GAS DYNAMICS

GAS DYNAMICS

M. Halûk Aksel

and

O. Cahit Eralp

Middle East Technical University
Ankara, Turkey

PRENTICE HALL
New York London Toronto Sydney Tokyo Singapore

First published 1994 by
Prentice Hall International (UK) Ltd
Campus 400, Maylands Avenue
Hemel Hempstead
Hertfordshire, HP2 7EZ
A division of
Simon & Schuster International Group

Typeset in 10/12 pt Plantin
by Mathematical Composition Setters, Salisbury, Wiltshire

Printed and bound in Great Britain by
Redwood Books, Trowbridge, Wiltshire

Library of Congress Cataloging-in-Publication Data

Aksel, Mehmet Halûk.
Gas dynamics/Mehmet Halûk Aksel and Osman Cahit Eralp.
p. cm.
Includes bibliographical references and index.
ISBN 0-13-497728-9 (pbk.)
1. Fluid dynamics. 2. Gas dynamics—Mathematics. 3. Engineering mathematics. I. Eralp, Osman Cahit. II. Title.
TA357.A388 1993
620.1'074—dc20 93–515
CIP

British Library Cataloguing in Publication Data

A catalogue record for this book is available from the British Library

ISBN 0-13-497728-9 (pbk)

1 2 3 4 5 96 95 94 93

This book is dedicated to Emine and Sacit Aksel without whose love, encouragement and support this text would not have been possible.

Contents

Preface

This text is developed for students as an introductory course to gas dynamics. It evolved from the lecture notes prepared by the authors and several of their colleagues for use in the courses 'Fluid Mechanics II' and 'Gas Dynamics' in the Mechanical Engineering Department of the Middle East Technical University.

Compressible fluid flow, or to use its more common name, gas dynamics, finds extensive applications in aerodynamics, turbomachinery, gas turbines and even in the most common engineering designs. Although in many of these cases two- or three-dimensional flows, unsteady flows and flows with chemical reactions are encountered, the basis of the topic can best be introduced by studying the simplified case of one-dimensional steady flow.

This text should be suitable for undergraduate students of mechanical and aeronautical engineering, and covers an important part of the compressible flow curriculum.

The intention of this text is to provide students with a good understanding of the physical behaviour of compressible fluid flow, and a good appreciation of the principles behind modern applications. The authors are aware of the fact that a student may study this text starting from any chapter. For this reason, the text is designed such that each chapter starts from the very basic principles. Thus, several important features of this course are intentionally repeated throughout the text. Even though there is no claim to complete originality in this text, it is believed that most topics are covered in quite a distinct manner. The material has been tested in the class over the last nine years. The problems provided at the end of each chapter are selected from the original problems which were mostly prepared for examinations and homework. Some classical problems, which may be found in several other textbooks on gas dynamics, are also given. Answers to all the problems are included, in order to give students a chance to check themselves. It is assumed that the reader has a basic knowledge of thermodynamics and fluid mechanics. References to other books are listed at the end of each chapter as extensions of each subject. The student is encouraged to use charts and tables to perform steady flow calculations. This avoids lengthy mathematical calculations.

In Chapter 1, the fundamental principles of fluid mechanics are discussed, and basic equations governing the motion of fluids are reviewed. In Chapter 2, the fundamental principles of thermodynamics are dealt with. Compressible flow and the phenomenon of wave propagation are discussed in Chapter 3. Chapter 4 is devoted to one-dimensional, steady and isentropic flow and the isentropic operation of converging–diverging nozzles. In Chapter 5, stationary, moving and reflected normal shock waves are discussed with emphasis on the applications such as supersonic wind tunnels, supersonic jet engine inlets, supersonic Pitot tube and shock tubes. One-dimensional adiabatic and isothermal flow with friction in constant-area ducts is discussed in Chapter 6. Chapter 7 is devoted to one-dimensional flow in constant-area ducts with simple heat transfer. In Chapter 8, oblique shock waves and Prandtl–Meyer expansion waves are discussed, and their applications are investigated in detail.

The authors wish to thank their colleagues Professor Dr Ahmet Ş. Üçer, Professor Dr Nuri Saryal, Associate Professor Dr Kahraman Albayrak and Associate Professor Dr Nafiz Alemdaroğlu for many helpful discussions during the teaching of this course over the last nine years.

The authors would also like to thank Mrs Emine Asli Aksel, who supported them by proofreading the text, criticizing it from the students' point of view and preparing the problems and their answers.

Finally, the authors acknowledge Mr Selçuk Tomaçoğlu for his effort in drawing the figures. The authors also appreciate the contributions of Classes 87 through 92 for correcting mistakes in the manuscript.

Dr M. Halûk Aksel
Dr O. Cahit Eralp
12 November 1992

Nomenclature

A cross-sectional area
A_C capture area
A_w wetted area
a speed of sound
C_D drag coefficient
C_d nozzle discharge coefficient
C_L lift coefficient
C_t nozzle thrust coefficient
C_V nozzle velocity coefficient
c chord length
c_p specific heat at constant pressure
c_v specific heat at constant volume
D hydraulic diameter, drag force
E total energy
e total energy per unit mass
F force
F_f frictional force
F_p pressure force
F_t thrust
f local coefficient of friction
$\bar{f}$ average coefficient of friction
G mass flux, modulus of elasticity in shear
g gravitational acceleration
H enthalpy
h enthalpy per unit mass
I impulse function
I_s specific impulse
i unit vector in x direction
j unit vector in y direction
KE kinetic energy
Kn Knudsen number
k specific heat ratio, unit vector in z direction
L length, lift force
L_{max} maximum duct length
M Mach number
$\bar{M}$ molecular mass
M_∞ flight Mach number
$M^\star$ Mach number referred to critical conditions
m mass
$\dot{m}$ mass flow rate
N any extensive property, number of molecules
n unit normal vector, normal direction
P linear momentum
PE potential energy
p pressure
Q heat transfer
$\dot{Q}$ rate of heat transfer
q heat transfer per unit mass
R gas constant
Re Reynolds number
$\bar{R}$ universal gas constant
r position
S entropy
s entropy per unit mass
T temperature
t time, tangential direction, unit tangential vector
U internal energy
u internal energy per unit mass

V velocity
V_x velocity in x direction
V_y velocity in y direction
V_z velocity in z direction
v volume per unit mass
W work
$\dot{W}$ rate of work (power)
$\dot{W}_{\text{flow}}$ rate of flow work
w work per unit mass
x x direction, displacement
y y direction
z z direction, elevation
α shear strain, wall angle, angle of attack, Riemann variable, slope of Rayleigh line
β Riemann variable
δ deflection angle
η any intensive property, efficiency
θ shock-wave angle
μ Mach angle
ν Prandtl–Meyer angle
ρ density
σ normal stress
τ thrust, shear stress
τ_w wall shear stress

$\mathbb{P}$ strength of a shock wave
$\forall$ volume
Δu° energy of combustion

SUBSCRIPTS

a after a moving shock wave, atmospheric
ai after an incident moving shock wave
ar after a reflected moving shock wave
atm atmospheric
b back, before a moving shock wave, burned products
bi before an incident moving shock wave
br before a reflected moving shock wave
cs control surface
d diffuser
dt diffuser throat
e exit
es isentropic exit
f frictional
g gas
H high temperature, differentiation along a Hugoniot curve
i inlet
irr irreversible
L lower, low temperature
max maximum
min minimum
n in the normal direction, nozzle
nt nozzle throat
p under constant pressure
R differentiation along a Rayleigh line
rev reversible
S system
s isentropic, shock
si incident shock wave
sr reflected shock wave
t throat, test section, in the tangential direction
U upper
u unburned reactant
v at constant volume
w wall
wr reflected wave
x before a stationary shock wave, in the x-direction
xi before an incident stationary shock wave
xr before a reflected stationary shock wave
y after a stationary shock wave in the y-direction
yi after an incident stationary shock wave

yr after a reflected stationary shock wave
z in the z-direction
0 stagnation stage
∞ free stream, flight

SUPERSCRIPTS

$\star$ critical state
$\star t$ critical state for isothermal flow

OPERATORS

d exact differential, infinitesimal
D/dt total derivative, derivative following a fluid
δ inexact differential
Δ difference
∇ gradient
$\boldsymbol{\nabla}$ divergence
$\partial/\partial t$ partial derivative, local derivative
Σ summation
lim limit

1 Fundamentals of fluid dynamics

1.1 INTRODUCTION

Gas dynamics is the science which investigates the motion of compressible fluids, particularly gases, and is a branch of the more general science of fluid dynamics. The basic difference between fluid dynamics and gas dynamics is that the latter deals with flows in which the changes in density that are brought about by the motion are appreciable when compared with the density at rest. Gas dynamics brings together several concepts and principles from other branches of science such as mechanics, thermodynamics, aerodynamics and chemical kinetics.

Four basic laws may be applied to study the motion of compressible fluids. These are as follows:

1 The law of conversation of mass or the continuity equation.
2 The law of conservation of momentum or Newton's second law of motion.
3 The first law of thermodynamics or the energy equation.
4 The second law of thermodynamics.

These laws are not dependent on the properties of a particular fluid. Therefore, in order to relate the motion to a particular fluid, it is necessary to use subsidiary laws in addition to these fundamental principles, such as the equation of state for a perfect gas.

The first two chapters are devoted to a review of the fundamental concepts of fluid dynamics and thermodynamics. In this chapter, a brief review of the classification of matter, the continuum concept, properties of the continuum, a description of the continuum, the classification of fluid flow and the application of conservation of mass and momentum to a material and/or control volume are presented.

1.2 SOLIDS AND FLUIDS

Matter may ideally be grouped into two classes: **solids** and **fluids**. This classification depends on the ease with which they may be deformed due to the different magnitudes of the intermolecular forces of attraction acting on them.

A solid is a state of matter in which the relative positions of molecules are fixed due to the very strong intermolecular forces of attraction. For this reason, solids

have definite shapes and definite volumes. The shape and volume of a solid can change only by a small amount when there is a large change in the external forces acting on it.

The word **fluid**, which is derived from the Latin word *fluidus*, means a substance having molecules that can easily change their relative positions. In fluids, the intermolecular forces of attraction are much weaker than those in solids, so the relative positions of molecules can change. Fluids can be classified further into two groups as **liquids** and **gases** with respect to the magnitudes of these intermolecular forces of attraction.

A liquid is a state of matter in which molecules are quite free to change their relative positions due to the medium intermolecular forces of attraction. For this reason, liquids do not have definite shapes but have definite volumes. Correspondingly, when external forces of small magnitude are applied to a liquid, the relative positions of the molecules may change by an amount which is not small, although the liquid maintains a relatively fixed volume.

A gas is a state of matter in which molecules are practically unrestricted due to very weak intermolecular forces of attraction. Therefore, gases have neither definite shapes nor definite volumes. As a direct consequence of this, the spacing of gas molecules can be changed by extremely small external forces so that gases expand to fill all the available space uniformly.

1.3 THE CONCEPT OF CONTINUUM

A gram of gas is composed of billions of molecules which are in constant motion and collision. The most general description of the motion of a gas would be the one in which the equations of motion are written for each separate molecule. This approach is used in statistical mechanics. Although it has some advantages, it is too cumbersome or even impossible for practical calculations. Besides, little is known about the intermolecular forces of attraction.

However, for most engineering problems, the principal interest is not the motion of individual molecules, but the gross behaviour of the gas as a continuous material. The treatment of a gas as a continuum is valid whenever the smallest volume of this gas contains enough molecules to take statistical averages.

The ratio of the molecular mean free path (the average distance travelled by molecules between two consecutive collisions) to a significant characteristic linear dimension related to the flow field, is the determining criterion for the validity of the continuum concept. This non-dimensional term is called the **Knudsen number**, *Kn*, and is defined as

$$Kn = \frac{\text{molecular mean free path}}{\text{characteristic linear dimension of the flow field}} \tag{1.1}$$

The continuum postulate is applicable to flows with a Knudsen number of less than about 0.01.

1.4 PROPERTIES OF CONTINUUM

1.4.1 Mass density

Consider a mass of fluid Δm, within a volume $\Delta \forall$, surrounding any arbitrary point P in a continuous fluid, as shown in Figure 1.1(a). Then the ratio $\Delta m/\Delta \forall$ represents the average mass density of the fluid within the volume $\Delta \forall$. The plot of $\Delta m/\Delta \forall$ versus $\Delta \forall$ is given in Figure 1.1(b) as $\Delta \forall$ is shrunk about point P. At first, the average density approaches an asymptotic value as the enclosed fluid becomes more and more homogeneous. However, when $\Delta \forall$ becomes very small, it contains relatively few molecules. For this reason, the average density fluctuates significantly with time, as molecules move into and out of the volume. Then, the minimum volume $\Delta \forall'$ can be introduced satisfying the continuum postulate; the fluid contained in this volume is known as the **fluid particle**. The definition of the mass density, ρ, at point P can be given as

$$\rho = \lim_{\Delta \forall \to \Delta \forall'} \Delta m/\Delta \forall \tag{1.2}$$

The density of a substance is a function of the temperature and the pressure. Fluids with constant density are called **incompressible fluids**. The density of liquids is only slightly affected by these properties, so that it is almost constant. Thus, liquids are incompressible fluids. However, the density of gases is strongly affected by the temperature and the pressure, and for this reason they are classified as **compressible fluids**.

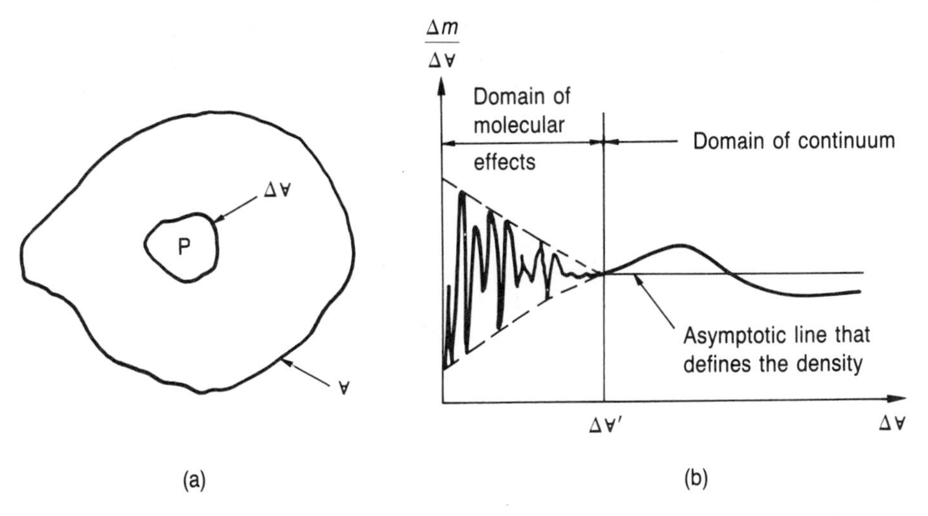

Figure 1.1 Definition of the mass density

For a homogeneous fluid, the average mass density is defined as the mass per unit volume,

$$\rho = m/\forall \tag{1.3}$$

and has units of kilograms per cubic metre (kg m^{-3}) in the SI system (the official abbreviation in all languages for the Système International d'Unités or International System of Units) of units.

1.4.2 Fluid velocity

The fluid velocity, $\mathbf{V}(x, y, z, t)$, at point $P(x, y, z)$ is defined as the instantaneous velocity of the centre of gravity of the volume $\Delta\forall'$. By definition

$$\mathbf{V} = \sum_{i=1}^{N} m_i \mathbf{V}_i \bigg/ \sum_{i=1}^{N} m_i = \frac{\text{total momentum contained in volume } \Delta\forall'}{\text{total mass inside volume } \Delta\forall'} \tag{1.4}$$

where m_i and $\mathbf{V}_i$ denote the mass and the velocity of the ith molecule, respectively, with N denoting the number of molecules inside $\Delta\forall'$, as shown in Figure 1.2.

Streamlines are imaginary lines drawn in the flow field so that at a given instant of time, they are tangential to the direction of flow or the velocity vector at every point of the flow field, as shown in Figure 1.3. As long as the streamlines indicate the direction of motion at each point, then there cannot be any flow across a streamline. For this reason, impermeable solid boundaries are naturally streamlines.

In time dependent flows, the velocity field is a function of time and the shape of the streamlines can vary from one instant of time to another. A fluid particle follows a streamline for only an infinitesimal time interval before shifting to another streamline. The position of a streamline at a given instant of time is known as the **instantaneous streamline**.

Two streamlines can only intersect when the fluid particle is at rest.

A **streamtube** is formed by a bundle of streamlines passing through a closed curve and is filled with the flowing fluid, as shown in Figure 1.4. As long as the boundaries

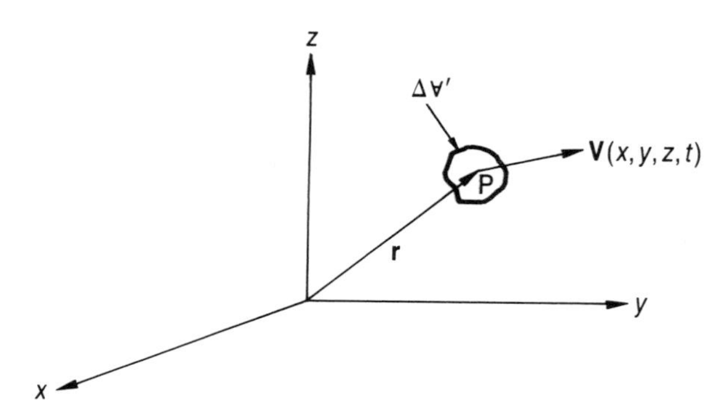

Figure 1.2 Definition of the fluid velocity at a point

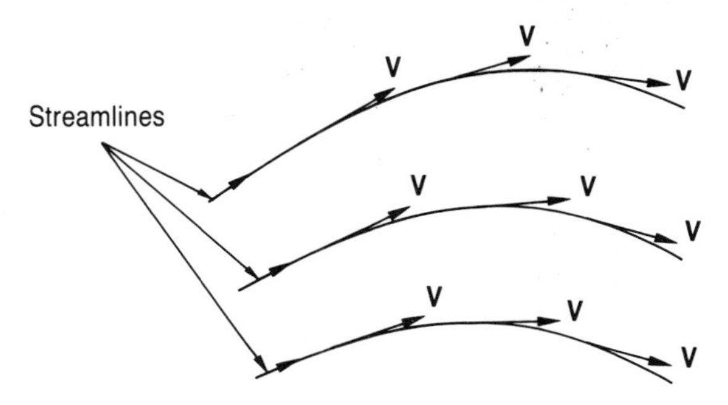

Figure 1.3 Streamlines

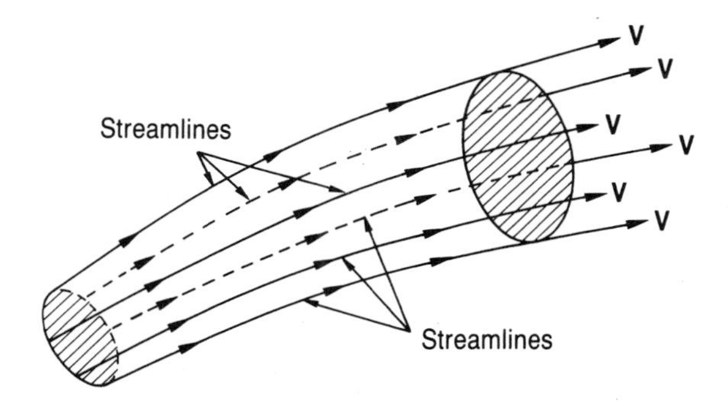

Figure 1.4 A streamtube

of the streamtube are formed by streamlines, then it behaves like a rigid tube through which the fluid flows. When the flow is time dependent, the shape of the streamtube will change from one instant of time to another, so that it is made up of ever-changing fluid particles. However, when the flow does not depend on time, a streamtube behaves like a real tube.

1.4.3 Forces acting on the body of a fluid

The external forces acting on the body of a fluid can be classified into two groups, namely the **body forces** and the **surface forces**.

A body force is distributed over the entire volume of a body of fluid and is defined as the force per unit mass of the fluid. It includes the forces which are caused by gravitational fields, magnetic fields, electrostatic fields, etc.

A surface force may have any orientation with respect to the surface of a body. These kinds of forces may be decomposed into a normal force acting perpendicular to the surface and a tangential or shear force acting parallel to the surface.

A continuum is in equilibrium if and only if the resultant of all forces acting on any part is zero. Consider a body of fluid in continuum under the action of several external forces, as shown in Figure 1.5(a). These are then transmitted as internal forces through the material composing the body. An insight into the internal forces at any part of the body may be obtained by imagining the body to be cut along a plane passing through point O, as shown in Figure 1.5(a). Since each half of the body is in equilibrium, then the resultant of the internal forces over the cut surface must be equal in magnitude and opposite in direction to the resultant of the external forces acting on the cut body. Then $\mathbf{F}_1, \mathbf{F}_2$ and $\mathbf{F}_3$ acting on part I of the body must be balanced by internal forces distributed over the cross-sectional area A of the cutting plane. If $\mathbf{F}$ denotes the resultant of the internal forces acting on the cross-sectional area, A, then it can be decomposed into its components as

$$\mathbf{F} = F_n\hat{\mathbf{n}} + F_t\hat{\mathbf{t}} \tag{1.5}$$

where F_n and F_t are the normal and tangential components of this force, respectively, and $\hat{\mathbf{n}}$ and $\hat{\mathbf{t}}$ are the unit vectors in the normal and tangential directions, respectively.

1.4.4 Stress

If $\Delta\mathbf{F} = \Delta\mathbf{F}_n\hat{\mathbf{n}} + \Delta F_t\hat{\mathbf{t}}$ is the resultant of the internal forces acting on a small-area element, ΔA, of Figure 1.5(b), then the normal stress, σ, and the shear stress, τ, can be defined as

$$\sigma = \lim_{\Delta A \to \Delta A'} \Delta F_n/\Delta A \tag{1.6a}$$

and

$$\tau = \lim_{\Delta A \to \Delta A'} \Delta F_t/\Delta A \tag{1.6b}$$

respectively. In Equations (1.6a) and (1.6b), $\Delta A'$ represents the minimum area for which the continuum postulate is valid.

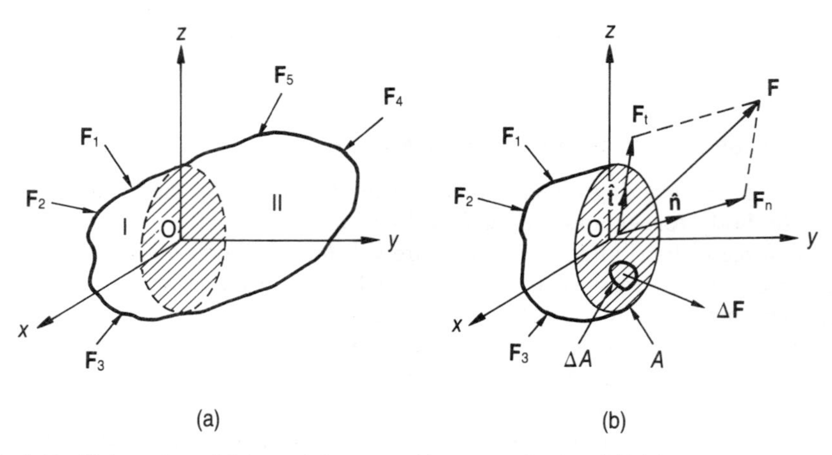

Figure 1.5 External and internal forces acting on a body of fluid

If the internal forces are uniformly distributed over the area, A, then the normal stress and the shear stress can be given as

$$\sigma = F_n / A \tag{1.7a}$$

and

$$\tau = F_t / A \tag{1.7b}$$

respectively. Stress has units of newtons per square metre ($\mathrm{N\,m^{-2}}$) in the SI system.

1.4.5 Viscosity

Consider the behaviour of a solid and a fluid under the action of a shear force, as shown in Figure 1.6. In Figure 1.6(a), a solid block is bonded between two parallel plates. When the shear force, F, is applied through the upper plate, side $\overline{AB}$ of the solid block is deformed, as shown by the dashed line $\overline{AB'}$. The angle, α, between the lines $\overline{AB}$ and $\overline{AB'}$ is known as the shear strain. From solid mechanics, it is known that the shear strain is directly proportional to the applied shear stress, τ, where $\tau = F/A$, and A is the area of contact of the upper plate with the solid block, provided that the elastic limit of the solid material is not exceeded. Then, by Hooke's law

$$\tau = G\alpha \tag{1.8}$$

where G is the modulus of elasticity in shear.

The same experiment may now be repeated by using a fluid between the parallel plates. The vertical fluid element that is shown in Figure 1.6(b) can be made visible with the aid of a dye at time t_0. The shear force, F, is applied such that the upper plate moves at a constant velocity of V_0. Under the action of this force, the fluid starts to deform, and continues to deform as long as the force is applied. The shape of the fluid element, at successive instants of time t_1 and t_2, is shown by the dashed lines in Figure 1.6(b). A better definition for a fluid may now be given.

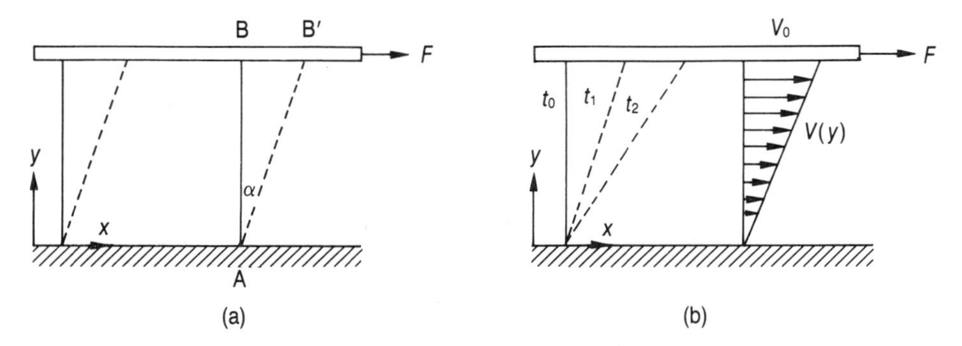

Figure 1.6 Behaviour of (a) a solid and (b) a fluid under the action of a constant shear force

A fluid is a substance that cannot sustain a shear stress when it is at rest, and it deforms continuously under the action of a shear stress, no matter how small the shear stress is. It is an experimental fact that the fluid particles which are in direct contact with the solid boundary have the same velocity as the boundary itself. In other words, the fluid particles stick to solid boundaries. This is known as the **no-slip condition.** Therefore, while the fluid has a finite velocity of V_0 at the upper boundary, the velocity is zero at the lower boundary. Thus, the velocity in the x direction, V_x, is only a function of the y direction, and it increases with increasing distance from the lower boundary, as shown in Figure 1.6(b).

During an infinitesimal time interval, dt, while the fluid element ABCD is translated to a new position A′EFD′, it is also distorted to the shape A′B′C′D′, as shown in Figure 1.7. The velocity of the fluid element at point $A(x,y)$ can now be expressed as $V_{Ax} = V_x(y)$, while the velocity at point $B(x, y+dy)$ is $V_{Bx} = V_x(y+dy)$. Then, V_{Bx} may be related to V_{Ax} by using the Taylor series expansion:

$$V_{Bx} = V_x + \frac{1}{1!}\frac{dV_x}{dy}\,dy + \frac{1}{2!}\frac{d^2V_x}{dy^2}(dy)^2 + \frac{1}{3!}\frac{d^3V_x}{dy^3}(dy)^3 + \cdots$$

In this equation, the second- and higher-order terms are neglected, since dy is infinitesimally small. Then

$$V_{Bx} = V_x + \frac{dV_x}{dy}\,dy$$

Hence, the distances $\overline{AA'}$ and $\overline{BB'}$ may now be expressed as

$$\overline{AA'} = V_{Ax}\,dt = V_x\,dt$$

and

$$\overline{BB'} = V_{Bx}\,dt = \left(V_x + \frac{dV_x}{dy}\,dy\right)dt$$

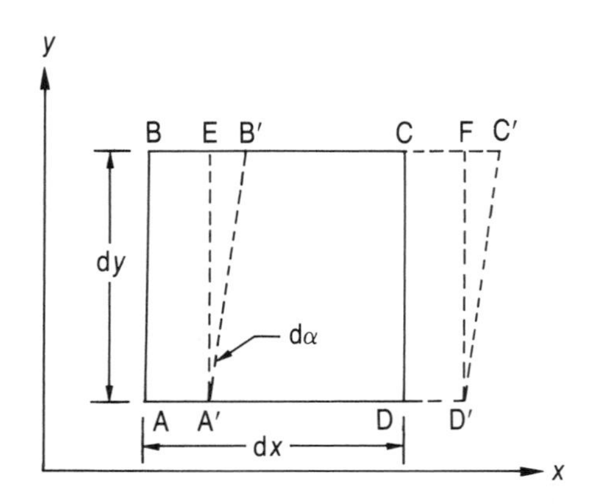

Figure 1.7 A fluid element under the action of a constant shear stress

respectively. Finally, the infinitesimal shear strain, $\mathrm{d}\alpha$, can be expressed as

$$\tan \mathrm{d}\alpha = \frac{\overline{\mathrm{EB}'}}{\overline{\mathrm{A}'\mathrm{E}'}} = \frac{\overline{\mathrm{BB}'} - \overline{\mathrm{AA}'}}{\overline{\mathrm{AB}}} = \frac{\mathrm{d}V_x}{\mathrm{d}y}\,\mathrm{d}t$$

But for small angles, $\tan \mathrm{d}\alpha \simeq \mathrm{d}\alpha$, and therefore the rate of change of the shear strain is

$$\frac{\mathrm{d}\alpha}{\mathrm{d}t} = \frac{\mathrm{d}V_x}{\mathrm{d}y}$$

For certain fluids, called **Newtonian fluids**, the shear stress on a surface tangential to the flow direction is proportional to the rate of change of shear strain or to the velocity gradient (the rate of change of velocity with respect to the normal direction of the surface). This relation is known as **Newton's law of viscosity**. Hence, for a flow in the x direction,

$$\tau = \mu \frac{\mathrm{d}V_x}{\mathrm{d}y} \tag{1.9}$$

where the coefficient of proportionality, μ, is known as the **absolute viscosity**. The viscosity of a fluid is a measure of its resistance to shear and angular deformation. Viscosity has units of newton seconds per square metre ($\mathrm{N\,s\,m^{-2}}$) or pascal seconds (Pa s).

Fluids which possess viscosity are known as **real** or **viscous fluids**, while fluids with zero viscosity are often referred to as **inviscid** or **ideal fluids**. The concept of the ideal fluid is imaginary, and is used in the analytical treatment of fluid flow phenomena. The mathematical analysis of a flow problem can be considerably simplified by assuming the fluid to be inviscid. But it is worth while to remember that all fluids that exist in nature are viscous to some extent.

Many viscous flows may be analyzed by dividing the flow field into two regions: one close to the solid boundaries and the other covering the rest of the flow field. Only in the thin region adjacent to the solid boundary, which is referred to as the **boundary layer**, are the viscous effects important, and the viscous forces are much more significant than the inertial forces. The velocity of the fluid at the wall relative to the solid boundary is zero and increases towards the main stream. Shear stresses in this zone are very high owing to the existence of extremely high velocity gradients at and near the solid boundaries. In the region outside the boundary layer, the influence of viscosity is negligible so that the inertial forces dominate the viscous ones. Therefore, the flow in this region may be considered inviscid.

1.4.6 Pressure

Pressure is defined as the normal component of the force acting on an area divided by that area. Therefore, it has the same magnitude as the normal stress. The surface force due to pressure is perpendicular to this surface and directed towards the surface.

To prove whether or not the pressure depends on the direction, consider a small body of inviscid fluid in the form of a prism, as shown in Figure 1.8. The prism is small enough so that the pressure over each face may be assumed constant. As long as the fluid is inviscid, there are no shear forces. Hence, in addition to the pressure forces, only body forces may act on the fluid element. Therefore, the summation of all forces acting on the fluid element must be equal to the product of its mass and acceleration. Then, for the y direction, it is possible to write

$$\Sigma F_y = p_y A_y - p_e A_e \cos \beta + F_{by} \left(\frac{A_y y_0}{3} \right) \rho = \frac{DV_y}{dt} \left(\frac{A_y y_0}{3} \right) \rho \tag{1.10}$$

where DV_y/dt is the acceleration in the y direction, A_y is the area of the prism in the xz plane, p_y is the average pressure on the area A_y, A_e is the area of the inclined surface, p_e is the average pressure on the area A_e, β is the angle between the normals of surfaces A_y and A_e, F_{by} is the component of the body force in the y direction, and y_0 is the intercept of the prism faces on the y axis. In Equation (1.10), the term $A_y y_0/3$ represents the volume of the prism, while the term $\rho F_{by}(A_y y_0/3)$ is the component of the body force in the y direction. Also, from the geometry of the prism

$$A_y = A_e \cos \beta \tag{1.11}$$

Now, substituting Equation (1.11) into Equation (1.10) and dividing through by A_y gives

$$p_y - p_e + \left(F_{by} - \frac{DV_y}{dt} \right) \rho \frac{y_0}{3} = 0 \tag{1.12}$$

Finally, considering successively smaller prisms, as the prism shrinks to a point, y_0 goes to zero and Equation (1.12) reduces to

$$p_y = p_e \tag{1.13a}$$

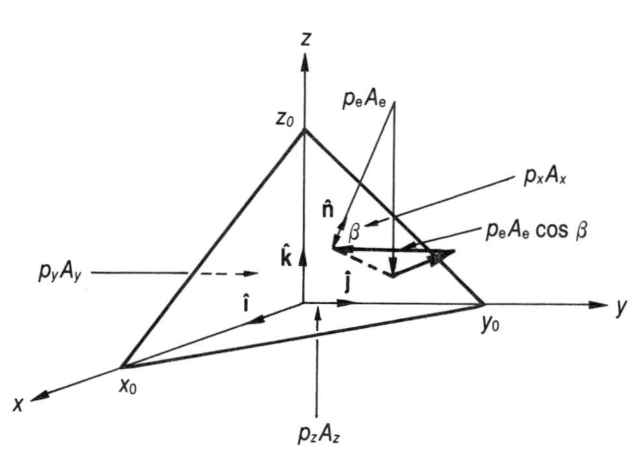

Figure 1.8 Pressure forces on a prism of inviscid fluid

If the same argument is repeated for the x and z directions, the relations

$$p_x = p_e \tag{1.13b}$$

and

$$p_z = p_e \tag{1.13c}$$

can be obtained. As a result, for an inviscid fluid, the pressure is equal in all directions, that is

$$p_x = p_y = p_z = p_e = p \tag{1.14}$$

Pressure has units of newtons per square metre ($N\,m^{-2}$) in the SI system which are also called pascals (Pa).

1.5 THE DESCRIPTION OF CONTINUUM

For each particle of a fluid in continuum, the properties can be defined instantaneously. As the fluid particle moves about in the flow field, its properties change from point to point and from time to time. The fluid flow can be described in the two ways below.

1.5.1 Material (Lagrangian) description

In this description, it is desirable to follow identified fluid particles in the course of time and to determine the variation of properties of these fluid particles. This approach is commonly used in dynamics.

All the particles of the fluid are identified by locating them at some reference instant of time, say $t = 0$. Now, consider the fluid particle, P, which is located at point A of the flow field in Figure 1.9, at time $t = 0$. In the course of time, the fluid particle will move in the flow field and will be located at point B at time t. Then, the path followed by the fluid particle may be specified by expressing its position vector, $\mathbf{r}_p$, as a function of time with respect to the Cartesian coordinate system, xyz, of Figure 1.9 as

$$\mathbf{r}_p(t) = x_p(t)\hat{\mathbf{i}} + y_p(t)\hat{\mathbf{j}} + z_p(t)\hat{\mathbf{k}} \tag{1.15}$$

where $x_p(t)$, $y_p(t)$ and $z_p(t)$ are coordinates of the fluid particle, P, at time t. The velocity, $\mathbf{V}_p$, of the fluid particle may then be obtained by differentiating its position vector with respect to time, that is

$$\mathbf{V}_p = V_{px}\mathrm{i} + V_{py}\mathrm{j} + V_{pz}\hat{\mathbf{k}} = \frac{d\mathbf{r}_p}{dt} = \frac{dx_p}{dt}\hat{\mathbf{i}} + \frac{dy_p}{dt}\hat{\mathbf{j}} + \frac{dz_p}{dt}\hat{\mathbf{k}} \tag{1.16}$$

where V_{px}, V_{py} and V_{pz} are the velocity components in the x, y and z directions, respectively.

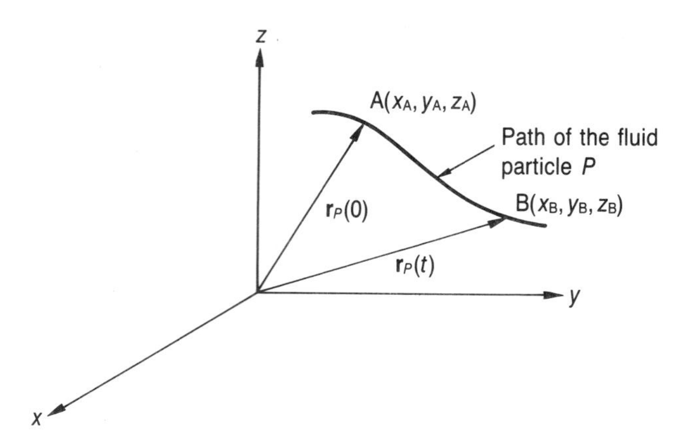

Figure 1.9 Path of a fluid particle

In this description, the velocity, the pressure and the density of the fluid particle can be specified as

$$\mathbf{V}_P = \mathbf{V}_P[x_P(t),\ y_P(t),\ z_P(t), t] \tag{1.17a}$$

$$p_P = p_P[x_P(t),\ y_P(t),\ z_P(t), t] \tag{1.17b}$$

and

$$\rho_P = \rho_P[x_P(t),\ y_P(t),\ z_P(t), t] \tag{1.17c}$$

In the above equations, t is the only independent variable, and x_P, y_P and z_P are the dependent variables used for locating the fluid particle, P, when t is known.

Of course, the motion of one fluid particle is inadequate to describe the entire flow field, so the motion of all fluid particles must be considered simultaneously. Therefore, this apparently very convenient method proves to be very troublesome when solutions to definite problems have to be found. Although it is very powerful, it can only be carried out for a few cases, and gives more information than one needs to know.

The collection of identified fluid particles is known as the **material volume** or the **closed system**. A closed system may change its shape or its position, but it always encloses the same fluid particles. The boundaries separating a closed system from its surroundings are known as **system boundaries**, and fluid particles cannot cross these boundaries.

1.5.2 Spatial (Eulerian) description

In this description, attention is focused on a fixed point in space, and the variation of properties is considered as fluid particles pass through this point.

As one moves in the flow field, the properties change from one point to another and from time to time. Therefore, the velocity, the pressure and the density of the flow field may be expressed as a function of space coordinates and time. In the Cartesian coordinate system,

$$\mathbf{V} = \mathbf{V}(x, y, z, t) \tag{1.18a}$$

$$p = p(x, y, z, t) \tag{1.18b}$$

and

$$\rho = \rho(x, y, z, t) \tag{1.18c}$$

Since the pressure and the density are scalar quantities, Equations (1.18b) and (1.18c) represent scalar fields. However, velocity is a vector quantity, and therefore Equation (1.18a) represents the magnitude as well as the direction of the velocity and is referred to as a vector field. In the above equations, x, y, z and t are independent variables.

There is certainly a relation between the Lagrangian and the Eulerian descriptions. Referring to Figure 1.9, at time $t = 0$, the fluid particle, P, is located at point A of the flow field. Therefore the properties at point A of the flow field are equivalent to the properties of the fluid particle, P, that is

$$\mathbf{V}(x_A, y_A, z_A, t = 0) = \mathbf{V}_P[x_P(t = 0), y_P(t = 0), z_P(t = 0), t = 0] \tag{1.19a}$$

$$p(x_A, y_A, z_A, t = 0) = p_P[x_P(t = 0), y_P(t = 0), z_P(t = 0), t = 0] \tag{1.19b}$$

and

$$\rho(x_A, y_A, z_A, t = 0) = \rho_P[x_P(t = 0), y_P(t = 0), z_P(t = 0), t = 0] \tag{1.19c}$$

But at a later time, the properties at point A can be represented by the properties of a different fluid particle passing through this point.

The volume, which is fixed in space relative to the coordinate system, is known as the **control volume** or the **open system**. The amount and the identity of the matter in the control volume may change with time, but the shape of the control volume is fixed. The boundaries which separate the control volume from its surroundings are known as the **control surface**.

It is an enormous task to keep track of the positions of all the fluid particles in a flow field, because, unlike the movement of solid bodies, their relative positions change continuously with time. For this reason, the Eulerian approach is generally favoured over the Lagrangian one.

1.6 METHODS FOR THE MATHEMATICAL FORMULATION OF FLUID FLOW

Whether the Lagrangian or the Eulerian approach is used for describing the fluid flow, it is always necessary to use either the **differential formulation** or the **integral formulation** for applying the basic laws to analyze the fluid motion.

1.6.1 Differential formulation

In the differential approach, detailed information of the flow field is required. If the Lagrangian description is used, then information about the motion of all fluid particles in the flow field is necessary. In the case of the Eulerian description, information about the fluid flow is sought at every point in the flow field. The basic laws are then applied to each fluid particle in the Lagrangian description or to an infinitesimal fluid element of the flow field in the case of the Eulerian description so that the resulting equations are differential equations. If it is possible to solve these differential equations, then detailed information (i.e. point by point) about the behaviour of the flow field is obtained.

1.6.2 Integral formulation

However, for most common fluid flow problems, the information sought does not require detailed knowledge of the flow field. In the integral approach, the basic laws are applied to a closed system in the case of the Lagrangian description or to a control volume in the case of the Eulerian description. They are formulated in terms of integral equations, which give information about the gross behaviour of the fluid flow and are usually easier to treat analytically.

1.7 RELATIONS BETWEEN THE MATERIAL AND THE SPATIAL DESCRIPTIONS

Whatever type of mathematical formulation used, there is always a relation between the material and the spatial descriptions of the flow field.

1.7.1 Relations in the differential formulation

The rate of change of any property of a fluid particle may be related to the rate of change of that property in the flow field at that point. To find this relation, the material description of any property, N, may be used. Similar to Equation (1.17), N may be expressed as

$$N = N[x(t), y(t), z(t), t] \tag{1.20}$$

Then, the total differential of the property, N, is

$$\mathrm{D}N = \frac{\partial N}{\partial x}\,\mathrm{d}x + \frac{\partial N}{\partial y}\,\mathrm{d}y + \frac{\partial N}{\partial z}\,\mathrm{d}z + \frac{\partial N}{\partial t}\,\mathrm{d}t$$

or dividing by $\mathrm{d}t$ yields

$$\frac{\mathrm{D}N}{\mathrm{d}t} = \frac{\partial N}{\partial x}\frac{\mathrm{d}x}{\mathrm{d}t} + \frac{\partial N}{\partial y}\frac{\mathrm{d}y}{\mathrm{d}t} + \frac{\partial N}{\partial z}\frac{\mathrm{d}z}{\mathrm{d}t} + \frac{\partial N}{\partial t}$$

Now, from Equation (1.16), $V_x = \mathrm{d}x/\mathrm{d}t$, $V_y = \mathrm{d}y/\mathrm{d}t$ and $V_z = \mathrm{d}z/\mathrm{d}t$, so that it is possible to obtain

$$\frac{\mathrm{D}N}{\mathrm{d}t} = \frac{\partial N}{\partial t} + V_x \frac{\partial N}{\partial x} + V_y \frac{\partial N}{\partial y} + V_z \frac{\partial N}{\partial z} \tag{1.21a}$$

or in vectorial form

$$\frac{\mathrm{D}N}{\mathrm{d}t} = \frac{\partial N}{\partial t} + \mathbf{V} \cdot \nabla N \tag{1.21b}$$

The differential operator, $\mathrm{D}/\mathrm{d}t$, is often referred to as the **total derivative**, the **substantial derivative**, the **material derivative** or the **derivative following the fluid**. It represents the rate of change of a property in the Lagrangian description.

The differential operator, $\partial/\partial t$, is often referred to as the **partial derivative**, the **local derivative** or the **spatial derivative**. It describes the rate of change of a property at a certain point in the flow field and is certainly related to the Eulerian description.

The term $\mathbf{V} \cdot \nabla N$ in Equation (1.21b) is known as the **convective derivative** of the property N, and it represents the change in this property from one point to another at a certain time. To have a better understanding of this derivative, assume that the property, N, does not change with time but changes from one point to another in the flow field. A group of curves of $N =$ constant is shown in Figure 1.10, and a fluid particle located at point A moves to point B during a time interval $\mathrm{d}t$. If the velocity of the fluid particle at point A is $\mathbf{V}$, then its displacement from A to B is $\mathbf{V}\,\mathrm{d}t$. The gradient of property N, that is ∇N, is perpendicular to the constant N curve passing through point A. The change in property N as the fluid particle moves from point A to B is equal to the scalar product of $\mathbf{V}\,\mathrm{d}t$ and ∇N. Therefore, the convective change of property N per unit time is $\mathbf{V} \cdot \nabla N$.

1.7.2 Relations in the integral formulation (Reynolds' transport theorem)

The properties of fluids may be divided into two general classes, namely the **extensive** and the **intensive** properties. An extensive property varies directly with the mass, and

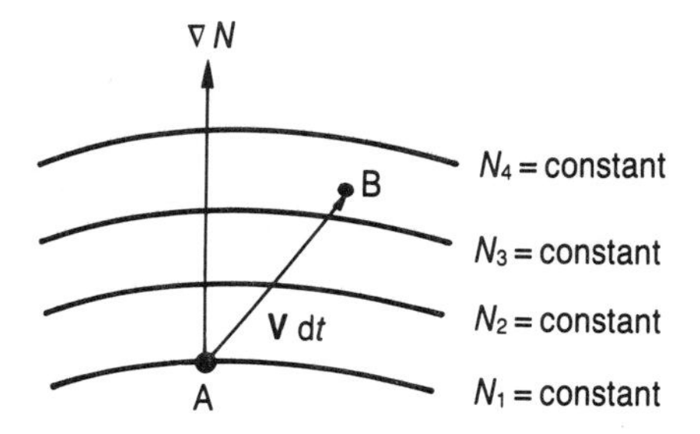

Figure 1.10 Convective change of property N of a fluid particle

examples of extensive properties are weight, volume and linear momentum. Clearly, changing the amount of mass directly changes the measure of these quantities. However, an intensive property does not depend on mass. Some examples of intensive properties are specific weight, volume per unit mass and linear momentum per unit mass (velocity). Furthermore, properties such as temperature and pressure, by their mass-independent nature, are already in the category of intensive properties. Now, if N designates any arbitrary extensive property, the corresponding intensive property (extensive property per unit mass) will be designated by η. Thus,

$$N = \int_{m_S} \eta \, dm \tag{1.22}$$

where m_S is the mass contained in the system. But as $dm = \rho \, d\forall$, then

$$N = \int_{\forall_S} \eta\rho \, d\forall \tag{1.23}$$

with ρ being the density and $\forall_S$ the volume of the system.

Consider an arbitrary flow field $\forall(x, y, z, t)$ as seen from an arbitrary reference frame xyz. The boundaries of the system of fluid particles at time t, which are shown in Figure 1.11(a), coincide with the boundaries of the control volume as indicated by the dashed lines. After a time interval of Δt, that is at time $t + \Delta t$, the positions of the system and the control volume are as shown in Figure 1.11(b). Now, the system occupies regions B and C, and the mass within region A enters into the control volume, while the mass within region C leaves the control volume. The main objective of this analysis is to relate the rate of change of any extensive property N of the system to variations of this property associated with the control volume. From the definition of a derivative, the rate of change of any extensive property in a system may be

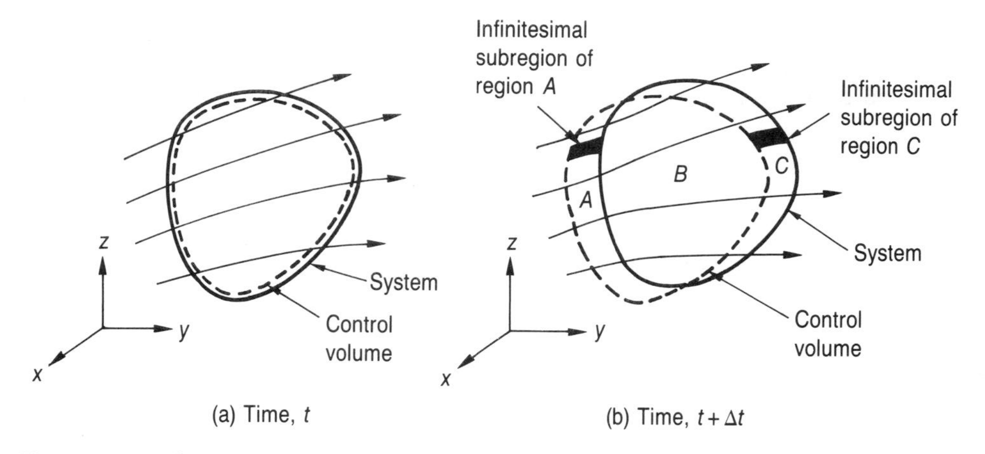

Figure 1.11 System and control volume configuration

given as

$$\frac{DN}{dt}=\frac{D}{dt}\int_{\forall_S}\rho\eta\,d\forall=\lim_{\Delta t\to 0}\frac{1}{\Delta t}\left(\int_{\forall_S}(\rho\eta\,d\forall)_{t+\Delta t}-\int_{\forall_S}(\rho\eta\,d\forall)_t\right) \tag{1.24}$$

As long as the system and the control volume coincide at time t, then

$$\int_{\forall_S}(\rho\eta\,d\forall)_t=\int_{\forall}(\rho\eta\,d\forall)_t \tag{1.25}$$

where $\forall$ represents the volume of the control volume. Also, at time $t+\Delta t$

$$\int_{\forall_S}(\rho\eta\,d\forall)_{t+\Delta t}=\int_{\forall}(\rho\eta\,d\forall)_{t+\Delta t}+\int_{\forall_C}(\rho\eta\,d\forall)_{t+\Delta t}-\int_{\forall_A}(\rho\eta\,d\forall)_{t+\Delta t} \tag{1.26}$$

where $\forall_A$ and $\forall_C$ represent the volumes of regions A and C, respectively. Substituting Equations (1.25) and (1.26) into Equation (1.24) and using the rule that the limit of a sum is equal to the sum of the limits, it is possible to write

$$\frac{D}{dt}\int_{\forall_S}\rho\eta\,d\forall=\lim_{\Delta t\to 0}\frac{1}{\Delta t}\left(\int_{\forall}(\rho\eta\,d\forall)_{t+\Delta t}-\int_{\forall}(\rho\eta\,d\forall)_t\right)$$
$$+\lim_{\Delta t\to 0}\frac{1}{\Delta t}\left(\int_{\forall_C}(\rho\eta\,d\forall)_{t+\Delta t}\right)-\lim_{\Delta t\to 0}\frac{1}{\Delta t}\left(\int_{\forall_A}(\rho\eta\,d\forall)_{t+\Delta t}\right) \tag{1.27}$$

The first limit on the right-hand side of Equation (1.27) represents the rate of change of an extensive property in a control volume, that is

$$\frac{\partial}{\partial t}\int_{\forall}\rho\eta\,d\forall=\lim_{\Delta t\to 0}\frac{1}{\Delta t}\left(\int_{\forall}(\rho\eta\,d\forall)_{t+\Delta t}-\int_{\forall}(\rho\eta\,d\forall)_t\right) \tag{1.28}$$

It is possible to express the second limit as

$$\lim_{\Delta t\to 0}\frac{1}{\Delta t}\left(\int_{\forall_C}(\rho\eta\,d\forall)_{t+\Delta t}\right)=\lim_{\Delta t\to 0}\frac{1}{\Delta t}(N_C)_{t+\Delta t}$$

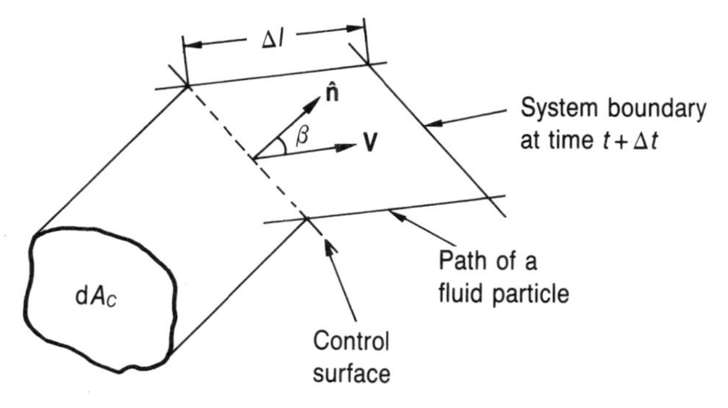

Figure 1.12 Enlarged view of an infinitesimal subregion of region C of Figure 1.11

To evaluate $(N_C)_{t+\Delta t}$, consider the enlarged view of the infinitesimal subregion of region C in Figure 1.12. dA_C represents the infinitesimal area of the control surface at the interface of the control volume and region C, $\hat{\mathbf{n}}$ is the unit normal vector of this area, β is the angle between the unit normal vector and the velocity vector at the interface, and Δl is the distance travelled by the fluid particles in a time interval of Δt. Then, the mass of this infinitesimal fluid element is $\rho \, \Delta l \, dA_C \cos \beta$ so that the value of the extensive property in this infinitesimal region is

$$d(N_C)_{t+\Delta t} = \rho\eta \, \Delta l \cos \beta \, dA_C$$

Integrating over the interface yields

$$(N_C)_{t+\Delta t} = \int_{\forall_C} \rho\eta \, \Delta l \cos \beta \, dA_C$$

so that the second limit becomes

$$\lim_{\Delta t \to 0} \frac{1}{\Delta t} (N_C)_{t+\Delta t} = \lim_{\Delta t \to 0} \int_{\forall_C} \rho\eta \frac{\Delta l}{\Delta t} \cos \beta \, dA_C$$

But as Δt approaches zero

$$\lim_{\Delta t \to 0} \Delta l / \Delta t = V$$

and hence

$$\lim_{\Delta t \to 0} \frac{1}{\Delta t} (N_C)_{t+\Delta t} = \int_{A_C} \rho\eta V \cos \beta \, dA_C$$

Since the mass is flowing out of the control volume, the angle, β, is always less than 90° and $\cos \beta$ is then always positive. Also $V \cos \beta = \mathbf{V} \cdot \hat{\mathbf{n}}$ so that

$$\lim_{\Delta t \to 0} \frac{1}{\Delta t} \left(\int_{\forall_C} (\rho\eta \, d\forall)_{t+\Delta t} \right) = \int_{A_C} \rho\eta (\mathbf{V} \cdot \hat{\mathbf{n}}) \, dA_C \qquad (1.29)$$

To evaluate the third limit on the right-hand side of Equation (1.27), the enlarged view of the infinitesimal subregion of region A in Figure 1.13 may be used. dA_A

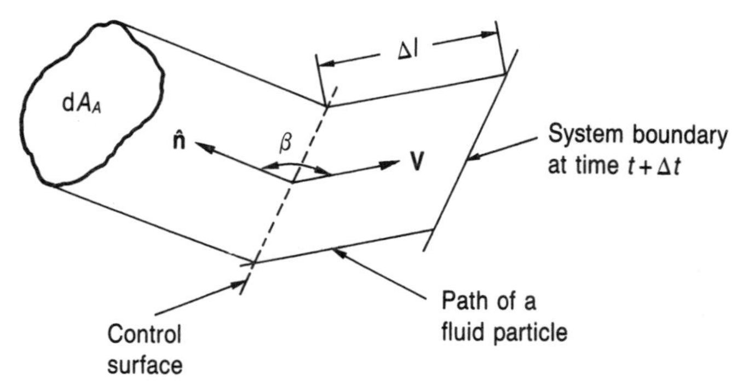

Figure 1.13 Enlarged view of an infinitesimal subregion of region A of Figure 1.11

represents the infinitesimal area of the control volume at the interface of the control volume and region A. The same analysis as that used in the evaluation of the second limit can also be used for the third limit with only one difference. In this case, since the mass is flowing out from the control volume, the angle, β, is always greater than 90° and $\cos\beta$ is then always negative. Therefore, the third limit may be expressed as

$$\lim_{\Delta t \to 0} \frac{1}{\Delta t}\left(\int_{\forall_A} (\rho\eta \, \mathrm{d}\forall)_{t+\Delta t}\right) = -\int_{A_A} \rho\eta(\mathbf{V}\cdot\hat{\mathbf{n}})\,\mathrm{d}A_A \tag{1.30}$$

Substituting Equations (1.28), (1.29) and (1.30) into Equation (1.27) yields

$$\frac{D}{\mathrm{d}t}\int_{\forall_S} \rho\eta\,\mathrm{d}\forall = \frac{\partial}{\partial t}\int_{\forall} \rho\eta\,\mathrm{d}\forall + \int_{A_C} \rho\eta(\mathbf{V}\cdot\hat{\mathbf{n}})\,\mathrm{d}A_C + \int_{A_A} \rho\eta(\mathbf{V}\cdot\hat{\mathbf{n}})\,\mathrm{d}A_A \tag{1.31}$$

The entire control surface, A, of Figure 1.11 consists of three different surfaces, that is

$$A = A_A + A_C + A_r$$

where the control surface A_r is characterized by either $\beta = 90^\circ$ or $\mathbf{V} = 0$, so that there is no flow through this surface. Therefore, the last two integrals of Equation (1.31) may be represented by a single integral over the control surface to yield

$$\frac{\mathrm{D}N}{\mathrm{d}t} = \frac{D}{\mathrm{d}t}\int_{\forall_S} \rho\eta\,\mathrm{d}\forall = \frac{\partial}{\partial t}\int_{\forall} \rho\eta\,\mathrm{d}\forall + \int_{A} \rho\eta(\mathbf{V}\cdot\hat{\mathbf{n}})\,\mathrm{d}A \tag{1.32}$$

Equation (1.32) is known as **Reynolds' transport theorem**. The term on the left-hand side of this equation represents the total rate of change of any extensive property of the system, the first term on the right-hand side is the rate of change of the arbitrary extensive property within the control volume, and the second term on the right-hand side is the net rate of efflux of the extensive property through the control surface.

1.8 CLASSIFICATION OF FLUID FLOW

It is possible to classify the fluid flow into several different types by using different viewpoints.

1.8.1 Steady and unsteady flows

If the properties of a flow field do not change with time, then the flow is said to be **steady**. Mathematically, the definition of a steady flow can be stated as

$$\partial\eta/\partial t = 0 \tag{1.33}$$

where η represents any fluid property. Therefore, in a steady flow, the properties may vary from point to point in the flow field, but they remain constant with time at any point. In an **unsteady flow**, one or more variables at a point do change with time.

It should be noted that, although in a steady flow, the partial derivatives with respect to time are zero, that is $\partial/\partial t = 0$, and the total derivatives with respect to time, which also include the convective derivatives, are non-zero, that is $\mathrm{D}/\mathrm{d}t \neq 0$.

To understand the physical difference between a steady and an unsteady flow, consider the motion of the body in Figure 1.14 with respect to an observer on the ground. The body is moving at a constant speed, V_0, into a fluid which is at rest. At time t_0, the fluid particle, which is initially at rest, is well ahead of the body as shown in Figure 1.14(a). In Figure 1.14(b), the fluid particle is accelerated by the approaching body at time $t_0 + \Delta t$. It is then decelerated at time $t_0 + 2\,\Delta t$ as the body passes near the fluid particle, as indicated in Figure 1.14(c). Hence, the motion of the fluid particle and the body itself are dependent on time with respect to an observer on the ground.

Now, if the same motion is investigated by an observer moving with the body, the flow becomes steady, as shown in Figure 1.15. In this case, the fluid is travelling towards the stationary body with a velocity equal and opposite to that of the body. Fluid particles A, B and C pass through the control volume, which is fixed relative to an observer moving with the body, as shown in Figures 1.15(a), 1.15(b) and 1.15(c) at times t_0, $t_0 + \Delta t$ and $t_0 + 2\,\Delta t$, respectively. Also, the position of particle A at time $t_0 + \Delta t$ and the positions of particles A and B at time $t_0 + 2\Delta t$ are indicated by squares in Figures 1.15(b) and 1.15(c), respectively. Therefore, the velocity and the other properties of fluid particles that pass through a specific location are

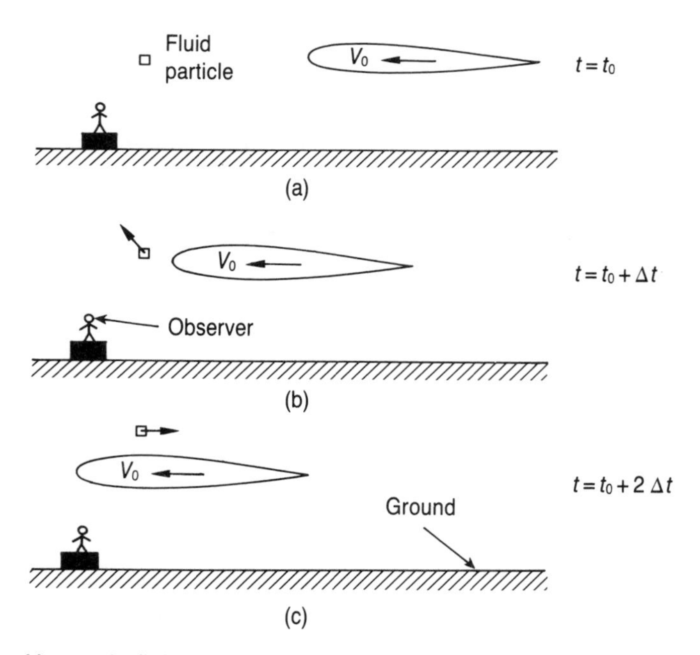

Figure 1.14 Unsteady fluid flow about a body as seen by an observer on the ground

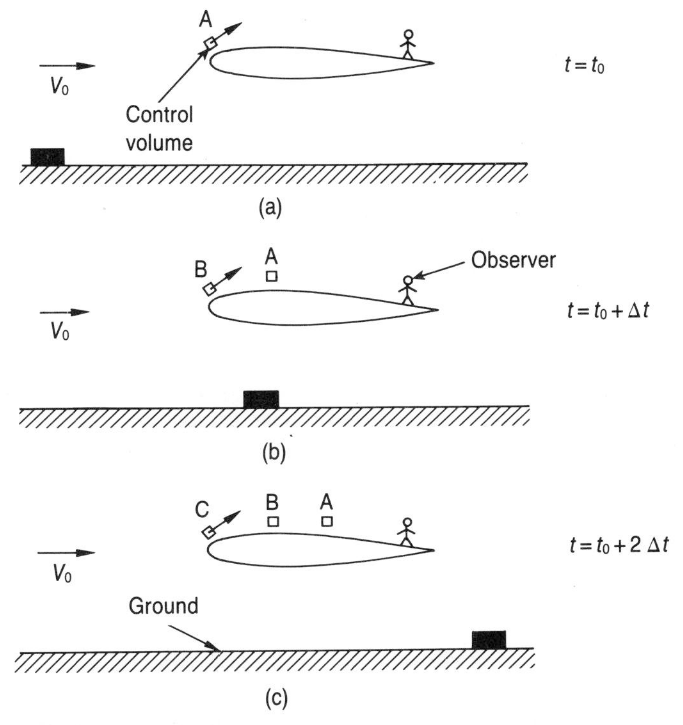

Figure 1.15 Steady fluid flow about a body as seen by an observer moving with the body

independent of time, and the flow is steady relative to a coordinate system which is attached to the body. Hence, whether or not the flow is steady depends on the choice of the coordinate system.

1.8.2 One-dimensional flow

In general, compressible flow is three dimensional in the sense that the properties such as the pressure, the density and the three components of the velocity may vary in all mutually perpendicular directions. As long as the information given in this book forms a basis for the study of the more advanced topics of two or three-dimensional compressible fluid flows, the discussion is limited to one-dimensional and quasi-one-dimensional flows.

No approximation is involved in the case of a streamtube. The flow through an infinitesimal streamtube is obviously one-dimensional.

To understand the basic idea behind this one-dimensional approximation, consider the steady flow of a compressible viscous fluid between two parallel plates, as shown in Figure 1.16. The plates have infinite length in the z direction. As long as a real fluid has to satisfy the no-slip condition at the solid boundaries, the velocity in the

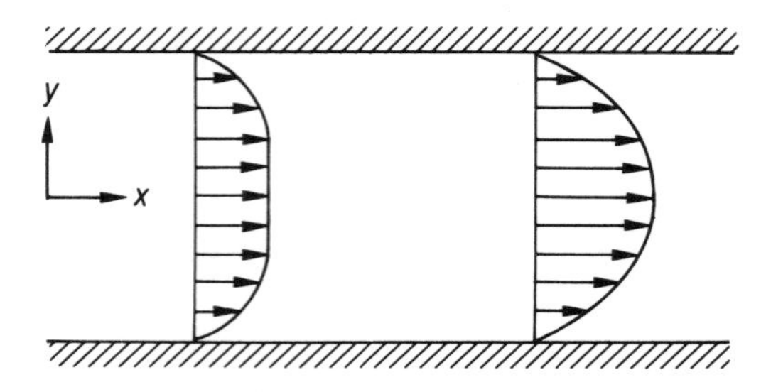

Figure 1.16 Steady flow of a compressible and viscous fluid between two parallel plates

flow direction (x direction) may be expressed as a function of x and y. Although the velocity profiles at two different sections are similar to each other, their variation may be due to external effects such as friction or heat transfer.

Although the static pressure over a cross-section may safely be assumed constant, the flow of real fluids cannot be completely one dimensional since the velocity at a solid boundary must be zero with respect to the solid boundary. In many cases, for the sake of simplicity, the properties at a cross-section can be approximated by their average values and are said to be uniform over that cross-section. As a result of this approximation, the two-dimensional steady flow field of Figure 1.16 becomes one dimensional, as indicated in Figure 1.17. The average properties vary only from one cross-section to another so that the velocity in the flow direction or any other related property is only a function of one space variable, that is x.

The one-dimensional approximation gives good results whenever the following occur:

1 The rate of change of the cross-sectional area along the duct axis is small.
2 The radius of curvature of the duct axis is very large compared with the diameter of the duct.
3 The velocity and the temperature profiles remain approximately unchanged from one section to the other along the duct.

One-dimensional flows in which there is a change of area are often referred to as

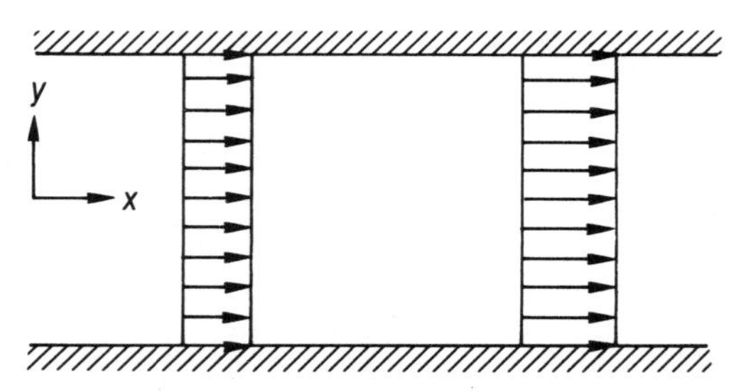

Figure 1.17 One-dimensional steady flow of a compressible and viscous fluid between two parallel plates

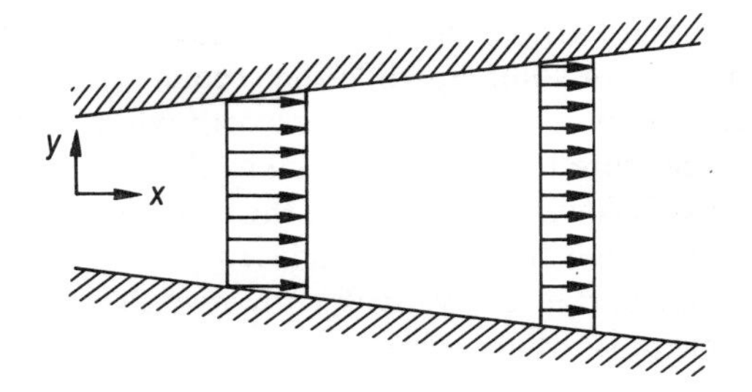

Figure 1.18 Quasi-one-dimensional steady flow of a compressible and viscous fluid

quasi-one-dimensional flows. An example of this kind of flow is the steady flow of a compressible viscous fluid between two parallel plates, as shown in Figure 1.18.

If there is a rapid change in the cross-sectional area of the duct, the one-dimensional approximation is no longer valid in the vicinity of this change of area. However, these cases, such as the flow around a sudden enlargement or contraction, may easily be treated with the one-dimensional approximation by applying the governing equations for the flow sufficiently far away both upstream and downstream of the discontinuity.

1.9 CONSERVATION OF MASS

The application of the principle of conservation of mass to a fluid flow yields an equation which is referred to as the **continuity equation**.

As long as a system is composed of the same quantity of matter at all times, the conservation of mass simply states that the mass of a system is constant. Therefore, the rate of change of the mass of a system is zero, that is

$$\frac{\mathrm{D}m_{\mathrm{S}}}{\mathrm{d}t}=0 \tag{1.34}$$

where

$$m_{\mathrm{S}}=\int_{m_{\mathrm{S}}}\mathrm{d}m \tag{1.35}$$

But $\mathrm{d}m=\rho\,\mathrm{d}\forall$, so that

$$m_{\mathrm{S}}=\int_{\forall_{\mathrm{S}}}\rho\,\mathrm{d}\forall \tag{1.36}$$

Now, if Equation (1.34) is compared with Equation (1.32), then it is possible to find that $N=m_{\mathrm{S}}$ and $\eta=1$. The application of Reynolds' transport theorem to Equation (1.34) yields

$$\frac{\mathrm{D}m_{\mathrm{S}}}{\mathrm{d}t}=\frac{\mathrm{D}}{\mathrm{d}t}\int_{\forall_{\mathrm{S}}}\rho\,\mathrm{d}\forall=\frac{\partial}{\partial t}\int_{\forall}\rho\,\mathrm{d}\forall+\int_{A}\rho(\mathbf{V}\cdot\hat{\mathbf{n}})\,\mathrm{d}A=0 \tag{1.37}$$

In Equation (1.37), the first term on the right-hand side represents the rate of change of the mass within the control volume, while the second term represents the net rate of mass efflux through the control surface. It is worth while to reiterate at this point that the velocity, $\mathbf{V}$, is measured relative to the control surface. Also, the second term is positive when the mass is flowing out through the control surface and negative when the mass is flowing in through the control surface.

For a steady flow, the partial derivatives with respect to time are zero, that is $\partial/\partial t = 0$; then Equation (1.37) reduces to

$$\int_A \rho(\mathbf{V} \cdot \hat{\mathbf{n}})\, \mathrm{d}A = 0 \tag{1.38}$$

Now, consider one-dimensional steady flow through a streamtube. The chosen control volume coincides with the portion of the streamtube shown in Figure 1.19. The fluid flows in through area A_1 of the control surface and out through area A_2 of the control surface. No flow can take place through other surfaces of the control volume, since they are formed by streamlines and $\mathbf{V} \cdot \hat{\mathbf{n}} = 0$ on those surfaces. Therefore, Equation (1.38) reduces to

$$\int_{A_1} \rho_1(\mathbf{V}_1 \cdot \hat{\mathbf{n}}_1)\, \mathrm{d}A + \int_{A_2} \rho_2(\mathbf{V}_2 \cdot \hat{\mathbf{n}}_2)\, \mathrm{d}A = 0$$

As long as the flow is one dimensional, cross-sectional areas A_1 and A_2 are perpendicular to velocities $\mathbf{V}_1$ and $\mathbf{V}_2$, respectively. Then $\mathbf{V}_1 \cdot \hat{\mathbf{n}}_1 = -V_1$ and $\mathbf{V}_2 \cdot \hat{\mathbf{n}}_2 = V_2$. Hence

$$-\int_{A_1} \rho_1 V_1\, \mathrm{d}A + \int_{A_2} \rho_2 V_2\, \mathrm{d}A = 0$$

For one-dimensional flow, the properties are uniform over each cross-section. Then the mass flow rate $\dot{m}$ is

$$\dot{m} = \rho_1 A_1 V_1 = \rho_2 A_2 V_2 \tag{1.39}$$

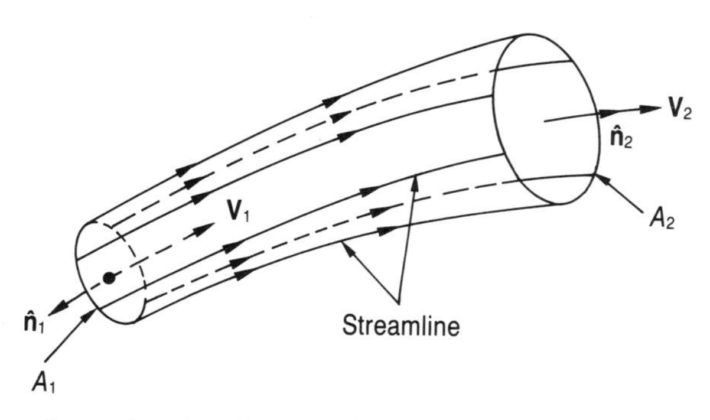

Figure 1.19 One-dimensional and steady flow through a streamtube

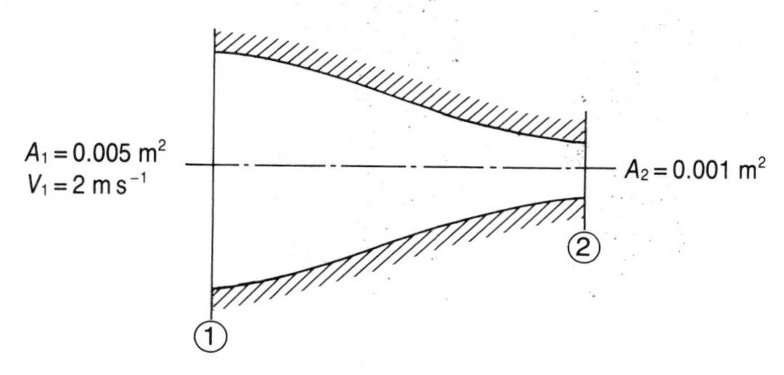

Figure 1.20 Sketch for Example 1.1

Example 1.1
A fluid flows steadily through a converging channel, as shown in Figure 1.20. The cross-sectional areas of sections 1 and 2 are 0.005 m^2 and 0.001 m^2, respectively. The velocity is 2 $m\,s^{-1}$ at section 1. Assuming uniform flow at both sections, determine the velocity at section 2 if

a the fluid is incompressible, and
b the densities at sections 1 and 2 are 1.2 $kg\,m^{-3}$ and 0.8 $kg\,m^{-3}$, respectively.

Solution
(a) For one-dimensional and steady flow of an incompressible fluid, the density is constant, so the continuity equation (1.39) becomes

$$V_2 = \frac{A_1 V_1}{A_2} = \frac{(0.005\ m^2)(2\ m\,s^{-1})}{(0.001\ m^2)} = \mathbf{10\ m\,s^{-1}}$$

Therefore, the velocity increases as the cross-sectional area decreases in the one-dimensional flow of an incompressible fluid.

(b) If the flow is one-dimensional and steady, then the continuity equation (1.39) states that

$$V_2 = \frac{\rho_1 A_1 V_1}{\rho_2 A_2} = \frac{(1.2\ kg\,m^{-3})(0.005\ m^2)(2\ m\,s^{-1})}{(0.8\ kg\,m^{-3})(0.001\ m^2)} = \mathbf{15\ m\,s^{-1}}$$

1.10 CONSERVATION OF MOMENTUM

In this section, an integral formulation of Newton's second law of motion or the conservation of linear momentum which is suitable for application to an inertial control volume is derived. An inertial control volume is a control volume which does not accelerate relative to a stationary frame of reference, that is an inertial coordinate system.

Newton's second law of motion states that the sum of the external forces acting on an inertial system of fluid particles is equal to the rate of change of linear momentum of the system of fluid particles, such that

$$\Sigma \mathbf{F} = \frac{D\mathbf{P}_S}{dt} \tag{1.40}$$

where $\Sigma \mathbf{F}$ is the sum of the external forces acting on the inertial system of fluid particles and $\mathbf{P}_S$ is the linear momentum of the system of fluid particles, such that

$$\mathbf{P}_S = \int_{m_S} \mathbf{V}\, dm \tag{1.41}$$

But $dm = \rho\, d\forall$, so that

$$\Sigma \mathbf{F} = \frac{D\mathbf{P}_S}{dt} = \frac{D}{dt} \int_{\forall_S} \rho \mathbf{V}\, d\forall \tag{1.42}$$

Now, if Equation (1.42) is compared with Equation (1.32), then it is possible to find out that the linear momentum is an extensive property and the velocity is an intensive property, that is $N = \mathbf{P}_S$ and $\eta = \mathbf{V}$. The application of Reynolds' transport theorem to Equation (1.42) yields

$$\Sigma \mathbf{F} = \frac{D}{dt} \int_{\forall_S} \rho \mathbf{V}\, d\forall = \frac{\partial}{\partial t} \int_{\forall} \rho \mathbf{V}\, d\forall + \int_A \rho \mathbf{V}(\mathbf{V} \cdot \hat{\mathbf{n}})\, dA \tag{1.43}$$

This equation is known as the **equation for the conservation of linear momentum** or simply the **linear momentum equation**. The first term on the right-hand side of Equation (1.43) represents the rate of change of linear momentum within the control volume, while the second term represents the net rate of linear momentum efflux through the control surface.

For a steady flow, the partial derivatives with respect to time are zero, that is $\partial/\partial t = 0$. Equation (1.43) reduces to

$$\Sigma \mathbf{F} = \int_A \rho \mathbf{V}(\mathbf{V} \cdot \hat{\mathbf{n}})\, dA \tag{1.44}$$

Now, consider one-dimensional steady flow through a streamtube. The chosen control volume coincides with the portion of the streamtube shown in Figure 1.19. The fluid flows in through area A_1 of the control surface and out through area A_2. No flow can take place through other surfaces of the control volume, since they are formed by streamlines and $\mathbf{V} \cdot \hat{\mathbf{n}} = 0$ on those surfaces. Therefore, Equation (1.44) reduces to

$$\Sigma \mathbf{F} = \int_{A_1} \rho_1 \mathbf{V}_1(\mathbf{V}_1 \cdot \hat{\mathbf{n}}_1)\, dA + \int_{A_2} \rho_2 \mathbf{V}_2(\mathbf{V}_2 \cdot \hat{\mathbf{n}}_2)\, dA$$

As long as the flow is one-dimensional, the cross-sectional areas A_1 and A_2 are perpendicular to velocities $\mathbf{V}_1$ and $\mathbf{V}_2$, respectively. Then $\mathbf{V}_1 \cdot \hat{\mathbf{n}}_1 = -V_1$ and $\mathbf{V}_2 \cdot \hat{\mathbf{n}}_2 = V_2$. Also, for one-dimensional flows, it is possible to express the vector equation as a scalar one. Hence,

$$\Sigma F = -\int_{A_1} \rho_1 V_1^2 \, dA + \int_{A_2} \rho_2 V_2^2 \, dA$$

For one-dimensional flows, the properties are uniform over each cross-section. Thus

$$\Sigma F = \rho_2 A_2 V_2^2 - \rho_1 A_1 V_1^2 \tag{1.45}$$

or using the continuity equation (1.39)

$$\Sigma F = \dot{m}(V_2 - V_1) \tag{1.46}$$

Example 1.2

A compressible and viscous fluid is flowing steadily through a converging duct, as shown in Figure 1.21. The cross-sectional areas of sections 1 and 2 are 0.01 m^2 and 0.005 m^2, respectively. The velocity, the density and the pressure at section 1 are 50 $m\,s^{-1}$, 1.5 $kg\,m^{-3}$ and 200 kPa, respectively, while the velocity and the pressure at section 2 are 150 $m\,s^{-1}$ and 150 kPa, respectively. Find the force exerted by the fluid on the converging duct.

Solution

As long as the flow is steady, the mass flow rate is constant across each cross-section. Then the density at section 2 may be evaluated as

$$\rho_2 = \frac{\rho_1 A_1 V_1}{V_2 A_2} = \frac{(1.5 \text{ kg m}^{-3})(0.01 \text{ m}^2)(50 \text{ m s}^{-1})}{(150 \text{ m s}^{-1})(0.005 \text{ m}^2)} = 1 \text{ kg m}^{-3}$$

The force exerted on the fluid may be evaluated by applying the momentum equation (1.45) as

$$F_{\text{fluid}} + p_1 A_1 - p_2 A_2 = \rho_2 V_2^2 A_2 - \rho_1 V_1^2 A_1$$

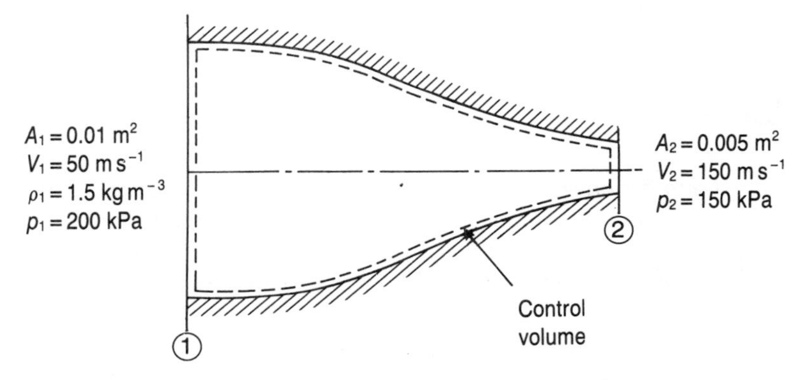

Figure 1.21 Sketch for Example 1.2

or

$$F_{\text{fluid}} = (1\ \text{kg m}^{-3})(150\ \text{m s}^{-1})^2(0.005\ \text{m}^2) - (1.5\ \text{kg m}^{-3})(50\ \text{m s}^{-1})^2(0.01\ \text{m}^2)$$

$$- (200\,000\ \text{N m}^{-2})(0.01\ \text{m}^2) + (150\,000\ \text{N m}^{-2})(0.005\ \text{m}^2) = -1175\ \text{N}$$

However, the force exerted on the duct by the fluid is equal in magnitude but opposite in direction:

$$F_{\text{duct}} = -F_{\text{fluid}} = \mathbf{1175\ N}$$

FURTHER READING

Bertin, J. J. and Smith, M. L. (1979) *Aerodynamics for Engineers*, Prentice Hall: Englewood Cliffs, NJ.

Daugherty, R. L., Franzini, J. B. and Finnemore, E. J. (1985) *Fluid Mechanics with Engineering Applications*, 8th edn, McGraw-Hill: New York.

Fox, R. W. and McDonald, A. T. (1985) *Introduction to Fluid Mechanics*, 3rd edn, Wiley: New York.

Owczarek, J. A. (1968) *Introduction to Fluid Mechanics*, International Textbook: Scranton, PA.

Prandtl, L. and Tietjens, O. G. (1957) *Fundamentals of Hydro- and Aero-mechanics*, Dover: New York.

Sabersky, R. H., Acosta, A. J. and Hauptmann, E. G. (1971) *Fluid Flow*, 2nd edn, Macmillan: New York.

Shames, I. H. (1982) *Mechanics of Fluids*, 2nd ed, McGraw-Hill: New York.

Streeter, V. L. and Wylie, E. B. (1983) *Fluid Mechanics*, McGraw-Hill: New York.

Vennard, J. K. and Street, R. L. (1982) *Elementary Fluid Mechanics*, 7th edn, Wiley: New York.

PROBLEMS

1.1 For quasi-one-dimensional and steady flow of a gas, the density, the cross-sectional area and the velocity at the inlet of a converging duct are 2 kg m^{-3}, 0.75 m^2 and 25 m s^{-1}, respectively. If the density and the cross-sectional area at the exit are 1.6 kg m^{-3} and 0.25 m^2, respectively, determine the exit velocity.

1.2 The flow area and the flow conditions at the inlet (station 1) and at the exit (station 2) of the gas turbine engine are shown in the figure. Some of the incoming air is discharged at an intermediate station by a bleed valve. The amount of bleed air is $\dot{m}_b = 0.1\dot{m}_1$. In the engine, the jet fuel is injected into the air and burned, increasing the temperature and the pressure of the air, as well as changing its composition. Assume that the inlet and the exit gases are perfect with gas constant 287.1 J kg^{-1} K^{-1}. Find the amount of fuel that must be injected, assuming steady and uniform flow. (METU Middle East Technical University 1985)

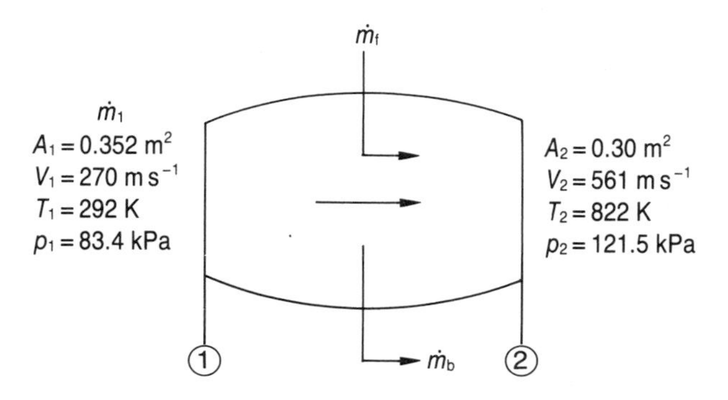

Problem 1.2

1.3 The velocity and density of a one-dimensional flow at sections 1 and 2 are shown in the figure below. If the properties are uniform over each cross-section, then determine whether

a the flow is compressible or not, and
b the flow is steady or not.

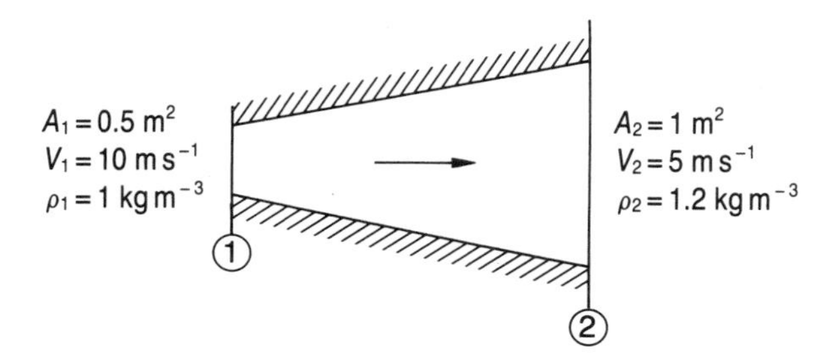

Problem 1.3

1.4 Oxygen escapes from a large tank with a volume of 2 m^3 through a small hole with a cross-sectional area of 1 mm^2. The mass flow rate of oxygen flowing through the hole is given as $\dot{m} = 0.68pA/\sqrt{(RT)}$, where p is the pressure in the tank, A is the cross-sectional area of the hole, R is the gas constant and T is the temperature in the tank. The escape of oxygen is so slow that the temperature in the tank remains constant at 15 °C. Determine the time required for the pressure in the tank to decrease from 600 kPa to 300 kPa, if the gas constant for oxygen is 259.8 J kg^{-1} K^{-1}.

1.5 A tank with a volume of 0.5 m^3 contains air at a pressure of 500 kPa and a temperature of 25 °C. At time $t = 0$, a hole with a cross-sectional area of 100 mm^2 is pierced so that the air escapes from the tank at a velocity of 346 m s^{-1}. Assuming that the properties in the tank are uniform, determine the instantaneous rate of change of the density in the tank at time $t = 0$. The gas constant for air is 287.1 J kg^{-1} K^{-1}.

1.6 A compressible and viscous fluid is flowing steadily through a diverging duct, as shown in the figure below. The cross-sectional areas of sections 1 and 2 are 0.001 m^2 and 0.005 m^2, respectively. The velocity, density and pressure at section 1 are 150 m s^{-1}, 2.5 kg m^{-3} and 200 kPa, respectively, while the velocity and pressure at section 2 are 50 m s^{-1} and 50 kPa, respectively. Find the force exerted by the fluid on the diverging duct.

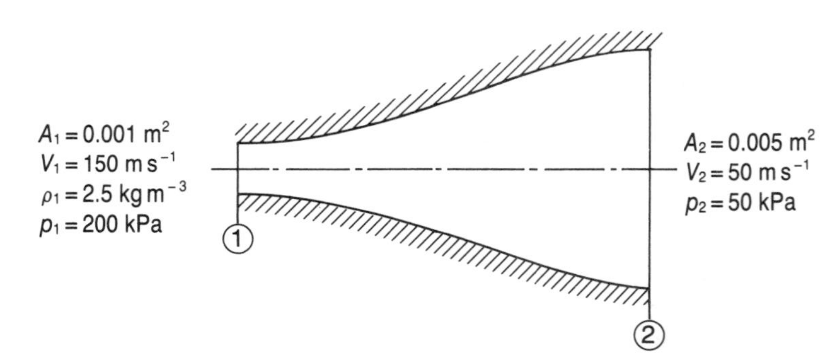

Problem 1.6

1.7 The pressure, density and velocity of the burned gases at the exit plane of the nozzle of a rocket motor, which is shown in the figure below, are p_e, ρ_e and V_e, respectively. The entire motor is surrounded by atmospheric pressure p_{atm}. The velocities of the entering propellant and the air as well as any forces due to the pressure in the feed pipes may be neglected. The flow may be assumed to be uniform at the exit of the nozzle. Determine the net thrust of the rocket motor. (METU 1983)

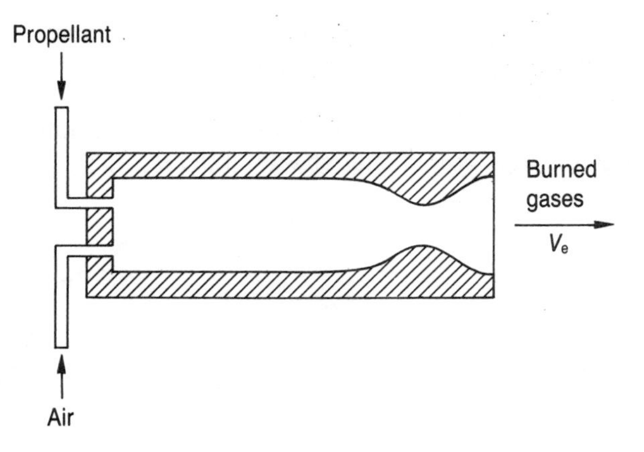

Problem 1.7

1.8 A water jet pump, which is shown in the figure below, has a jet cross-sectional area of 0.01 m^2 and a jet speed of 25 m s^{-1}. The secondary stream of water surrounding the jet has a speed of 5 m s^{-1}. The total cross-sectional area of the duct, which is equivalent to the sum of the cross sectional areas of the jet and the secondary stream, is 0.1 m^2. Downstream of the duct, the jet and the secondary stream are completely mixed, and the flow leaves the jet pump in a uniform stream. The pressures of the jet and the secondary stream are the same at the inlet of the jet pump. Determine

a the speed at the exit of the jet pump, and
b the pressure rise across the jet pump.

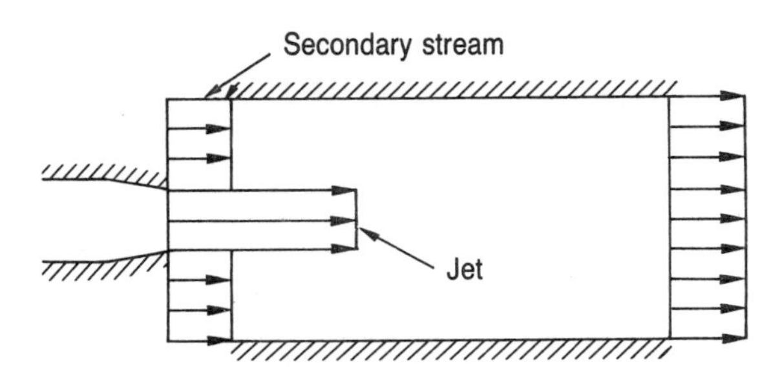

Problem 1.8

1.9 The relative velocity of air at the intake of an F-16 (Fighting Falcon) engine is 100 m s^{-1} on landing at an airport. The pressure and density at the intake are 100 kPa and 1.25 kg m^{-3}, respectively. The cross-sectional area of the air intake is 2 m^2. The exhaust gases leaving the engine have a pressure of 100 kPa and a density of 0.5 kg m^{-3}. The exit cross-sectional area of the engine is 0.65 m^2. The increase in the mass flow rate due to the fuel injection is 4 per cent. At the exit of the engine, there is a conical thrust reversal mechanism, as shown in the figure above, to deflect the jet of exhaust gases from the engine back through 120°. Determine the thrust on the F-16 engine. Neglect frictional effects. (METU 1986)

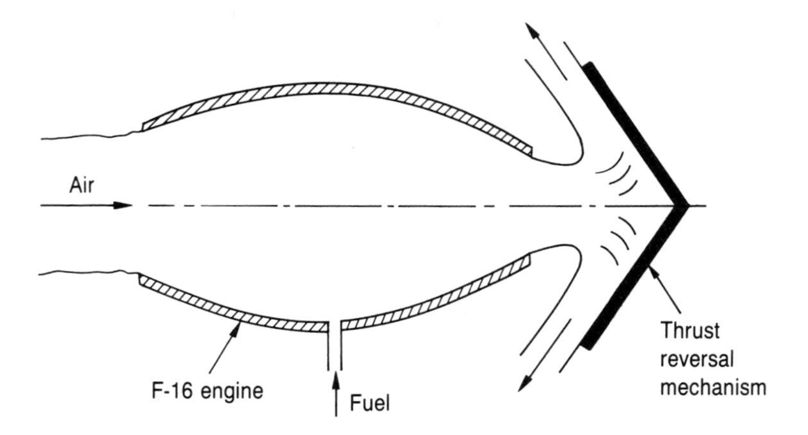

Problem 1.9

2 Fundamentals of thermodynamics

2.1 INTRODUCTION

Thermodynamics is a science which deals with heat and work, and those properties of substances that are related to heat and work. As in all branches of science, the basis of thermodynamics is experimental observation. In thermodynamics, these observations are formulated in terms of certain basic laws, which are known as the first and second laws of thermodynamics. In addition to these laws, there is the zeroth law of thermodynamics.

2.2 PROPERTIES AND STATE OF A SUBSTANCE

A **state** can be identified or described by certain observable properties, such as the pressure, temperature or density. In a given state, each of the properties of a substance has only one definite value, and these properties always have the same value for a given state independently of how the substance arrived at that state. A **property** can now be defined as the quantity that depends on the state of the system and is independent of how the system arrived at the given state. Conversely, a state is identified or described by the properties.

2.3 PROCESSES AND CYCLES

There is always a change in the state of a system whenever one or more properties of the system change. In Figure 2.1, the gas, which is confined in a piston–cylinder arrangement, can be considered as a thermodynamic system. When one or more weights on the piston are removed, the gas expands while the piston rises. As a result, a change in state occurs as the pressure decreases and the volume per unit mass increases. Finally, mechanical equilibrium will be reached and the gas will be at a new state. The path of successive states through which the system passes is called a **process**.

As long as the properties describe the state of a system when it is in equilibrium, there is a problem in describing the states of a system during an actual process which

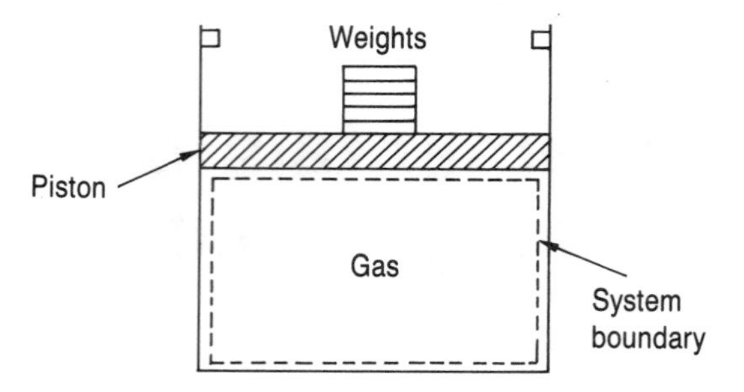

Figure 2.1 Change of state in a thermodynamic system

occurs when there is no equilibrium. This problem may be eliminated by defining a **quasi-equilibrium process** for which the deviation from thermodynamic equilibrium is infinitesimal. In this case, all the states that the system passes during a quasi-equilibrium process may be considered as quasi-equilibrium states. Therefore, if the weights are small enough and are taken off one by one, the process may be considered as quasi-equilibrium.

However, if all the weights are removed at once, the piston will rise quickly and hit the stops. Therefore, the process will be a **non-equilibrium process**. For non-equilibrium processes, the description of a system can only be made just before the process occurs and just after the process is completed and equilibrium is again restored. However, no information can be given about the states through which the system passes and the rate at which the process occurs.

A system undergoes a thermodynamic **cycle** when it goes through a number of processes and finally returns to the initial state. Therefore, when a cycle is completed, the final properties of the system are the same as the initial ones.

2.4 EQUALITY OF TEMPERATURE

Although an exact definition of temperature is quite difficult, everybody is aware of it. One is first aware of temperature as a sense of hotness and coldness. Also, when a hot and a cold body are brought together, the hot body becomes colder and the cold body becomes warmer. If these bodies remain in contact for a certain time, they will have the same sense of hotness and coldness. However, the sense of hotness and coldness is very unreliable. Sometimes, bodies of different materials that are at the same temperature might appear to have different temperatures.

Because of these difficulties in defining temperature, one might define the equality of temperature. Now, consider a hot body and a cold body, each of which is in contact with a mercury-in-glass thermometer. When these two bodies are brought into contact, the mercury column of the thermometer in contact with the hot body drops, while that in contact with the cold body rises. After a period of time, there will be no further changes and the level of the two mercury columns will be the same. In this case, one might say that the two bodies have equality of temperature.

2.5 THE ZEROTH LAW OF THERMODYNAMICS

Now, consider two bodies and one thermometer. Suppose that one of the bodies is brought into contact with the thermometer until the equality of temperature is established. Now, suppose the same thermometer is removed from the first body and brought into contact with the second one. If there is no change in the level of the mercury column, then both bodies are in thermal equilibrium.

The zeroth law of thermodynamics states that when two bodies have equality of temperature with a third body, then they in turn have equality of temperature with each other.

2.6 WORK

In mechanics, work, W, is usually defined as a force, F, acting through a displacement, x, with the displacement being in the direction of the force, that is

$$W = \int_1^2 F \, \mathrm{d}x \tag{2.1}$$

However, in thermodynamics, it is quite advantageous to relate this definition of work to the thermodynamic concepts of systems, properties and processes. Work is done by a system, if its only effect on the surroundings could be the raising of a weight. One should note that this definition does not state that a weight was actually raised. Work done by a system is considered positive, while work done on the system is negative.

To understand the definition of work clearly, consider a system formed by a battery and a motor, as shown in Figure 2.2(a). Suppose that the motor drives a fan. In this case, one should consider the question of whether the work crosses the boundaries of the system or not. To answer this question, the fan of Figure 2.2(a) may be replaced by a weight and pulley system, as shown in Figure 2.2(b). In this case, as the motor turns, the weight is raised. Therefore, the only effect external to the system is just the raising of a weight. Now, if the boundaries of the system are changed to include only the battery, then one may wonder whether or not the electrical energy crossing the system boundaries is doing work. When the bearing and electrical losses are negligible, the only external effect is the raising of a weight. Hence, when there is a flow of electricity across the system boundaries, there is also work done on or by the system.

In Figure 2.3, a gas is confined in a piston–cylinder arrangement. If one of the infinitesimal weights on the piston is removed, the piston will then rise upwards by an infinitesimal distance, $\mathrm{d}x$. If the pressure of the gas is p and the cross-sectional

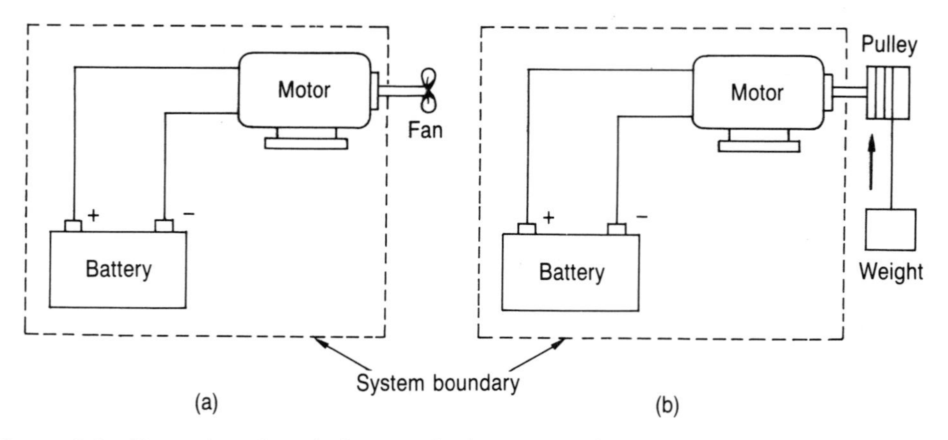

Figure 2.2 Examples of work done at the boundary of a system

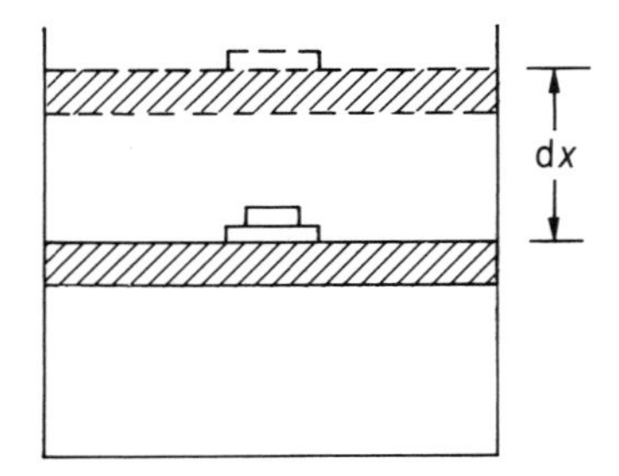

Figure 2.3 Work done by moving the boundary of a system in a quasi-equilibrium process

area of the piston is A, then the force on the piston is pA. Hence, the infinitesimal work done is

$$\delta W = pA \, dx$$

But $d\forall = A \, dx$, so that

$$\delta W = p \, d\forall \tag{2.2}$$

The work done by moving the system boundaries in a quasi-equilibrium process may be found by integrating Equation (2.2) if the relation between p and $\forall$ is known. For this reason, consider the compression of a gas from state 1 to state 2 in a piston–cylinder arrangement, as shown in Figure 2.4. If the compression is a quasi-equilibrium process, then each point on the curve connecting states 1 and 2 shows the actual states of the system, as long as the deviation from equilibrium is

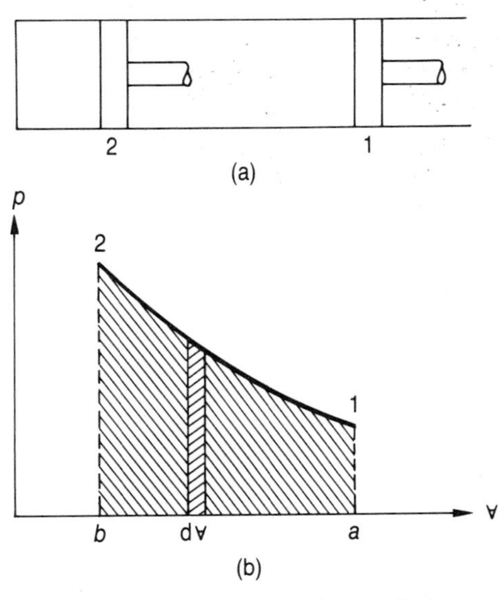

Figure 2.4 Pressure–volume diagram showing the work done at the moving boundary during a quasi-equilibrium compression process

infinitesimal. Then, the work done, ${}_1W_2$, during the compression process from state 1 to state 2 may be found by integrating Equation (2.2) as

$$ {}_1W_2 = \int_1^2 \delta W = \int_1^2 p \, \mathrm{d}\forall \tag{2.3} $$

From Figure 2.4(b), it is quite clear that the work done during the process is represented by the area under curve 1–2 of the p–$\forall$ diagram, that is by the area a–1–2–b–a.

A further investigation of the p–$\forall$ diagram of Figure 2.4(b) shows that it is possible to go from state 1 to state 2 along many different quasi-equilibrium paths, such as A, B and C, as shown in Figure 2.5. As long as the area under the curve represents the work for each process, then the amount of work involved for each process is not only a function of the end states of the process, but also a function of the path between the two states. Hence, work is a path function or δW is an inexact differential in mathematical language.

At this point, it is necessary to make a distinction between point and path functions, or exact and inexact differentials. Thermodynamic properties are point functions since there is a definite value of each property at every different state corresponding to a point on the p–$\forall$ diagram. Therefore, the differentials of point functions are exact differentials and their integration is simply

$$ \int_1^2 \mathrm{d}\forall = \forall_2 - \forall_1 \tag{2.4} $$

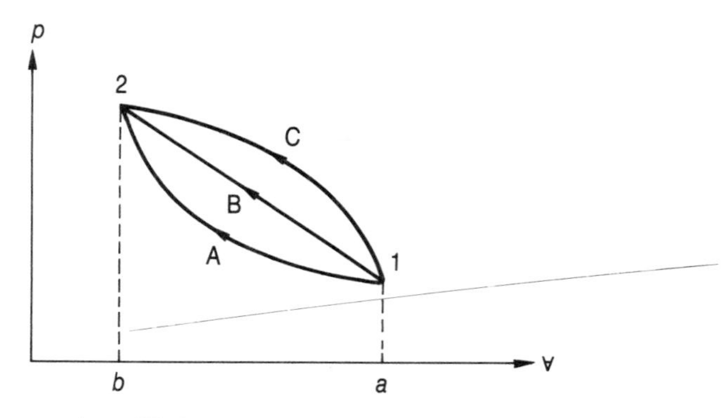

Figure 2.5 Various quasi-equilibrium processes between two given states

as in the case of the total volume. Therefore, the change in volume depends only on the initial and final states. However, on the other hand, work is a path function and the work done during a quasi-equilibrium process between two given states depends on the path followed. The differentials of path functions are inexact differentials and the symbol δ is used to designate them in contrast to d for exact differentials. Hence,

$$\int_1^2 \delta W = {}_1W_2 \tag{2.5}$$

It would be more precise to use the notation $({}_1W_2)_A$ for the work done during process A from state 1 to state 2. Also, one should note that it is not possible to talk about the work at state 1 or 2.

Example 2.1

A gas is confined in a cylinder, as shown in Figure 2.6. The cylinder is fitted with a piston on which there are a number of small weights. The initial volume and

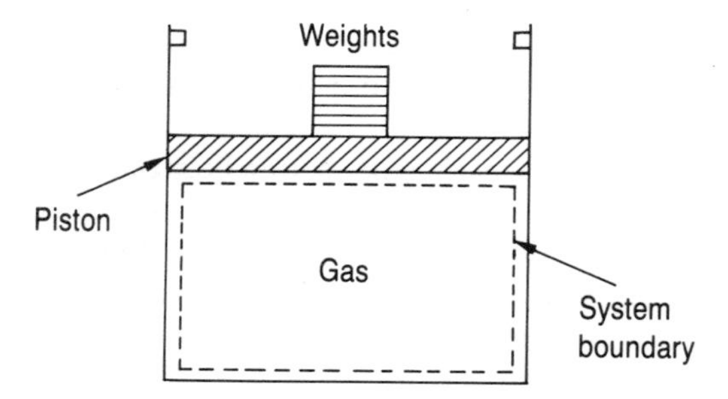

Figure 2.6 Sketch for Example 2.1

the pressure of the gas are 0.05 m^3 and 150 kPa, respectively. For a final volume of 0.1 m^3, find the work done by the system

a if the gas is heated by a Bunsen burner placed under the cylinder,
b if the gas heated by a Bunsen burner and weights are removed at a rate such that $p\forall$ = constant,
c if the weights are removed at a rate so that $p\forall^{1.3}$ = constant during the heating process, and
d if the piston is held in place by means of a pin so that the volume is constant.

Solution
(a) As the gas is heated by a Bunsen burner, the gas expands against a constant pressure due to the weight of the piston and the additional small weights. Therefore, Equation (2.3) becomes

$$_1W_2 = \int_1^2 p\, \mathrm{d}\forall = p \int_1^2 \mathrm{d}\forall = p(\forall_2 - \forall_1)$$

Substituting in the numerical values gives

$$_1W_2 = (150\,000 \text{ N m}^{-2})(0.1 \text{ m}^3 - 0.05 \text{ m}^3) = \mathbf{7500 \text{ J}}$$

(b) When $p\forall = p_1\forall_1 = p_2\forall_2$ = constant, the work done by the system can be found as

$$_1W_2 = \int_1^2 p\, \mathrm{d}\forall = p_1\forall_1 \int_1^2 \mathrm{d}\forall/\forall = p_1\forall_1 \ln(\forall_2/\forall_1)$$

Substituting in the numerical values gives

$$_1W_2 = (150\,000 \text{ N m}^{-2})(0.05 \text{ m}^3)\ln(0.1 \text{ m}^3/0.05 \text{ m}^3) = \mathbf{5199 \text{ J}}$$

(c) During the process $p\forall^{1.3} = p_1\forall_1^{1.3} = p_2\forall_2^{1.3}$ = constant, the work done by the system can be found as

$$_1W_2 = \int_1^2 p\, \mathrm{d}\forall = p_1\forall_1^{1.3} \int_1^2 \frac{\mathrm{d}\forall}{\forall^{1.3}} = -\frac{p_1\forall_1^{1.3}}{0.3}\left[\forall_2^{-0.3} - \forall_1^{-0.3}\right]$$

Substituting in the numerical values gives

$$_1W_2 = -\frac{(150\,000 \text{ N m}^{-2})(0.05 \text{ m}^3)^{1.3}}{0.3}\left[(0.1 \text{ m}^3)^{-0.3} - (0.05 \text{ m}^3)^{-0.3}\right] = \mathbf{4694 \text{ J}}$$

(d) When the volume is constant, $\mathrm{d}\forall = 0$ and the work done by the system is zero, that is

$$_1W_2 = \mathbf{0}$$

The processes that the system undergoes in parts (a), (b), (c) and (d) are indicated by the curves 1–2*a*, 1–2*b*, 1–2*c* and 1–2*d*, respectively, on the p–$\forall$ diagram of Figure 2.7. The areas 1–2*a*–*e*–*f*–1, 1–2*b*–*e*–*f*–1 and 1–2*c*–*e*–*f*–1

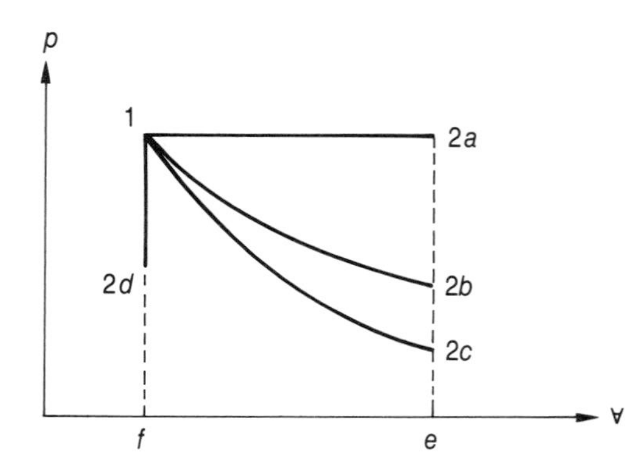

Figure 2.7 *p*–∀ diagram for Example 2.1

under the curves represent the work done by the system for the processes 1–2*a*, 1–2*b* and 1–2*c*, respectively. For the process 1–2*d*, there is no area under this curve so that the work done is zero.

2.7 HEAT

Heat is defined as the form of energy that is transferred across the boundary of a system at a given temperature to another system or surroundings at a lower temperature due to the presence of a temperature difference between them. One should note that a body never contains heat and heat can only be identified as it crosses the boundary.

Heat transfer is a transient phenomenon. When a hot body and a cold body are brought together, there will be heat transfer from the hot one to the cold one until temperature equilibrium is reached. After this point, there will no longer be heat transfer since there is no temperature difference.

Heat transfer to a system is considered to be positive, while that from a system is negative. It will be indicated by Q.

A process in which there is no heat transfer is known as an **adiabatic** process.

From a mathematical point of view, heat is also a path function like work and is an inexact differential. Therefore, the amount of heat transfer during a process from state 1 to state 2 depends on the path followed during the process, and can be given as

$$\int_1^2 \delta Q = {}_1Q_2 \tag{2.6}$$

2.8 COMPARISON OF HEAT AND WORK

At this point, one should note that there are many similarities between work and heat.

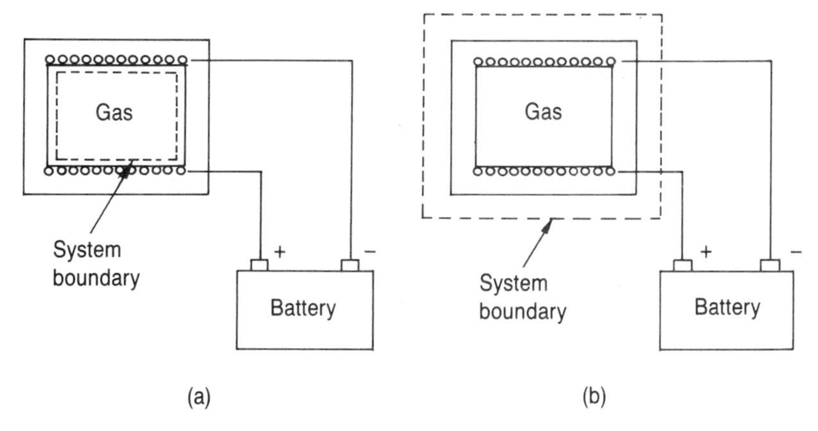

Figure 2.8 Difference between heat and work

These similarities can be summarized as follows:

1 Systems never possess work or heat.
2 Both work and heat are boundary phenomena. They are only observed when either or both cross through the system boundaries during a process, and they both represent the energy crossing the system boundaries. Therefore, they are both transient phenomena.
3 Heat and work are both path functions and so they are inexact differentials.

To illustrate the difference between heat and work, consider a gas which is confined in a rigid container. Resistance coils, which are connected to a battery, are wrapped around this container. As the electric current flows through the resistance coils, the temperature of the gas increases. Now, one may wonder whether the work or the heat crosses the system boundaries. In Figure 2.8(a), the system consists of only the gas in the container. As the electric current flows through the resistance coils, the temperature of the container walls increases. As a result, there will be heat transfer from the container to the gas since the temperature of the container is higher than the temperature of the gas. Now, consider another system which includes both the container and the resistance coils, as shown in Figure 2.8(b). In this case, electricity is crossing the system boundaries which is nothing but work.

2.9 THE FIRST LAW OF THERMODYNAMICS

In this section, the first law of thermodynamics is introduced for a system undergoing a cycle, and then modified for a system undergoing a change of state. Finally, the first law of thermodynamics for a control volume is obtained with the aid of Reynolds' transport theorem.

2.9.1 The first law of thermodynamics for a system undergoing a cycle

The first law of thermodynamics states that when a system undergoes a cycle, the cyclic integral of work is equal to the cyclic integral of heat.

To illustrate this law, consider a system of gas in a container, as shown in Figure 2.9(a), undergoing a cycle which is composed of two different processes. In the first process, as the weight is lowered, the paddle turns and work is done on the system. The system may now be returned to its initial state by transferring heat from the system to the surroundings, as shown in Figure 2.9(b).

Therefore, this law can be expressed mathematically as

$$\oint \delta Q_S = \oint \delta W_S \tag{2.7}$$

where the symbol $\oint \delta Q_s$, is known as the cyclic integral of the heat transfer undergoing the cycle, and $\oint \delta W_s$ is known as the cyclic integral of the work representing the net work during the cycle.

2.9.2 The first law of thermodynamics for a change in the state of a system

In the previous section, the first law of thermodynamics for a system during a cycle was discussed. However, most frequently, one usually deals with a process rather than a cycle. For this reason, the first law of thermodynamics for a system that undergoes a change of state must be considered. Now, consider a system which undergoes a cycle from state 1 to state 2 by process A and returns from state 2 to state 1 by process B, as shown in Figure 2.10. Then, the first law of thermodynamics for this cycle is

$$\int_{1A}^{2A} \delta Q_S + \int_{2B}^{1B} \delta Q_S = \int_{1A}^{2A} \delta W_S + \int_{2B}^{1B} \delta W_S$$

Next, consider another cycle changing from state 1 to state 2 by process A, and

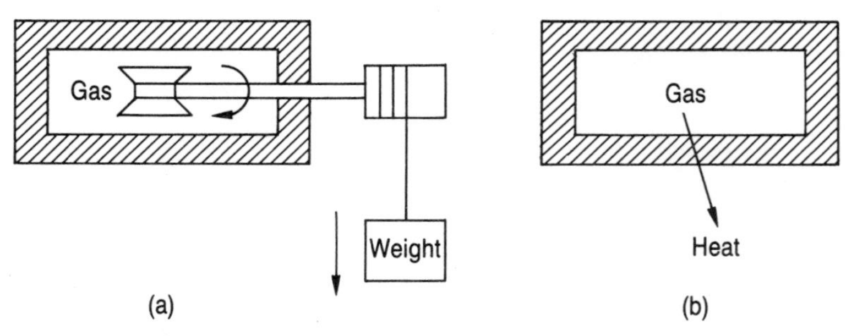

Figure 2.9 A system undergoing a cycle

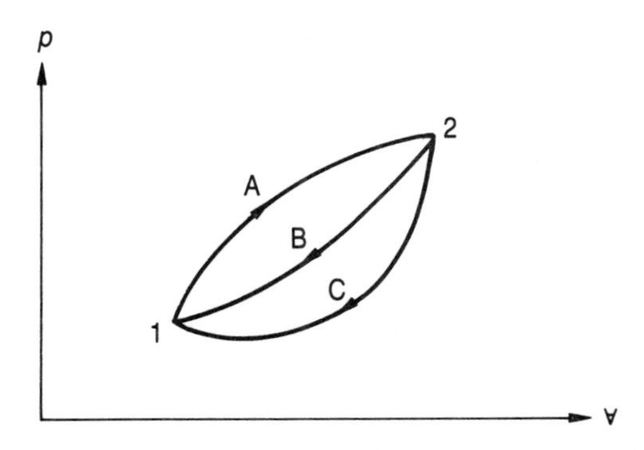

Figure 2.10 Demonstration of the existence of the energy as a thermodynamic property

returning to state 1 by process C. Then, the first law of thermodynamics for this cycle is

$$\int_{1A}^{2A} \delta Q_S + \int_{2C}^{1C} \delta Q_S = \int_{1A}^{2A} \delta W_S + \int_{2C}^{1C} \delta W_S$$

Now, subtracting the second equation from the first

$$\int_{2B}^{1B} \delta Q_S - \int_{2C}^{1C} \delta Q_S = \int_{2B}^{1B} \delta W_S - \int_{2C}^{1C} \delta W_S$$

or by rearranging

$$\int_{2B}^{1B} (\delta Q_S - \delta W_S) = \int_{2C}^{1C} (\delta Q_S - \delta W_S)$$

Since B and C are arbitrary processes between states 1 and 2, the quantity $(\delta Q_S - \delta W_S)$ is the same for all processes between states 1 and 2. Therefore, $(\delta Q_S - \delta W_S)$ depends only on the initial and final states and not on the path followed between the two states. Hence, $(\delta Q_s - \delta W_S)$ is a point function and a differential of a property of a system; it is referred to as the energy of a system. Then

$$\delta Q_S - \delta W_S = \mathrm{D}E_S \tag{2.8}$$

The property, E_S, of the system represents all the energy of the system at a given state. This energy represents the mechanical and thermal energies of a system when the other forms of energy such as chemical energy etc. are disregarded. In this case, it may be present in a variety of forms, such as kinetic or potential energy with respect to the chosen coordinate frame, energy associated with the motion and position of molecules, energy associated with the structure of the atom or in any other forms. In thermodynamics, it is convenient to consider the kinetic and potential energies separately from the other forms of energy, and all these other forms of energy may then be considered as a single property referred to as the internal energy of a system, U_S. Hence

$$E_S = U_S + (KE)_S + (PE)_S \tag{2.9}$$

where $(KE)_S$ and $(PE)_S$ represent the kinetic and the potential energies of the system, respectively.

Frequently, it is desirable to express the first law of thermodynamics as a rate equation. In this case, it is possible to consider either the instantaneous or the average rate at which the energy crosses the system boundaries in the form of work and heat and the rate at which the energy of the system changes. For this reason, one can divide Equation (2.8) by an infinitesimal time interval $\mathrm{d}t$ to obtain

$$\dot{Q}_S - \dot{W}_S = \mathrm{D}E_S/\mathrm{d}t \tag{2.10}$$

where $\dot{Q}_S = \delta Q_S/\mathrm{d}t$ is the rate of heat transfer and $\dot{W}_S = \delta W_S/\mathrm{d}t$ is the power. In Equation (2.10), the rate of heat transfer, $\dot{Q}_S$, is taken to be positive when heat is added to the system from the surroundings, and the rate of work, $\dot{W}_S$, is positive when work is done by the system on the surroundings.

2.9.3 The first law of thermodynamics for a control volume

It is possible to express the total energy possessed by the system, E_S, as

$$E_S = \int_{m_S} e \, \mathrm{d}m = \int_{\forall_S} \rho e \, \mathrm{d}\forall \tag{2.11}$$

where e is the total energy per unit mass. Then, it is possible to relate Equation (2.10) to a control volume by using Reynolds' transport theorem (1.32) as

$$\dot{Q}_S - \dot{W}_S = \frac{\mathrm{D}E_S}{\mathrm{d}t} = \frac{\mathrm{D}}{\mathrm{d}t} \int_{\forall_S} \rho e \, \mathrm{d}\forall = \frac{\partial}{\partial t} \int_{\forall} \rho e \, \mathrm{d}\forall + \int_A \rho e (\mathbf{V} \cdot \hat{\mathbf{n}}) \, \mathrm{d}A \tag{2.12}$$

The total energy per unit mass may be given as

$$e = u + \frac{V^2}{2} + gz \tag{2.13}$$

where u is the internal energy per unit mass, g is the gravitational acceleration and z is the elevation. In this case, it is possible to express the first law of thermodynamics as

$$\dot{Q}_S - \dot{W}_S = \frac{\partial}{\partial t} \int_{\forall} \rho \left(u + \frac{V^2}{2} + gz \right) \mathrm{d}\forall + \int_A \rho \left(u + \frac{V^2}{2} + gz \right) (\mathbf{V} \cdot \hat{\mathbf{n}}) \, \mathrm{d}A \tag{2.14}$$

Now, introducing a new intensive property, enthalpy, as

$$h = u + \frac{p}{\rho} \tag{2.15}$$

in this case Equation (2.14) becomes

$$\dot{Q}_S - \dot{W}_S = \frac{\partial}{\partial t}\int_{\forall} \rho\left(u + \frac{V^2}{2} + gz\right) \mathrm{d}\forall + \int_A \rho\left(h - \frac{p}{\rho} + \frac{V^2}{2} + gz\right)(\mathbf{V}\cdot\hat{\mathbf{n}})\ \mathrm{d}A \quad (2.16)$$

When mass flows across the boundaries of a control volume, it is convenient to divide the work into two parts: (i) the work necessary to push the mass across the system boundaries which is often referred to as the flow work, W_{flow}, and (ii) all the other work including the shaft work, electric work, magnetic work and the shear work, W. At this point, one should note that the flow work is positive in pushing the mass out of the control volume and negative when the surroundings push the mass into the control volume. The rate of flow work may now be determined by referring to Figure 2.11. The force acting normal to the infinitesimal area, $\mathrm{d}A$, of the control surface is $\mathrm{d}\mathbf{F}_n = p\ \mathrm{d}A\hat{\mathbf{n}}$. The rate of flow work can now be obtained by multiplying the component of $\mathrm{d}F_n$ in the direction of the velocity V by the magnitude of the velocity. Hence

$$\delta\dot{W}_{\text{flow}} = \mathrm{d}F_n V \cos\beta = \mathrm{d}\mathbf{F}_n \cdot \mathbf{V} = p(\mathbf{V}\cdot\hat{\mathbf{n}})\ \mathrm{d}A \quad (2.17)$$

Since $(\mathbf{V}\cdot\hat{\mathbf{n}})$ is positive for the mass leaving the control volume and negative for the mass entering the control volume, the rate of flow work $\delta\dot{W}_{\text{flow}}$ is also positive for the mass leaving the control volume and negative for the mass entering the control volume. This is in agreement with the sign convention that the work done by the system is positive and the work done on the system is negative. The rate of flow work can now be evaluated by integrating $\delta\dot{W}_{\text{flow}}$ over the control surface, A, as

$$\dot{W}_{\text{flow}} = \int \delta\dot{W}_{\text{flow}} = \int_A p(\mathbf{V}\cdot\hat{\mathbf{n}})\ \mathrm{d}A \quad (2.18)$$

Then, it is possible to express the rate of work done by the system as

$$\dot{W}_S = \dot{W} + \dot{W}_{\text{flow}} = \dot{W} + \int_A p(\mathbf{V}\cdot\hat{\mathbf{n}})\ \mathrm{d}A \quad (2.19)$$

Also, the rate of heat transfer to a control volume, $\dot{Q}$, is equal to the rate of heat

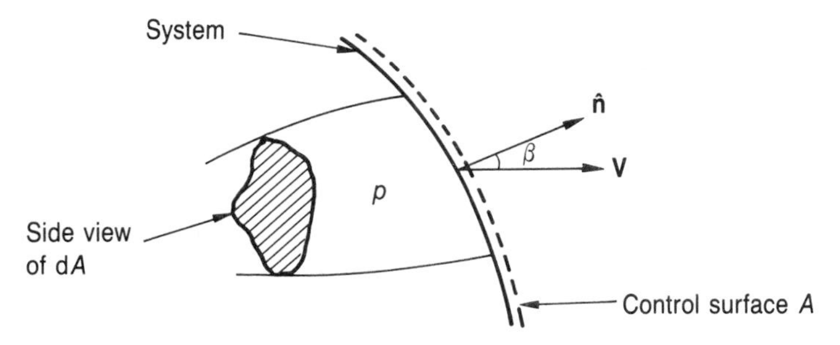

Figure 2.11 Determination of the flow work

transfer to the mass occupying the control volume, that is $\dot{Q}_S = \dot{Q}$. In this case, the first law of thermodynamics (2.16) takes the following form:

$$\dot{Q} - \dot{W} = \frac{\partial}{\partial t} \int_{\forall} \rho \left(u + \frac{V^2}{2} + gz \right) \mathrm{d}\forall + \int_A \rho \left(h + \frac{V^2}{2} + gz \right) (\mathbf{V} \cdot \hat{\mathbf{n}}) \, \mathrm{d}A \quad (2.20)$$

For a steady flow, the partial derivatives with respect to time are zero, that is $\partial / \partial t = 0$, and Equation (2.20) reduces to

$$\dot{Q} - \dot{W} = \int_A \rho \left(h + \frac{V^2}{2} + gz \right) (\mathbf{V} \cdot \hat{\mathbf{n}}) \, \mathrm{d}A \quad (2.21)$$

Now, consider one-dimensional steady flow through a streamtube. The chosen control volume coincides with the portion of the streamtube shown in Figure 1.19. The fluid flows in through area A_1 of the control surface and flows out through area A_2 of the control surface. No flow can take place through other surfaces of the control volume, since they are formed by streamlines and $\mathbf{V} \cdot \hat{\mathbf{n}} = 0$ on those surfaces. Therefore, Equation (2.21) reduces to

$$ {}_1\dot{Q}_2 - {}_1\dot{W}_2 = \int_{A_1} \rho_1 \left(h_1 + \frac{V_1^2}{2} + gz_1 \right) (\mathbf{V}_1 \cdot \hat{\mathbf{n}}_1) \, \mathrm{d}A + \int_{A_2} \rho_2 \left(h_2 + \frac{V_2^2}{2} + gz_2 \right) (\mathbf{V}_2 \cdot \hat{\mathbf{n}}_2) \, \mathrm{d}A $$

As long as the flow is one dimensional, the cross-sectional areas A_1 and A_2 are perpendicular to velocities $\mathbf{V}_1$ and $\mathbf{V}_2$, respectively. Then $\mathbf{V}_1 \cdot \hat{\mathbf{n}}_1 = -V_1$ and $\mathbf{V}_2 \cdot \hat{\mathbf{n}}_2 = V_2$. Thus

$$ {}_1\dot{Q}_2 - {}_1\dot{W}_2 = \int_{A_2} \rho_2 V_2 \left(h_2 + \frac{V_2^2}{2} + gz_2 \right) \mathrm{d}A - \int_{A_1} \rho_1 V_1 \left(h_1 + \frac{V_1^2}{2} + gz_1 \right) \mathrm{d}A $$

For one-dimensional flows, the properties are uniform over each cross-section. Then

$$ {}_1\dot{Q}_2 - {}_1\dot{W}_2 = \rho_2 V_2 \left(h_2 + \frac{V_2^2}{2} + gz_2 \right) A_2 - \rho_1 V_1 \left(h_1 + \frac{V_1^2}{2} + gz_1 \right) A_1 $$

or using the continuity equation (1.39)

$$ {}_1\dot{Q}_2 - {}_1\dot{W}_2 = \dot{m} \left[\left(h_2 + \frac{V_2^2}{2} + gz_2 \right) - \left(h_1 + \frac{V_1^2}{2} + gz_1 \right) \right] \quad (2.22a)$$

or

$$ {}_1q_2 - {}_1w_2 = \left(h_2 + \frac{V_2^2}{2} + gz_2 \right) - \left(h_1 + \frac{V_1^2}{2} + gz_1 \right) \quad (2.22b)$$

where ${}_1q_2$ is the heat transfer per unit mass and ${}_1w_2$ the work per unit mass.

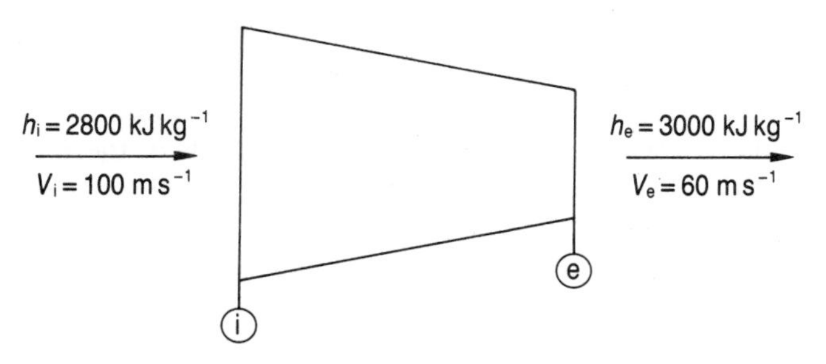

Figure 2.12 Sketch for Example 2.2

Example 2.2
Steam enters a compressor with a velocity of 100 m s^{-1} and with an enthalpy of 2800 kJ kg^{-1}, as shown in Figure 2.12. It is discharged from the compressor with a velocity and an enthalpy of 60 m s^{-1} and 3000 kJ kg^{-1}, respectively. The compressor receives a power input of 25 kW from the turbine. If the mass flow rate of the steam is 0.1 kg s^{-1}, find the heat loss to the surroundings. Neglect the changes in potential energy.

Solution
When the changes in the potential energy are negligible, the first law of thermodynamics (2.22b) becomes

$$ {}_iq_e + h_i + \frac{V_i^2}{2} = h_e + \frac{V_e^2}{2} + {}_iw_e $$

or substituting in the numerical values

$$ {}_iq_e + 2\,800\,000 \text{ J kg}^{-1} + \frac{(100 \text{ m s}^{-1})^2}{2} = 3\,000\,000 \text{ J kg}^{-1} + \frac{(60 \text{ m s}^{-1})^2}{2} - \frac{25\,000 \text{ J s}^{-1}}{0.1 \text{ kg s}^{-1}} $$

or

$$ {}_iq_e = -53\,200 \text{ J kg}^{-1} $$

Then, the rate of heat transfer is

$$ {}_i\dot{Q}_e = \dot{m}_i q_e = (0.1 \text{ kg s}^{-1})(-53\,200 \text{ J kg}^{-1}) = \mathbf{-5.32\ kW} $$

2.10 THE SECOND LAW OF THERMODYNAMICS

The first law of thermodynamics states that the cyclic integral of heat is equal to the cyclic integral of work for a system undergoing a cycle. However, the first law of thermodynamics places no restriction on the direction of flow of heat and work. For example, when heat is transferred from a system and an equal amount of work is done on the system in a cycle, then the first law of thermodynamics is satisfied. Similarly, when the directions of heat and work are reversed, the first law of thermodynamics is also satisfied. However, from previous experience, one should be aware of the fact that any proposed cycle which does not violate the first law of thermodynamics does not guarantee that the cycle will actually occur. Hence, a cycle only occurs whenever the first and the second laws of thermodynamics are satisfied at the same time. In other words, the second law of thermodynamics implies that certain processes proceed only in a certain direction but not in the opposite direction. For example, a hot cup of tea cools down owing to heat transfer to the surroundings. However, the reverse process is not possible.

2.10.1 Heat engines and heat pumps

Now, consider the system that was introduced during the discussion of the first law of thermodynamics, as shown in Figure 2.13. Work is first done on the system by turning the paddle as the weight is lowered. Then the cycle is completed by transferring heat to the surroundings. However, one should realize that the reverse cycle is impossible from previous experience. In other words, when heat is transferred to the system in the direction of the dashed arrow, the temperature of the gas increases. However, the paddle cannot be turned and the weight cannot be raised. From this discussion, it is possible to conclude that a system can operate in a cycle in which both the heat transfer and work are negative. However, it cannot operate in a cycle in which both the heat transfer and work are positive.

Now, consider that heat is transferred from a system at a higher temperature to another one at a lower temperature. Although such a process is possible, the reverse process, in which heat is transferred from a system at a lower temperature to another one at a higher temperature, is impossible. Therefore, it can be concluded that a cycle cannot be completed with heat transfer only.

As a result of these two illustrations, it is now possible to consider the heat engine and the refrigerator. The latter is often referred to as the heat pump. A heat engine is a system which operates in a cycle with net positive work and net positive heat

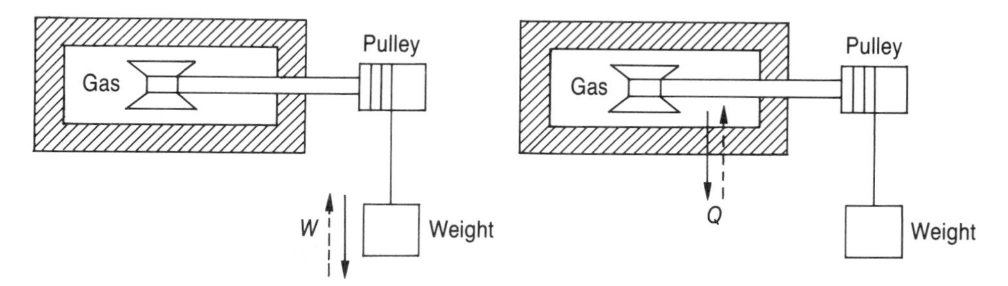

Figure 2.13 A system that undergoes a cycle involving work and heat

transfer. By using a heat pump, it is possible to transfer heat from a low-temperature body to a high-temperature one by using the necessary work required to do this.

A heat engine is a device that operates in a thermodynamic cycle and does a certain amount of net positive work as a result of heat transfer from a body at a higher temperature to another one at a lower temperature. The substance in the heat engine to and from which the heat is transferred is known as the **working substance** or the **working fluid**. A simple example of a heat engine is a simple steam power plant, as shown in Figure 2.14. During the cycle in the heat engine, an amount of heat, Q_H, is transferred from the products of combustion to the working fluid in the boiler so that the water will be in a superheated state. The superheated steam is then expanded in the turbine producing a certain amount of work. Some of this work is used to drive the pump to increase the pressure of the condensed water. Certainly, there is always net work delivered during the cycle. After the turbine, an amount of heat, Q_L, is rejected to the low-temperature body which is usually the cooling water in the condenser. Therefore, a simple steam power plant is a heat engine which does a certain amount of work as heat is transferred to and from the working fluid during a cycle.

In order to transfer heat from a low-temperature body to another one at a higher temperature, one may use a heat pump or refrigerator. A vapour compression refrigeration cycle is shown in Figure 2.15. The typical working fluid is a refrigerant, such as ammonia or Freon-12. Heat is transferred to the working fluid from the low-temperature body in the evaporator where the temperature and pressure of the working fluid are low. Work is done on the refrigerant as it is compressed in the compressor to a higher pressure. Heat is then transferred from the working fluid to

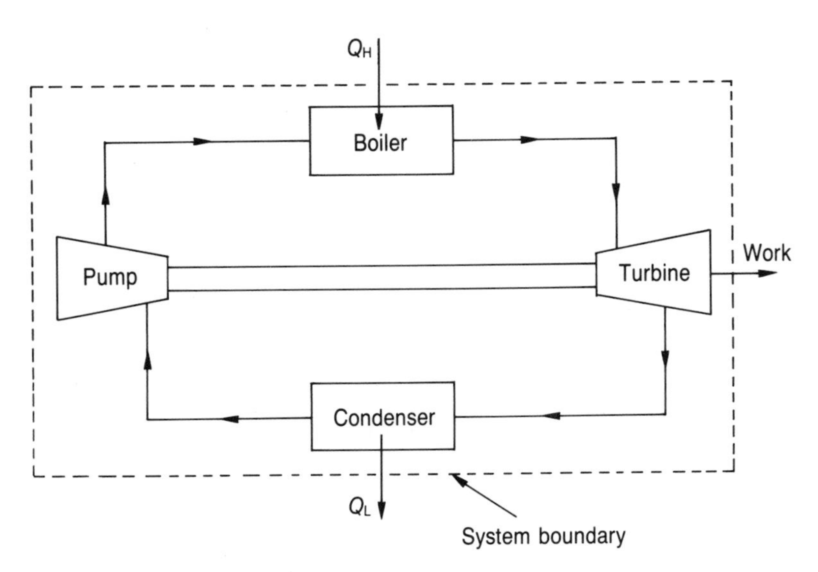

Figure 2.14 A simple heat engine

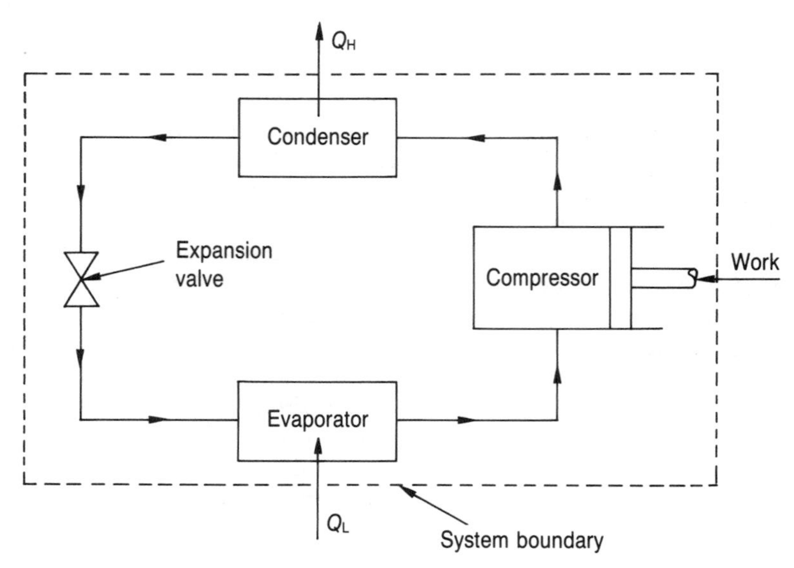

Figure 2.15 A simple heat pump

a high-temperature body in the condenser where the temperature and pressure of the working fluid are high. The pressure of the refrigerant drops as it flows through the expansion valve. Therefore, a heat pump transfers heat from a body at a lower temperature to another one at a higher temperature as a result of work input in a cycle.

2.10.2 Statement of the second law of thermodynamics

There are two classical statements of the second law of thermodynamics. These are the Kelvin–Planck and the Clausius statements:

1 *The Kelvin–Planck statement*: It is impossible to construct a device operating in a thermodynamic cycle which produces no effect other than the raising of a weight and the exchange of heat with a single reservoir. This statement is related to the discussion of a heat engine. It describes the impossibility of constructing a heat engine operating in a cycle which receives a certain amount of heat from a high-temperature body and does an equal amount of work. Such a heat engine is only possible when some of the heat is transferred to a low-temperature body. Therefore, a heat engine must always involve two different temperature levels and it is impossible to build a heat engine with 100 per cent thermal efficiency.
2 *The Clausius statement*: It is impossible to construct a device operating in a thermodynamic cycle which produces no other effect than the transfer of heat from a body at a low temperature to another one at a higher temperature. This statement is related to the discussion of the heat pump or refrigerator. It describes the impossibility of constructing a refrigerator which operates without any work input.

There are three important observations about these two statements:

1 Both are negative statements and it is impossible to prove them. However, every experiment till now verified the second law of thermodynamics directly or indirectly and no experiment that has ever been conducted contradicts the second law of thermodynamics. Hence, the second law of thermodynamics is based on experimental evidence.
2 Both are equivalent. Two statements are equivalent whenever the truth of one statement implies the truth of the other or the violation of one statement implies the violation of the other. Therefore, the violation of the Kelvin–Planck statement implies the violation of the Clausius statement.
3 The second law of thermodynamics implies the impossibility of constructing a perpetual motion machine of the second kind. A perpetual motion machine of the first kind produces mass and energy from nothing and violates the first law of thermodynamics, while a perpetual motion machine of the second kind violates the second law of thermodynamics.

2.10.3 Reversible and irreversible processes

In the previous section, the impossibility of a 100 per cent efficient heat engine was discussed. However, one may wonder about the maximum possible efficiency which may be obtained by defining an ideal process or a reversible process.

A **reversible process** for a system is one that can be reversed after it has taken place without leaving any effect on the system and the surroundings.

To illustrate the difference between a reversible and an irreversible process, consider a gas at high pressure which is confined in a cylinder–piston arrangement, as shown in Figure 2.16. The piston is held in place by means of a pin. When the pin is removed, the piston will rise abruptly and hit the stops. During this process, work is certainly done by the gas on the surroundings. In order to restore the system to its initial state, work can be done on the system so that the gas is compressed until the pin is inserted into the piston. At this point, one should note that work is done

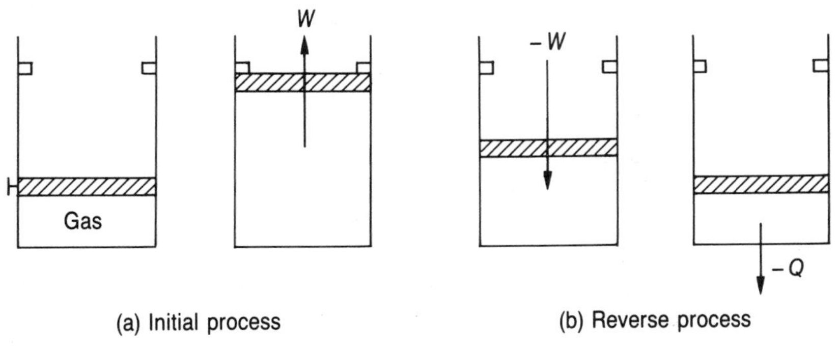

Figure 2.16 An irreversible process

by the gas against atmospheric pressure during the initial process, while work is done on the gas against the pressure on the face of the piston which is greater than atmospheric pressure during the reverse process. Therefore, the work done on the system during the reverse process is greater than the work done by the system during the initial process. During the compression process, the gas is heated up so that heat must be transferred to the surroundings in order to restore the system to its original state. At the end, although the system is restored to its original state, there are changes in the surroundings due to the net work done on the system and heat transfer to the surroundings. Hence, the initial process is an irreversible one and it cannot be reversed without leaving any effect on the surroundings.

Next, consider the gas confined in a cylinder by means of a piston, as indicated in Figure 2.17. The piston is loaded with a number of small weights. When the weights are taken off horizontally one at a time, the gas expands and work is done on the remaining weights. During the reverse process, there is always a small weight on the platform which is at the same level as the piston so that these weights can be placed on the piston without requiring any work. Hence, the process becomes reversible as the weights become smaller and smaller.

Friction, unrestrained expansion, heat transfer through a finite temperature difference and mixing of different substances render processes irreversible.

2.10.4 A Carnot cycle

Consider a heat engine which is operating between a high-temperature and a low-temperature reservoir in a cycle in which every process is reversible. As long as every process in the thermodynamic cycle is reversible, then the cycle itself is reversible. In this case, if the cycle is reversed, the heat engine becomes a refrigerator. The most efficient cycle operating between two constant-temperature reservoirs is known as the Carnot cycle.

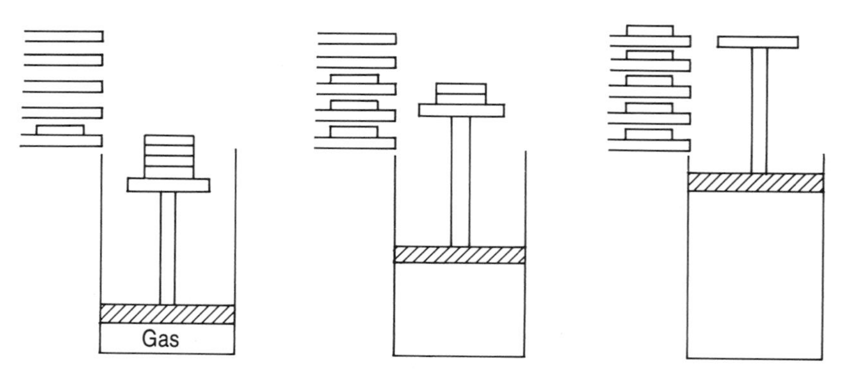

Figure 2.17 A reversible process

In Figure 2.18, a simple steam power plant, which is operating in a Carnot cycle, is shown. Assuming that the working fluid is a pure substance, such as water, the basic processes in a Carnot cycle can be summarized as follows:

1 Heat is transferred from the high-temperature reservoir to the water in the boiler. In order to have a reversible heat transfer, the temperature of the water must only be infinitesimally lower than the temperature of the reservoir. As long as the temperature of the reservoir is constant, then the temperature of the water should also be constant. This can be easily accomplished, since the change of phase from liquid to vapour at constant pressure is isothermal for a pure substance. Therefore, the first process in the Carnot cycle is the reversible isothermal heat transfer from the high-temperature reservoir to the working fluid.
2 The next process in the Carnot cycle is the adiabatic expansion process in the turbine. As long as all the processes in a Carnot cycle are reversible, then during this reversible adiabatic process the temperature of the working fluid decreases from that of the high-temperature reservoir to that of the low-temperature reservoir.
3 Then, the heat is rejected from the working fluid to the low-temperature reservoir in a reversible isothermal process. In order to achieve this, the temperature of the working fluid is infinitesimally higher than the temperature of the low-temperature reservoir. During this isothermal process, some of the steam is condensed.

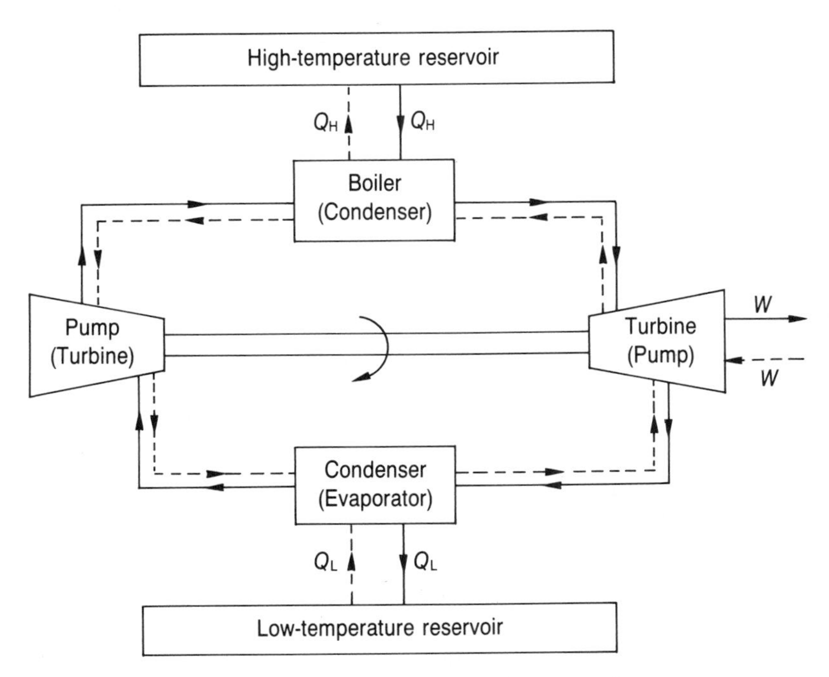

Figure 2.18 A simple power plant operating in a Carnot cycle

4 The cycle is then completed by a reversible adiabatic process in which the temperature of the working fluid increases from that of the low-temperature reservoir to that of the high-temperature reservoir. This can be accomplished by compressing the mixture of liquid and vapour from the condenser, which is very impractical. For this reason, the working fluid is completely condensed in the condenser and then handled by a pump in a completely liquid phase.

As long as the heat engine operating in a Carnot cycle is reversible, every process can be reversed. In this case, the heat engine operates as a refrigerator, as indicated by the dashed lines and parentheses in Figure 2.18.

2.10.5 The Clausius inequality

The Clausius inequality is a corollary and consequence of the second law of thermodynamics which is valid for all possible cycles, including reversible and irreversible heat engines and refrigerators, and given by

$$\oint \delta Q/T \leqslant 0 \tag{2.23}$$

To demonstrate the validity of the Clausius inequality, first consider a Carnot cycle operating between two reservoirs at temperatures T_H and T_L, as shown in Figure 2.19. For this cycle, the cyclic integral of heat transfer is greater than zero:

$$\oint \delta Q = Q_H - Q_L > 0$$

Now, introducing Kelvin's definition of absolute temperature as

$$\frac{Q_H}{T_H} = \frac{Q_L}{T_L} \tag{2.24}$$

and noting that T_H and T_L are constant, the Clausius inequality becomes

$$\oint \frac{\delta Q}{T} = \frac{Q_H}{T_H} - \frac{Q_L}{T_L} = 0$$

Now, if the cyclic integral of δQ is made to approach zero by making T_H approach T_L while the cycle remains reversible, then the cyclic integral of $\delta Q/T$ is zero. Therefore, for all reversible engine cycles

$$\oint \delta Q \geqslant 0 \quad \text{and} \quad \oint \delta Q/T = 0$$

Next, consider an irreversible cyclic heat engine operating between the same temperatures T_H and T_L as the reversible heat engine of Figure 2.19 and receiving

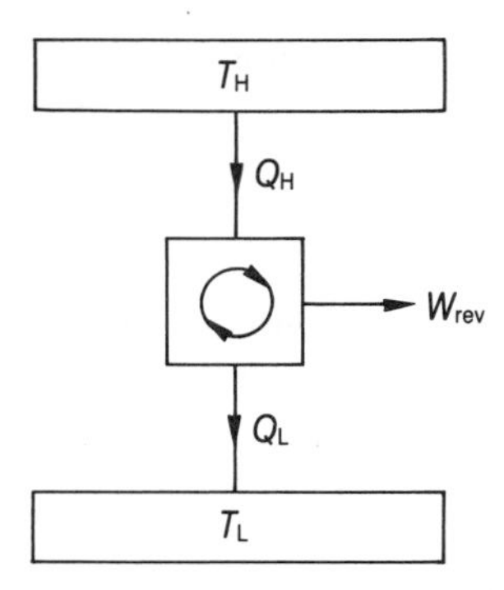

Figure 2.19 A Carnot cycle

the same quantity of heat, Q_H. Certainly, the work done by the irreversible engine is less than the work done by the reversible engine, that is

$$W_{irr} = Q_H - (Q_L)_{irr} < W_{rev} = Q_H - (Q_L)_{rev}$$

or

$$(Q_L)_{irr} > (Q_L)_{rev}$$

As a result, for the irreversible cyclic engine

$$\oint \delta Q = Q_H - (Q_L)_{irr} > 0$$

$$\oint \frac{\delta Q}{T} = \frac{Q_H}{T_H} - \frac{(Q_L)_{irr}}{T_L} < 0$$

Now, suppose that the heat engine becomes more and more irreversible while Q_H, T_H and T_L are fixed. Then, the cyclic integral of δQ approaches zero, while the cyclic integral of $\delta Q/T$ attains a progressively larger negative value. Therefore, for all irreversible engine cycles

$$\oint \delta Q \geqslant 0 \quad \text{and} \quad \oint \delta Q/T < 0$$

The demonstration of the Clausius inequality can also be proved by carrying out a similar analysis for both a reversible and an irreversible refrigeration cycle. As a summary, the equality of Equation (2.23) holds for reversible cycles and the inequality holds for irreversible cycles.

2.10.6 Entropy

Consider a system which undergoes a reversible process from state 1 to state 2 along path A. The cycle is then completed by another reversible process from state 2 to state 1 along path B. Hence

$$\oint \frac{\delta Q}{T} = \int_{1A}^{2A} \frac{\delta Q}{T} + \int_{2B}^{1B} \frac{\delta Q}{T} = 0$$

Now, consider another reversible cycle with the same initial process but with the cycle completed along path C. Then, for this cycle

$$\oint \frac{\delta Q}{T} = \int_{1A}^{2A} \frac{\delta Q}{T} + \int_{2C}^{1C} \frac{\delta Q}{T} = 0$$

Subtracting the second equation from the first:

$$\int_{2B}^{1B} \frac{\delta Q}{T} = \int_{2C}^{1C} \frac{\delta Q}{T}$$

At this point, one may conclude that $\int \delta Q/T$ is the same for all reversible paths between states 2 and 1. Therefore, it is independent of the path and only a function of the end states. Hence, it is a property. This property is known as the **entropy**, S_S, of the system and is defined as

$$DS_S = (\delta Q/T)_{rev} \qquad (2.25)$$

Consider a system that is undergoing a cycle which is formed by two reversible processes A and B. Then, for this cycle

$$\oint \frac{\delta Q}{T} = \int_{1A}^{2A} \frac{\delta Q}{T} + \int_{2B}^{1B} \frac{\delta Q}{T} = 0$$

Next, consider another cycle which is formed by a reversible process A and an irreversible process C:

$$\oint \frac{\delta Q}{T} = \int_{1A}^{2A} \frac{\delta Q}{T} + \int_{2C}^{1C} \frac{\delta Q}{T} < 0$$

Subtracting the second equation from the first:

$$\int_{2B}^{1B} \frac{\delta Q}{T} > \int_{2C}^{1C} \frac{\delta Q}{T}$$

As long as the process along path B is reversible

$$\int_{2B}^{1B} \frac{\delta Q}{T} = \int_{2B}^{1B} DS_S = \int_{2C}^{1C} DS_S$$

Therefore

$$\int_{2C}^{1C} DS_S > \int_{2C}^{1C} \frac{\delta Q}{T}$$

Hence, for an irreversible process

$$DS_S > (\delta Q/T)_{irr} \tag{2.26}$$

Finally, for the very general case

$$DS_S \geqslant \delta Q/T \tag{2.27}$$

2.10.7 The second law of thermodynamics for a system

Now, the rate of change of entropy for a system may be obtained from Equation (2.27) as

$$\frac{DS_S}{dt} \geqslant \frac{\dot{Q}}{T} \tag{2.28}$$

where the total entropy, S_S, of the system is given by

$$S_S = \int_{m_S} s \, dm = \int_{\forall_S} \rho s \, d\forall \tag{2.29}$$

with s being the entropy per unit mass. The inequality in Equation (2.28) holds for irreversible processes, while the equality holds for reversible processes.

2.10.8 The second law of thermodynamics for a control volume

By using Reynolds' transport theorem, it is possible to write the second law of thermodynamics for a control volume as

$$\frac{\dot{Q}}{T} \leqslant \frac{DS_S}{dt} = \frac{D}{dt} \int_{\forall_S} \rho s \, d\forall = \frac{\partial}{\partial t} \int_{\forall} \rho s \, d\forall + \int_A \rho s (\mathbf{V} \cdot \hat{\mathbf{n}}) \, dA \tag{2.30}$$

For a steady flow, the partial derivatives with respect to time are zero, that is $\partial/\partial t = 0$, and Equation (2.30) reduces to

$$\frac{\dot{Q}}{T} \leqslant \int_A \rho s (\mathbf{V} \cdot \hat{\mathbf{n}}) \, dA \tag{2.31}$$

Now, consider one-dimensional steady flow through a streamtube. The chosen control volume coincides with the portion of the streamtube shown in Figure 1.19. The fluid flows in through area A_1 of the control surface and flows out through area A_2 of the control surface. No flow can take place through other surfaces of the control volume, since they are formed by streamlines and $\mathbf{V} \cdot \hat{\mathbf{n}} = 0$ on those surfaces. Therefore, Equation (2.31) reduces to

$$\frac{\dot{Q}}{T} \leqslant \int_{A_1} \rho_1 s_1 (\mathbf{V}_1 \cdot \hat{\mathbf{n}}_1) \, dA + \int_{A_2} \rho_2 s_2 (\mathbf{V}_2 \cdot \hat{\mathbf{n}}_2) \, dA$$

As long as the flow is one-dimensional, the cross-sectional areas A_1 and A_2 are perpendicular to velocities $\mathbf{V}_1$ and $\mathbf{V}_2$, respectively. Then $\mathbf{V}_1 \cdot \hat{\mathbf{n}}_1 = -V_1$ and $\mathbf{V}_2 \cdot \hat{\mathbf{n}}_2 = V_2$. Thus

$$\frac{\dot{Q}}{T} \leqslant \int_{A_2} \rho_2 s_2 V_2 \, dA - \int_{A_1} \rho_1 s_1 V_1 \, dA$$

For one-dimensional flows, the properties are uniform over each cross-section. Then

$$\frac{\dot{Q}}{T} \leqslant \rho_2 s_2 V_2 A_2 - \rho_1 s_1 V_1 A_1$$

or using the continuity equation (1.39)

$$\frac{\dot{Q}}{T} \leqslant \dot{m}(s_2 - s_1) \tag{2.32}$$

When the process is adiabatic, there is no heat transfer, so Equation (2.32) reduces to

$$s_2 \geqslant s_1 \tag{2.33}$$

When the thermodynamic process is also reversible and adiabatic, which is known as an isentropic process, Equation (2.32) simplifies to

$$s_1 = s_2 \tag{2.34}$$

2.11 EQUATION OF STATE

A gas is usually a collection of particles involving molecules, atoms, ions, etc., which are in a random motion. There is a force field around these particles due to their electronic nature. In general, the force field of one particle is affected by the force fields of neighbouring particles. For this reason, these force fields are often referred to as intermolecular force fields. The intermolecular force varies with the distance between the particles. In general, the intermolecular force is a weak force of attraction when the particles are far away from each other, but it becomes a strong force of repulsion as they come closer. Therefore, these intermolecular forces affect the motion of the particles and certainly the thermodynamic properties of gases.

However, for most compressible flow applications, gas particles are, on average, far away from each other. When the average distance between the molecules is greater than ten molecular diameters, the molecules exert a very weak attractive force on each other. Therefore, for many engineering applications, the effect of the intermolecular forces on the properties of a gas may be neglected. **A perfect gas can now be defined as a gas in which the intermolecular forces are negligible.**

When the intermolecular forces are negligible, the equation of state for a perfect gas can be derived by using the theoretical concepts of statistical mechanics or kinetic theory as

$$p\forall = mRT \tag{2.35a}$$

or

$$p = \rho RT \tag{2.35b}$$

where R is the gas constant of a particular gas and related to the universal gas constant, $\bar{R}$, by

$$\bar{R} = R\bar{M} = 8314\ \mathrm{J\,kg^{-1}\ mol^{-1}\ K^{-1}} \tag{2.36}$$

with $\bar{M}$ being the molecular mass.

At low pressures less than 100 kPa and at high temperatures greater than 273 K, the deviation of the value of $p/(\rho RT)$ from unity is less than 1 per cent. However, at low temperatures and high pressures, the molecules are packed together more closely so that the intermolecular forces are dominant. In this case, a gas is often referred to as a **real gas**.

The internal energy and the enthalpy can be related to the temperature by using the following relations:

$$\left(\frac{\partial u}{\partial T}\right)_{\mathrm{v}} = c_{\mathrm{v}} \tag{2.37}$$

and

$$\left(\frac{\partial h}{\partial T}\right)_{\mathrm{p}} = c_{\mathrm{p}} \tag{2.38}$$

where c_{v} and c_{p} are the specific heats at constant volume and constant pressure, respectively. Then, the specific heat ratio, k, may be defined as

$$k = c_{\mathrm{p}}/c_{\mathrm{v}} \tag{2.39}$$

For a perfect gas, the specific heats at constant volume and constant pressure can be taken as constants. It is then possible to express Equations (2.37) and (2.38) as

$$u = c_{\mathrm{v}}T \tag{2.40}$$

and

$$h = c_{\mathrm{p}}T \tag{2.41}$$

Substitution of Equations (2.35b), (2.40) and (2.41) into Equation (2.15) yields the following relation:

$$R = c_p - c_v \tag{2.42}$$

or, using Equation (2.39), it is possible to obtain

$$c_p = \frac{kR}{k-1} \tag{2.43}$$

and

$$c_v = \frac{R}{k-1} \tag{2.44}$$

The following general relations between thermodynamic properties may be recalled:

$$T\,ds = dh - \frac{dp}{\rho} \tag{2.45}$$

and

$$T\,ds = du - RT\frac{d\rho}{\rho} \tag{2.46}$$

Using Equations (2.35b), (2.40) and (2.41), it is possible to express Equations (2.45) and (2.46) as

$$ds = c_p\frac{dT}{T} - R\frac{dp}{p} \tag{2.47}$$

and

$$ds = c_v\frac{dT}{T} - R\frac{d\rho}{\rho} \tag{2.48}$$

Integrating Equations (2.47) and (2.48) between two states yields

$$s_2 - s_1 = c_p \ln\left(\frac{T_2}{T_1}\right) - R\ln\left(\frac{p_2}{p_1}\right) \tag{2.49}$$

and

$$s_2 - s_1 = c_v \ln\left(\frac{T_2}{T_1}\right) - R\ln\left(\frac{\rho_2}{\rho_1}\right) \tag{2.50}$$

For an isentropic process, the entropy is constant and it is possible to express Equations (2.49) and (2.50) with the aid of Equations (2.43) and (2.44) as

$$p/\rho^k = \text{constant} \tag{2.51}$$

FURTHER READING

Anderson, J. D. Jr (1982) *Modern Compressible Flow with Historical Perspective*, McGraw-Hill: New York.

John, J. E. A. (1984) *Gas Dynamics*, 2nd edn, Allyn & Bacon: Boston, MA.

Shapiro, A. H. (1953) *The Dynamics and Thermodynamics of Fluid Flow*, Vol. 1, Ronald Press: New York.

Van Wylen, G. J. and Sonntag, R. E. (1973) *Fundamentals of Classical Thermodynamics*, 2nd edn, Wiley: New York.

Zucrow, M. J. and Hoffman, J. D. (1976) *Gas Dynamics*, Vol. 1, Wiley: New York.

PROBLEMS

2.1 Oxygen is confined in a piston–cylinder arrangement. There are a number of small weights on the piston. The initial pressure and the volume of the gas are 150 kPa and 0.03 m^3, respectively. For a final volume of 0.09 m^3, find the work done by the system

a if the gas is heated by a Bunsen burner placed under the cylinder,
b if the gas is heated by a Bunsen burner and at the same time weights are removed at a rate such that $p\forall$ = constant, and
c if the weights are removed at a rate so that $p\forall^{1.2}$ = constant during the heating process.

2.2 A gas is confined in a piston–cylinder arrangement, as shown in the figure below. The initial volume and pressure of the gas inside the cylinder are 0.1 m^3 and 120 kPa, respectively. At this state, the pressure balances both atmospheric pressure and the weight of the piston. At the same time, although the spring touches the piston it exerts no force on the piston. The gas is then heated until the volume is doubled. The final pressure of the gas

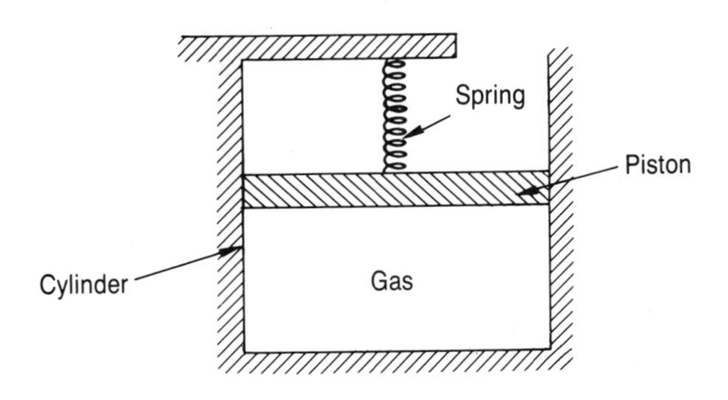

Problem 2.2

is 300 kPa. The spring force is proportional to the displacement from the original position during the process.

a Show the process on a p–$\forall$ diagram.
b Determine the work done by the system.
c What percentage of this work is done against the spring?

2.3 Nitrogen is contained in a cylinder fitted with a piston. There are some weights on the piston. The initial volume of nitrogen is 0.1 m^3 while the initial pressure and temperature are 420 kPa and 5 °C, respectively. The nitrogen is then heated so that the final volume is 0.2 m^3. Calculate the work done and the heat transfer during the process. Assume constant specific heats.

2.4 Steam enters a turbine with an enthalpy per unit mass of 3200 kJ kg^{-1} and a velocity of 60 m s^{-1}. At the exit, the enthalpy per unit mass and the velocity of the steam are 2700 kJ kg^{-1} and 170 m s^{-1}, respectively. For a mass flow rate of 7.5 kg s^{-1}, the power output of the turbine is 4000 kW. Determine the rate of heat transfer.

2.5 Air at an initial pressure of 100 kPa and an initial temperature of 17 °C is compressed to a pressure and temperature of 500 kPa and 117 °C, respectively. Under steady flow conditions, the power required for the compression process is 5 kW, while the heat loss is 5000 J kg^{-1} during this process. Neglecting the changes in the kinetic and the potential energies, determine the mass flow rate. Assume constant specific heats.

2.6 An initially evacuated tank with a volume of 0.5 m^3 is connected to a pipe by means of a valve, as shown in the figure below. Air is flowing through the pipe at a pressure of 900 kPa and a temperature of 350 K. The valve is then opened allowing air to enter the tank until the final pressure is 500 kPa. Find the mass and the temperature of the air in the tank when the valve is closed. Assume that the process is adiabatic with constant specific heats.

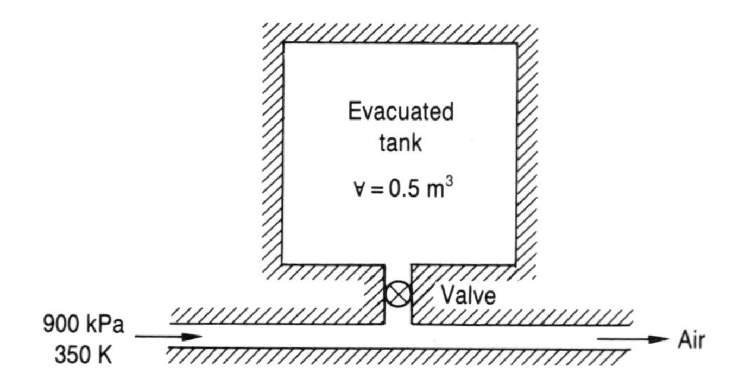

Problem 2.6

2.7 A tank with a volume of 0.1 m^3 contains carbon dioxide at a pressure and temperature of 800 kPa and 400 K, respectively. A valve which is connected to the tank is then opened until the pressure in the tank drops to 400 kPa. During the process, heat is added to carbon dioxide within the tank to keep it at constant temperature. Determine the amount of heat used during the process. Assume constant specific heats.

2.8 A piston–cylinder arrangement is attached through a valve to a pipe inside which air is flowing steadily, as shown in the figure below. Initially, the volume of air inside the cylinder is 0.01 m^3 and it is at a temperature and pressure of 40 °C and 100 kPa, respectively. When the valve is opened, the volume of air in the cylinder increases to 0.02 m^3. The temperature and the pressure of the supply line are 100 °C and 600 kPa, respectively. Find the final temperature of the air and the mass of the air that entered the cylinder. Assume that the process is adiabatic and there is no friction between the cylinder and the piston. The specific heats may be taken as constant.

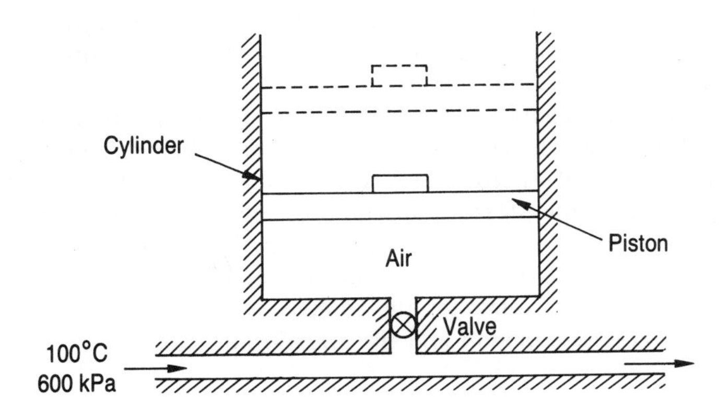

Problem 2.8

2.9 Two adiabatic tanks are connected by a valve. Tank A contains 0.25 m^3 of air at a pressure of 5 MPa and a temperature of 127 °C. Tank B contains 2 m^3 of air at a pressure of 100 kPa and a temperature of 27 °C. The valve is then opened until the pressure in tank A drops to 2.5 MPa. Assuming constant specific heats, at this instant determine

a the temperature in tank A,
b the mass left in tank A,
c the temperature and the pressure in tank B, and
d the total change in entropy for the process.

3 Introduction to compressible flow

3.1 INTRODUCTION

When a fluid is moving at low velocities, the variations in the pressure are usually small, and the flow can be considered as incompressible. As a result, the density can be assumed to be constant. In this case, the energy equation is decoupled so that the continuity and the momentum equations can be solved independently for the velocity and the pressure.

In this chapter, compressible flow is studied, which implies appreciable variations in the density throughout the flow field. Compressibility becomes quite an important factor, especially in high-speed flows. Large changes in the velocity result in considerable changes in the pressure for compressible flows. These pressure changes are then accompanied by significant variations in both the temperature and the density. Therefore, two additional variables are encountered in compressible flows which require two more relations for the complete solution of the flow field. These equations are the energy equation and the equation of state.

3.2 WAVE PROPAGATION IN COMPRESSIBLE MEDIA

In Figure 3.1, a gas is confined in a long tube with a piston at the left-hand side. The piston is then given a sudden push to the right. At first, a layer of gas piles up next to the piston and is compressed, but the remainder of the gas is unaffected. Then, the compression wave created by the piston moves through the gas, and eventually all the gas is able to feel the movement of the piston. If the pressure pulse given to the gas is infinitesimally small, then the wave is called a **sound wave**, and the resultant compression wave moves at the **speed of sound**.

When the medium is incompressible, no changes in the density can be allowed. If the piston is moved towards the right, then there will be no piling up of the fluid or no density changes at any point in the fluid. Thus, all the fluid would have to move instantaneously with the piston. Therefore, the speed of wave propagation in an incompressible fluid is infinite, and a disturbance created at any point is sensed instantaneously at all other points in the fluid. However, no medium is truly incompressible. The more compressible the substance through which the wave

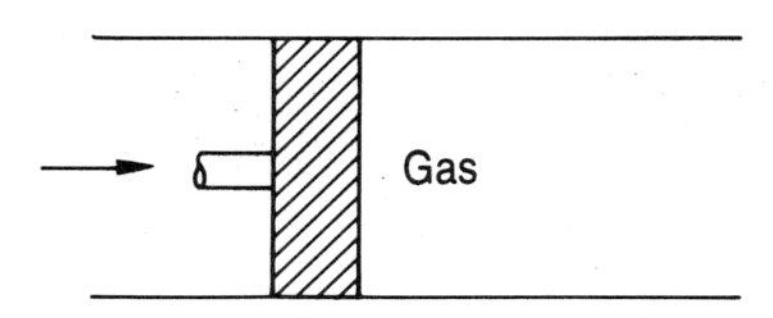

Figure 3.1 Piston moving into a stationary gas

propagates, the smaller will be the speed of sound in that substance. For example, the speed of sound in water is greater than the speed of sound in air.

3.3 SPEED OF SOUND

If an infinitesimal disturbance is created by an infinitesimal acceleration of the piston in Figure 3.2(a), then a compression wave propagates with the speed of sound, a, into the stationary gas. Owing to this infinitesimal acceleration, the piston acquires a steady velocity of $\mathrm{d}V$ to the right, and the gas between the piston and the sound wave moves at the piston velocity. As a result of the compression, the pressure and the density next to the piston are infinitesimally greater than the pressure and the density of the gas in front of the wave, which is at rest.

Unfortunately, the flow of Figure 3.2(a) appears unsteady to a stationary observer, observing the motion of the sound wave from a fixed point on the ground. To simplify the analysis, the flow can be reduced to one of steady motion by imagining that the observer travels with the steady speed of the wave, as shown in Figure 3.2(b). In this case, the fluid flows steadily from right to left, and as it passes through the stationary wave its velocity is reduced from a to $a - \mathrm{d}V$. At the same time, the pressure changes from p to $p + \mathrm{d}p$, while the density changes from ρ to $\rho + \mathrm{d}\rho$. This transformation is a **dynamic transformation** because only the dynamic properties of the fluid, such as the fluid velocity, are affected. The static properties such as the static temperature and static pressure are unaffected by this transformation.

The basic equations may now be applied to the differential control volume shown in Figure 3.2(b).

(a) Continuity equation

For steady flow through the control volume, the mass flow rate, $\dot{m}$, is constant. Also, the flow can be assumed to be one dimensional so that the continuity equation (1.39) becomes

$$\dot{m} = \rho a A = (\rho + \mathrm{d}\rho)(a - \mathrm{d}V)A$$

or, expanding,

$$\cancel{\rho a A} = \cancel{\rho a A} - \rho A\ \mathrm{d}V + aA\ \mathrm{d}\rho - A\ \underbrace{\cancel{\mathrm{d}\rho}}_{\simeq 0}\ \mathrm{d}V$$

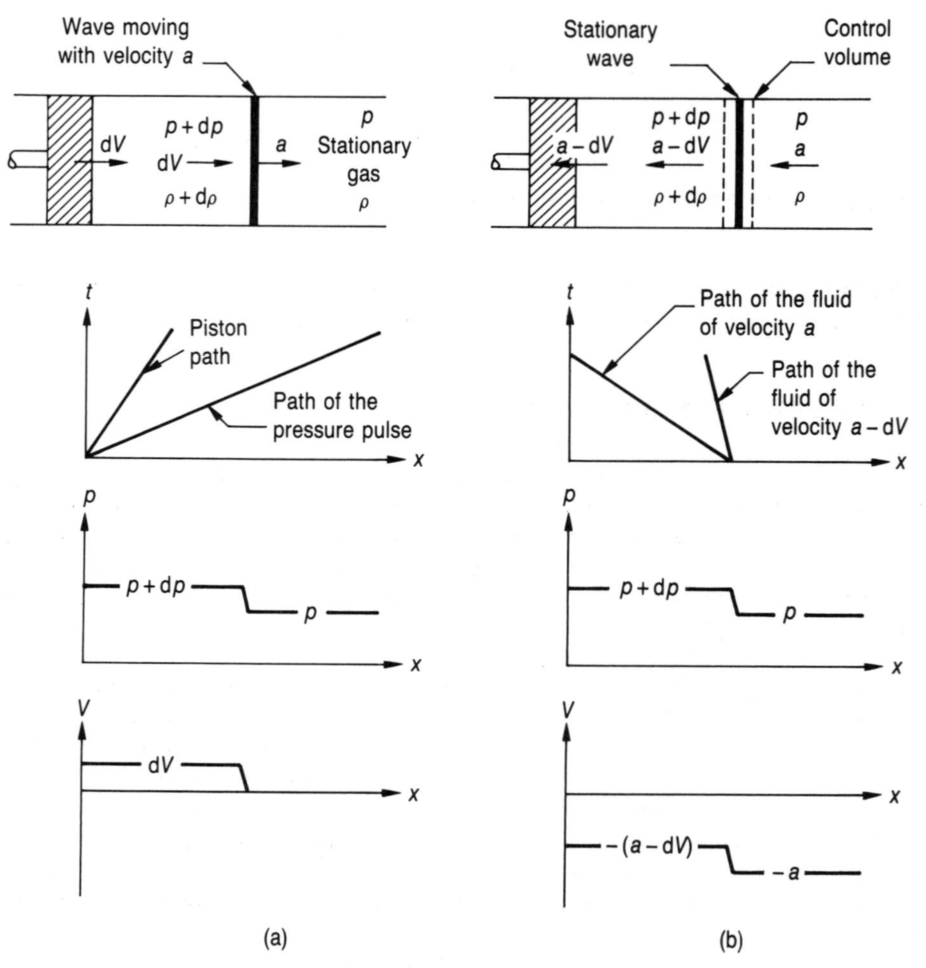

Figure 3.2 Propagation of an infinitesimal pressure disturbance: (a) observer at rest; (b) observer moving with the wave at a velocity of *a*

The product of the differentials, $d\rho\, dV$, may be neglected when compared with the infinitesimal differentials $d\rho$ or dV. Hence

$$dV = \frac{a}{\rho}\, d\rho \qquad (3.1)$$

(b) Momentum equation

As long as the compression wave is very thin, the shear forces on the control volume are negligibly small compared with the pressure forces. In this case, the application of the momentum equation (1.46) to the control volume yields

$$(p + \mathrm{d}p)A - pA = \dot{m}a - \dot{m}(a - \mathrm{d}V)$$

Using the continuity equation (3.1)

$$A\ \mathrm{d}p = \dot{m}\ \mathrm{d}V = \rho a A\ \mathrm{d}V$$

so that

$$\mathrm{d}V = \frac{1}{\rho a}\ \mathrm{d}p \tag{3.2}$$

(c) Energy equation

As long as the sound wave is very thin and the motion is very rapid, the heat transfer between the control volume and the surroundings can be neglected, and the process can be considered as **adiabatic**. As a result, the first law of thermodynamics (2.22b) can be applied to the control volume to yield

$$(h + \mathrm{d}h) + \frac{(a - \mathrm{d}V)^2}{2} = h + \frac{a^2}{2}$$

or

$$\cancel{h} + \mathrm{d}h + \cancel{\frac{a^2}{2}} - a\ \mathrm{d}V + \overbrace{\cancel{\frac{(\mathrm{d}V)^2}{2}}}^{\simeq 0} = \cancel{h} + \cancel{\frac{a^2}{2}}$$

Then

$$\mathrm{d}h = a\ \mathrm{d}V \tag{3.3}$$

(d) The second law of thermodynamics

Till now, both the heat transfer and the frictional effects have been neglected so that the flow through the control volume is adiabatic and reversible, that is isentropic. This can also be proved with the thermodynamic relation, $T\,\mathrm{d}s = \mathrm{d}h - \mathrm{d}p/\rho$. When Equations (3.2) and (3.3) are substituted into this relation, it is possible to obtain

$$\mathrm{d}s = 0 \tag{3.4}$$

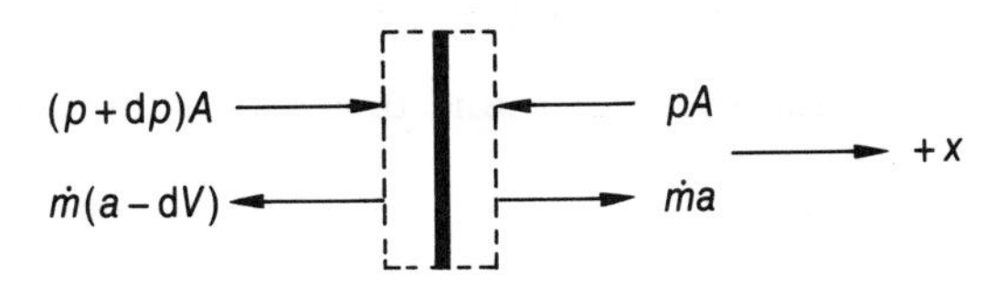

Figure 3.3 Pressure forces and rates of linear momentum on the control volume moving with the wave

Therefore, across an infinitesimal pressure wave, the entropy is constant and the flow is isentropic.

To obtain the equation for the speed of the sound wave, one may combine Equations (3.1) and (3.2) to yield

$$a^2 = \mathrm{d}p/\mathrm{d}\rho$$

But, to evaluate the derivative of a thermodynamic property, the property to be held constant during the differentiation must be specified. For the present case, the entropy is constant. Thus

$$a^2 = (\partial p/\partial \rho)_s \tag{3.5}$$

For an isentropic process, the pressure and the density are related by

$$p/\rho^k = \text{constant}$$

as shown previously. Taking the logarithm of both sides

$$\ln p - k \ln \rho = \ln(\text{constant})$$

and differentiating

$$\frac{\mathrm{d}p}{p} - k\frac{\mathrm{d}\rho}{\rho} = 0$$

Therefore,

$$\left(\frac{\partial p}{\partial \rho}\right)_s = k\frac{p}{\rho}$$

and

$$a = \sqrt{\left(k\frac{p}{\rho}\right)} \tag{3.6}$$

(e) Equation of state

If the equation of state for a perfect gas (2.35b) is substituted into Equation (3.6)

$$a = \sqrt{(kRT)} \tag{3.7a}$$

or using Equation (2.36) for eliminating the gas constant

$$a = \sqrt{\left(\frac{k\bar{R}T}{\bar{M}}\right)} \tag{3.7b}$$

Hence, for a perfect gas, the speed of sound depends on the temperature and the molecular mass.

Example 3.1

Calculate the speed of sound in air at 25 °C. The gas constant and the specific heat ratio for air are 287.1 J kg^{-1} K^{-1} and 1.4, respectively.

Solution

The speed of sound can be obtained from Equation (3.7a) as

$$a = \sqrt{(kRT)} = \sqrt{[(1.4)(287.1\ \text{J kg}^{-1}\,\text{K}^{-1})(298\ \text{K})]} = \mathbf{346.1\ m\,s^{-1}}$$

3.4 MACH NUMBER

When there is a large relative speed between a body and the compressible fluid surrounding it, the variation of its density with the speed influences the properties of the flow field. The dimensionless criterion of the flow phenomena is called the **Mach number**, M, and is defined as the ratio of the velocity of the fluid, V, to the local speed of sound, a, as

$$M = V/a \tag{3.8}$$

The physical significance of the Mach number can easily be understood by squaring Equation (3.8) and introducing Equation (3.7a) for the speed of sound in a perfect gas as

$$M^2 = \frac{V^2}{a^2} = \frac{V^2}{kRT} \propto \frac{\text{directed kinetic energy}}{\text{random kinetic energy}} \tag{3.9}$$

The velocity, V, is a measure of the directed motion of the gas particles, while $V^2/2$ measures the kinetic energy per unit mass of the directed flow. The speed of sound for a gas is proportional to the square root of its temperature, which is also proportional to the random velocity of gas particles. Therefore, a^2 is a measure of the kinetic energy per unit mass associated with the random motion of the gas molecules. The Mach number may then be regarded as the ratio of the kinetic energy of the directed fluid flow to the kinetic energy of the random molecular motion.

There are large differences in flow patterns for compressible flows. General behaviour of the flow depends on whether the velocity is greater or less than the speed of sound. When the Mach number is less than unity, the flow is termed **subsonic**. However, the flow is known as **supersonic** when the Mach number is greater than unity. Finally, when the Mach number equals unity, the flow is identified as **sonic**.

The flow can usually be assumed to be **incompressible** when the flow speed is small compared with the speed of sound, that is when $M \leqslant 0.3$.

A **transonic flow** is defined as a flow having regions in which the flow velocity changes from subsonic to supersonic, that is when $0.9 \leqslant M \leqslant 1.1$. A **hypersonic flow** is a supersonic flow at high Mach numbers and is usually defined as a flow whose Mach number is greater than 5. The treatment of hypersonic flows is somewhat different from that of low-Mach-number supersonic flows since the real-gas effects become important.

Example 3.2

An aircraft is travelling at a velocity of 500 m s^{-1} at an altitude where the pressure and the density are 70 kPa and 0.909 kg m^{-3}, respectively. Meanwhile, the wind is blowing in the opposite direction at a velocity of 35 m s^{-1}. Find

a the Mach number of the wind,
b the Mach number of the aircraft, if there were no wind, and
c the Mach number of the aircraft relative to the moving air.

Solution
The temperature of the air may be evaluated from the equation of state (2.35b) for a perfect gas as,

$$T = \frac{p}{\rho R} = \frac{70\,000 \text{ N m}^{-2}}{(287.1 \text{ J kg}^{-1}\text{K}^{-1})(0.909 \text{ kg m}^{-3})} = 268.2 \text{ K}$$

The speed of sound may now be evaluated by using Equation (3.7a):

$$a = \sqrt{(kRT)} = \sqrt{[(1.4)(287.1 \text{ J kg}^{-1}\text{K}^{-1})(268.2 \text{ K})]} = 328.3 \text{ m s}^{-1}$$

(a) The Mach number of the wind can be found via Equation (3.8) as

$$M_{\text{wind}} = \frac{V_{\text{wind}}}{a} = \frac{35 \text{ m s}^{-1}}{328.3 \text{ m s}^{-1}} = \mathbf{0.1066}$$

(b) One can again use Equation (3.8) to determine the Mach number of the aircraft when there is no wind as

$$M_{\text{aircraft}} = \frac{V_{\text{aircraft}}}{a} = \frac{500 \text{ m s}^{-1}}{328.3 \text{ m s}^{-1}} = \mathbf{1.523}$$

(c) The velocity of the aircraft with respect to the blowing wind is then

$$V_{\text{rel}} = V_{\text{wind}} + V_{\text{aircraft}} = 35 \text{ m s}^{-1} + 500 \text{ m s}^{-1} = 535 \text{ m s}^{-1}$$

Therefore, the relative Mach number is

$$M_{\text{rel}} = \frac{V_{\text{rel}}}{a} = \frac{535 \text{ m s}^{-1}}{328.3 \text{ m s}^{-1}} = \mathbf{1.630}$$

3.5 PRESSURE DISTURBANCES IN A COMPRESSIBLE FLUID

When an object moves through a gas, it pushes the neighbouring gas out of the way creating a pressure pulse which propagates into the exterior gas. This is very similar to the impulsive motion of the piston in Figure 3.2 which causes the propagation of a compression wave down through the tube. To understand the differences between the physical natures of incompressible and compressible flows, one may consider a point source of infinitesimal pressure disturbances. Then, any pressure disturbance created by the source propagates in all directions at the speed of sound, a, relative to the gas. At any time t, the location of the wave from the disturbance emitted at time t_0, will be represented by a sphere with a radius $a(t - t_0)$, whose centre is coincident with the location of the disturbance at time t_0.

3.5.1 Stationary point source

A sound pattern propagates uniformly in all directions. At the instant Δt after emission, any given sound pulse is located at radius $a\ \Delta t$ from the source. At the instant $2\ \Delta t$, of course, the radius is $a(2\ \Delta t)$. Any given waves are spherical in shape and all the waves are concentric spheres as shown in Figure 3.4.

3.5.2 Subsonic point source

Now, consider the case where a point source moves at a uniform subsonic speed, V, in an isothermal homogeneous fluid, as illustrated in Figure 3.5. If the pressure disturbances created by the moving body are small, then it may be assumed that the sonic speed remains constant. Initially, the point source is at point A_0. At $t = \Delta t$, the point source moves to A_1, and the distance between points A_0 and A_1 is $V(\Delta t)$. At $t = 2\ \Delta t$ and $t = 3\ \Delta t$, the point source is at points A_2 and A_3 which are at distances of $V(2\ \Delta t)$ and $V(3\ \Delta t)$ respectively from the initial location of the point source. The pressure disturbances that were created by the point source at points A_0, A_1 and A_2 as observed at $t = 3\ \Delta t$ are spheres of radii $a(3\ \Delta t)$, $a(2\ \Delta t)$ and $a(\Delta t)$ centred at points A_0, A_1 and A_2, respectively. Consequently, if a body moves in a compressible fluid with a subsonic speed, the fluid ahead of the body becomes aware of the presence of the body, since the body emits disturbance signals ahead of itself.

The circles shown can be imagined to be the amplitude peaks of a sinusoidal tone, and the same qualitative picture holds for a moving source of a continuous sound. If the source moves at a uniform speed, then the whole pattern is carried along with the emitter. Thus, a stationary observer would hear more peaks per unit time as the source approaches than after it has passed. This is known as the **Doppler effect**.

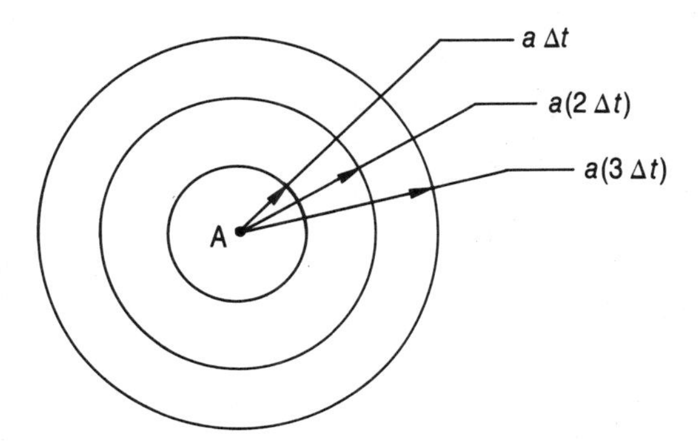

Figure 3.4 Propagation of sound waves from a stationary point source

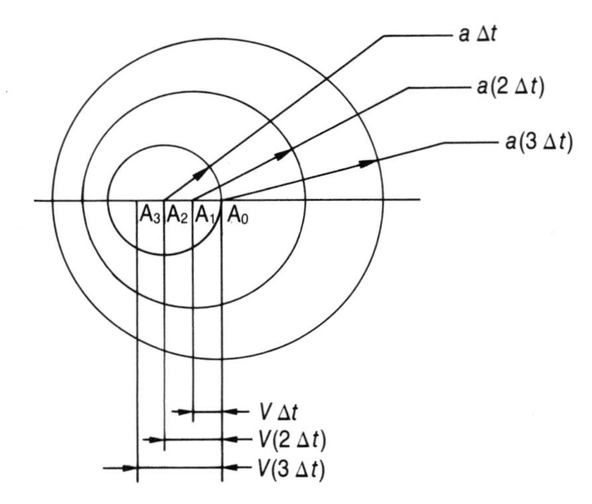

Figure 3.5 Pressure disturbances produced by a point source moving at subsonic speed

3.5.3 Sonic point source

Consider the case where a body moves at sonic speed, that is $V = a$, as shown in Figure 3.6. During any time interval, Δt, pressure disturbances move a distance of $a\,\Delta t$ as the point source moves an equal distance of $V\,\Delta t$. The locus of the leading surfaces of all waves will be a plane passing through the source, perpendicular to the path of

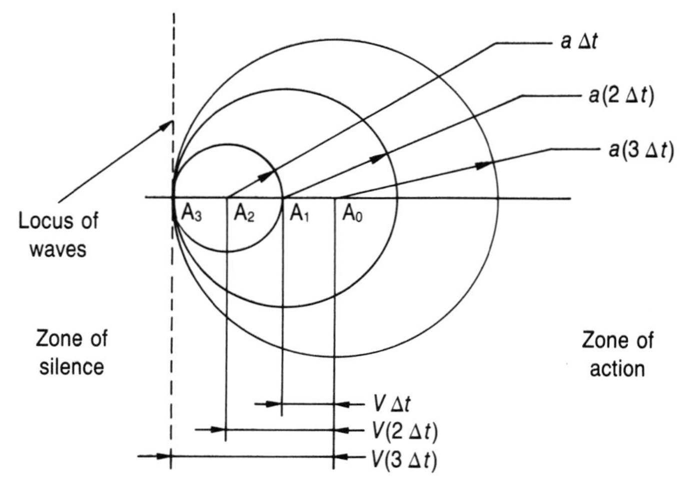

Figure 3.6 Pressure disturbances produced by a point source moving at sonic speed

the motion. No sound wave can travel in front of the source. Consequently, an observer in front of the source will not hear any sound.

3.5.4 Supersonic point source

Now, consider the case where the body moves at a uniform speed, V, which is greater than the speed of sound, as illustrated in Figure 3.7. Then, the pressure disturbance created by the body lags behind the body that created the disturbance, and the disturbance wave cannot overtake the moving body. Consequently, in all of the successive positions A_1, A_2 and A_3, the moving body is ahead of the disturbance waves that it has produced. Therefore, the effect of pressure changes produced by a body cannot reach points ahead of the body, and this fact is known as the **rule of forbidden signals**.

In this case, the locus of the leading surfaces of the waves will be a cone with the body at its apex. This cone is known as the **Mach cone**. Then, all the disturbances are confined inside the Mach cone extending downstream from the moving body, and this region is called the **zone of action**. The region outside the Mach cone, extending upstream from the moving body, is not aware of any disturbance and is known as the **zone of silence**. The half angle of the Mach cone is called the **Mach angle**, μ, and from the geometry of Figure 3.7

$$\sin \mu = \frac{a(3\ \Delta t)}{V(3\ \Delta t)} = \frac{a(2\ \Delta t)}{V(2\ \Delta t)} = \frac{a(\Delta t)}{V(\Delta t)} = \frac{a}{V} = \frac{1}{M} \tag{3.10}$$

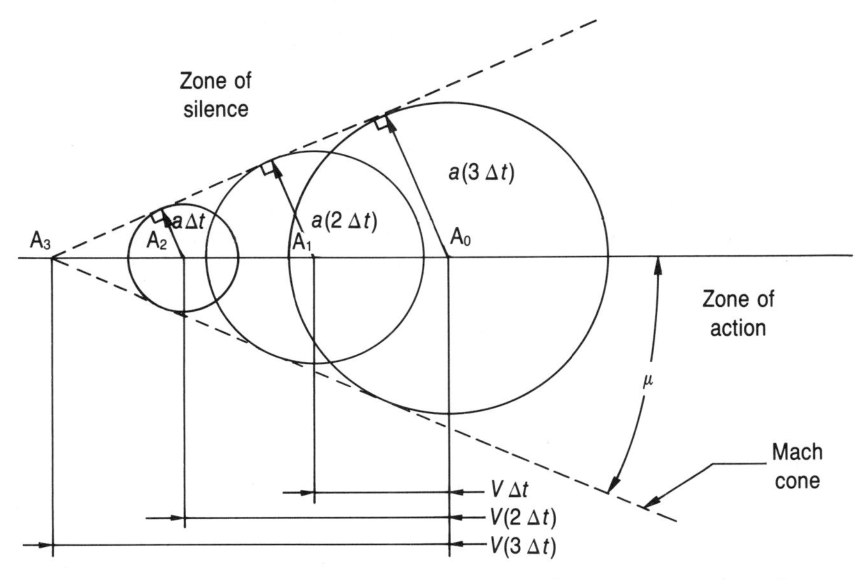

Figure 3.7 Pressure disturbances produced by a point source moving at supersonic speed

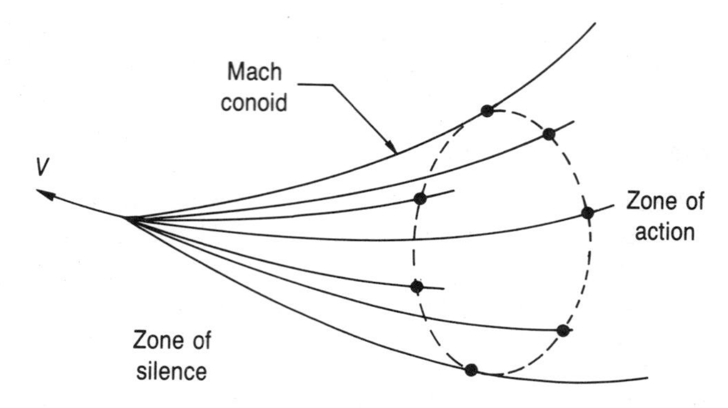

Figure 3.8 The Mach conoid created by a disturbance moving at a variable velocity in a non-uniform region

The closeness of the circles representing the different pressure pulses is a measure of the intensity of the pressure disturbances at each point of the flow field. In supersonic flows, the pressure disturbances are largely concentrated in the neighbourhood of the Mach cone that forms the outer limit of the zone of action. This is known as the **rule of concentrated action.**

The Mach cone is usually a three-dimensional conoid, as shown in Figure 3.8, when the source of disturbance moves at a variable velocity into a non-uniform region where the properties of the fluid change from one point to another. The Mach conoid becomes a right circular one when the disturbance is moving at a constant velocity into a uniform region.

Example 3.3

A supersonic aircraft flies at an altitude of 3000 m at a constant velocity of 650 m s^{-1}. The aircraft passes directly over an observer who is stationary on the ground. How much time elapses, after the aircraft has passed over the observer, before the observer hears the sound of the aircraft? Assume that the average speed of sound is 325 m s^{-1}.

Solution

The physical situation is shown in Figure 3.9. The observer located on the ground at point G will first hear the sound wave generated by the aircraft at point A. During the time interval required for the sound wave to travel from point A to G, the aircraft travels from point A to B. The Mach angle, μ, can be evaluated from Equation (3.10) as

$$\mu = \sin^{-1}\left(\frac{1}{M}\right) = \sin^{-1}\left(\frac{a}{V}\right) = \sin^{-1}\left(\frac{325 \text{ m s}^{-1}}{650 \text{ m s}^{-1}}\right) = 30^\circ$$

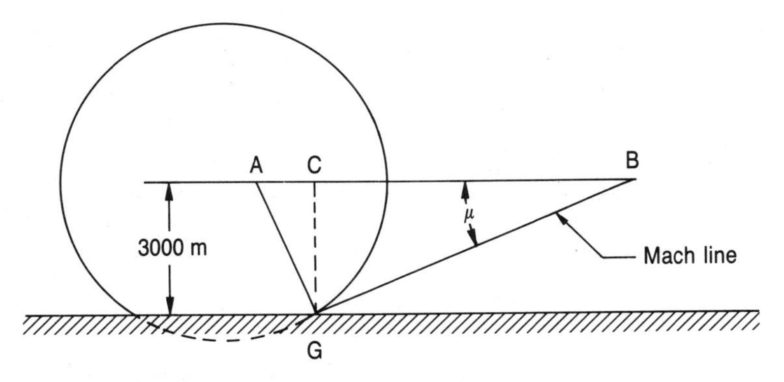

Figure 3.9 Sketch for Example 3.3

The distance from point C to point B is

$$\overline{CB} = \frac{\overline{GC}}{\tan \mu} = \frac{3000 \text{ m}}{\tan 30^\circ} = 5196 \text{ m}$$

The time required for the aircraft to travel from point C to point B is

$$t = \frac{\overline{CB}}{V} = \frac{5196 \text{ m}}{650 \text{ m s}^{-1}} = \mathbf{7.994 \text{ s}}$$

FURTHER READING

Fox, R. W. and McDonald, A. T. (1985) *Introduction to Fluid Mechanics*, 3rd edn., Wiley: New York.

John, J. E. A. (1984) *Gas Dynamics*, 2nd edn, Allyn & Bacon: Boston, MA.

Owczarek, J. A. (1964) *Fundamentals of Gas Dynamics*, International Textbook: Scranton, PA.

Shapiro, A. H. (1953) *The Dynamics and Thermodynamics of Fluid Flow*, Vol. 1, Ronald Press: New York.

Zucrow, M. J. and Hoffman, J. D. (1976) *Gas Dynamics*, Vol. 1, Wiley: New York.

PROBLEMS

3.1 Find the speed of sound in air at an altitude of 1500 m where the pressure and the temperature are 83 kPa and 5 °C, respectively.

3.2 Find the temperature of the air if a sound wave travels in the air with a speed of 400 m s^{-1}.

3.3 When the temperature of nitrogen is low and its pressure is high, its properties are often represented by the van der Waals equation of state as

$$p = \frac{\rho RT}{1 - b\rho} - a\rho^2$$

where a and b are constants and $a = 0.1790\ \mathrm{N\,m^4\,kg^{-2}}$ and $b = 0.001\,397\ \mathrm{m^3\,kg^{-1}}$ for nitrogen. Determine the speed of sound in nitrogen at a temperature of $-50\ ^\circ\mathrm{C}$ and a pressure of 2 MPa by using

a the equation of state for a perfect gas, and
b the van der Waals equation of state.

3.4 An aircraft is flying with the same Mach number at elevations of 5000 m and 10 000 m. If the aircraft flies 50 $\mathrm{m\,s^{-1}}$ faster at the altitude of 5000 m, find its Mach number.

3.5 An aircraft is flying at a speed of 960 $\mathrm{km\,h^{-1}}$ through air at a pressure of 82 kPa and a temperature of 0 °C. Calculate the Mach number and the Mach angle for the aircraft.

3.6 A projectile is fired into air in which the pressure is 340 kPa and the density is 4.5 $\mathrm{kg\,m^{-3}}$. It can be observed that a Mach cone is originated from the projectile with a total angle of 20°. What is the speed of the projectile with respect to the air?

3.7 A particle is moving in air at sea level where the temperature is 15 °C. The two disturbance spheres created by the particle at points A and B are shown in the figure below. Determine the Mach number and the velocity of the particle.

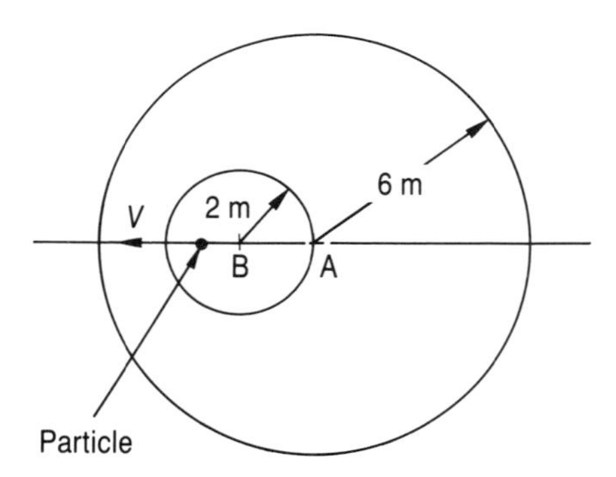

Problem 3.7

3.8 A particle which is moving in air at a temperature of 20 °C creates two disturbance spheres at points A and B, as shown in the figure below. Determine the Mach number and the velocity of the particle. Also, find the Mach angle.

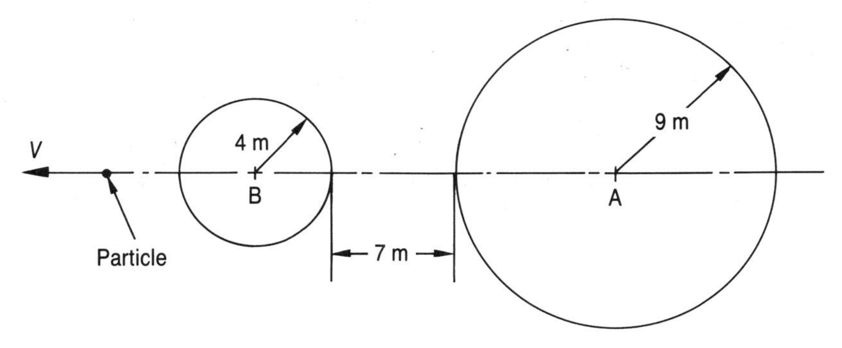

Problem 3.8

3.9 An aircraft at an altitude of 2000 m passes over the head of an observer on the ground. The Mach number of the aircraft is 1.5. Assuming an isothermal atmosphere with a temperature of 10 °C, find the speed of the aircraft. How long after it passes directly overhead does its Mach cone pass a point on the ground?

3.10 A supersonic aircraft flies at an altitude of 5000 m at a constant velocity of 750 $\mathrm{m\,s^{-1}}$. The aircraft passes directly over an observer who is stationary on the ground. How much time elapses, after the aircraft has passed over the observer, before the observer hears the sound of the aircraft? Assume that the atmosphere is isothermal with a temperature of 20 °C.

3.11 An observer on the ground hears the sound of a supersonic aircraft 7 s after it has passed directly overhead. In the same time interval, the aircraft travels a distance of 8 km. Find the altitude of the aircraft if the average temperature of the atmosphere is 5 °C.

3.12 A supersonic aircraft flies at an altitude of 2000 m. Determine its velocity if the sound is heard 4 s after its passage over the head of an observer. Assume that the air has an average temperature of 10 °C.

3.13 A supersonic aircraft flies at a constant altitude of 3000 m. An observer on the ground hears the noise generated by the aircraft, which is 3000 m away horizontally, after 12 s. Assuming that the atmosphere is isothermal, determine

a the Mach number of the aircraft,
b the velocity of the aircraft,
c the distance travelled before the observer hears the noise, and
d the temperature of the isothermal atmosphere. (METU 1987)

3.14 Show that, for a perfect gas, the fractional change in pressure across an infinitesimal pressure pulse is given by the equation

$$\frac{dp}{p} = k\frac{dV}{a}$$

and that the fractional change in the temperature is given by the equation

$$\frac{dT}{T} = (k-1)\,\frac{dV}{a}$$

3.15 A compression pulse changes the velocity of the air over which it passes by 3 m s^{-1}. Calculate the pressure rise and the temperature change across the pulse, if the air is initially at a temperature and pressure of 15 °C and 101.3 kPa, respectively.

3.16 When a reciprocating compressor is connected to suction and discharge pipes, as shown in the figure below, it creates an unsteady flow within the cylinder and its piping system. The rate of change of pressure in the cylinder is given as dp_c/dt and the rate of change of volume is $d\forall_c/dt$. Show that the speed of sound in the gas in the cylinder changes according to the equation

$$\frac{da_c}{dt} = \frac{a_c}{2}\left\{\frac{1}{p_c}\frac{dp_c}{dt} + \frac{1}{\forall_c}\frac{d\forall_c}{dt} - \frac{1}{m_c}\left[\left(\frac{dm}{dt}\right)_{in} - \left(\frac{dm}{dt}\right)_{out}\right]\right\}$$

It is assumed that the properties in the cylinder are uniform at any instant of time. (METU 1991)

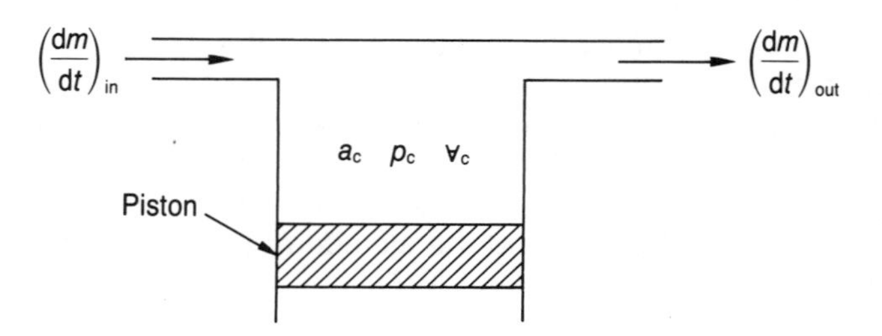

Problem 3.16

4 Isentropic flow

4.1 INTRODUCTION

From a one-dimensional point of view, the fluid properties in compressible flows are affected by three important factors: (i) the changes in the cross-sectional area; (ii) the friction; and (iii) the heat transfer. In most practical engineering problems, these effects occur simultaneously. Owing to the complexity of the analysis of such flows, the effects of these factors are studied one at a time, while the others are neglected.

In this chapter, the effects of friction and heat transfer are neglected, and steady, one-dimensional adiabatic and reversible flows with changes in area are studied. These isentropic flows are usually called **simple area-change flows**, and they have two important applications, namely (i) in ducts and (ii) in streamtubes.

The flow in pipes and ducts is very often adiabatic. In the case of nozzles and diffusers, the duct is usually very short, so frictional effects can be disregarded. Then the flow, as a first approximation, is reversible and therefore isentropic. The main function of nozzles and diffusers is, respectively, to accelerate and to decelerate the fluid as efficiently as possible. Therefore, an isentropic process provides a useful standard for comparing the performance of actual nozzles and diffusers.

For external flows around bodies and internal flows through passages, there are streamtubes which lie completely outside the boundary layer. In these streamtubes, the effects of heat transfer and viscosity are negligible. Therefore, the flow in these streamtubes may be considered as isentropic.

4.2 GOVERNING EQUATIONS

Consider steady, one-dimensional isentropic flow through an arbitrary duct as shown in Figure 4.1. The governing equations are given below for the control volume shown.

(a) Continuity equation

For steady and uniform flow at each cross-section, the mass flow rate is constant. Thus

$$\dot{m} = \rho_1 V_1 A_1 = \rho_2 V_2 A_2 \tag{4.1}$$

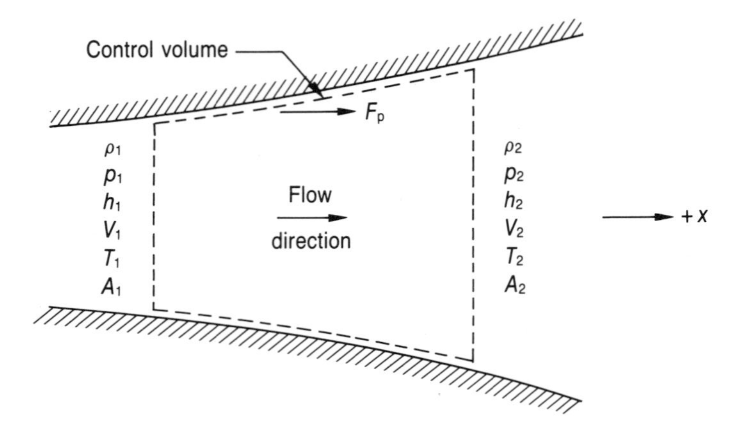

Figure 4.1 Control volume for steady, one-dimensional isentropic flow

(b) Momentum equation

Since the flow is frictionless, the pressure forces acting on the control volume of Figure 4.1 for the isentropic flow must be balanced by the rate of change of momentum across the control volume, that is

$$F_p + p_1 A_1 - p_2 A_2 = \rho_2 A_2 V_2^2 - \rho_1 A_1 V_1^2 \tag{4.2}$$

where F_p is the pressure force acting on the side walls of the control volume.

(c) Energy equation

When there is no heat transfer between the control volume and its surroundings

$$h_1 + \frac{V_1^2}{2} = h_2 + \frac{V_2^2}{2} \tag{4.3}$$

(d) The second law of thermodynamics

For isentropic flows

$$s_2 = s_1 \tag{4.4}$$

(e) Equation of state

If the working fluid is a perfect gas, then

$$p = \rho R T \tag{4.5}$$

4.3 STAGNATION CONDITIONS

The **stagnation conditions** are extremely important for defining a reference state for compressible flows. The **stagnation state** is defined as the state where the flow velocity is zero.

Most of the practical flows start from a large reservoir where the velocity of the fluid is negligible, as shown in Figure 4.2(a). Therefore, the gas in the reservoir is stagnant. The fluid must then be accelerated from an infinitely large reservoir in order to attain the actual state of the fluid flow. Alternatively, the stagnation state can be attained by decelerating the fluid to zero speed, as shown in Figure 4.2(b).

Obviously, the nature of the deceleration or the acceleration determines the final stagnation state. In order to obtain a unique stagnation state, a restriction is put on the deceleration process by keeping the entropy constant. Then, the **stagnation state is the state attained by a fluid when it is decelerated to zero speed isentropically, or the state from which a fluid is accelerated isentropically to its actual state.** The stagnation state will be designated by subscript zero.

When the energy equation is applied to the fluid flow in Figure 4.2, then

$$h_0 = h + \frac{V^2}{2} \tag{4.6}$$

where h_0 is the stagnation enthalpy. The stagnation enthalpy has the same value for all states, which are reachable **adiabatically** from a given static state, whether or not the deceleration process is reversible. The stagnation state may be fixed by using the second law of thermodynamics as

$$s_0 = s \tag{4.7}$$

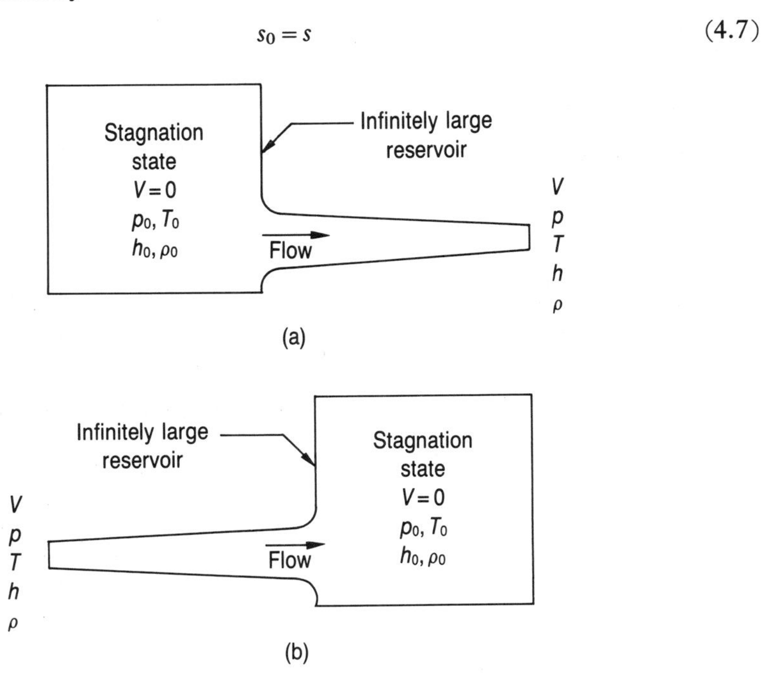

Figure 4.2 Stagnation conditions

The declaration process, which is governed by Equations (4.6) and (4.7), may be illustrated on a **Mollier diagram**, as shown in Figure 4.3.

For a perfect gas, $h = c_p T$, and then Equation (4.6) becomes

$$T_0 = T + \frac{V^2}{2c_p} \tag{4.8a}$$

or using $c_p = kR/(k-1)$

$$T_0 = T + \frac{k-1}{2kR} V^2 \tag{4.8b}$$

where T_0 is the stagnation temperature and has the same value whether or not the acceleration or the deceleration process is reversible. The process which is now governed by Equations (4.7) and (4.8a) may be illustrated on a T–s diagram (temperature–entropy diagram), as shown in Figure 4.4.

One should note that the stagnation pressure, p_0, has the same value for all states that are reachable isentropically from a given state.

The concept of stagnation state as a reference state is independent of the type of fluid being considered, as well as the actual flow under investigation. The actual flow process may involve effects such as heat transfer and friction along the flow path. At every point along the actual flow path, there is a local stagnation state which is defined by an imaginary isentropic deceleration. Therefore, in an actual flow, the stagnation conditions may vary from one point to another.

Example 4.1

An aircraft is flying at an altitude where the air temperature is 17 °C. The

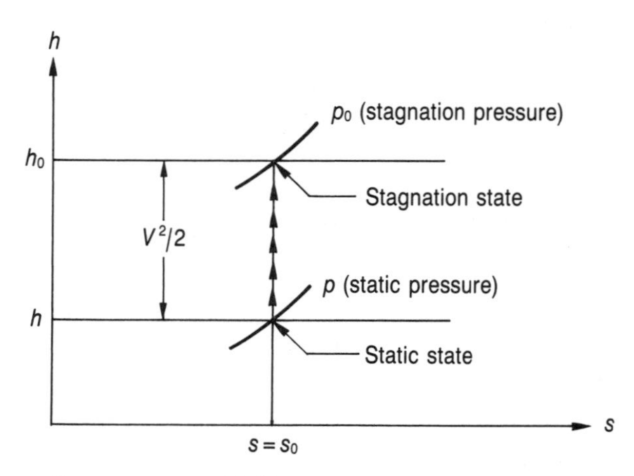

Figure 4.3 Stagnation state on a Mollier diagram

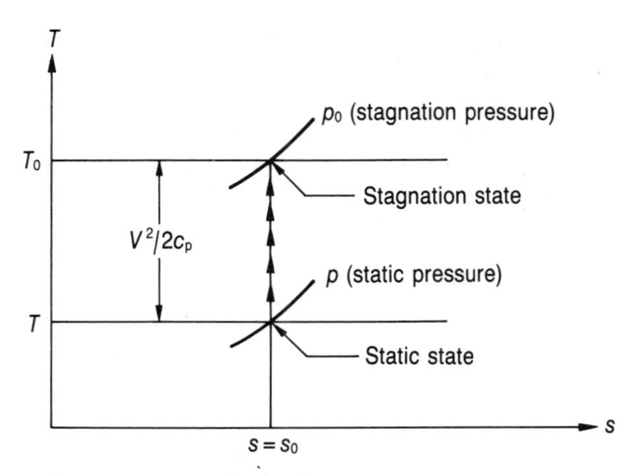

Figure 4.4 Stagnation state on a T–s diagram

temperature at the nose of the aircraft is found to be 57 °C, where the velocity of the air relative to the aircraft is zero. Determine

a the velocity of the aircraft, and
b the Mach number of the aircraft.

Solution
When the aircraft travels into still air, as shown in Figure 4.5(a), the flow is unsteady, since the position of the aircraft is changing with respect to an observer at a fixed point on the ground. To simplify the analysis, the flow can be

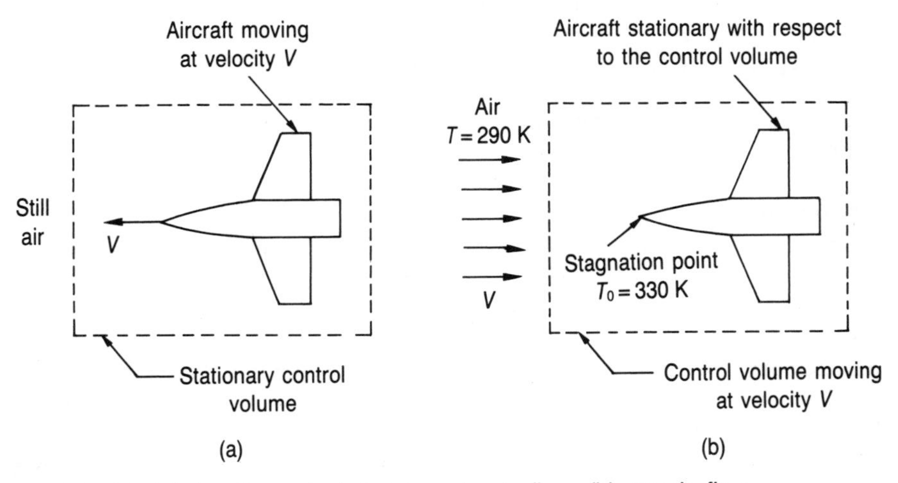

Figure 4.5 Sketch for Example 4.1: (a) unsteady flow; (b) steady flow

seen as steady, by considering a control volume which is travelling with the aircraft. In this case, the air is travelling towards the stationary aircraft with a velocity equal and opposite to that of the aircraft. Then the temperature of the air is the static temperature $T = 290$ K. The temperature at the nose of the stationary aircraft corresponds to the stagnation temperature $T_0 = 330$ K, since at that point the velocity of air is zero.

(a) Equation (4.8b) can then be rearranged for the velocity as

$$V = \sqrt{\left(\frac{2kR}{k-1}(T_0 - T)\right)}$$

$$V = \sqrt{\left(\frac{(2)(1.4)(287.1\ \mathrm{J\,kg^{-1}\,K^{-1}})}{1.4-1}(330\ \mathrm{K} - 290\ \mathrm{K})\right)} = \mathbf{283.5\ m\,s^{-1}}$$

(b) The speed of sound in the air can now be evaluated by using Equation (3.7a):

$$a = \sqrt{(kRT)} = \sqrt{[(1.4)(287.1\ \mathrm{J\,kg^{-1}\,K^{-1}})(290\ \mathrm{K})]} = 341.4\ \mathrm{m\,s^{-1}}$$

Then the Mach number of the aircraft, from Equation (3.8), is

$$M = \frac{V}{a} = \frac{283.5\ \mathrm{m\,s^{-1}}}{341.4\ \mathrm{m\,s^{-1}}} = \mathbf{0.8304}$$

4.4 CHARACTERISTIC SPEEDS OF GAS DYNAMICS

In the analysis of gas dynamic processes, there are several different characteristic speeds which are used for reference purposes. Some of these reference speeds are dimensional, while others are non-dimensional.

4.4.1 Dimensional characteristic speeds

There are three important dimensional speeds, which are representative of the overall properties of the flow. These are discussed below.

(a) Stagnation speed of sound

This is the speed of sound at the stagnation state and for a perfect gas it can be defined as

$$a_0 = \sqrt{(kRT_0)} \tag{4.9}$$

For an adiabatic flow, the stagnation speed of sound remains constant.

(b) Maximum speed

A gas can attain its maximum speed when it is hypothetically expanded to zero pressure. The static temperature corresponding to this state is also zero. Therefore, the maximum speed of a gas represents the speed corresponding to the complete transformation of the kinetic energy associated with the random motions of the gas molecules (thermal energy) into directed kinetic energy. The equation for maximum velocity is obtained by setting the temperature to zero in Equation (4.8b):

$$V_{max} = \sqrt{\left(\frac{2k}{k-1} RT_0\right)} \tag{4.10}$$

(c) Critical speed of sound

This is the speed of sound at the sonic state of a perfect gas, where $M = 1$, and can be given as

$$V^* = a^* = \sqrt{(kRT^*)} \tag{4.11}$$

The critical state, where the flow is sonic, is denoted by a superscript asterisk. Now, Equation (4.8b) becomes

$$V^* = \sqrt{\left(\frac{2k}{k-1} R(T_0 - T^*)\right)}$$

Eliminating T^* by using the definition of V^* in Equation (4.11), it is possible to obtain

$$V^* = a^* = \sqrt{\left(\frac{2k}{k+1} RT_0\right)} \tag{4.12}$$

The relation between these three reference speeds can be obtained via Equations (4.9), (4.10) and (4.11) as

$$\frac{a^*}{a_0} = \sqrt{\left(\frac{2}{k+1}\right)} \tag{4.13}$$

$$\frac{V_{max}}{a_0} = \sqrt{\left(\frac{2}{k-1}\right)} \tag{4.14}$$

$$\frac{V_{max}}{a^*} = \sqrt{\left(\frac{k+1}{k-1}\right)} \tag{4.15}$$

When the definition of the speed of sound for a perfect gas is introduced into the adiabatic energy equation (4.8b), then

$$V^2 + \frac{2}{k-1} a^2 = \text{constant}$$

From the definition of maximum velocity, this constant is equal to V_{max}^2 when the speed of sound or the temperature is equal to zero, that is

$$V^2 + \frac{2}{k-1} a^2 = V_{max}^2$$

Using Equations (4.14) and (4.15),

$$V^2 + \frac{2}{k-1}\, a^2 = V_{\max}^2 = \frac{2}{k-1}\, a_0^2 = \frac{k+1}{k-1}\, a^{*2} \tag{4.16a}$$

Equation (4.16a) is the kinematic form of the steady, adiabatic energy equation, and it can be rearranged as

$$\frac{V^2}{V_{\max}^2} + \frac{a^2}{a_0^2} = 1 \tag{4.16b}$$

Equation (4.16b) is the equation of an ellipse, and it is known as the **steady flow adiabatic ellipse**, as shown in Figure 4.6. All the points on this ellipse have the same total energy. Each point differs from the others owing to the relative proportions of the thermal and kinetic energies, and thus corresponds to different Mach numbers. By differentiating Equation (4.16a), and using the definition of the Mach number, the Mach number at any point can be related the slope of the ellipse as

$$\mathrm{M} = -\frac{2}{k-1}\frac{\mathrm{d}a}{\mathrm{d}V}$$

Thus the change of the slope from point to point indicates how the changes in the Mach number are related to the changes in the speed of sound and the velocity. Therefore, it is a direct comparison of the relative magnitudes of the thermal and kinetic energies. As observed from Figure 4.6 for low-Mach-number flows, the changes in the Mach number are mainly due to the changes in velocity. However, at high-Mach-number flows, they are due to the changes in the speed of sound, and the compressibility effects are dominant. One important point to note in Figure 4.6 is the case corresponding to $M \lesssim 0.3$, where the changes in the speed of sound (the compressibility effects) are negligibly small. These flows are considered to be **incompressible**.

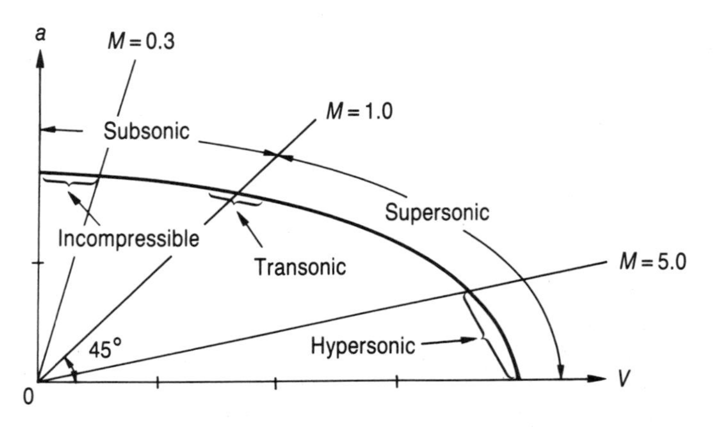

Figure 4.6 Steady flow adiabatic ellipse showing different types of compressible flow

4.4.2 Non-dimensional characteristic speeds

There are two important non-dimensional characteristic speeds, namely the Mach number and the Mach number referred to critical conditions.

(a) Mach number

This is a very convenient dimensionless characteristic speed and it is defined in Chapter 3, but it has two important disadvantages:

1 The Mach number is not only proportional to the velocity, but it is also a function of the temperature due to the definition of the speed of sound.
2 The Mach number tends to infinity at high velocities.

Thus a dimensionless velocity which eliminates these two disadvantages must be defined.

(b) Mach number referred to critical conditions

This can be obtained by dividing the flow velocity, V, by the critical speed of sound, a^*, as

$$M^* = V/a^* \tag{4.17}$$

There is a unique relation between M and M^* for adiabatic flows. Using the definitions of M and M^* from Equations (3.8) and (4.17)

$$M^{*2} = \frac{V^2}{a^{*2}} = \frac{V^2}{a^2}\frac{a^2}{a^{*2}} = M^2\frac{a^2}{a^{*2}} \tag{4.18}$$

If the first and last terms of Equation (4.16a) are divided by a^*, then the following relation can be obtained:

$$\frac{V^2}{a^{*2}} + \frac{2}{k-1}\frac{a^2}{a^{*2}} = \frac{k+1}{k-1}$$

A relation between M^* and M can be obtained by substituting Equations (4.17) and (4.18) into the above expression as

$$M^{*2} = M^2\left(\frac{k+1}{2} - \frac{k-1}{2}M^{*2}\right)$$

This can be solved for M^* and M to yield

$$M^{*2} = \frac{[(k+1)/2]M^2}{1+[(k-1)/2]M^2} \tag{4.19a}$$

and

$$M^2 = \frac{[2/(k+1)]M^{*2}}{1-[(k-1)/(k+1)]M^{*2}} \tag{4.19b}$$

From Equation (4.19a), it is possible to observe that, as M tends to infinity, M^* approaches a constant value of $\sqrt{[(k+1)/(k-1)]}$. Figure 4.7 illustrates the relative magnitudes of M and M^* in subsonic and supersonic flows.

Example 4.2
At a point in a flow passage, the velocity and temperature of the air are 802 m s^{-1} and 400 K, respectively.

- **a** Calculate the stagnation speed of sound.
- **b** Calculate the critical speed of sound.
- **c** Calculate the Mach number and the Mach number referred to critical conditions at the given state.
- **d** If the air is accelerated in a suitable flow passage, find the maximum possible velocity that can be attained. Determine the Mach number and the Mach number referred to critical conditions corresponding to the state of maximum velocity.
- **e** Obtain the kinematic form of the adiabatic energy equation and sketch the steady flow adiabatic ellipse.

Solution
(a) The stagnation temperature of the air may be evaluated by using Equation (4.8b) as

$$T_0 = T + \frac{k-1}{2kR}V^2 = 400\ \text{K} + \frac{1.4-1}{(2)(1.4)(287.1\ \text{J kg}^{-1}\,\text{K}^{-1})}(802\ \text{m s}^{-1})^2 = 720.1\ \text{K}$$

Then the stagnation speed of sound from Equation (4.9) is

$$a_0 = \sqrt{(kRT_0)} = \sqrt{[(1.4)(287.1\ \text{J kg}^{-1}\,\text{K}^{-1})(720.1\ \text{K})]} = \mathbf{538\ m\,s^{-1}}$$

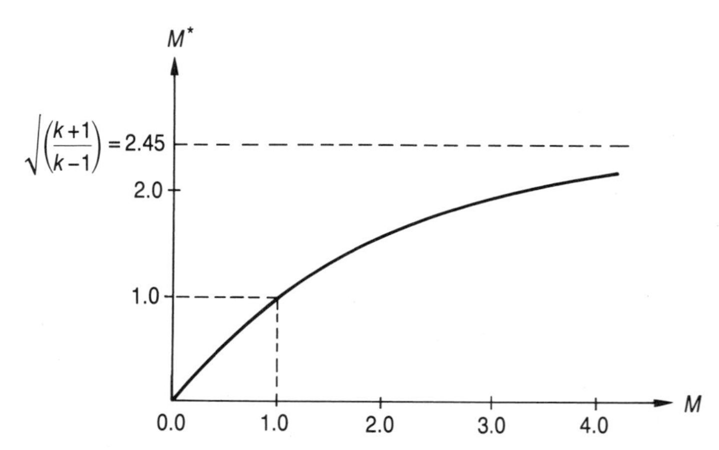

Figure 4.7 Relative magnitudes for M and M^* (for $k = 1.4$)

In this state, $M = 0$ and $M^* = 0$, since the fluid is stagnant with $V = 0$.

(b) The critical speed of sound from Equation (4.12) is

$$a^* = \sqrt{\left(\frac{2k}{k+1} RT_0\right)} = \sqrt{\left(\frac{(2)(1.4)}{1.4+1} (287.1\ \mathrm{J\,kg^{-1}\,K^{-1}})(720.1\ \mathrm{K})\right)} = \mathbf{491.1\ m\,s^{-1}}$$

At the critical state $M = 1$ and $M^* = 1$, since $a^* = V^*$.

(c) The speed of sound at the given state is

$$a = \sqrt{(kRT)} = \sqrt{[(1.4)(287.1\ \mathrm{J\,kg^{-1}\,K^{-1}})(400\ \mathrm{K})]} = 401\ \mathrm{m\,s^{-1}}$$

Then the Mach number is

$$M = \frac{V}{a} = \frac{802\ \mathrm{m\,s^{-1}}}{401\ \mathrm{m\,s^{-1}}} = \mathbf{2.0}$$

and the Mach number referred to critical conditions from Equation (4.17) is

$$M^* = \frac{V}{a^*} = \frac{802\ \mathrm{m\,s^{-1}}}{491.1\ \mathrm{m\,s^{-1}}} = \mathbf{1.633}$$

(d) The maximum velocity that can be attained by the air can be evaluated by using Equation (4.10) as

$$V_{max} = \sqrt{\left(\frac{2k}{k-1} RT_0\right)} = \sqrt{\left(\frac{(2)(1.4)}{1.4-1} (287.1\ \mathrm{J\,kg^{-1}\,K^{-1}})(720.1\ \mathrm{K})\right)} = \mathbf{1203\ m\,s^{-1}}$$

As the velocity of the air approaches its maximum value, the temperature and the speed of sound approach zero. Hence, the Mach number goes to infinity, that is $M \to \infty$.

The Mach number referred to critical conditions from Equation (4.17) is

$$M^* = \frac{V_{max}}{a^*} = \frac{1203\ \mathrm{m\,s^{-1}}}{491.1\ \mathrm{m\,s^{-1}}} = \mathbf{2.450}$$

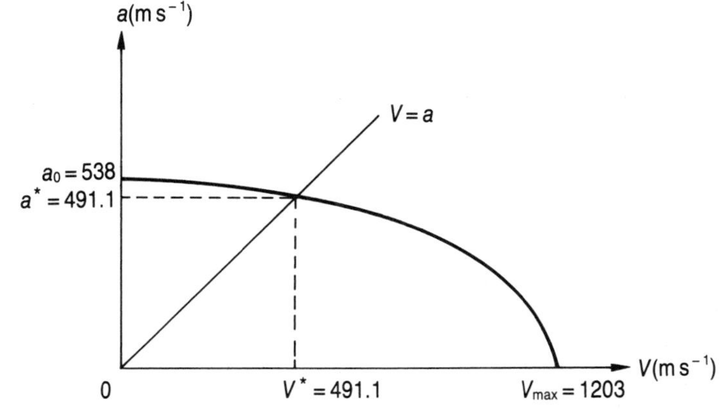

Figure 4.8 Steady-state adiabatic ellipse for Example 4.2

(e) The kinematic form of the adiabatic energy equation is obtained via Equation (4.16b) as

$$\frac{V^2}{V_{max}^2}+\frac{a^2}{a_0^2}=1 \quad \text{or} \quad \frac{V^2}{(1203)^2}+\frac{a^2}{(538)^2}=1$$

This steady-state adiabatic ellipse is shown in Figure 4.8.

4.5 EFFECTS OF AREA VARIATION ON FLOW PROPERTIES IN ISENTROPIC FLOW

Consider a one-dimensional compressible flow in an infinitesimal duct with variable area, as shown in Figure 4.9. The infinitesimal variation of the cross-sectional area of the duct causes infinitesimal changes in the flow properties. These changes may be related by using the basic equations.

(a) Continuity equation

For steady and uniform flow at each cross-section, the mass flow rate is constant. Therefore,

$$\dot{m}=\rho AV=(\rho+\mathrm{d}\rho)(A+\mathrm{d}A)(V+\mathrm{d}V)$$

or

$$\cancel{\rho A V}=\cancel{\rho A V}+\rho V\,\mathrm{d}A+AV\,\mathrm{d}\rho+\underbrace{V\,\mathrm{d}\rho\,\mathrm{d}A}_{\simeq 0}+\rho A\,\mathrm{d}V+\underbrace{\rho\,\mathrm{d}A\,\mathrm{d}V}_{\simeq 0}+\underbrace{A\,\mathrm{d}\rho\,\mathrm{d}V}_{\simeq 0}+\underbrace{\mathrm{d}\rho\,\mathrm{d}A\,\mathrm{d}V}_{\simeq 0}$$

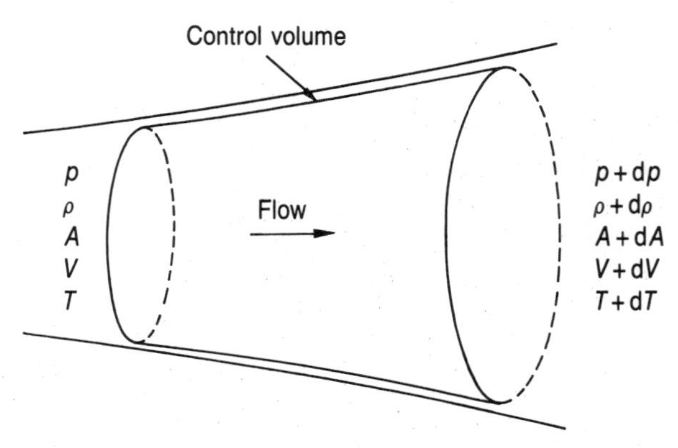

Figure 4.9 Compressible flow in an infinitesimal duct with an area change

Neglecting the products of differentials and dividing by ρAV, it is possible to obtain

$$\frac{dV}{V}+\frac{d\rho}{\rho}+\frac{dA}{A}=0 \tag{4.20}$$

Equation (4.20) can also be obtained by taking the logarithm of $\dot{m}=\rho AV$ and then by differentiating (logarithmic differentiation).

(b) Momentum equation

For frictionless flows, the pressure forces acting on the control volume must be balanced by the rate of change of momentum across the control volume, that is

$$pA-(p+dp)(A+dA)+dF_p=\dot{m}(V+dV)-\dot{m}V$$

where dF_p is the infinitesimal pressure force acting in the x direction on the side walls of the control volume. On these side walls, the average pressure is $p+dp/2$, and the cross-sectional area of the side walls perpendicular to the flow direction is dA, so $dF_p=(p+dp/2)\,dA$. Substituting the pressure force on the side walls and using the continuity equation as $\dot{m}=\rho AV$, the momentum equation becomes

$$pA-(p+dp)(A+dA)+(p+\tfrac{1}{2}\,dp)\,dA=\rho AV(V+dV)-\rho AV^2$$

or

$$\cancel{pA}-\cancel{pA}-A\,dp-\cancel{p\,dA}-\underbrace{dp\,dA}_{\simeq 0}+\cancel{p\,dA}+\underbrace{\tfrac{1}{2}\,dp\,dA}_{\simeq 0}=\cancel{\rho AV^2}+\rho AV\,dV-\cancel{\rho AV^2}$$

Cancelling and neglecting the higher-order differentials

$$dp=-\rho V\,dV \tag{4.21a}$$

or

$$\frac{dp}{\rho}+d\left(\frac{V^2}{2}\right)=0 \tag{4.21b}$$

A relation between the infinitesimal pressure, dp, and the infinitesimal area element, dA, can be obtained by eliminating dV with the aid of Equations (4.20) and (4.21a) as

$$\frac{dA}{A}=\frac{dp}{\rho V^2}\left(1-\frac{V^2}{dp/d\rho}\right)$$

But recalling Equation (3.5), $a^2=dp/d\rho=(\partial p/\partial\rho)_s$, for an isentropic process and using the definition of the Mach number, $M=V/a$, it is possible to obtain

$$\frac{dA}{A}=\frac{dp}{\rho V^2}(1-M^2) \tag{4.22}$$

A relation between the infinitesimal area element, $\mathrm{d}A$, and the infinitesimal velocity change, $\mathrm{d}V$, can be obtained by combining Equations (4.21a) and (4.22) as

$$\frac{\mathrm{d}A}{A} = -\frac{\mathrm{d}V}{V}(1 - M^2) \tag{4.23}$$

4.5.1 Nozzles

Nozzles are ducts with a variable cross-sectional area and are used to accelerate the flow, that is $\mathrm{d}V > 0$. As may be observed from Equation (4.21a), the pressure decreases in the direction of flow so that $\mathrm{d}p < 0$.

For subsonic flows ($M < 1$), Equation (4.23) states that the changes in the velocity and the area are inversely proportional. The cross-sectional area must then decrease to cause a velocity increase in the direction of flow. Therefore, for the acceleration of a subsonic flow, a **converging nozzle** is required, as shown in Figure 4.10(a).

Conversely, if a compressible fluid enters a passage with a supersonic velocity ($M > 1$), to accelerate the flow, the cross-sectional area must increase in the flow direction, as indicated by Equation (4.23). Hence, for supersonic acceleration, a **diverging nozzle** is necessary, as shown in Figure 4.10(b).

4.5.2 Diffusers

Diffusers are ducts of variable cross-sectional area for decelerating the flow so that $\mathrm{d}V < 0$. Then, Equation (4.21a) indicates that the pressure increases in the direction of flow, that is $\mathrm{d}p > 0$.

If the flow is subsonic ($M < 1$), it follows from Equations (4.22) and (4.23) that the cross-sectional area must increase in the direction of flow, such that $\mathrm{d}A > 0$. Therefore, for the deceleration of subsonic flows, a **diverging diffuser** is required, as shown in Figure 4.11(a).

Conversely, if the flow is supersonic ($M > 1$), to obtain diffusion the cross-sectional area must decrease in the flow direction. Hence, a **converging diffuser** is required for the deceleration of supersonic flows, as illustrated in Figure 4.11(b).

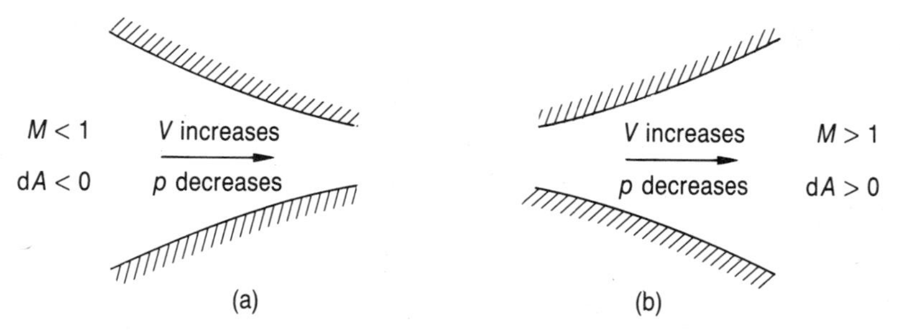

Figure 4.10 Effect of area change on steady, one-dimensional isentropic flow in nozzles: (a) subsonic flow; (b) supersonic flow

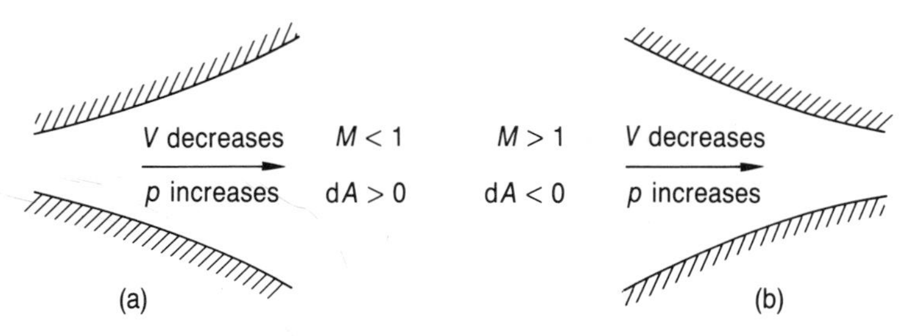

Figure 4.11 Effect of area change on steady, one-dimensional isentropic flow in diffusers: (a) subsonic flow; (b) supersonic flow

4.5.3 Choking in isentropic flow

First, it is necessary to obtain a relation between the changes in the velocity and the Mach number. From Figure 4.4, one should note that the temperature decreases as the velocity increases. For all compressible fluids, the speed of sound decreases as the temperature decreases. Therefore, when the velocity increases, the speed of sound decreases and the Mach number increases. Hence, the changes in the velocity and the Mach number are in the same direction.

In a diverging passage, shown in Figure 4.12(a) the cross-sectional area increases in the flow direction, that is $dA > 0$. Then the Mach number decreases, that is $M_2 < M_1$, if M_1 is subsonic, but it increases, that is $M_2 > M_1$, if M_1 is supersonic.

However, in a converging passage, shown in Figure 4.12(b) the cross-sectional area decreases in the flow direction, that is $dA < 0$. Then the Mach number increases, that is $M_2 > M_1$, if M_1 is subsonic, but it decreases, that is $M_2 < M_1$, if M_1 is supersonic.

When $M = 1$, inspection of Equation (4.23) shows that $dA/dV = 0$, which means that the flow cross-sectional area must pass through a maximum or minimum at $M = 1$. To illustrate this, suppose that the Mach number reaches unity either during the acceleration of a subsonic flow or during the deceleration of a supersonic flow in a converging passage. After this point, the cross-sectional area can either decrease or increase in the flow direction.

The case corresponding to decreasing flow area, after $M_1 = 1$, is given in Figure 4.13(a). If M decreases, then M_2 will be less than 1.0 and $dV < 0$. In this case, Equation (4.23) indicates that $dA > 0$, which contradicts the initial assumption that the flow area is decreasing in the flow direction. On the other hand, if M increases after $M_1 = 1$, then M_2 will be greater than 1.0 and $dV > 0$. This again contradicts the initial assumption, since Equation (4.23) again indicates that $dA > 0$. Therefore, a sonic flow cannot enter a converging passage and still satisfy the governing equations for steady, one-dimensional isentropic flow. This contradiction is due to a phenomenon known as **choking**. Therefore, the Mach number always changes towards unity in a converging passage whether the flow is initially subsonic or supersonic. When the Mach number is unity, the area can no longer decrease without

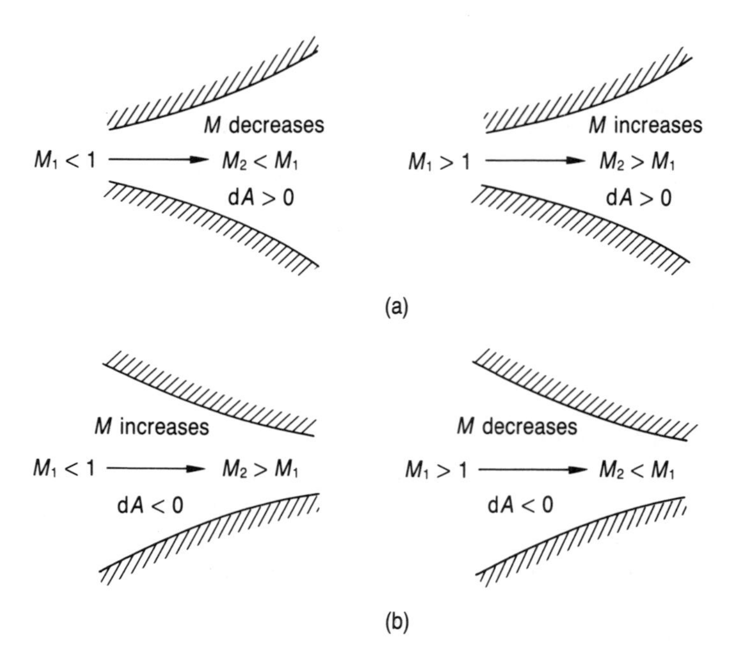

Figure 4.12 Effect of area change on the Mach number in (a) a diverging passage; (b) a converging passage

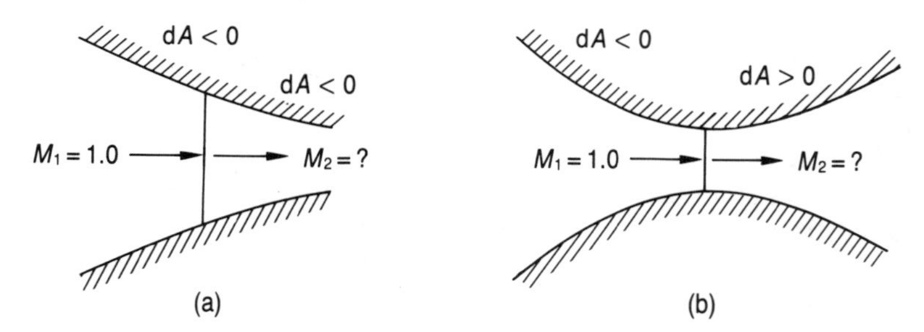

Figure 4.13 Limiting conditions in steady, one-dimensional isentropic compressible flow: (a) sonic flow entering a converging passage; (b) sonic flow entering a diverging passage

violating the governing equations for steady, one-dimensional isentropic flow and the flow is known as a **choked flow**.

The other possibility to be examined is illustrated in Figure 4.13(b). If M decreases, then M_2 will be less than 1.0 and $dV < 0$. In this case, Equation (4.23) indicates that $dA > 0$, which is in agreement with the assumption that M decreases. Conversely, if M increases, then M_2 becomes greater than 1.0 and $dV > 0$. This, again, is in agreement with the assumption, since Equation (4.23) indicates that $dA > 0$.

Therefore, a sonic flow can enter a diverging passage and keep satisfying the governing equations for steady, one-dimensional isentropic flow. The Mach number in the diverging passage may be either subsonic or supersonic depending on the downstream physical boundary conditions at the exit section of the flow passage. These downstream boundary conditions will be discussed in Section 4.8.

Therefore, if the mass flow rate is fixed, there is a minimum cross-sectional area required to pass this flow, and this phenomenon is known as **choking. For a given reduction in area, there is a maximum initial Mach number which can be maintained in a subsonic flow, while there is a minimum initial Mach number which can be maintained steadily in a supersonic flow**. At either of these limiting conditions, the flow at the exit section is sonic.

Physically, the phenomenon of choking can be illustrated by using the duct with adjustable walls shown in Figure 4.14. Now, suppose that for a subsonic flow A_1, p_1, T_1 and M_1 are known at the inlet section so that the mass flow rate is fixed by these parameters. Since the walls of the duct are flexible, it is possible to adjust the exit flow area, A_2. If the exit flow area, A_2, is equal to the inlet flow area, A_1, then the conditions at the exit section will be identical to those at the inlet section. A slight reduction in the exit area increases the Mach number, M_2, while it decreases the pressure, p_2, and the temperature, T_2. Further reductions in A_2 are possible until the exit Mach number, M_2, reaches unity. There, the flow is choked. After the flow is choked, there is no way of reducing the exit area, A_2, further without changing the steady-state conditions at the inlet. If A_2 is reduced beyond the limiting value, then the inlet Mach number, M_1, will decrease and settle at another steady-state value after a transient period of wave propagation. In this case, the mass flow rate will also be reduced.

Therefore, when $M = 1$, the flow cross-sectional area must pass through a minimum. Then, for a **subsonic nozzle**, only a converging section is necessary, as shown in Figure 4.15(a). However, for a **supersonic nozzle**, which continuously accelerates an initially subsonic flow to a supersonic one, the duct must consist of a converging section followed by a diverging one, as illustrated in Figure 4.15(b). This

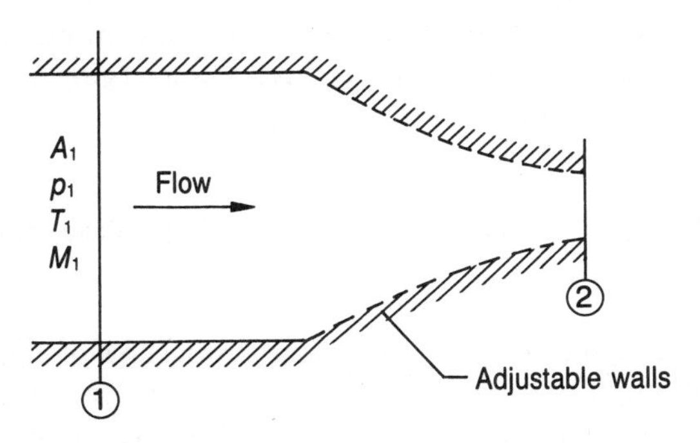

Figure 4.14 Choking of flow in a duct with adjustable walls

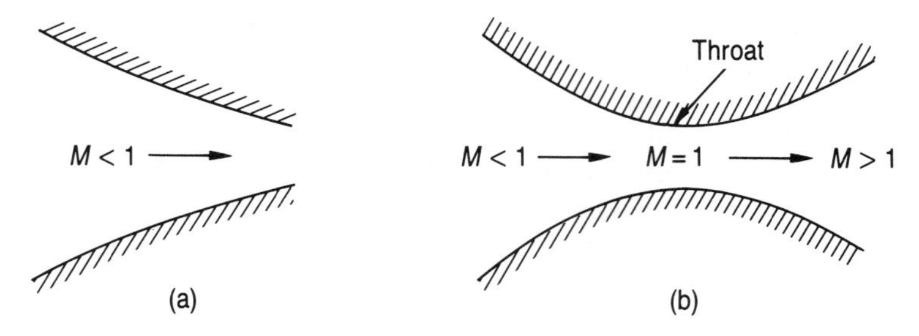

Figure 4.15 Nozzle types for subsonic and supersonic flows: (a) subsonic nozzle; (b) supersonic nozzle

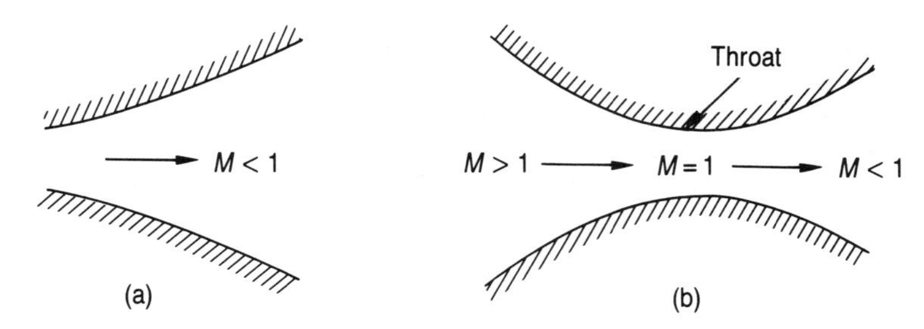

Figure 4.16 Diffuser types for subsonic and supersonic flows: (a) subsonic diffuser; (b) supersonic diffuser

is known as the **de Laval nozzle**. The minimum cross-sectional area of a converging–diverging nozzle is called the **throat**.

Similarly, when $M = 1$, the flow cross-sectional area must pass through a minimum in diffusers. This may be proved by using the same argument that is used for nozzles. Therefore, only a diverging section is required for a **subsonic diffuser**, as shown in Figure 4.16(a). However, a **supersonic diffuser**, which continuously decelerates the flow speed from supersonic to subsonic, requires a converging section followed by a diverging one, as shown in Figure 4.16(b).

4.6 RELATIONS FOR THE ISENTROPIC FLOW OF A PERFECT GAS

The equations governing the one-dimensional isentropic flow of a perfect gas are given in Section 4.2. Although any isentropic flow problem can be solved using these equations, the solution can be simplified by rearranging them in a more suitable form. These relations are derived in this section.

4.6.1 Stagnation temperature ratio

Equation (4.8b) is an alternative form of the adiabatic energy equation, and can be rearranged by using the definitions of the speed of sound and the Mach number from Equations (3.7a) and (3.8), respectively. Then, the following equation may be obtained for the ratio of the stagnation temperature to the static temperature as

$$\frac{T_0}{T} = 1 + \frac{k-1}{2} M^2 \tag{4.24}$$

When $M = 1$, Equation (4.24) becomes

$$\frac{T^*}{T_0} = \frac{2}{k+1} \tag{4.25}$$

4.6.2 Stagnation pressure ratio

For an isentropic process, p/ρ^k is constant. Then

$$\frac{p}{p_0} = \left(\frac{\rho}{\rho_0}\right)^k \tag{4.26}$$

But for a perfect gas $p/(\rho T) = p_0/(\rho_0 T_0) = R$, and therefore

$$\frac{p_0}{p} = \left(\frac{T_0}{T}\right)^{k/(k-1)} \tag{4.27}$$

The ratio of the stagnation pressure to the static pressure can now be obtained by combining Equations (4.24) and (4.27) as

$$\frac{p_0}{p} = \left(1 + \frac{k-1}{2} M^2\right)^{k/(k-1)} \tag{4.28}$$

When $M = 1$, this ratio reduces to

$$\frac{p^*}{p_0} = \left(\frac{2}{k+1}\right)^{k/(k-1)} \tag{4.29}$$

4.6.3 Stagnation density ratio

The ratio of the stagnation density to the static density can be obtained by using Equations (4.26) and (4.28) as

$$\frac{\rho_0}{\rho} = \left(1 + \frac{k-1}{2} M^2\right)^{1/(k-1)} \tag{4.30}$$

At the critical state, that is when $M = 1$, Equation (4.30) becomes

$$\frac{\rho^*}{\rho_0} = \left(\frac{2}{k+1}\right)^{1/(k-1)} \tag{4.31}$$

4.6.4 Mass flow rate per unit area

For a perfect gas, $p=\rho RT$ and the continuity equation can be transformed to

$$\frac{\dot{m}}{A}=\rho V=\frac{pV}{RT}$$

Now, substituting $a=\sqrt{(kRT)}$ and $M=V/a$ yields

$$\frac{\dot{m}}{A}=\frac{pV}{RT}=\frac{\sqrt{(k)}pV}{\sqrt{(kRT)}\sqrt{(RT)}}=\sqrt{\left(\frac{k}{RT_0}\right)}pM\sqrt{\left(\frac{T_0}{T}\right)}$$

Introducing Equation (4.24)

$$\frac{\dot{m}}{A}=\sqrt{\left(\frac{k}{RT_0}\right)}pM\sqrt{\left(1+\frac{k-1}{2}M^2\right)} \qquad (4.32)$$

This equation is valid for adiabatic and/or isentropic flows because, until now, only the continuity equation, the adiabatic energy equation and the equation of state for a perfect gas have been used.

At this point, the static pressure, p, may be eliminated by using the isentropic relation (4.28). Then

$$\frac{\dot{m}}{A}=\sqrt{\left(\frac{k}{RT_0}\right)}p_0M\left(1+\frac{k-1}{2}M^2\right)^{-(k+1)/[2(k-1)]} \qquad (4.33a)$$

or in dimensionless form

$$\frac{\dot{m}\sqrt{(RT_0)}}{Ap_0}=\sqrt{(k)}M\left(1+\frac{k-1}{2}M^2\right)^{-(k+1)/[2(k-1)]} \qquad (4.33b)$$

In Equation (4.33a), the Mach number, M, can be eliminated by using Equation (4.28) for the isentropic stagnation pressure ratio. Then

$$\frac{\dot{m}}{A}=\sqrt{\left(\frac{k}{RT_0}\right)}p_0\sqrt{\left\{\frac{2}{k-1}\left[\left(\frac{p}{p_0}\right)^{2/k}-\left(\frac{p}{p_0}\right)^{(k+1)/k}\right]\right\}} \qquad (4.34)$$

The plot of $\dot{m}/A$ versus p/p_0 is given in Figure 4.17.

The condition of the maximum mass flow rate per unit area is obtained by differentiating Equation (4.34) with respect to p/p_0 and setting this derivative to zero. In this case, $\dot{m}/A$ is maximum when

$$\frac{p}{p_0}=\left(\frac{2}{k+1}\right)^{k/(k-1)}=\frac{p^*}{p_0}$$

Therefore, the maximum mass flow rate per unit area is obtained for the critical pressure ratio, that is when $M=1$. The same result can also be obtained by differentiating Equation (4.33a) with respect to the Mach number, M, and setting this

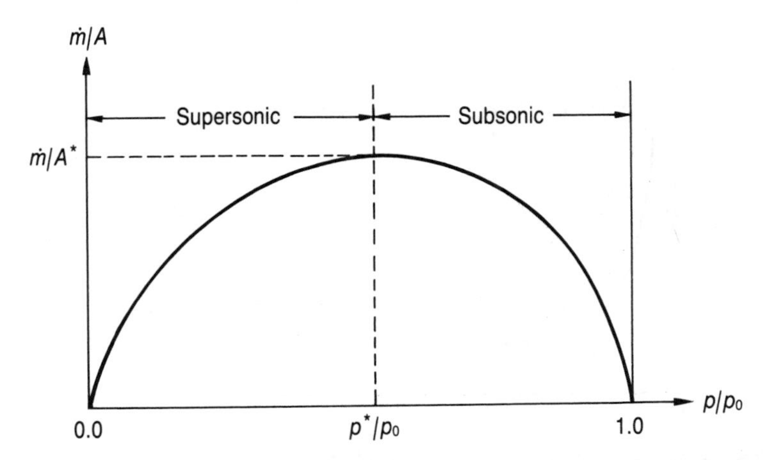

Figure 4.17 Variation of the mass flow rate per unit area with p/p_0 for isentropic flow

derivative to zero, which yields $M = 1$. An equation for the maximum mass flow rate per unit area may be obtained by setting the Mach number to unity in Equation (4.33a) as

$$\left(\frac{\dot{m}}{A}\right)_{\max} = \frac{\dot{m}}{A^*} = \sqrt{\left(\frac{k}{RT_0}\right)} p_0 \left(\frac{2}{k+1}\right)^{(k+1)[/2(k-1)]} \tag{4.35}$$

Therefore, for a given gas, the maximum mass flow rate per unit area depends on the ratio $p_0/\sqrt{T_0}$.

4.6.5 The area ratio

It is very convenient to introduce a dimensionless area ratio by dividing Equation (4.35) by Equation (4.33a):

$$\frac{A}{A^*} = \frac{1}{M}\left[\frac{2}{k+1}\left(1 + \frac{k-1}{2}M^2\right)\right]^{(k+1)/[2(k-1)]} \tag{4.36}$$

The area ratio, A/A^*, is plotted with respect to the Mach number in Figure 4.18, by using Equation (4.36). This plot shows that the area ratio is minimum and has a value of unity at the critical point, where $M = 1$. For other values of the Mach number, the area ratio is always greater than unity, and for any given value of the area ratio, there always corresponds two values of the Mach number, one for subsonic flow and the other for supersonic flow.

4.6.6 Impulse function

It is sometimes convenient to define a quantity called the **impulse function** for problems involving jet propulsion, which may be given as

$$I = pA + \rho AV^2 \tag{4.37}$$

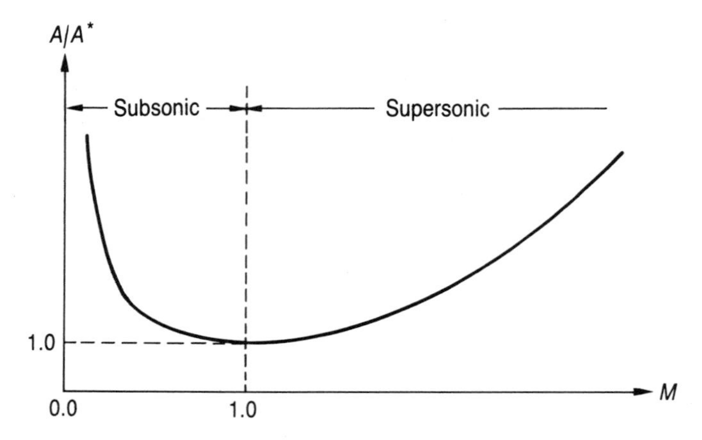

Figure 4.18 Area ratio as a function of the Mach number

The momentum equation (4.2) may now be rearranged by using Equation (4.37) as

$$\tau = I_1 - I_2 = (p_1 A_1 + \rho_1 A_1 V_1^2) - (p_2 A_2 + \rho_2 A_2 V_2^2) \tag{4.38}$$

where τ is the **net thrust** and is defined as the net force exerted by the stream on the internal solid surfaces which the fluid wets, acting in the direction opposite to the direction of the flow. It includes the forces due to the pressure and the viscous stresses on the duct walls. Equation (4.38) is valid whether the flow is adiabatic or non-adiabatic, and whether it is reversible or irreversible.

For a perfect gas, it is convenient to express the impulse function in terms of the Mach number. Then,

$$I = pA + \frac{p}{RT} AV^2 = pA\left(1 + \frac{kV^2}{a^2}\right) = pA(1 + kM^2) \tag{4.39}$$

For isentropic flow, a dimensionless impulse function is formed by writing

$$\frac{I}{p_0 A^*} = \frac{p}{p_0} \frac{A}{A^*} (1 + kM^2) \tag{4.40}$$

where p/p_0 and A/A^* are functions of M, and are given by Equations (4.28) and (4.36), respectively.

Another way of forming the dimensionless impulse function is by evaluating Equation (4.40) at $M = 1$ as

$$\frac{I^*}{p_0 A^*} = \frac{p^*}{p_0} (1 + k)$$

and dividing Equation (4.40) by $I^*/(p_0 A^*)$. Then

$$\frac{I}{I^*} = \frac{p}{p_0} \frac{A}{A^*} \frac{p_0}{p^*} \frac{1 + kM^2}{1 + k}$$

Now, the ratios p/p_0, A/A^* and p_0/p^* may be substituted by using Equations (4.28), (4.36) and (4.29), respectively. Hence

$$\frac{I}{I^*} = \frac{1 + kM^2}{M\sqrt{\{2(k+1)\,[1 + \frac{1}{2}(k-1)M^2]\}}} \tag{4.41}$$

4.7 WORKING CHART AND WORKING TABLE FOR ISENTROPIC FLOW

Formulas derived thus far introduce quite tedious numerical calculations, and their solutions often involve a trial-and-error procedure. The numerical calculations can be facilitated by introducing a working chart and a working table. The following equations are the property ratios for the steady, isentropic and one-dimensional flow of a perfect gas that were derived in the previous sections.

$$M^{*2} = \frac{[(k+1)/2]\,M^2}{1 + [(k-1)/2]\,M^2} \tag{4.19a}$$

$$\frac{T_0}{T} = 1 + \frac{k-1}{2}\,M^2 \tag{4.24}$$

$$\frac{p_0}{p} = \left(1 + \frac{k-1}{2}\,M^2\right)^{k/(k-1)} \tag{4.28}$$

$$\frac{\rho_0}{\rho} = \left(1 + \frac{k-1}{2}\,M^2\right)^{1/(k-1)} \tag{4.30}$$

$$\frac{A}{A^*} = \frac{1}{M}\left[\frac{2}{k+1}\left(1 + \frac{k-1}{2}\,M^2\right)\right]^{(k+1)/[2(k-1)]} \tag{4.36}$$

$$\frac{I}{I^*} = \frac{1 + kM^2}{M\sqrt{\{2(k+1)\,[1 + \frac{1}{2}(k-1)M^2]\}}} \tag{4.41}$$

$$\frac{\dot{m}\sqrt{(RT_0)}}{Ap_0} = \sqrt{(k)}\,M\left(1 + \frac{k-1}{2}\,M^2\right)^{-(k+1)/[2(k-1)]} \tag{4.33b}$$

4.7.1 Working chart for isentropic flow

The right-hand sides of the above equations are all functions of the Mach number, M, and the specific heat ratio, k. The dimensionless ratios on the left-hand sides of these equations can then be represented in graphical form with the Mach number as the independent variable for a fixed value of k. Such a working chart for steady, one-dimensional isentropic flow of a perfect gas is given in Figure 4.19 for $k = 1.4$.

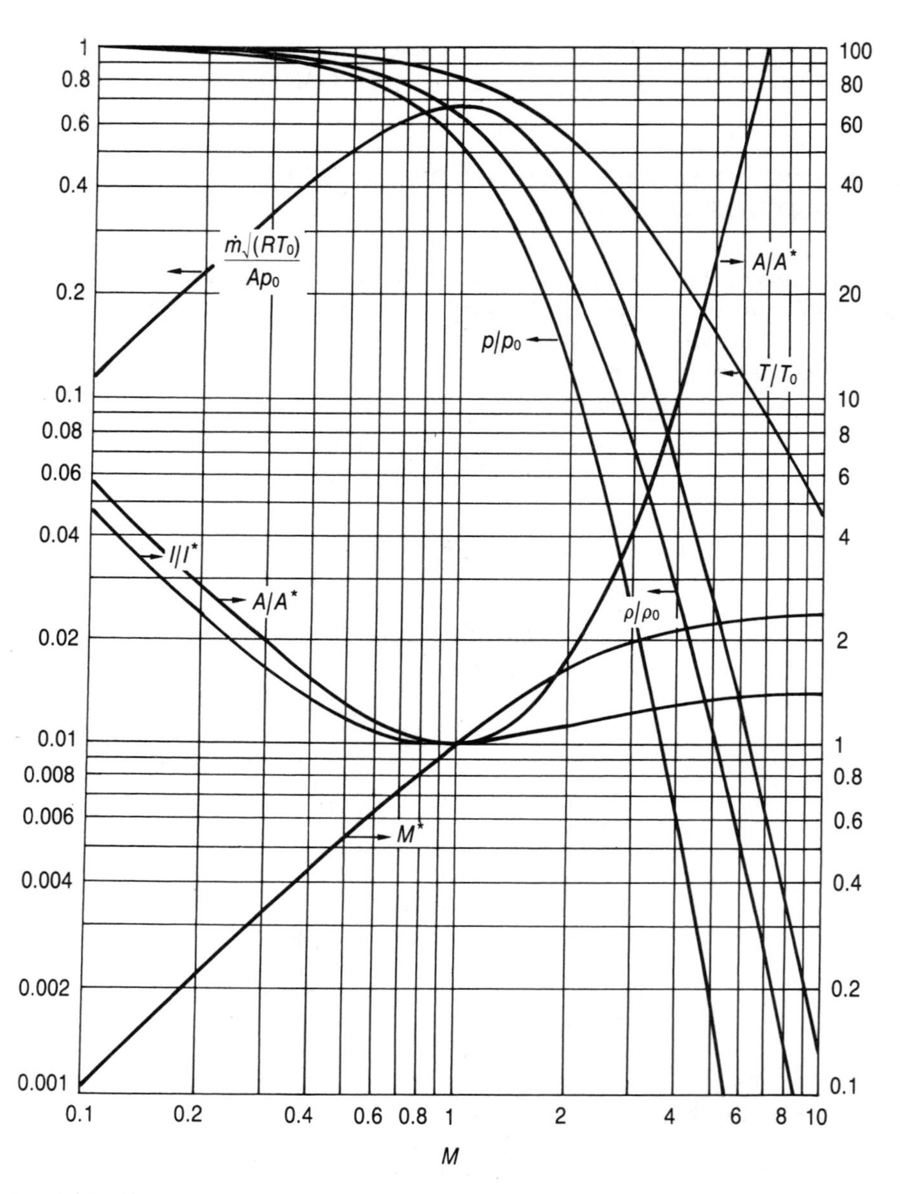

Figure 4.19 Working chart for isentropic flow ($k = 1.4$)

4.7.2 Working table for isentropic flow

The dimensionless ratios can also be represented in tabular form with the Mach number as the independent variable. For $k = 1.4$, the values of these non-dimensional ratios are given in Appendix C as a function of the Mach number. For convenience, the critical values of these non-dimensional ratios are given in Table 4.1.

Example 4.3
An airstream flows in a converging duct, as shown in Figure 4.20(a) from a cross-sectional area of 0.001 176 m^2 to a cross-sectional area of 0.001 057 m^2, where the pressure, temperature and velocity are 150 kPa, 125 °C and 304 m s^{-1}, respectively.

a Determine the remaining fluid properties at section 2.
b Determine the fluid properties at section 1.
c Show the process on a $T-s$ diagram.
d Find the mass flow rate.

Assume steady, one-dimensional isentropic flow.

Solution
(a) The speed of sound at section 2 is

$$a_2 = \sqrt{(kRT_2)} = \sqrt{[(1.4)(287.1\ \text{J kg}^{-1}\ \text{K}^{-1})(398\ \text{K})]} = 400\ \text{m s}^{-1}$$

Then, the Mach number is

$$M_2 = \frac{V_2}{a_2} = \frac{304\ \text{m s}^{-1}}{400\ \text{m s}^{-1}} = \mathbf{0.76}$$

Table 4.1 The limiting values of isentropic ratios for $k = 1.4$

Ratio	$M = 0$	$M = 1$	$M \to \infty$
M^*	0.000	1.000	2.450
T/T_0	1.000	0.8333	0.000
p/p_0	1.000	0.5283	0.000
ρ/ρ_0	1.000	0.6339	0.000
A/A^*	∞	1.000	∞
I/I^*	∞	1.000	1.429
$\dot{m}\sqrt{(RT_0)}/(Ap_0)$	0.000	0.6847	0.000

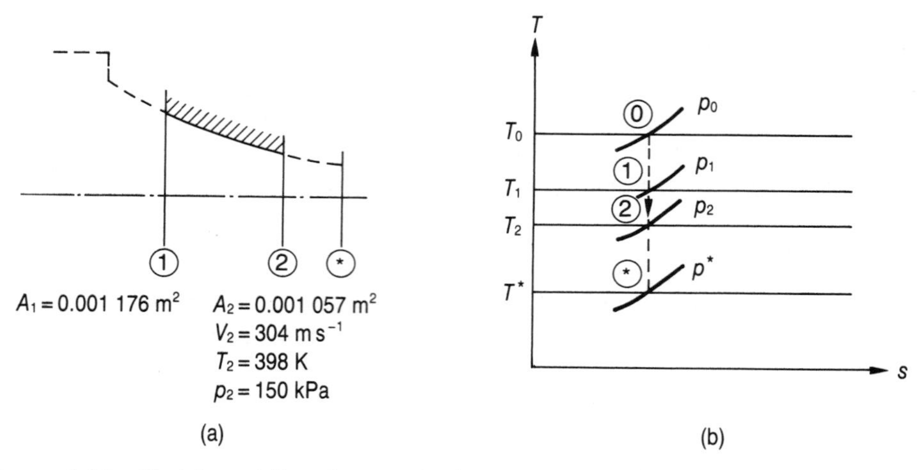

Figure 4.20 Sketch and T–s diagram for Example 4.3

Also, the density may be evaluated from the equation of state for a perfect gas:

$$\rho_2 = \frac{p_2}{RT_2} = \frac{150\ 000\ \mathrm{N\,m^{-2}}}{(287.1\ \mathrm{J\,kg^{-1}\,K^{-1}})(398\ \mathrm{K})} = \mathbf{1.313\ kg\,m^{-3}}$$

The stagnation properties, from which the airstream is accelerated to the condition at section 2, may be evaluated by using the isentropic tables in Appendix C at $M_2 = 0.76$ as

$$p_2/p_{02} = 0.6821 \quad \text{or} \quad p_{02} = 150\ \mathrm{kPa}/0.6821 = \mathbf{219.9\ kPa}$$

$$T_2/T_{02} = 0.8964 \quad \text{or} \quad T_{02} = 398\ \mathrm{K}/0.8964 = \mathbf{444.0\ K}$$

$$\rho_2/\rho_{02} = 0.7609 \quad \text{or} \quad \rho_{02} = 1.313\ \mathrm{kg\,m^{-3}}/0.7609 = \mathbf{1.726\ kg\,m^{-3}}$$

(b) If the isentropic flow in the converging duct were to be maintained with the duct extended imaginarily, as shown in Figure 4.20, the flow would be accelerated to $M = 1$, with the area at this point being equal to the critical area, A^*. Then, at $M_2 = 0.76$ from Appendix C,

$$A_2/A^* = 1.057 \quad \text{or} \quad A^* = 0.001\ 057\ \mathrm{m^2}/1.057 = 0.001\ \mathrm{m^2}$$

Hence

$$A_1/A^* = 0.001\ 176\ \mathrm{m^2}/0.001\ \mathrm{m^2} = 1.176$$

Corresponding to an area ratio of 1.176, there exist two solutions for the Mach number, one for subsonic flow with $M = 0.61$ and the other for supersonic flow with $M = 1.50$. But in a converging duct, only subsonic speeds are possible, so

$$M_1 = \mathbf{0.61}$$

Also for isentropic flow, the stagnation conditions do not change from one state to the other. Therefore,

$$p_{01} = p_{02} = \mathbf{219.9\ kPa}$$

$$T_{01} = T_{02} = \mathbf{444.0\ K}$$

$$\rho_{01} = \rho_{02} = \mathbf{1.726\ kg\,m^{-3}}$$

From Appendix C, with $M_1 = 0.61$,

$$p_1/p_{01} = 0.7778 \quad \text{or} \quad p_1 = (0.7778)(219.9\ \text{kPa}) = \mathbf{171.0\ kPa}$$

$$T_1/T_{01} = 0.9307 \quad \text{or} \quad T_1 = (0.9307)(444.0\ \text{K}) = \mathbf{413.2\ K}$$

$$\rho_1/\rho_{01} = 0.8357 \quad \text{or} \quad \rho_1 = (0.8357)(1.726\ \text{kg}\,\text{m}^{-3}) = \mathbf{1.442\ kg\,m^{-3}}$$

The speed of sound at section 1 may now be evaluated as

$$a_1 = \sqrt{(kRT_1)} = \sqrt{[(1.4)(287.1\ \text{J}\,\text{kg}^{-1}\,\text{K}^{-1})(413.2\ \text{K})]} = 407.5\ \text{m}\,\text{s}^{-1}$$

and the velocity as

$$V_1 = M_1 a_1 = (0.61)(407.5\ \text{m}\,\text{s}^{-1}) = \mathbf{248.6\ m\,s^{-1}}$$

(c) The temperature–entropy diagram for the acceleration process is shown in Figure 4.20(b)

(d) The mass flow rate may be computed from the continuity equation at section 1 or at section 2. For example, at section 1

$$\dot{m} = \rho_1 A_1 V_1 = (1.442\ \text{kg}\,\text{m}^{-3})(0.001\ 176\ \text{m}^2)(248.6\ \text{m}\,\text{s}^{-1}) = \mathbf{0.4216\ kg\,s^{-1}}$$

Alternatively, the mass flow rate at section 1 or at section 2 may be calculated by using the isentropic tables in Appendix C. For example, at section 2,

$$\frac{\dot{m}\sqrt{(RT_{02})}}{A_2 p_{02}} = 0.6478 \quad \text{or} \quad \dot{m} = \frac{(0.6478)(0.001\ 057\ \text{m}^2)(219\ 900\ \text{N}\,\text{m}^{-2})}{\sqrt{[(287.1\ \text{J}\,\text{kg}^{-1}\,\text{K}^{-1})(444.0\ \text{K})]}}$$

$$= \mathbf{0.4217\ kg\,s^{-1}}$$

Example 4.4

A supersonic nozzle has throat and exit cross-sectional areas of 0.09 m^2 and 0.15 m^2, respectively, as shown in Figure 4.21(a). The mass flow rate of the air is 14 $kg\,s^{-1}$. The supply pressure and temperature at the nozzle inlet, where the velocity is negligible, are 68 kPa and 40 °C, respectively. Assuming steady, one-dimensional isentropic flow

a determine the fluid properties at the throat,
b determine the fluid properties at the exit of the nozzle, and
c show the process on a *T–s* diagram.

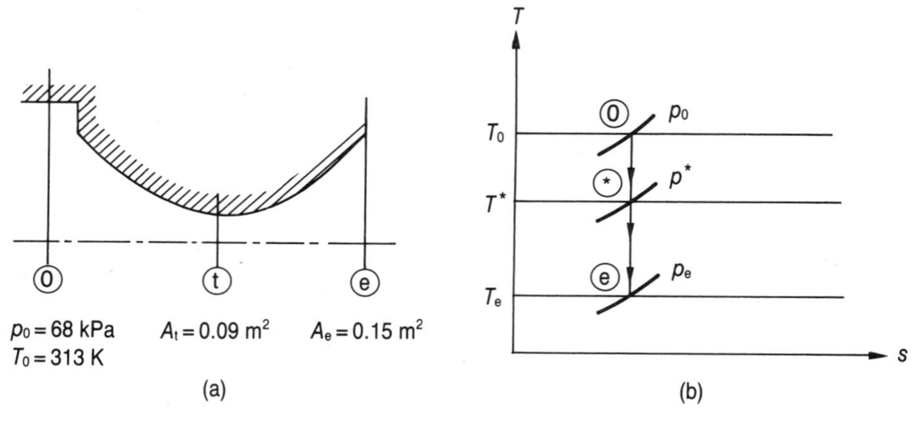

Figure 4.21 Sketch and T–s diagram for Example 4.4

Solution

The pressure and the temperature at the nozzle inlet, where the velocity is negligible, represent the stagnation pressure and temperature, that is $p_0 = 68$ kPa and $T_0 = 313$ K. Then the stagnation density is

$$\rho_0 = \frac{p_0}{RT_0} = \frac{(68\ 000\ \text{N m}^{-2})}{(287.1\ \text{J kg}^{-1}\ \text{K}^{-1})(313\ \text{K})} = 0.7567\ \text{kg m}^{-3}$$

(a) The nozzle under consideration is a supersonic one, and therefore the Mach number is unity at the throat. Also, the stagnation conditions in an isentropic flow do not change from one state to the other. Then, from Appendix C,

$$p^*/p_0 = 0.5283 \quad \text{or} \quad p^* = (0.5283)(68\ \text{kPa}) = \mathbf{35.92\ kPa}$$

$$T^*/T_0 = 0.8333 \quad \text{or} \quad T^* = (0.8333)(313\ \text{K}) = \mathbf{260.8\ K}$$

$$\rho^*/\rho_0 = 0.6339 \quad \text{or} \quad \rho^* = (0.6339)(0.7567\ \text{kg m}^{-3}) = \mathbf{0.4797\ kg\ m^{-3}}$$

The speed of sound at the throat is

$$a^* = \sqrt{(kRT^*)} = \sqrt{[(1.4)(287.1\ \text{J kg}^{-1}\ \text{K}^{-1})(260.8\ \text{K})]} = 323.8\ \text{m s}^{-1}$$

and the velocity at the throat, where $M = 1$, is

$$V^* = a^* = \mathbf{323.8\ m\ s^{-1}}$$

(b) The non-dimensional mass flow rate at the exit of the nozzle can be evaluated as

$$\frac{\dot{m}\sqrt{(RT_0)}}{A_e p_0} = \frac{(14\ \text{kg s}^{-1})\sqrt{[(287.1\ \text{J kg}^{-1}\ \text{K}^{-1})(313\ \text{K})]}}{(0.15\ \text{m}^2)(68\ 000\ \text{N m}^{-2})} = 0.4115$$

The Mach number at the exit of the nozzle corresponding to this non-dimensional mass flow rate can be obtained from Appendix C as

$$M_e = 1.98$$

The properties at the cross-sectional area where $M_e = 1.98$ can be obtained via Appendix C as

$$p_e/p_0 = 0.1318 \quad \text{or} \quad p_e = (0.1318)(68 \text{ kPa}) = \mathbf{8.962 \text{ kPa}}$$

$$T_e/T_0 = 0.5605 \quad \text{or} \quad T_e = (0.5605)(313 \text{ K}) = \mathbf{175.4 \text{ K}}$$

$$\rho_e/\rho_0 = 0.2352 \quad \text{or} \quad \rho_e = (0.2352)(0.7567 \text{ kg m}^{-3}) = \mathbf{0.178 \text{ kg m}^{-3}}$$

$$M_e^* = V_1/a^* = 1.624 \quad \text{or} \quad V_e = (1.624)(323.8 \text{ m s}^{-1}) = \mathbf{525.9 \text{ m s}^{-1}}$$

(c) The $T-s$ diagram for the acceleration process of this example is shown in Figure 4.21(b).

4.8 ISENTROPIC OPERATION OF NOZZLES

In this section, the isentropic performances of converging and converging–diverging nozzles under varying pressure ratios are discussed.

4.8.1 Converging nozzles

Consider a converging nozzle which is fed from a large reservoir where the conditions are stagnant due to the negligible flow velocity, as shown in Figure 4.22(a). This converging passage discharges into another reservoir where the back pressure, p_b, is controllable by means of a valve. The pressure at the exit plane of the nozzle is denoted by p_e. The possible isentropic flow patterns can now be investigated by changing the back pressure. The effects of variations in the back pressure on the distribution of the pressure in the passage, on the mass flow rate and on the exit plane pressure are given in Figures 4.22(b), 4.22(c) and 4.22(d), respectively.

(a) No-flow condition ($p_b = p_0$)

To begin with, suppose that the valve is closed as shown by flow pattern 1 of Figure 4.22(b). In this case, the pressure is constant throughout the converging nozzle, such that $p_b = p_e = p_0$, and there is no flow.

(b) Subcritical flow regime ($p^* < p_b < p_0$)

If the valve at the discharge reservoir is now opened slightly, then there will be a flow with a constantly decreasing pressure through the nozzle, as indicated by flow pattern 2 of Figure 4.22(b). Since the flow is subsonic at the exit plane, the exit pressure, p_e, must be the same as the back pressure p_b. To prove this statement, assume that p_e is larger than p_b; then the stream would expand upon leaving the nozzle. However,

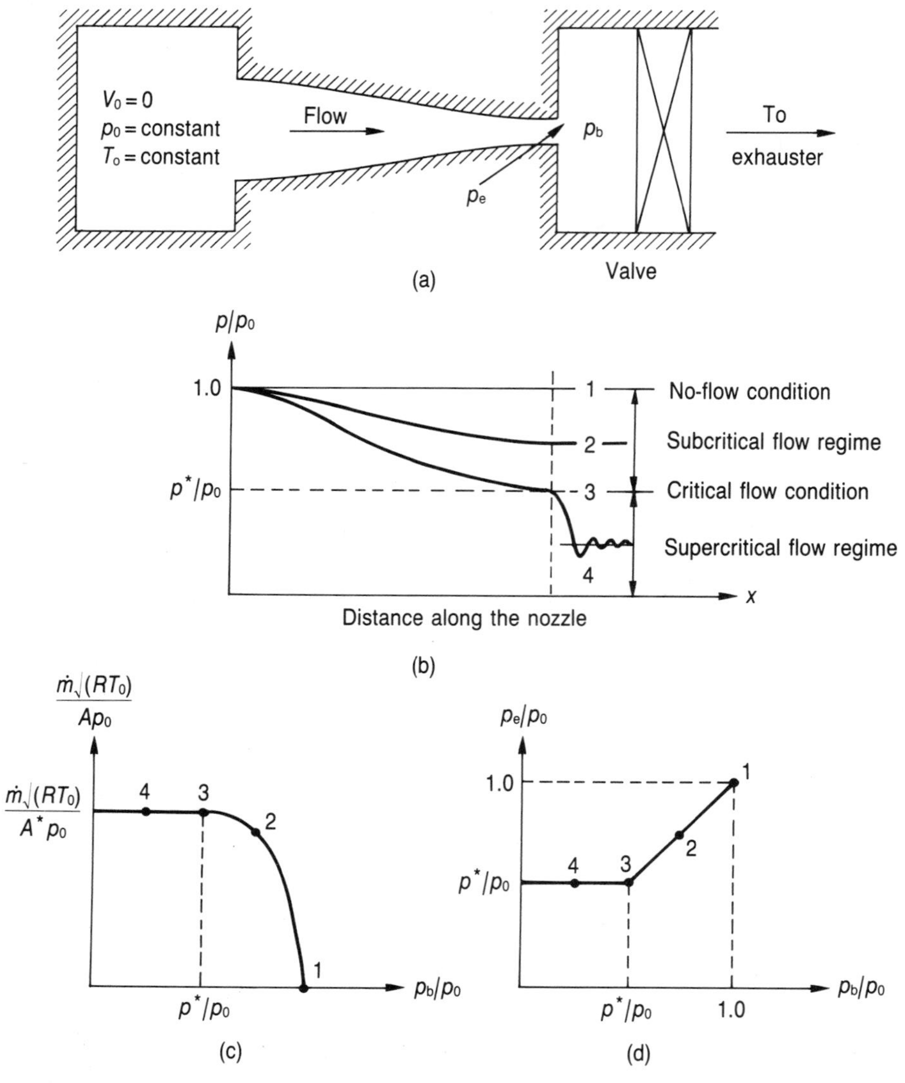

Figure 4.22 Operation of a converging nozzle at various back pressures

such an increase in area will further increase the stream pressure. This contradicts the fact that p_b is the final pressure to be reached by the stream, since the stream pressure is greater than p_b after expansion. Conversely, if p_e is less than p_b, then the stream would contract when it leaves the nozzle. Such a contraction will further decrease the stream pressure, which again contradicts the fact that p_b is the final pressure of the stream leaving the nozzle.

When the back pressure, p_b, is equal to the exit pressure, p_e, the jet issues as a cylindrical parallel stream. The surface of this jet is then gradually retarded by the surrounding gases so that a mixing zone is produced in which the velocity of the jet finally drops to that of the surroundings, as shown in Figure 4.23(a).

Further reductions in p_b will increase the mass flow rate and will change the pressure distribution without a significant change in the performance, until the critical flow condition is reached.

(c) Critical flow condition ($p_b = p^*$)

When the critical flow condition is reached, as shown by flow pattern 3 of Figure 4.22(b), then the converging nozzle is choked. The Mach number at the exit plane is then unity with p_b/p_0 equal to the critical pressure ratio p^*/p_0.

(d) Supercritical flow regime ($p_b < p^*$)

Further reductions in the back pressure by opening the valve cannot produce further changes in the conditions within the nozzle, as shown by flow pattern 4 of Figure 4.22(b). This is due to the fact that the value of p_e/p_0 cannot be made less than the critical pressure ratio p^*/p_0, unless there is a throat upstream of the exit section. Therefore, in this regime, the pressure distribution within the nozzle, the mass flow rate and the exit pressure are all identical to the corresponding quantities in the critical flow condition, as shown in Figures 4.22(b), 4.22(c) and 4.22(d), respectively. The stream just outside the nozzle expands from the critical pressure to the back pressure. The pressure distribution and the expansion process outside the nozzle cannot be investigated using one-dimensional gas dynamics.

In this regime, the pressure of the jet leaving the nozzle is the critical pressure p^*, which is larger than the back pressure. The sudden reduction in the pressure causes the gaseous jet to expand in an explosive fashion. The gas particles accelerate radially, creating a pressure reduction in the core of the jet so that the flow is reversed. This phenomenon is periodic and the jet becomes thinner in some sections and thicker in others, as shown in Figure 4.23(b).

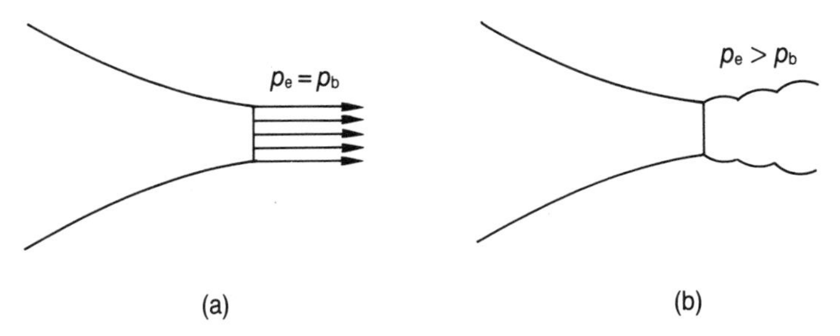

Figure 4.23 Discharge jet shapes for subcritical and supercritical flows in converging nozzles

Example 4.5
A converging nozzle with an exit cross-sectional area of 0.001 m^2 is operated with air at a back pressure of 69.5 kPa, as shown in Figure 4.24(a). The nozzle is fed from a large reservoir where the stagnation pressure and temperature are 100 kPa and 60 °C, respectively. Determine the Mach number and the temperature at the nozzle exit. Also, find the mass flow rate through the nozzle. Assume one-dimensional steady isentropic flow.

Solution
The first step is to check the converging nozzle for choking. Then

$$p_b/p_0 = 69.5 \text{ kPa}/100 \text{ kPa} = 0.6950 > p^*/p_0 = 0.5283$$

Therefore the flow is not choked, and it is in the subcritical flow regime. Thus

$$p_e = p_b = 69.50 \text{ kPa}$$

From Appendix C, corresponding to $p_e/p_0 = 0.695$,

$$M_e = \mathbf{0.74}$$

$$T_e/T_0 = 0.9013 \quad \text{or} \quad T_e = (0.9013)(333 \text{ K}) = \mathbf{300.1\ K}$$

and

$$\frac{\dot{m}\sqrt{(RT_0)}}{A_e p_0} = 0.6410 \quad \text{or} \quad \dot{m} = \frac{(0.6411)(0.001 \text{ m}^2)(100\,000 \text{ N m}^{-2})}{\sqrt{[(287.1 \text{ J kg}^{-1} \text{K}^{-1})(333 \text{ K})]}}$$

$$= \mathbf{0.2073\ kg\ s^{-1}}$$

The expansion process is shown in Figure 4.24(b) on a T–s diagram.

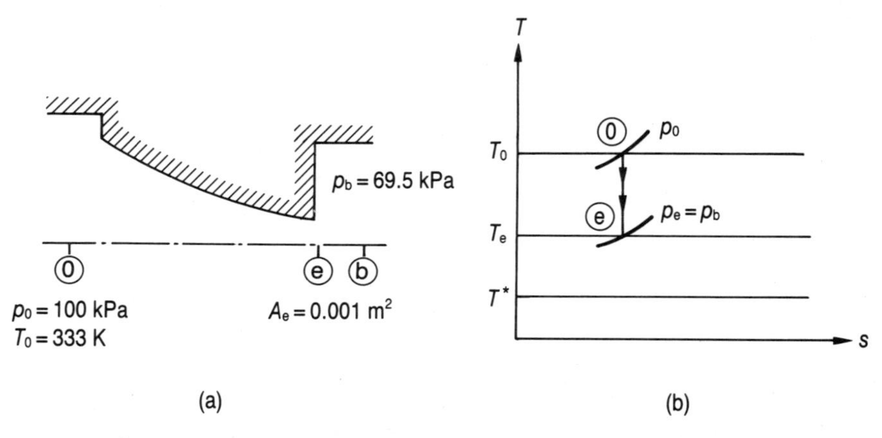

Figure 4.24 Sketch and T–s diagram for Example 4.5

Example 4.6
Air, which is flowing isentropically through a converging nozzle, discharges to a pressure of 40 kPa, as shown in Figure 4.25(a). At a section where the pressure is 179 kPa, the temperature is 39 °C and the air velocity is 177 m s^{-1}. If the mass flow rate is 4 kg s^{-1}, determine the pressure, the temperature and the cross-sectional area at the nozzle exit.

Solution
In order to check whether or not the nozzle is choked, the stagnation properties must be evaluated. Thus, the speed of sound at section 1 is

$$a_1 = \sqrt{(kRT_1)} = \sqrt{[(1.4)(287.1\ \mathrm{J\,kg^{-1}\,K^{-1}})(312\ \mathrm{K})]} = 354.1\ \mathrm{m\,s^{-1}}$$

and the Mach number is

$$M_1 = \frac{V_1}{a_1} = \frac{177\ \mathrm{m\,s^{-1}}}{354.1\ \mathrm{m\,s^{-1}}} = 0.4999$$

The stagnation conditions may now be evaluated by using Appendix C. At $M_1 = 0.50$

$$p_1/p_0 = 0.8430 \quad \text{or} \quad p_0 = 179\ \mathrm{kPa}/0.8430 = 212.3\ \mathrm{kPa}$$

$$T_1/T_0 = 0.9524 \quad \text{or} \quad T_0 = 312\ \mathrm{K}/0.9524 = 327.6\ \mathrm{K}$$

Then

$$p_b/p_0 = 40\ \mathrm{kPa}/212.3\ \mathrm{kPa} = 0.1884 < p^*/p_0 = 0.5283$$

so the flow is choked. For the choked flow

$$M_e = 1.0$$

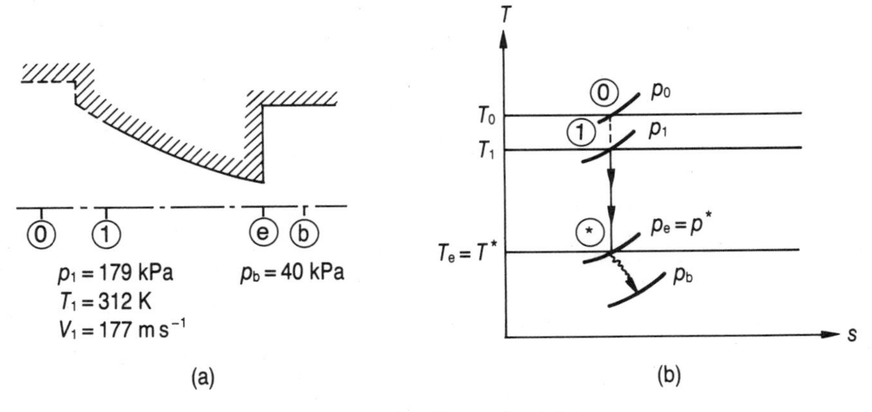

Figure 4.25 Sketch and T–s diagram for Example 4.6

The process is shown in Figure 4.25(b) on a $T-s$ diagram. To calculate the cross-sectional area at section 1, the density must first be evaluated as

$$\rho_1 = \frac{p_1}{RT_1} = \frac{(179\,000\ \mathrm{N\,m^{-2}})}{(287.1\ \mathrm{J\,kg^{-1}\,K^{-1}})(312\ \mathrm{K})} = 1.998\ \mathrm{kg\,m^{-3}}$$

and from the continuity equation

$$A_1 = \frac{\dot{m}}{\rho_1 V_1} = \frac{4\ \mathrm{kg\,s^{-1}}}{(1.998\ \mathrm{kg\,m^{-3}})(177\ \mathrm{m\,s^{-1}})} = 0.011\,31\ \mathrm{m^2}$$

The nozzle exit area is then evaluated by using the area ratio for $M_1 = 0.50$ from Appendix C:

$$A_1/A^* = A_1/A_e = 1.340 \quad \text{or} \quad A_e = 0.011\,31\ \mathrm{m^2}/1.340 = \mathbf{0.008\,440\ m^2}$$

Similarly, the pressure and the temperature at the nozzle exit for $M_e = 1$ are

$$p_e/p_0 = 0.5283 \quad \text{or} \quad p_e = (0.5283)(212.3\ \mathrm{kPa}) = \mathbf{112.2\ kPa}$$

$$T_e/T_0 = 0.8333 \quad \text{or} \quad T_e = (0.8333)(327.6\ \mathrm{K}) = \mathbf{273.0\ K}$$

Example 4.7

Air is stored in a tank of 2 m^3 in volume at a pressure of 800 kPa and a temperature of 300 K, as shown in Figure 4.26. Air is discharged through a converging nozzle with an exit area of 12 cm^2. For a back pressure of 100 kPa, find the time required for the tank pressure to drop to 125 kPa. Assume isentropic, quasi-steady flow through the nozzle, where the steady flow equations are applicable at each instant of time. Also, there is enough heat transfer to the tank so that the temperature of the air in the tank is constant.

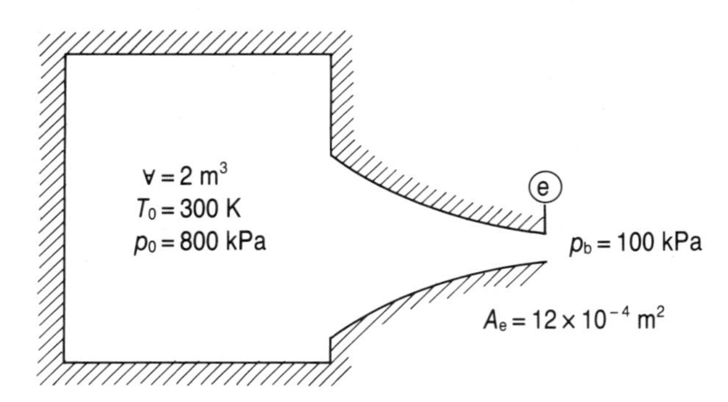

Figure 4.26 Sketch for Example 4.7

Solution
The flow in the nozzle is initially choked, since

$$p_b/p_0 = 100\text{ kPa}/800\text{ kPa} = 0.125 < p^*/p_0 = 0.5283$$

and it will remain choked until

$$p_0 = p_b/0.5283 = 100\text{ kPa}/0.5283 = 189.3\text{ kPa}$$

Afterwards, the flow will be in the subcritical flow regime when
189.3 kPa $> p_0 \geqslant$ 125 kPa. For this reason, it is necessary to investigate the flow in two regimes:

1 *Supercritical flow regime (800 kPa $> p_0 >$ 189.3 kPa)*: When the flow is choked, the Mach number is unity at the exit area. The continuity equation then becomes

$$\frac{dm}{dt} = \rho_e A_e V_e = \rho^* A^* V^*$$

For sonic conditions, from Appendix C,

$$p^*/p_0 = 0.5283 \quad T^*/T_0 = 0.8333 \quad \text{and} \quad \rho^*/\rho_0 = 0.6339$$

But the velocity at the throat is

$$V^* = \sqrt{(kRT^*)} = \sqrt{[(1.4)(287.1\text{ J kg}^{-1}\text{K}^{-1})(0.8333)(300\text{ K})]} = 317\text{ m s}^{-1}$$

Also, by using the equation of state for a perfect gas, $p = \rho RT$,

$$\rho^* = 0.6339\rho_0 = 0.6339\frac{p_0}{RT_0} = \frac{0.6339}{(287.1\text{ J kg}^{-1}\text{K}^{-1})(300\text{ K})}p_0$$

$$= 7.360 \times 10^{-6}\, p_0$$

Then the mass flow rate becomes

$$\frac{dm}{dt} = (7.360 \times 10^{-6}\, p_0)(12 \times 10^{-4}\text{ m}^2)(317.0\text{ m s}^{-1}) = 2.8 \times 10^{-6}\, p_0$$

At any time, the mass of air which escaped from the tank is equal to the difference between the initial mass of air in the tank m_i and the present mass of air in the tank, that is

$$m = m_i - \frac{p_0 \forall}{RT_0}$$

Differentiating

$$\frac{dm}{dt} = -\frac{\forall}{RT_0}\frac{dp_0}{dt} = -\frac{2\text{ m}^3}{(287.1\text{ J kg}^{-1}\text{K}^{-1})(300\text{ K})}\frac{dp_0}{dt} = -2.322 \times 10^{-5}\frac{dp_0}{dt}$$

and eliminating $\mathrm{d}m/\mathrm{d}t$

$$2.8 \times 10^{-6}\, p_0 + 2.322 \times 10^{-5} \frac{\mathrm{d}p_0}{\mathrm{d}t} = 0$$

The above differential equation can be solved by separating the variables as

$$\mathrm{d}t = -8.293 \frac{\mathrm{d}p_0}{p_0}$$

and integrating

$$t_1 = -8.293 \int_{8.0 \times 10^5}^{1.893 \times 10^5} \frac{\mathrm{d}p_0}{p_0} = 8.293 \ln\left(\frac{8 \times 10^5}{1.893 \times 10^5}\right) = 11.95 \text{ s}$$

2 *Subcritical flow regime (189.3 kPa > $p_0 \geqslant$ 125 kPa)*: In the subcritical regime, the pressure at the nozzle exit is equal to the back pressure. Then Equation (4.34) may be rearranged as

$$\frac{\mathrm{d}m}{\mathrm{d}t} = \sqrt{\left(\frac{k}{RT_0}\right)}\, p_0 A_e \sqrt{\left\{\frac{2}{k-1}\left[\left(\frac{p_b}{p_0}\right)^{2/k} - \left(\frac{p_b}{p_0}\right)^{(k+1)/k}\right]\right\}}$$

or

$$\frac{\mathrm{d}m}{\mathrm{d}t} = \sqrt{\left(\frac{(1.4)}{(287.1 \text{ J kg}^{-1}\text{K}^{-1})(300 \text{ K})}\right)}\, p_0(12 \times 10^{-4} \text{ m}^2)$$

$$\times \sqrt{\left\{\frac{2}{1.4-1}\left[\left(\frac{p_b}{p_0}\right)^{2/1.4} - \left(\frac{p_b}{p_0}\right)^{(1.4+1)/1.4}\right]\right\}}$$

Hence

$$\frac{\mathrm{d}m}{\mathrm{d}t} = 1.082 \times 10^{-5}\, p_0 \sqrt{\left[\left(\frac{p_b}{p_0}\right)^{10/7} - \left(\frac{p_b}{p_0}\right)^{12/7}\right]}$$

But $\mathrm{d}m/\mathrm{d}t = -2.322 \times 10^{-5}\, \mathrm{d}p_0/\mathrm{d}t$ from the supercritical regime. Therefore

$$\frac{\mathrm{d}(p_0/p_b)}{\mathrm{d}t} = -0.466 \frac{p_0}{p_b} \sqrt{\left[\left(\frac{p_0}{p_b}\right)^{-10/7} - \left(\frac{p_0}{p_b}\right)^{-12/7}\right]}$$

or separating the variables

$$-0.466\, \mathrm{d}t = \frac{\mathrm{d}(p_0/p_b)}{(p_0/p_b)^{2/7}\sqrt{[1 - (p_0/p_b)^{-2/7}]}}$$

The integration may be facilitated by introducing the transformation

$$\xi = \sqrt{\left[1 - \left(\frac{p_0}{p_b}\right)^{-2/7}\right]}$$

Then

$$-0.066\,57\ \mathrm{d}t = \frac{\mathrm{d}\xi}{(1-\xi^2)^{7/2}}$$

The supercritical flow in the converging nozzle transforms into a subcritical one when $t = 11.95$ s. At this instant, $p_0 = 189.3$ kPa and $\xi = \sqrt{[1-(189.3\ \text{kPa}/100\ \text{kPa})^{-2/7}]} = 0.4083$. It is desirable to evaluate the time elapsed when $p_0 = 125$ kPa. In this case, $\xi = \sqrt{[1-(125\ \text{kPa}/100\ \text{kPa})^{-2/7}]} = 0.2485$. Therefore, the above equation may now be integrated to yield

$$\int_{11.95\ \mathrm{s}}^{t_2} -0.066\,57\ \mathrm{d}t = \int_{0.4083}^{0.2485} \frac{\mathrm{d}\xi}{(1-\xi^2)^{7/2}}$$

From integration tables (Carmichael and Smith, 1962)

$$\int \frac{\mathrm{d}\xi}{(1-\xi^2)^r} = \frac{\xi}{2(r-1)(1-\xi^2)^{r-1}} + \frac{(2r-3)}{2(r-1)} \int \frac{\mathrm{d}\xi}{(1-\xi^2)^{r-1}}$$

Hence

$$-0.066\,57(t_2 - 11.95\ \mathrm{s}) = \left[\frac{\xi(15-20\xi^2+8\xi^4)}{15(1-\xi^2)^{5/2}}\right]_{0.4083}^{0.2485} = -0.2425$$

and the overall time required for the tank pressure to drop from 800 kPa to 125 kPa may now be evaluated as

$$t = t_2 = \mathbf{15.59\ s}$$

4.8.2 Converging–diverging nozzles

Consider a similar experiment to the one described in the previous section, except now a converging–diverging nozzle is to be used, as shown in Figure 4.27(a). This nozzle is fed from a stagnant reservoir and discharges into another reservoir, where the back pressure, p_b, is controlled by a valve. The pressure at the exit plane of the converging–diverging nozzle is again denoted by p_e. The effects of the variation of the back pressure on the distribution of the pressure in the passage, on the exit plane pressure and on the mass flow rate are given in Figures 4.27(b), 4.27(c) and 4.27(d), respectively.

(a) No-flow condition ($p_b = p_0$)

When the valve in the discharge reservoir is closed, the pressure is constant throughout the converging–diverging nozzle such that $p_b = p_e = p_t = p_0$ and, certainly, there is no flow. This condition is indicated by flow pattern 1 in Figure 4.27(b).

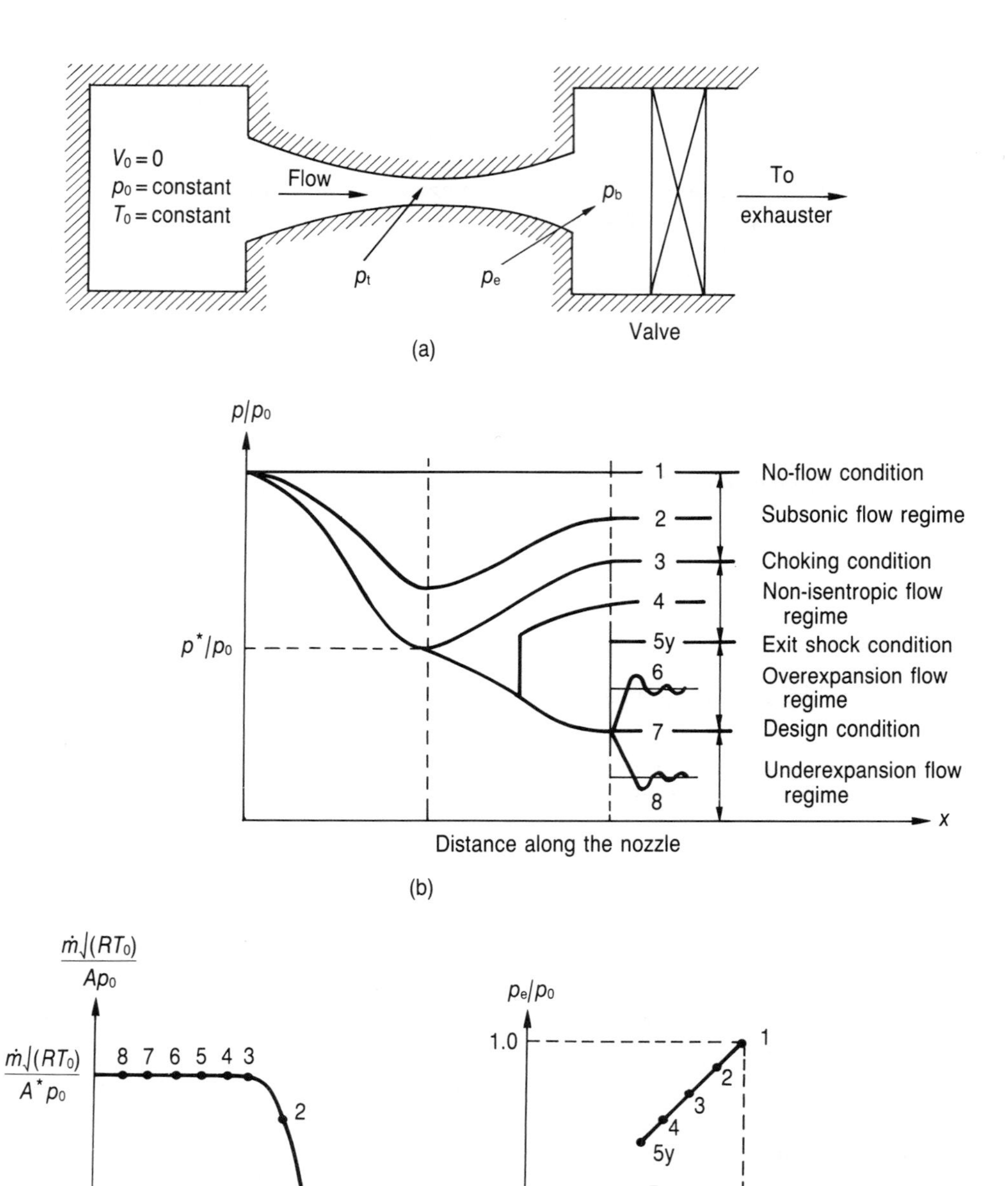

Figure 4.27 Operation of a converging–diverging nozzle at various back pressures

(b) Subsonic flow regime ($p_{e3} < p_b < p_0$)

When the back pressure in the discharge reservoir is slightly decreased by opening the valve, then the flow is subsonic throughout the converging–diverging nozzle, as shown in flow pattern 2 of Figure 4.27(b). The static pressure of the fluid decreases from the entrance to the throat, where it reaches its minimum value, and then increases in the diverging section. Therefore, the converging part of the nozzle acts as a subsonic nozzle, while the diverging part acts as a subsonic diffuser and the passage behaves like a conventional Venturi tube. The mass flow rate and the throat pressure are sensitive to the back pressure. The exit pressure is identical to the back pressure since the argument given in Section 4.8.1 is still valid. Therefore the jet leaves the nozzle as a cylindrical parallel stream, as indicated in Figure 4.28(a).

(c) Choking condition ($p_b = p_{e3}$)

In flow pattern 3 of Figure 4.27(b), the flow is choked at the throat where the Mach number is unity. The throat pressure is equal to the critical pressure, and the pressure at the exit is equal to the back pressure. The flow is subsonic at every point except at the throat so that the diverging part again acts as a subsonic diffuser. The jet again leaves the converging–diverging nozzle as a cylindrical stream, as shown in Figure 4.28(b), since $p_e = p_b$.

(d) Non-isentropic flow regime ($p_{e5y} < p_b < p_{e3}$)

Flow pattern 4 of Figure 4.27(b) is a typical representation of this flow regime, which is not isentropic. A non-isentropic phenomenon, known as a shock, transforms the supersonic flow in the diverging section into a subsonic one. This flow regime will be discussed in Chapter 5, since it is not possible to deal with it using isentropic flow theory.

(e) Exit plane shock condition ($p_b = p_{e5y}$)

In this case, the shock phenomenon moves to the exit plane of the converging–diverging nozzle as indicated by flow pattern 5 of Figure 4.27(b). This condition will again be investigated in Chapter 5.

(f) Overexpansion flow regime ($p_{e7} < p_b < p_{e5y}$)

A typical representation of this flow regime is given by flow pattern 6 of Figure 4.27(b). The flow is sonic at the throat where the Mach number is unity, and supersonic in the entire diverging section of the nozzle. The mass flow rate is invariant with respect to the back pressure, since the flow is choked at the throat. The gas is overexpanded at the exit plane, because the pressure at the exit plane is lower than the back pressure. The compression, which occurs outside the nozzle, involves non-isentropic oblique compression waves, as shown in Figure 4.28(e), which cannot be treated with one-dimensional flow, while the flow throughout the nozzle is isentropic.

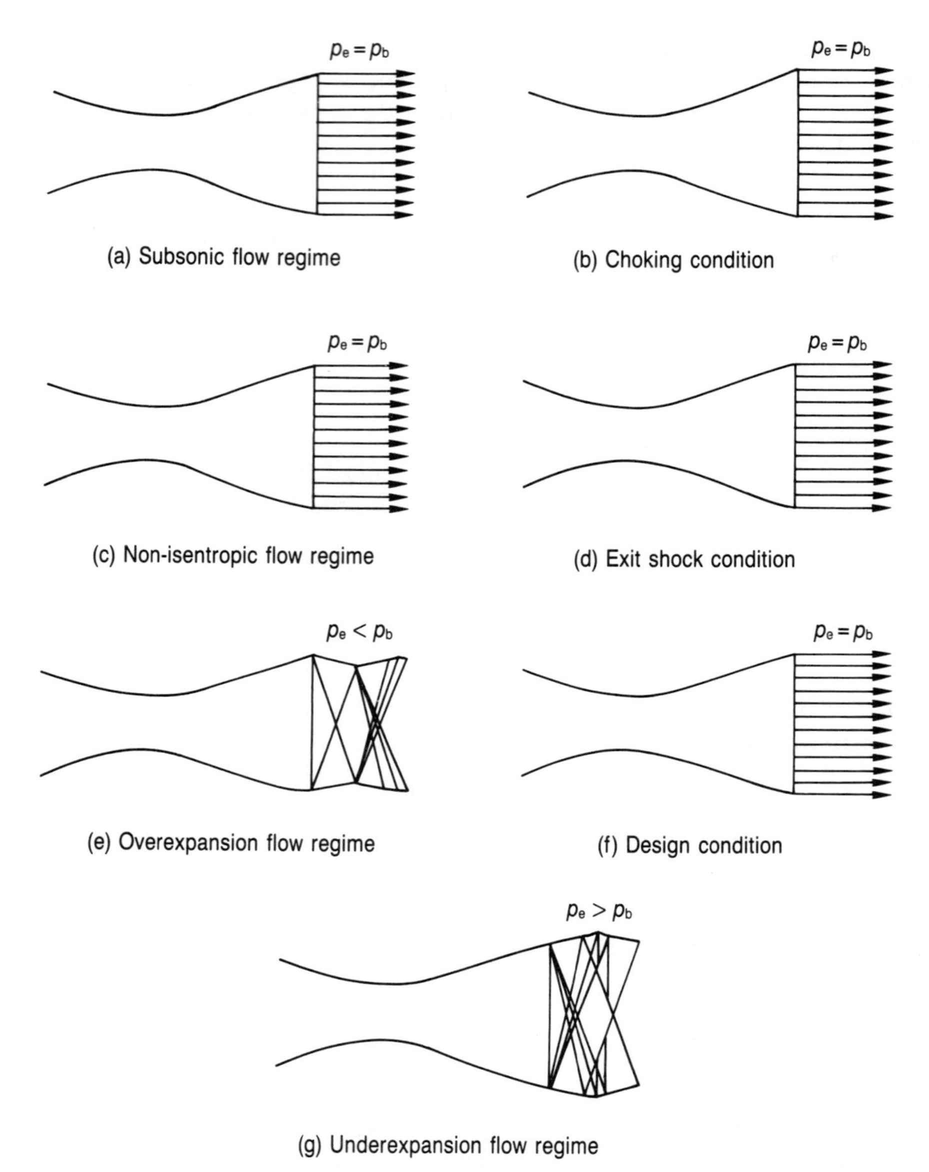

Figure 4.28 Discharge jet shapes in converging–diverging nozzles

(g) Design condition ($p_b = p_{e7}$)

Flow pattern 7 of Figure 4.27(b) represents the condition for which the converging–diverging nozzle is actually designed. The flow is entirely isentropic within and outside the nozzle such that the exit plane pressure is identical with the back pressure. Owing to the choking at the throat, the flow is supersonic in the entire diverging section of the nozzle. As long as $p_e = p_b$, the shape of the jet leaving the nozzle is cylindrical, as indicated in Figure 4.28(f).

(h) Underexpansion flow regime ($p_b < p_{e7}$)

A typical flow pattern for this regime is indicated by flow pattern 8 in Figure 4.27(b). The flow is entirely isentropic within the nozzle, but the expansion from the exit plane pressure to the back pressure, which occurs outside the nozzle, is in the form of non-isentropic oblique expansion waves, as shown in Figure 4.28(g). This expansion process cannot be treated with one-dimensional flow. The mass flow rate is constant and is not affected by the variations in the back pressure, since the flow is choked at the throat.

Example 4.8

A converging–diverging nozzle with a throat area of 0.0035 m^2 is attached to a very large tank of air in which the pressure is 125 kPa and the temperature is 47 °C, as shown in Figure 4.29(a). The nozzle exhausts to the atmosphere with a pressure of 100 kPa. If the mass flow rate is 0.9 kg s^{-1}, determine the exit area and the Mach number at the throat. The flow in the nozzle is isentropic.

Solution

First, it is necessary to check whether or not the converging–diverging nozzle is choked. For this reason, the non-dimensional mass flow rate at the throat may be calculated as

$$\frac{\dot{m}\sqrt{(RT_0)}}{A_t p_0} = \frac{(0.9\ \text{kg s}^{-1})\sqrt{[(287.1\ \text{J kg}^{-1}\ \text{K}^{-1})(320\ \text{K})]}}{(0.0035\ \text{m}^2)(125\,000\ \text{N m}^{-2})} = 0.6235$$

which is less than the critical value of the non-dimensional mass flow rate, since

$$\frac{\dot{m}\sqrt{(RT_0)}}{A_t p_0} = 0.6235 < \frac{\dot{m}\sqrt{(RT_0)}}{A^* p_0} = 0.6847$$

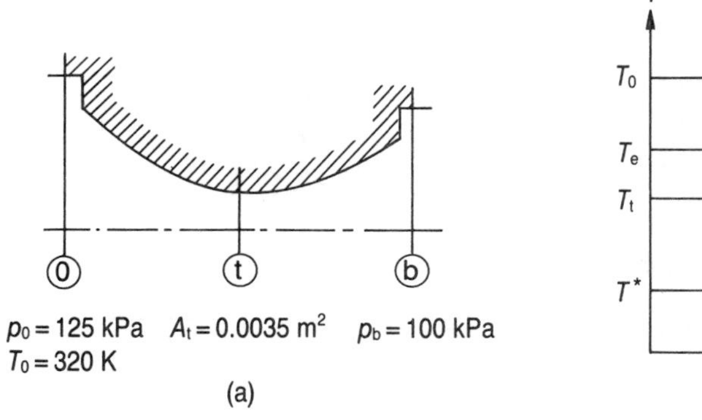

T, T_0, T_e, T_t, T^*, s, 0, e, t, p_0, p_e, p_t (b)

Figure 4.29 Sketch and *T–s* diagram for Example 4.8

Therefore the nozzle is not choked and the only possible isentropic pattern is in the subsonic flow regime. Hence

$$p_e = p_b = 100 \text{ kPa}$$

From Appendix C, corresponding to $p_e/p_0 = 100 \text{ kPa}/125 \text{ kPa} = 0.8$,

$$M_e = 0.57$$

$$\frac{\dot{m}\sqrt{(RT_0)}}{A_e p_0} = 0.5584 \quad \text{and} \quad A_e = \frac{(0.9 \text{ kg s}^{-1})\sqrt{[(287.1 \text{ J kg}^{-1} \text{ K}^{-1})(320 \text{ K})]}}{(0.5584)(125\,000 \text{ N m}^{-2})}$$

$$= \mathbf{0.003\,908 \text{ m}^2}$$

Finally, the Mach number at the throat corresponding to $\dot{m}\sqrt{(RT_0)}/(A_t p_0) = 0.6235$ can be obtained via Appendix C as

$$M_t = \mathbf{0.69}$$

The process is shown on the T–s diagram of Figure 4.29(b). Air is first accelerated in the converging part of the nozzle so that the pressure decreases from p_0 to p_t. However, the flow in the entire nozzle is subsonic and the diverging part acts as a diffuser. Therefore, the flow is decelerated in the diverging part, increasing the pressure from p_t to p_e.

Example 4.9

At a point upstream of the throat in a converging–diverging nozzle, the velocity, temperature and pressure are 172 m s^{-1}, 22 °C and 200 kPa, respectively, as shown in Figure 4.30. If the nozzle, operating at its design condition, has an exit

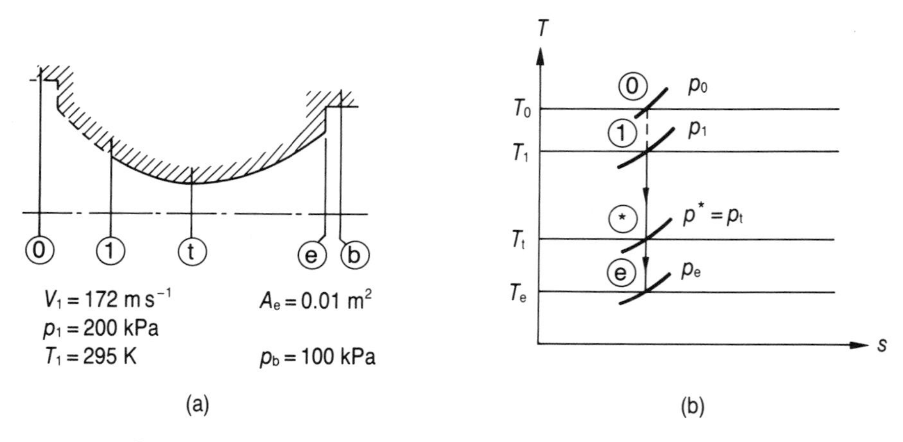

Figure 4.30 Sketch and T–s diagram for Example 4.9

area of 0.01 m^2 and discharges to the atmosphere with a pressure of 100 kPa, determine the mass flow rate and the nozzle throat area.

Solution

If the converging–diverging nozzle is operating at its design condition, then the flow within and outside the nozzle is entirely isentropic. Also, the flow is choked at the throat. At section 1, the speed of sound is

$$a_1 = \sqrt{(kRT_1)} = \sqrt{[(1.4)(287.1\ \mathrm{J\,kg^{-1}\,K^{-1}})(295\ \mathrm{K})]} = 344.3\ \mathrm{m\,s^{-1}}$$

and the Mach number is

$$M_1 = \frac{V_1}{a_1} = \frac{172\ \mathrm{m\,s^{-1}}}{344.3\ \mathrm{m\,s^{-1}}} = 0.4996$$

Then, the stagnation conditions corresponding to $M_1 = 0.50$ may be evaluated by using Appendix C:

$$p_1/p_0 = 0.8430 \quad \text{or} \quad p_0 = 200\ \mathrm{kPa}/0.8430 = 237.3\ \mathrm{kPa}$$

and

$$T_1/T_0 = 0.9524 \quad \text{or} \quad T_0 = 295\ \mathrm{K}/0.9524 = 309.7\ \mathrm{K}$$

At the design condition, $p_e = p_b$, then

$$p_e/p_0 = 100\ \mathrm{kPa}/237.3\ \mathrm{kPa} = 0.4214$$

and corresponding to $p_e/p_0 = 0.4214$ from Appendix C

$$M_e = 1.18$$

$$\frac{\dot{m}\sqrt{(RT_0)}}{A_e p_0} = 0.6681 \quad \text{or} \quad \dot{m} = \frac{(0.6681)(0.01\ \mathrm{m^2})(237\,300\ \mathrm{N\,m^{-2}})}{\sqrt{[(287.1\ \mathrm{J\,kg^{-1}\,K^{-1}})(309.7\ \mathrm{K})]}}$$

$$= \mathbf{5.317\ kg\,s^{-1}}$$

$$A_e/A^* = 1.025 \quad \text{or} \quad A^* = 0.01\ \mathrm{m^2}/1.025 = 9.756 \times 10^{-3}\ \mathrm{m^2}$$

As the flow is choked at the throat,

$$A_t = A^* = \mathbf{9.756 \times 10^{-3}\ m^2}$$

The process is shown on the T–s diagram of Figure 4.30(b). Air is accelerated to a sonic speed in the converging part of the nozzle so that the pressure decreases from p_0 to p_t. In the diverging part, the flow is accelerated from sonic speed to supersonic speeds, while the pressure is decreased from p_t to p_e.

4.9 PERFORMANCE OF REAL NOZZLES

The performance of real nozzles differs slightly from that computed by isentropic flow relations owing to the frictional effects. Since departures from isentropic flow are usually small, the usual design procedure is based on the use of isentropic flow formulas which are then modified by empirically determined coefficients. These coefficients are the nozzle efficiency, the nozzle velocity coefficient and the nozzle discharge coefficient.

In order to define these coefficients, consider a converging–diverging nozzle which is supplied with a gas at a stagnation pressure of p_{0i} and a stagnation temperature of T_{0i}, as shown in Figure 4.31(a). The gas expands adiabatically but with increasing entropy to the exit state, e, as shown in Figure 4.31(b). If it had been expanded without friction to the same final pressure, then the end state would have been es.

The **nozzle efficiency**, η_n, may now be defined as the ratio of the exit kinetic energy to the kinetic energy which would be obtained by expanding the gas isentropically to the same final pressure. Therefore,

$$\eta_n = \frac{V_e^2/2}{V_{es}^2/2} \tag{4.42}$$

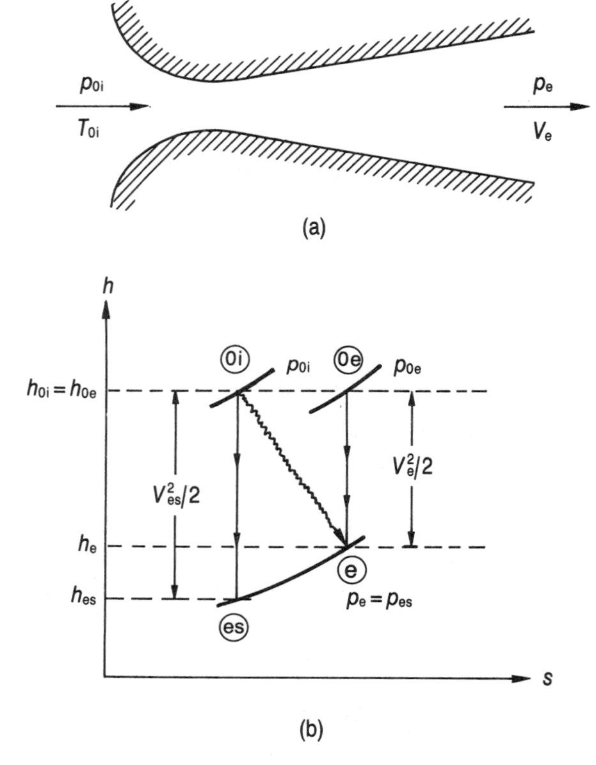

Figure 4.31 Performance of a real nozzle

Occasionally, the term **nozzle velocity coefficient**, C_V, is used for denoting the square root of the nozzle efficiency and may be given as

$$C_V = \sqrt{\eta_n} = V_e / V_{es} \tag{4.43}$$

The **nozzle discharge coefficient**, C_d, is defined as the ratio of the actual mass flow rate, $\dot{m}$, to the isentropic mass flow rate, $\dot{m}_s$, which would be obtained by expanding the gas isentropically to the same final pressure. Hence

$$C_d = \dot{m} / \dot{m}_s \tag{4.44}$$

The nozzle efficiency, the nozzle velocity coefficient and the nozzle discharge coefficient are interrelated. One may start to derive this relation by applying the first law of thermodynamics to the isentropic flow between the inlet state, 0i, and the isentropic exit state, es, as

$$h_{0i} = h_{es} + \frac{V_{es}^2}{2}$$

But, for a perfect gas, $h = c_p T$, so that

$$V_{es}^2 = 2c_p T_{0i}\left(1 - \frac{T_{es}}{T_{0i}}\right)$$

However, for an isentropic process

$$\frac{T_{es}}{T_{0i}} = \left(\frac{p_{es}}{p_{0i}}\right)^{(k-1)/k}$$

so that

$$V_{es}^2 = 2c_p T_{0i}\left[1 - \left(\frac{p_{es}}{p_{0i}}\right)^{(k-1)/k}\right] \tag{4.45}$$

Similarly, one might consider the imaginary isentropic process between the actual exit state, e, and its stagnation state, 0e. Application of the first law of thermodynamics to this process yields

$$h_{0e} = h_e + \frac{V_e^2}{2}$$

However, $h = c_p T$ for a perfect gas, so that

$$V_e^2 = 2c_p T_{0e}\left(1 - \frac{T_e}{T_{0e}}\right)$$

As long as the process is isentropic, then

$$\frac{T_e}{T_{0e}} = \left(\frac{p_e}{p_{0e}}\right)^{(k-1)/k}$$

and it is possible to obtain

$$V_e^2 = 2c_p T_{0e}\left[1 - \left(\frac{p_e}{p_{0e}}\right)^{(k-1)/k}\right] \tag{4.46}$$

Since the process within the nozzle is adiabatic, then the stagnation temperature remains the same during this process, that is $T_{0e} = T_{0i}$. Now, Equations (4.45) and (4.46) may be substituted into Equation (4.42) to yield

$$\frac{p_e}{p_{0e}} = \left\{1 - \eta_n\left[1 - \left(\frac{p_{es}}{p_{0i}}\right)^{(k-1)/k}\right]\right\}^{k/(k-1)} \tag{4.47}$$

The mass flux for the isentropic flow between states 0i and es may be obtained via Equation (4.34) as

$$\frac{\dot{m}_s}{A_e} = \sqrt{\left(\frac{k}{RT_{0i}}\right)}p_{0i}\sqrt{\left\{\frac{2}{k-1}\left[\left(\frac{p_{es}}{p_{0i}}\right)^{2/k} - \left(\frac{p_{es}}{p_{0i}}\right)^{(k+1)/k}\right]\right\}} \tag{4.48}$$

Similarly, the actual mass flux may be obtained by considering the imaginary isentropic process between states 0e and e as

$$\frac{\dot{m}}{A_e} = \sqrt{\left(\frac{k}{RT_{0e}}\right)}p_{0e}\sqrt{\left\{\frac{2}{k-1}\left[\left(\frac{p_e}{p_{0e}}\right)^{2/k} - \left(\frac{p_e}{p_{0e}}\right)^{(k+1)/k}\right]\right\}} \tag{4.49}$$

Now, by dividing Equation (4.49) by Equation (4.48) and noting that $p_e = p_{es}$, the nozzle discharge coefficient may be obtained as

$$C_d = \frac{(p_{es}/p_{0i})^{(k-1)/k}}{(p_e/p_{0e})^{(k-1)/k}}\sqrt{\left(\frac{1-(p_e/p_{0e})^{(k-1)/k}}{1-(p_{es}/p_{0i})^{(k-1)/k}}\right)}$$

or eliminating p_e/p_{0e} by using Equation (4.47)

$$C_d = \frac{\sqrt{\eta_n}(p_{es}/p_{0i})^{(k-1)/k}}{1-\eta_n[1-(p_{es}/p_{0i})^{(k-1)/k}]} \tag{4.50}$$

Example 4.10

Air is flowing in a converging–diverging nozzle which has an efficiency of 95 per cent, as shown in Figure 4.32(a). At the inlet of the nozzle, the stagnation temperature and the stagnation pressure of the air are 500 K and 700 kPa, respectively. The ratio of the exit cross-sectional area to the throat cross-sectional area is 2.0.

a Determine the Mach number, pressure, temperature, mass flux and velocity at the exit of the nozzle for the isentropic operation.

b Determine the actual values of the pressure, velocity, temperature, Mach number and mass flux at the exit of the nozzle.

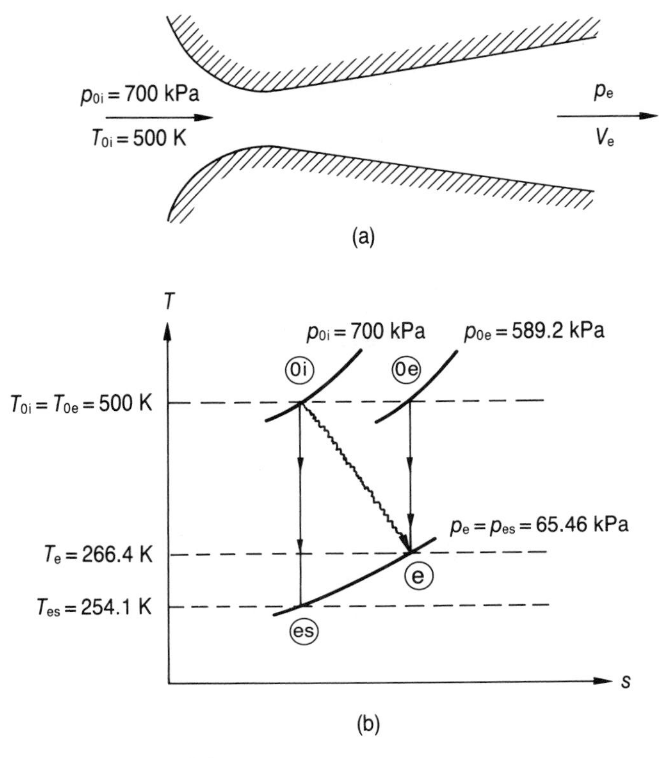

Figure 4.32 Sketch for Example 4.10

c Determine the discharge coefficient of the nozzle.
d Determine the velocity coefficient of the nozzle.

Solution

(a) One should note that the sonic condition exists at the throat of the nozzle for isentropic operation at the design condition. Then the Mach number, pressure, temperature and mass flux at the exit of the nozzle may be determined from Appendix C by using the area ratio, $A_e/A_t = A_e/A^* = 2.0$ as

$$M_e = \mathbf{2.20}$$

$$p_{es}/p_{0i} = 0.093\,52 \quad \text{or} \quad p_{es} = (0.093\,52)(700 \text{ kPa}) = \mathbf{65.46 \text{ kPa}}$$

$$T_{es}/T_{0i} = 0.5081 \quad \text{or} \quad T_{es} = (0.5081)(500 \text{ K}) = \mathbf{254.1 \text{ K}}$$

$$\frac{\dot{m}_s\sqrt{(RT_{0i})}}{A_e p_{0i}} = 0.3415 \quad \text{or} \quad \frac{\dot{m}_s}{A_e} = \frac{(0.3415)(700\,000 \text{ N m}^{-2})}{\sqrt{[(287.1 \text{ J kg}^{-1}\text{ K}^{-1})(500 \text{ K})]}}$$

$$= \mathbf{630.9 \text{ kg m}^{-2}\text{ s}^{-1}}$$

The speed of sound at the exit may now be evaluated as

$$a_{es} = \sqrt{(kRT_{es})} = \sqrt{[(1.4)(287.1\ \text{J kg}^{-1}\ \text{K}^{-1})(254.1\ \text{K})]} = 319.6\ \text{m s}^{-1}$$

so that the exit velocity is

$$V_{es} = M_{es} a_{es} = (2.20)(319.6\ \text{m s}^{-1}) = \mathbf{703.1\ m\ s^{-1}}$$

(b) As long as the isentropic and actual processes expand to the same exit pressure, then

$$p_e = p_{es} = \mathbf{65.46\ kPa}$$

Now, the actual exit velocity may be evaluated by using Equation (4.42) as

$$V_e = \sqrt{(\eta_n V_{es}^2)} = \sqrt{[(0.95)(703.1\ \text{m s}^{-1})^2]} = \mathbf{685.3\ m\ s^{-1}}$$

Also, the stagnation temperature remains the same during an adiabatic process, so $T_{0e} = T_{0i} = 500$ K. Now, applying the adiabatic energy equation to the imaginary process between the actual exit state, e, and its stagnation state, 0e,

$$T_e = T_{0e} - \frac{V_e^2}{2c_p} = 500\ \text{K} - \frac{(685.3\ \text{m s}^{-1})^2}{(2)(1005\ \text{J kg}^{-1}\ \text{K}^{-1})} = \mathbf{266.4\ K}$$

At this point, it is possible to determine the actual values of the exit Mach number and the exit mass flux from Appendix C by using $T_e/T_{0e} = 266.4\ \text{K}/500\ \text{K} = 0.5328$ as

$$M_e = \mathbf{2.09}$$

$$p_e/p_{0e} = 0.1111 \quad \text{or} \quad p_{0e} = 65.46\ \text{kPa}/0.1111 = 589.2\ \text{kPa}$$

$$\frac{\dot{m}\sqrt{(RT_{0e})}}{A_e p_{0e}} = 0.3760 \quad \text{or} \quad \frac{\dot{m}}{A_e} = \frac{(0.3760)(589\,200\ \text{N m}^{-2})}{\sqrt{[(287.1\ \text{J kg}^{-1}\ \text{K}^{-1})(500\ \text{K})]}}$$

$$= \mathbf{584.7\ kg\ m^{-2}\ s^{-1}}$$

(c) The discharge coefficient of the nozzle may now be obtained from Equation (4.44) as

$$C_d = \frac{\dot{m}/A_e}{\dot{m}_s/A_e} = \frac{584.7\ \text{kg m}^{-2}\ \text{s}^{-1}}{630.9\ \text{kg m}^{-2}\ \text{s}^{-1}} = \mathbf{0.9268}$$

(d) The velocity coefficient of the nozzle may be evaluated by using Equation (4.43) as

$$C_V = \sqrt{\eta_n} = \sqrt{0.95} = \mathbf{0.9747}$$

4.10 THRUST OF A ROCKET ENGINE

A rocket engine, which generates gases steadily at a stagnation pressure of p_0 and a stagnation temperature of T_0 as the solid propellant burns, is shown in Figure 4.33.

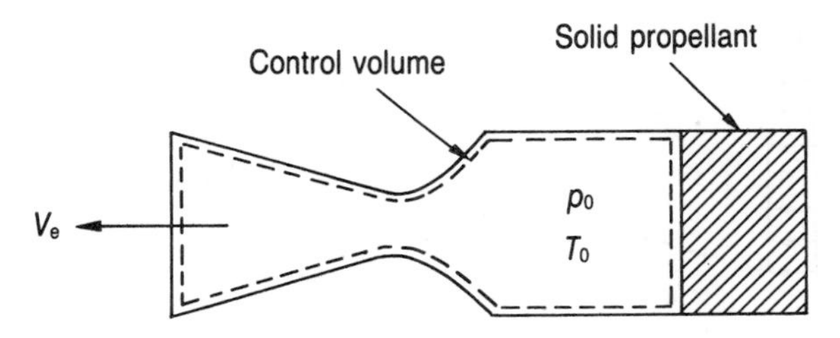

Figure 4.33 Sketch of a rocket engine

The converging–diverging nozzle has a throat area of A_t and an exit area of A_e. The generated gases discharge to the atmosphere at a pressure of p_a.

Most rocket engines generate gases at about 3500 kPa and operate in atmospheres with pressures of 101.325 kPa or less. Therefore, such a reduction in pressure is only possible by a converging–diverging nozzle.

Experimental data verify that the isentropic flow equations predict the thrust produced by such a rocket engine within an error of a few per cent.

The force, F_c, acting on the control volume of Figure 4.33 may be found by applying the momentum equation as

$$F_c = \dot{m}V_e + p_e A_e$$

The net thrust, F_t, acting on the rocket engine may now be obtained by considering the free-body diagrams of the control volume and the rocket engine, which are shown in Figures 4.34(a) and 4.34(b) respectively. Hence

$$F_t = F_c - p_a A_e = \dot{m}V_e + A_e(p_e - p_a) \tag{4.51}$$

The above equation may be non-dimensionalized by dividing it by $p_0 A_t$ as

$$\frac{F_t}{p_0 A_t} = \frac{\dot{m}V_e}{p_0 A_t} + \frac{A_e}{A_t}\left(\frac{p_e}{p_0} - \frac{p_a}{p_0}\right) \tag{4.52}$$

For the isentropic operation of the nozzle at its design condition, the flow is choked at the throat and the mass flux at the throat may be obtained from Equation (4.35) as

$$\frac{\dot{m}}{A_t} = \sqrt{\left(\frac{k}{RT_0}\right)} p_0 \left(\frac{2}{k+1}\right)^{(k+1)/[2(k-1)]} \tag{4.53}$$

Now, the first law of thermodynamics may be applied to the isentropic process between the stagnation state and the exit state to yield

$$h_0 = h_e + \frac{V_e^2}{2}$$

However, for a perfect gas $h = c_p T = kRT/(k-1)$, so that

$$V_e = \sqrt{\left[\frac{2kRT_0}{k-1}\left(1 - \frac{T_e}{T_0}\right)\right]}$$

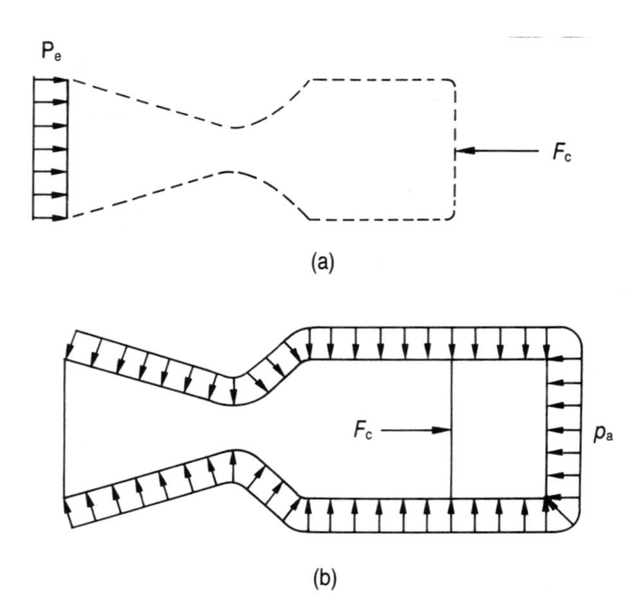

Figure 4.34 Free-body diagrams of (a) the control volume and (b) the rocket

Also, for an isentropic process,

$$\frac{T_e}{T_0} = \left(\frac{p_e}{p_0}\right)^{(k-1)/k}$$

and

$$V_e = \sqrt{\left\{\frac{2kRT_0}{k-1}\left[1 - \left(\frac{p_e}{p_0}\right)^{(k-1)/k}\right]\right\}} \tag{4.54}$$

The mass flux at the exit of the nozzle may be obtained via Equation (4.34) in the following form:

$$\frac{\dot{m}}{A_e} = \sqrt{\left(\frac{k}{RT_0}\right)} p_0 \sqrt{\left\{\frac{2}{k-1}\left[\left(\frac{p_e}{p_0}\right)^{2/k} - \left(\frac{p_e}{p_0}\right)^{(k+1)/k}\right]\right\}} \tag{4.55}$$

Now, by dividing Equation (4.53) by Equation (4.55) it is possible to obtain the following equation:

$$\frac{A_e}{A_t} = \left(\frac{2}{k+1}\right)^{(k+1)/[2(k-1)]} \Bigg/ \sqrt{\left\{\frac{2}{k-1}\left[\left(\frac{p_e}{p_0}\right)^{2/k} - \left(\frac{p_e}{p_0}\right)^{(k+1)/k}\right]\right\}} \tag{4.56}$$

At this point, Equations (4.53), (4.54) and (4.56) may be substituted into Equation (4.52) to yield

$$\frac{F_t}{p_0 A_t} = \left(\frac{2}{k+1}\right)^{(k+1)/[2(k-1)]} \left[\!\left[\sqrt{\left\{\frac{2k^2}{k-1}\left[1 - \left(\frac{p_e}{p_0}\right)^{(k-1)/k}\right]\right\}} + \left(\frac{p_e}{p_0} - \frac{p_a}{p_0}\right) \Big/ \sqrt{\left\{\frac{2}{k-1}\left[\left(\frac{p_e}{p_0}\right)^{2/k} - \left(\frac{p_e}{p_0}\right)^{(k+1)/k}\right]\right\}} \right]\!\right] \quad (4.57)$$

Equation (4.56) indicates that the pressure ratio, p_e/p_0, depends only on the area ratio or the nozzle geometry. Then, the thrust of a rocket with a nozzle of given size and geometry depends only on the stagnation pressure, p_0, and the pressure ratio, p_a/p_0, and is independent of the stagnation temperature, T_0.

To find the maximum thrust, one may differentiate Equation (4.57) with respect to p_e/p_0, and set this derivative to zero. In this case, the thrust is maximum when the area ratio is chosen in such a way to make the pressure at the exit plane exactly equal to atmospheric pressure, that is

$$p_e = p_a \quad (4.58)$$

However, this result may be obtained more easily by physical reasoning. The net thrust on the rocket is the resultant of static pressure forces on all the surfaces of the rocket engine. Suppose that there is a certain exit area in the diverging part of the nozzle for which $p_e = p_a$, as shown in Figure 4.35. If the nozzle is continued beyond this point, then the pressure inside the nozzle will drop further. Therefore, the added

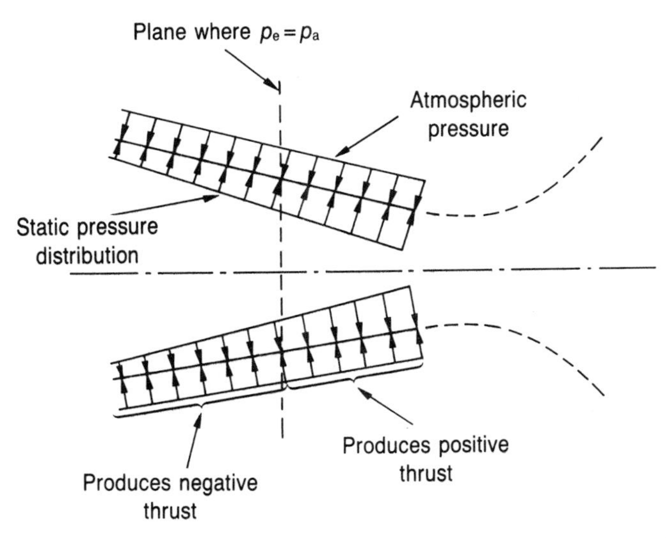

Figure 4.35 Pressure distribution on the internal and external surfaces of the diverging portion of the nozzle

piece of divergent nozzle will produce negative thrust, since the internal pressure on this added piece is less than the external pressure. By similar reasoning, it follows that cutting off a piece of the nozzle upstream of the plane where $p_e = p_a$ would act to reduce the thrust, because, on this piece, the internal pressure is greater than the external pressure. Hence, it is possible to conclude that the thrust is maximum when $p_e = p_a$.

In this case, the equation for the maximum thrust may be obtained from Equation (4.56) as

$$\frac{(F_t)_{max}}{p_0 A_t} = \left(\frac{2}{k+1}\right)^{(k+1)/[2(k-1)]} \sqrt{\left\{\frac{2k^2}{k-1}\left[1 - \left(\frac{p_e}{p_0}\right)^{(k-1)/k}\right]\right\}} \qquad (4.59)$$

Two performance criteria are commonly employed to compare the performance of propulsive nozzles in rocket engines. These are the **specific impulse**, I_s and the **thrust coefficient**, C_t, which are defined as

$$I_s = F_t / \dot{m} \qquad (4.60)$$

and

$$C_t = \frac{F_t}{p_0 A_t} \qquad (4.61)$$

respectively.

Example 4.11
A solid-propellant rocket engine generates gases at a stagnation temperature of 3000 K and a stagnation pressure of 7000 kPa, as shown in Figure 4.36. The exit and throat areas of the nozzle are 0.024 m^2 and 0.006 m^2, respectively. The generated gases may be assumed to be perfect with a specific heat ratio of 1.4 and a gas constant of 320 $J\,kg^{-1}\,K^{-1}$. At an altitude of 30 000 m, determine

a the thrust developed,
b the specific impulse, and
c the thrust coefficient.

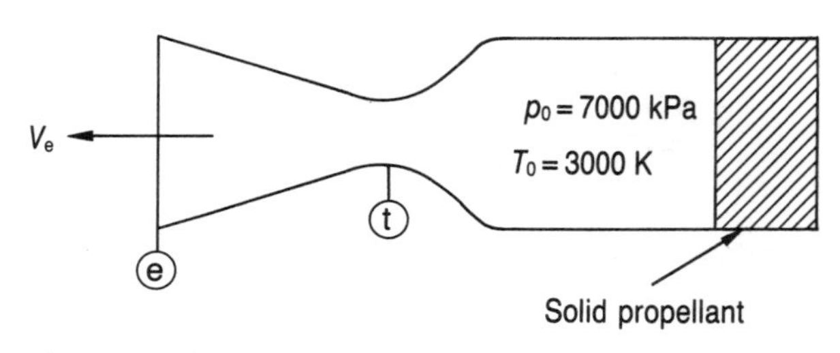

Figure 4.36 Sketch for Example 4.11

Solution

(a) For isentropic operation of the converging–diverging nozzle at its design condition, the flow is choked at the throat. Therefore, the properties at the exit section and the mass flow rate may be determined from Appendix C by using the area ratio $A_e/A_t = A_e/A^* = 0.024\ \text{m}^2/0.006\ \text{m}^2 = 4$ as

$$M_e = 2.94$$

$$p_e/p_0 = 0.029\,80 \quad \text{or} \quad p_e = (0.029\,80)(7000\ \text{kPa}) = 208.6\ \text{kPa}$$

$$T_e/T_0 = 0.3665 \quad \text{or} \quad T_e = (0.3665)(3000\ \text{K}) = 1100\ \text{K}$$

$$\frac{\dot{m}\sqrt{(RT_0)}}{A_e p_0} = 0.1712 \quad \text{or} \quad \dot{m} = \frac{(0.1712)(0.024\ \text{m}^2)(7 \times 10^6\ \text{N m}^{-2})}{\sqrt{[(320\ \text{J kg}^{-1}\,\text{K}^{-1})(3000\ \text{K})]}}$$

$$= 29.36\ \text{kg s}^{-1}$$

The speed of sound at the exit may be evaluated as

$$a_e = \sqrt{(kRT_e)} = \sqrt{[(1.4)(320\ \text{J kg}^{-1}\,\text{K}^{-1})(1100\ \text{K})]} = 702.0\ \text{m s}^{-1}$$

Then, the exit velocity is

$$V_e = M_e a_e = (2.94)(702.0\ \text{m s}^{-1}) = 2064\ \text{m s}^{-1}$$

Also, atmospheric pressure at an altitude of 30 000 m may be obtained from Appendix A as

$$p_a = 1197\ \text{Pa}$$

At this point, the thrust developed by the rocket engine may be evaluated from Equation (4.51) as

$$F_t = \dot{m}V_e + A_e(p_e - p_a) = (29.36\ \text{kg s}^{-1})(2064\ \text{m s}^{-1}) + (0.024\ \text{m}^2)(208\,600\ \text{N m}^{-2} - 1197\ \text{N m}^{-2}) = \mathbf{65.58\ kN}$$

(b) The specific impulse of the rocket engine may now be obtained from Equation (4.60) as

$$I_s = \frac{F_t}{\dot{m}} = \frac{65\,580\ \text{N}}{29.36\ \text{kg s}^{-1}} = \mathbf{2234\ N\,s\,kg^{-1}}$$

(c) The specific thrust of the rocket engine may be evaluated by using Equation (4.61) as

$$C_t = \frac{F_t}{p_0 A_t} = \frac{65\,580\ \text{N}}{(7 \times 10^6\ \text{N m}^{-2})(0.006\ \text{m}^2)} = \mathbf{1.561}$$

FURTHER READING

Carmichael, R. D. and Smith, E. R. (1962) *Mathematical Tables and Formulas*, Dover: New York.
Cambel, A. B. and Jennings, B. H. (1958) *Gas Dynamics*, Dover: New York.
Cohen, H., Rogers, G. F. C. and Saravanamuttoo, H. I. H. (1972) *Gas Turbine Theory*, 2nd edn, Longman: Harlow.
Fox, R. W. and McDonald, A. T. (1985) *Introduction to Fluid Mechanics*, 3rd edn, Wiley: New York.
John, J. E. A. (1984) *Gas Dynamics*, 2nd edn, Allyn & Bacon: Boston, MA.
Liepmann, H. W. and Roshko, A. (1957) *Elements of Gas Dynamics*, Wiley: New York.
Owczarek, J. A. (1964) *Fundamentals of Gas Dynamics*, International Textbook: Scranton, PA.
Shapiro, A. H. (1953) *The Dynamics and Thermodynamics of Fluid Flow*, Vol. 1, Ronald Press: New York.
Zucrow, M. J. and Hoffman, J. D. (1976) *Gas Dynamics*, Vol. 1, Wiley: New York.

PROBLEMS

4.1 The relation between the static and the stagnation conditions for any perfect gas is found to be

$$\frac{p_0}{p} = \left(1 + \frac{k-1}{2} M^2\right)^{k/(k-1)}$$

The same relation for an incompressible flow can be given with the aid of Bernoulli equation as

$$p_0 = p + \tfrac{1}{2}\rho V^2$$

a Show that both the compressible and the incompressible equations give the same result as the Mach number approaches zero.

b Find the approximate percentage error if the incompressible equation is used when $M = 0.3$ for air with $k = 1.4$

Hint: The binomial expansion theorem is given by

$$(1 + x)^n = 1 + nx + \frac{n(n-1)}{2!} x^2 + \cdots \quad \text{for } |x| < 1$$

4.2 Consider the steady, adiabatic flow of air through a long, straight pipe with a cross-sectional area of 0.05 m^2. At the inlet section, the air with a pressure of 200 kPa and a temperature of 50 °C has a velocity of 150 m s^{-1}. At the exit section, the air has a temperature of 300 K.

Determine the stagnation pressure and temperature at the inlet and exit sections, and show the process on a $T-s$ diagram.

4.3 The pressure at the nose of an aircraft in flight was found to be 120 kPa (the velocity of the air relative to the aircraft was zero at this point). Estimate the Mach number and the velocity of the aircraft if the pressure and temperature of the undisturbed air were 30 kPa and 225 K, respectively. Assume isentropic flow.

4.4 At a point in a flow passage, the velocity and temperature of the air are 200 m s^{-1} and 350 K, respectively.

a Calculate the stagnation speed of sound.
b Calculate the critical speed of sound.
c Calculate the Mach number and the Mach number referred to critical conditions at the given state.
d If the air is accelerated in a suitable passage, find the maximum possible velocity that can be attained. Determine the Mach number and the Mach number referred to critical conditions corresponding to the state of maximum velocity.
e Obtain the kinematic form of the adiabatic energy equation, and sketch the steady flow adiabatic ellipse.

4.5 A converging duct is fed with air from a large reservoir where the temperature and pressure are 400 K and 200 kPa, respectively. The Mach number at a point along the duct, where the cross-sectional area is 0.01 m^2, is 0.7. Determine the temperature, pressure, velocity and density at that point, and also calculate the mass flow rate. Assume isentropic flow throughout the nozzle.

4.6 Air is flowing isentropically through a converging duct which is fed from a large reservoir where the temperature and pressure are 350 K and 250 kPa, respectively. At a certain point along the duct, where the cross-sectional area is 0.005 m^2, the pressure is 150 kPa. Determine the Mach number, temperature and velocity at that point and also calculate the mass flow rate.

4.7 Air is flowing isentropically through a converging duct which is fed from a large reservoir where the temperature and pressure are 300 K and 400 kPa, respectively. The mass flow rate through the duct is 4 kg s^{-1}. Determine the Mach number, pressure, temperature and velocity at a point where the cross-sectional area is 0.01 m^2.

4.8 Air is flowing isentropically through a converging duct which is fed from a large reservoir where the temperature and pressure are 350 K and 200 kPa, respectively. At a certain point along the duct, where the cross-sectional area is 0.005 m^2, the velocity is 200 m s^{-1}. Determine the temperature, Mach number and pressure at that point, and also calculate the mass flow rate.

4.9 Air is supplied to a converging nozzle from a large reservoir where the temperature and pressure are 400 K and 150 kPa, respectively. At a certain cross-section where the area is 0.01 m^2, the pressure and velocity are 100 kPa and 120 $m\,s^{-1}$, respectively. Assuming adiabatic flow, find the Mach number at this cross-section and determine the mass flow rate. (METU 1987)

4.10 Helium flows adiabatically through a duct. At some section, where the area is 0.02 m^2, the velocity is 500 $m\,s^{-1}$ and the pressure is 100 kPa. Find the Mach number at the given section, and determine the mass flow rate if the stagnation temperature is 300 K. (METU 1985)

4.11 Air is supplied to a converging nozzle from a large reservoir where the temperature and pressure are 400 K and 100 kPa, respectively. At a certain cross-section, the temperature and pressure are measured to be 383.8 K and 63 kPa, respectively. Assuming adiabatic flow, find the Mach number at this cross-section and the mass flow rate per unit area. (METU 1985)

4.12 Air is flowing in a converging duct from a cross-sectional area of 0.0015 m^2, where the pressure, temperature and velocity are 200 kPa, 400 K and 100 $m\,s^{-1}$, respectively, to a cross-sectional area of 0.001 m^2. Determine the Mach number, temperature and pressure at the smaller cross-section, and also find the mass flow rate. The flow is isentropic throughout the nozzle.

4.13 A supersonic diffuser decelerates air isentropically from a Mach number of 3.0 to a Mach number of 1.5. If the static pressure and temperature at the diffuser inlet are 30 kPa and 250 K, respectively, calculate the static pressure rise in the diffuser and the ratio of the inlet to the exit cross-sectional area of the diffuser. Also find the mass flow rate per unit area at the inlet of the diffuser. Show the process on a $T-s$ diagram.

4.14 Air flows steadily and isentropically into an aircraft inlet at a mass flow rate of 100 $kg\,s^{-1}$. At a section where the cross-sectional area is 0.5 m^2, the temperature and pressure are 200 K and 20 kPa, respectively. Determine the velocity, pressure and cross-sectional area at a downstream section where the temperature is 350 K. Sketch the flow passage, and show the process on a $T-s$ diagram.

4.15 Air flows from a large tank where the pressure and temperature are 600 kPa and 800 K, respectively, through a converging nozzle with an exit area of 0.0005 m^2 and discharges to the atmosphere with a pressure of 100 kPa. The flow is isentropic throughout the nozzle. Determine the Mach number, temperature, pressure and velocity at the exit section, and also calculate the mass flow rate.

4.16 A converging nozzle is fed with air from a large reservoir where the temperature and pressure are 400 K and 170 kPa, respectively. The nozzle has an exit cross-sectional area of 0.001 m^2 and discharges into the atmosphere with a pressure of 100 kPa. The flow is isentropic throughout the nozzle. Determine the pressure, Mach number, temperature and velocity at the exit plane, and also find the mass flow rate.

4.17 Air with a stagnation pressure of 650 kPa and a stagnation temperature of 350 K is flowing isentropically through a converging nozzle. At a section within the nozzle, where the cross-sectional area is 2.6×10^{-3} m^2, the Mach number is 0.5. The nozzle discharges to a back pressure of 270 kPa. Determine the Mach number, pressure, temperature and cross-sectional area at the exit plane. Also find the mass flow rate.

4.18 Air with a stagnation temperature of 400 K and a stagnation pressure of 150 kPa is flowing isentropically through a converging nozzle. At a section within the nozzle, where the cross-sectional area is 0.005 m^2, the pressure is 125 kPa. The nozzle discharges to a back pressure of 100 kPa. Determine the pressure, Mach number, temperature and cross-sectional area at the exit plane. Also evaluate the mass flow rate.

4.19 A converging nozzle is fed with air from a large reservoir where the temperature and pressure are 400 K and 300 kPa, respectively. At a section within the nozzle, where the cross-sectional area is 0.005 m^2, the velocity is 200 m s^{-1}. The nozzle discharges to the atmosphere with a pressure of 100 kPa. Determine the Mach number, pressure, temperature and cross-sectional area at the exit plane. Also find the mass flow rate. The flow is isentropic throughout the nozzle.

4.20 A converging nozzle is fed with air from a large reservoir where the temperature and pressure are 350 K and 150 kPa, respectively. The flow in the nozzle is isentropic throughout, and the nozzle discharges to the atmosphere with a pressure of 100 kPa. At a section within the nozzle, where the cross-sectional area is 0.005 m^2, the velocity is 100 m s^{-1}. Determine the pressure, Mach number, temperature and cross-sectional area at the exit plane. Also find the mass flow rate.

4.21 A converging nozzle discharging air at 47.8 kg s^{-1} has an exit cross-sectional area of 0.1 m^2. The stagnation pressure and the stagnation temperature of the air are 240 kPa and 20 °C, respectively. Calculate the back pressure into which the nozzle is discharging. The flow is isentropic throughout. (METU 1986)

4.22 A converging nozzle is fed with a gas having a specific heat ratio of 1.33 and a gas constant of 460 J kg^{-1} K^{-1} from a large reservoir where the pressure and temperature are 250 kPa and 300 K, respectively. If the nozzle discharges to a region where the pressure is 100 kPa, determine the pressure and cross-sectional area at the exit of the nozzle. Assume that the flow is adiabatic and reversible with a mass flow rate of 10 kg s^{-1}. (METU 1987)

4.23 Consider the flow of air in a duct, which is shown in the figure below. The flow is adiabatic between sections 1 and 2, and isentropic between sections 2 and 3. The pressure, temperature and velocity at section 1 are 200 kPa, 320 K and 150 m s^{-1}, respectively. The drop in the stagnation pressure between sections 1 and 2 is 10 kPa. The nozzle discharges to a region where the pressure is 150 kPa. If the exit cross-sectional area of the duct is 0.001 m^2, determine the mass flow rate and the cross-sectional area at section 1. (METU 1987)

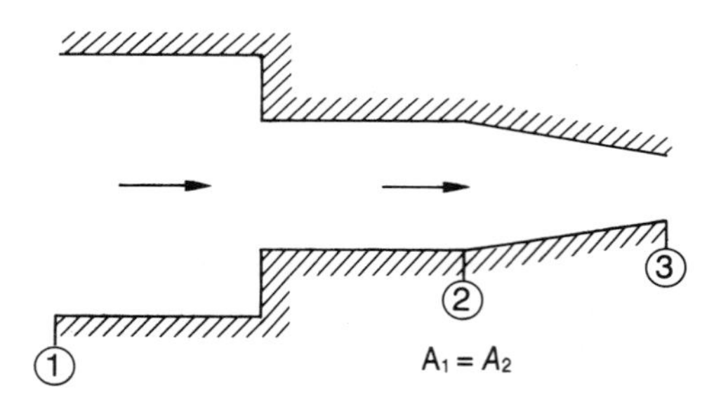

Problem 4.23

4.24 At the discharge of an air-conditioning duct, it is desirable to reduce the cross-sectional area of the duct, as shown in the figure below. The cross-sectional area of the smaller duct is 0.02 m^2. At section 1 of the larger duct, the pressure, stagnation temperature and velocity are 150 kPa, 327 K and 71 m s^{-1}, respectively. If the flow throughout the duct is isentropic and if it discharges to an ambient pressure of 100 kPa, determine the mass flow rate. (METU 1986)

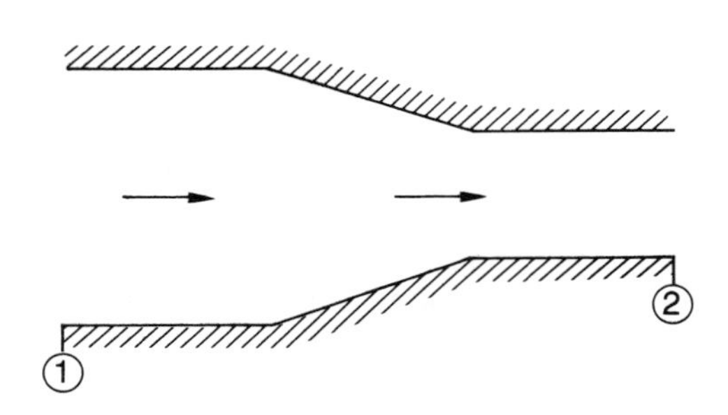

Problem 4.24

4.25 Air is supplied to a converging nozzle from a large reservoir, where the pressure and temperature are kept constant at 500 kPa and 300 K, respectively. If the nozzle discharges to the atmosphere with a pressure of 100 kPa, find the time required to discharge 2 kg of air. The exit area of the nozzle is 10 cm^2. Assume isentropic flow. (METU 1985)

4.26 Air is stored in a tank of 2 m^2 in volume at a pressure of 3 MPa and a temperature of 300 K. The gas is discharged through a converging nozzle with an exit cross-sectional area of 0.0012 m^2. For a back pressure of 101 kPa, find the time required for the tank pressure to drop to 300 kPa. Assume quasi-steady flow through the nozzle, where the steady flow equations are applicable at each instant of time. Also assume that the temperature of the air in the tank is constant.

4.27 Air is stored in a tank of 3 m^3 in volume at a pressure of 300 kPa and a temperature of 400 K. Air is discharged through a converging nozzle with an exit area of 0.001 m^2. Determine the time required for the tank pressure to drop to 125 kPa if the nozzle discharges to the atmosphere with a pressure of 100 kPa. Assume isentropic, quasi-steady flow through the nozzle, where the steady flow equations are applicable at each instant of time. Also, there is enough heat transfer to the tank so that the temperature of the air in the tank is constant.

4.28 An isentropic converging–diverging nozzle is operating at its design condition. It receives air from a large reservoir where the pressure and temperature are 200 kPa and 300 K, respectively. The exit and throat cross-sectional areas of the nozzle are 0.002 m^2 and 0.001 m^2, respectively. Determine the Mach number, temperature and pressure at the exit plane. Also find the back pressure and mass flow rate.

4.29 At a point upstream of the throat in a converging–diverging nozzle, the pressure, temperature and velocity of the air are 200 kPa, 400 K and 200 $m\,s^{-1}$, respectively. The exit and throat cross-sectional areas of the nozzle are 0.002 m^2 and 0.001 m^2, respectively. Determine the Mach number, temperature and pressure at the exit plane. Also find the back pressure and mass flow rate. The nozzle is operating at its design condition.

4.30 An isentropic converging–diverging nozzle receives air from a large reservoir, where the pressure and temperature are 110 kPa and 350 K, respectively. The exit and throat cross-sectional areas of the nozzle are 0.0015 m^2 and 0.001 m^2, respectively. The nozzle discharges to the atmosphere with a pressure of 100 kPa. Determine the pressure, the Mach number and temperature at the exit plane, and also find the mass flow rate.

4.31 At a point upstream of the throat of an isentropic converging–diverging nozzle, the pressure, temperature and velocity of the air are 105 kPa, 400 K and 100 $m\,s^{-1}$, respectively. The exit and throat cross-sectional areas of the nozzle are 0.0015 m^2 and 0.001 m^2, respectively. The nozzle discharges to the atmosphere with a pressure of 100 kPa. Determine the pressure, the Mach number and the temperature at the exit plane, and also calculate the mass flow rate.

4.32 An isentropic converging–diverging nozzle is fed with air from a large reservoir where the temperature and pressure are 500 K and 900 kPa, respectively. The ratio of the exit to the throat cross-sectional area is 1.25. The nozzle discharges to the atmosphere with a pressure of 100 kPa. Determine the Mach number, temperature, pressure and mass flux at the exit plane.

4.33 At a point upstream of an isentropic converging–diverging nozzle, the pressure, temperature and velocity of the air are 600 kPa, 400 K and 200 $m\,s^{-1}$, respectively. The ratio of the exit to the throat cross-sectional area is 1.5. The nozzle discharges to the atmosphere with a pressure of 100 kPa. Determine the Mach number, pressure, temperature and mass flux at the exit plane.

4.34 A large tank supplies helium to a converging–diverging nozzle. The pressure and temperature in the tank remain constant at 6 MPa and 1000 K, respectively. The flow throughout the nozzle is isentropic. The nozzle discharges to the atmosphere with a pressure of 100 kPa at an exit Mach number of 3. The exit cross-sectional area of the nozzle is 0.001 m^2. What is the mass flow rate through the nozzle?

4.35 Air is flowing in a converging–diverging nozzle which has an efficiency of 90 per cent. At the inlet of the nozzle, the stagnation temperature and the stagnation pressure are 350 K and 500 kPa, respectively. The ratio of the exit cross-sectional area to the throat cross-sectional area is 1.5.

a Determine the Mach number, pressure, temperature, mass flux and velocity at the exit of the nozzle for isentropic operation.
b Determine the actual values of the velocity, pressure, temperature, Mach number and mass flux at the exit of the nozzle.
c Determine the velocity coefficient of the nozzle.
d Determine the discharge coefficient of the nozzle.

4.36 A converging–diverging nozzle with a throat cross-sectional area of 6.5×10^{-4} m^2 discharges air into a duct with a cross-sectional area of 1.3×10^{-3} m^2. At the inlet of the nozzle, the stagnation temperature and the stagnation pressure are 500 K and 700 kPa, respectively. The nozzle has a discharge coefficient of 0.98.

a Determine the Mach number, pressure, temperature, mass flow rate and velocity at the exit of the nozzle for isentropic operation.
b Determine the actual values of the mass flow rate, pressure, velocity, temperature and the Mach number at the exit of the nozzle.
c Determine the efficiency of the nozzle.
d Determine the velocity coefficient of the nozzle.

4.37 A perfect gas is supplied to a converging–diverging nozzle from a large reservoir where the temperature and pressure are T_0 and p_0, respectively. The nozzle has an efficiency of η between the inlet section and any downstream section.

a Derive an expression for p/p_0 in terms of the nozzle efficiency, η, the specific heat ratio, k, and the Mach number, M, of the downstream section.
b Derive an expression for $\dot{m}\sqrt{(RT_0)}/(Ap_0)$ in terms of the nozzle efficiency, η, the specific heat ratio, k, and the Mach number, M, of the downstream section.

4.38 A solid-propellant rocket engine generates gases at a stagnation temperature of 2500 K and a stagnation pressure of 4000 kPa. The exit and throat cross-sectional areas of the nozzle are 0.020 m^2 and 0.005 m^2, respectively. The generated gases are perfect with a specific heat ratio of 1.4 and a gas constant of 320 J kg^{-1} K^{-1}. The flow is isentropic. For an altitude of 20 000 m, determine

a the thrust developed,
b the specific impulse, and
c the thrust coefficient.

4.39 A solid-propellant rocket engine generates gases at a stagnation temperature of 3000 K and a stagnation pressure of 7000 kPa. The exit and throat cross-sectional areas of the nozzle are 0.025 m^2 and 0.002 m^2, respectively. The generated gases are perfect with a specific heat ratio of 1.3 and a gas constant of 320 J kg^{-1} K^{-1}. The flow is isentropic. For an altitude of 30 000 m, determine

a the thrust developed,
b the specific impulse, and
c the thrust coefficient.

4.40 A gas turbine engine produces a thrust of 10 000 N while the gas is entering the engine at a velocity of 100 m s^{-1}, as shown in the figure below. The mass flow rate through the engine is 50 kg s^{-1}. The nozzle exit area is 0.2 m^2. The atmospheric pressure is 100 kPa and the nozzle is not choked. Find the Mach number at the exit of the nozzle. The specific heat ratio and the specific heat at constant pressure are 1.33 and 1150 J kg^{-1} K^{-1} for the exhaust gases, respectively.

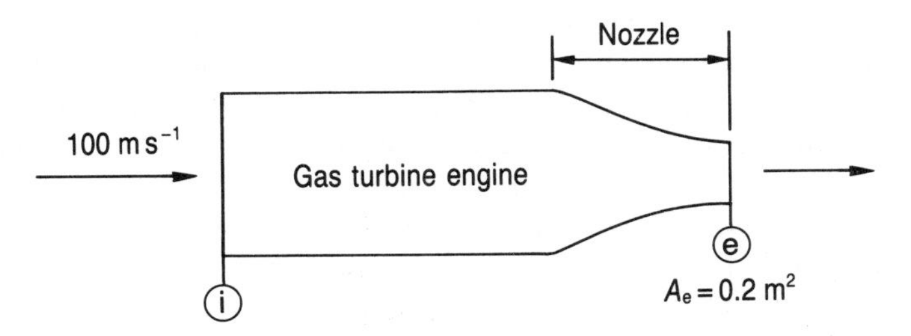

Problem 4.40

5 Normal shock waves

5.1 INTRODUCTION

A shock wave represents an abrupt change in fluid properties, in which finite variations in pressure, temperature and density occur over a shock thickness comparable with the mean free path of the gas molecules involved. Also, it will be shown that a shock wave can only be of the compression type, and a rarefaction shock which requires a decrease in entropy in an adiabatic process is not possible. It is known that supersonic flow adjusts to the presence of a body by means of the shock process, whereas subsonic flow can adjust by gradual changes in flow properties. Familiar examples are the phenomena associated with detonation waves and explosions. In all of these cases, the wave is very steep with a large pressure rise across the wave. Because of the large pressure gradient across the shock wave, the gas experiences a large increase in its density with a corresponding change in its refractive index. Therefore, the shock wave may be visualized by using a suitable apparatus, known as the **schlieren system**.

The compression of the gas across the shock wave is effectively instantaneous, and therefore it cannot be a reversible process. Because of the irreversibility of the shock process, the kinetic energy of the gas leaving the shock wave is smaller than that of the isentropic compression between the same pressure limits. The reduction in kinetic energy due to the shock wave appears as heating of the gas to a static temperature above that corresponding to the isentropic compression value. Consequently, as the gas flows through the shock wave, it experiences a decrease in its available energy and an increase in its entropy.

Experiments indicate that these discontinuities are often oblique to the flow direction, and they are referred to as **oblique shock waves**. However, with one-dimensional gas dynamics, it is only possible to treat the simplest case of a shock, where the changes in flow properties are only in the direction of the flow. Such shock waves are called **normal shock waves**.

Another way of classifying shock waves is according to their relative velocity. Thus, one may speak of **stationary shock waves** and **moving shock waves**. A moving shock wave may be rendered stationary by making the gas, through which it propagates travel at a speed equal to that of the shock wave but opposite in direction. This is very similar to the case of a sound wave which was rendered stationary in Figure 3.2.

5.2 PHYSICAL DESCRIPTION OF WAVE DEVELOPMENT

The development and the propagation of a single wave were discussed in Section 3.2. However, in this section, the development and the propagation of series of waves will be discussed. These waves may involve compression and expansion waves.

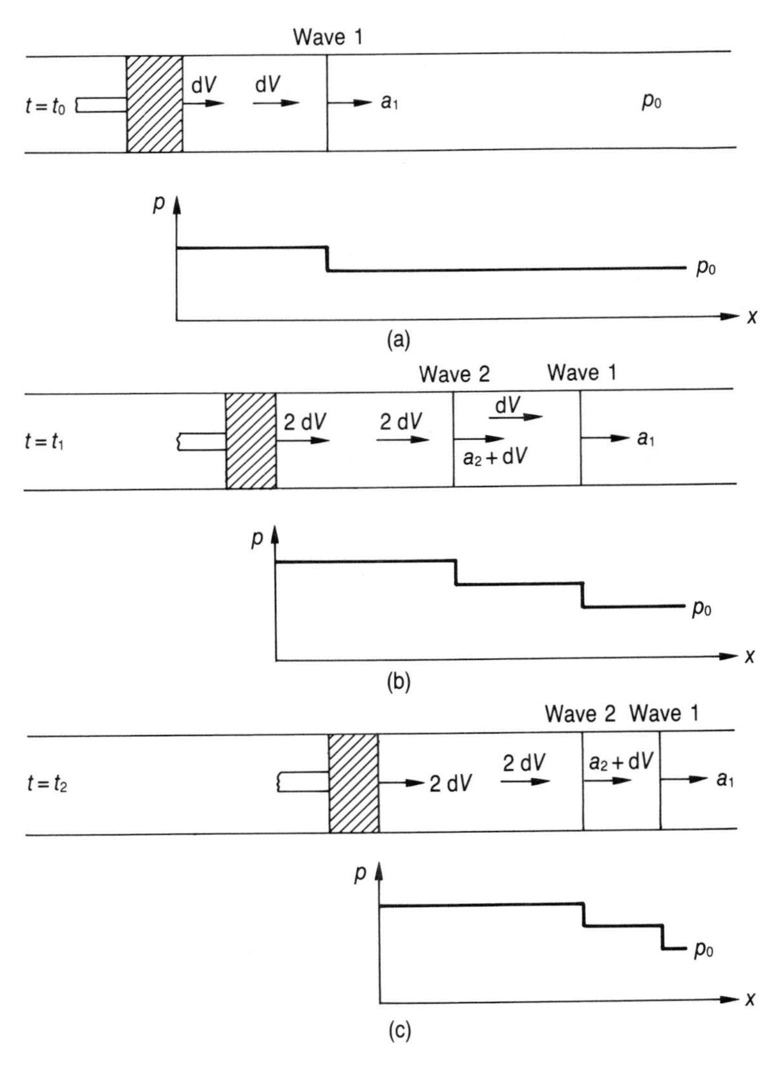

Figure 5.1 Development of a compression wave

5.2.1 Development of compression waves

It was shown in Section 3.3 that, when a piston in a tube is given a steady velocity to the right of magnitude dV (Figure 3.2), a sound wave travels ahead of the piston through the medium in the tube. This situation is also indicated in Figure 5.1(a). Suppose that the piston is now given a second increment of velocity, dV, causing a second wave to move into the compressed gas behind the first wave. The location of these waves and the pressure distribution at time $t = t_1$ are shown in Figure 5.1(b). The first and second waves, created by the acceleration of the piston, travel respectively at speeds of sound a_1 and a_2, with respect to the gas into which they are moving. Since the second wave is moving into a compressed gas with a slightly elevated temperature, then a_2 is greater than a_1. The first wave is moving into a stationary gas, so the absolute velocity of this wave is a_1. However, the second wave is moving into a gas which is already moving to the right at a velocity dV, and therefore the absolute velocity of the second wave is $a_2 + dV$. Hence, the second wave moves faster than the first and gradually gets closer, as shown in Figure 5.1(c), at time $t = t_2$.

Now, suppose that the piston is accelerated from the rest to a finite velocity increment of magnitude ΔV to the right. This finite velocity increment can be thought to consist of a large number of infinitesimal increments, each of magnitude dV. Figure 5.2 shows the velocity of the piston versus time with the incremental velocities superimposed. The corresponding location of the waves and the pressure distribution in tube, at $t = t_1$, are shown in Figure 5.3(a). The first wave of the series of waves created by the piston is called the **head**, while the last is known as the **tail**. As was demonstrated in Figure 5.1, the waves next to the piston tend to overtake those further down the tube, as shown in Figures 5.3(b) and 5.3(c) at $t = t_2$ and $t = t_3$, respectively. As time elapses, the compression wave tends to become infinitely steep at some point between $t = t_3$ and $t = t_4$. The present analysis then ceases to be valid, because the viscous and heat transfer effects are no longer negligible with such

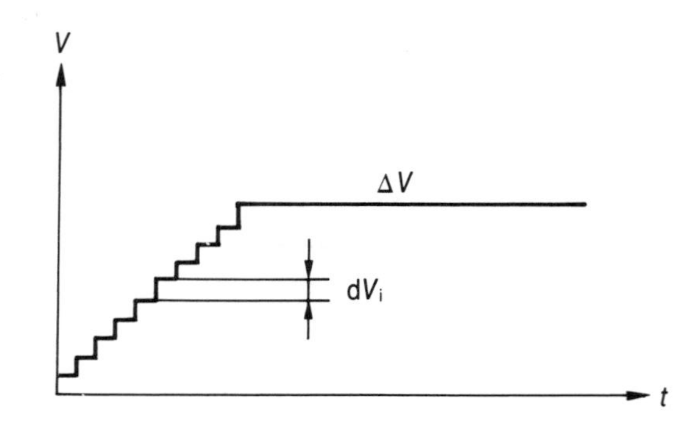

Figure 5.2 Velocity of the piston versus time

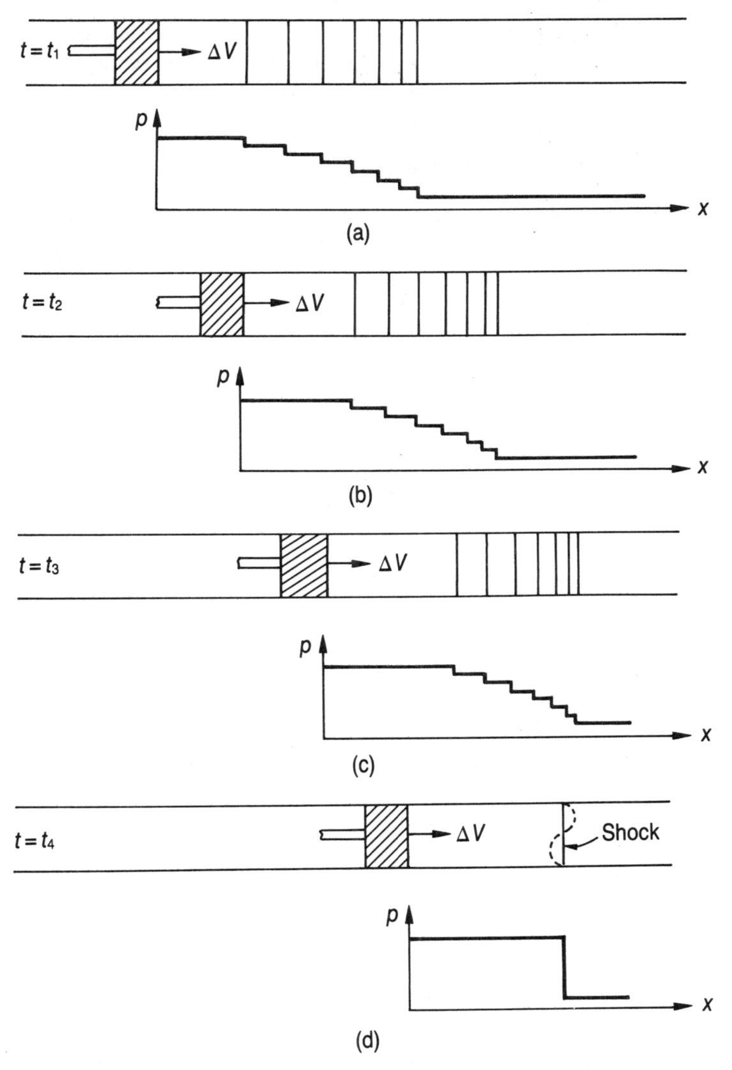

Figure 5.3 Steepening of compression waves and formation of a normal shock wave

extraordinary gradients in the velocity and the temperature along the wave. Indeed, if the isentropic analysis were continued up to $t = t_4$, the wave would topple over to the form shown by the dashed line in Figure 5.3(d). This is physically absurd, since it means that at the same time and at the same location, the fluid has simultaneously three different values of pressure, density and velocity. Therefore, viscous and heat conduction effects intervene soon after $t = t_3$, and produce a normal shock wave at $t = t_4$. The thickness of a normal shock wave is approximately

2.5×10^{-7} m, so the pressure and temperature gradients along the wave are very large, yet not truly infinite. Hence, a shock wave can be approximated by a discontinuity.

5.2.2 Development of expansion (rarefaction) waves

If the piston in Figure 5.4(a) is given an infinitesimal velocity of dV to the left, a weak expansion wave propagates to the right at the speed of sound, a_1, relative to the

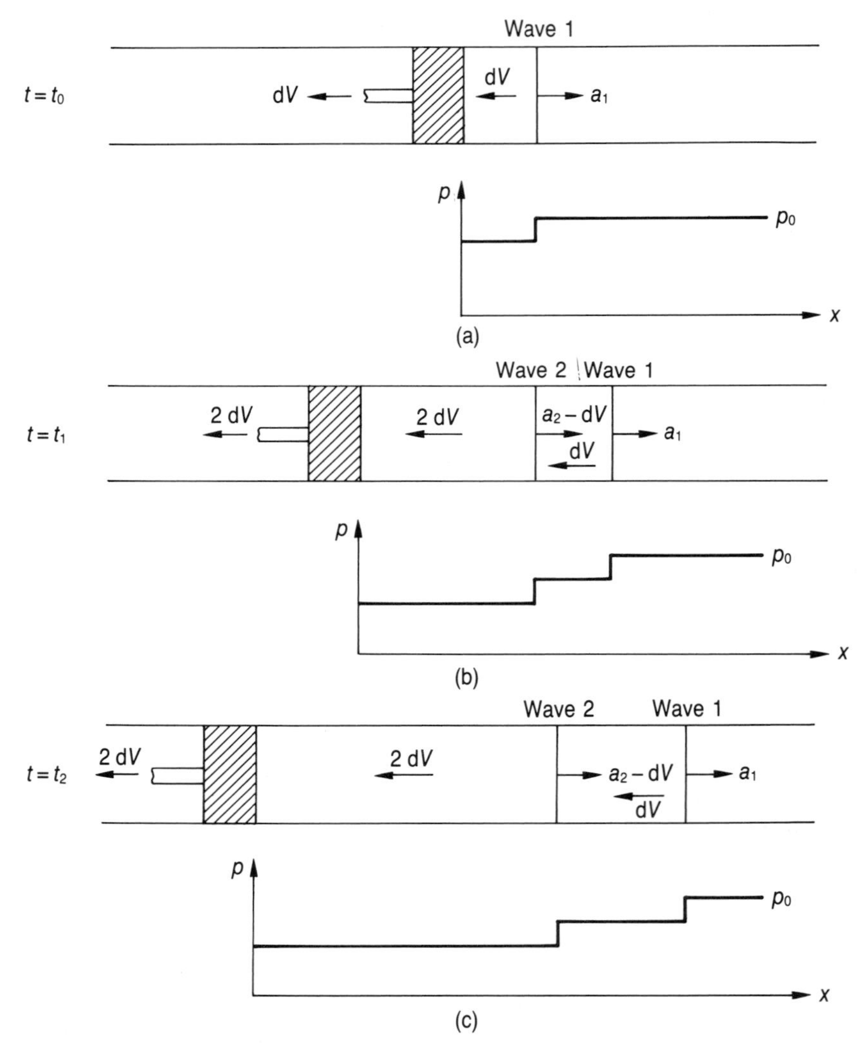

Figure 5.4 Development of expansion waves

stationary gas. At time $t = t_1$, when the piston is given a second infinitesimal velocity of dV to the left, a second expansion wave moves into the expanded gas behind the first wave, as shown in Figure 5.4(b). This second wave travels at the speed of sound, a_2, with respect to the gas into which it is moving. In this case, the second wave and the gas are moving in opposite directions, and therefore the wave has an absolute velocity of $a_2 - dV$ to the right. Furthermore, the second wave is travelling into the gas that has already been expanded and cooled, so that a_2 is smaller than a_1. Therefore, the second wave falls further and further behind the first one, as shown in Figure 5.4(c). In this manner, expansion waves spread out and are not able to reinforce each other. Therefore, the creation of a finite expansion shock wave is impossible.

5.3 GOVERNING EQUATIONS FOR THE FLOW ACROSS A NORMAL SHOCK WAVE

In this section, the governing equations for the flow across a normal shock wave are derived. Also, the mathematical definition of the normal shock wave and its direction is discussed.

5.3.1 Governing physical equations

The flow through a stationary normal shock wave may be analyzed by considering one-dimensional flow. In this analysis, the following assumptions are made:

1. The normal shock wave is perpendicular to the streamlines.
2. The normal shock takes place at a constant cross-sectional area, since the shock thickness is very small compared with the dimensions of the duct.
3. Although the shear stress is present within the boundary layer, it acts on a very small area since the shock wave is very thin. For this reason, frictional effects are negligible.
4. The flow process, including the shock wave, can be assumed to be adiabatic with no external work, and the effects of the body forces are negligible, since the shock wave is very thin.

Suppose that the conditions at the upstream and downstream sides of the normal shock wave, which is shown in Figure 5.5, are denoted by subscripts x and y, respectively. Then the governing physical equations for the stationary normal shock wave are as given below.

(a) Continuity equation

For steady and uniform flow, the mass flow rate upstream and downstream of the normal shock wave is constant. Then

$$\dot{m} = \rho_x V_x A_x = \rho_y V_y A_y$$

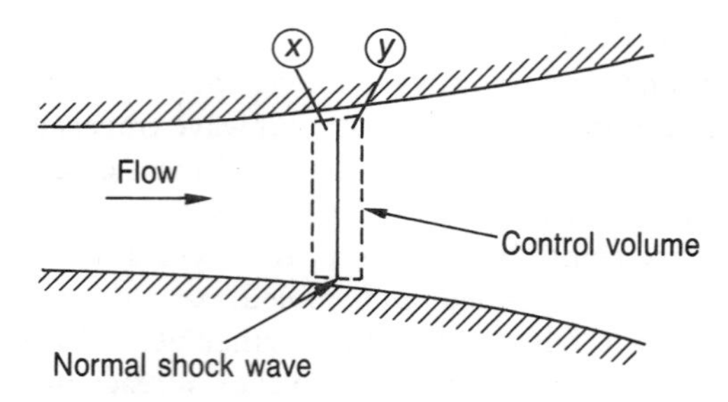

Figure 5.5 Control volume around a normal shock wave

Since $A_x = A_y = A$, the mass flow rate per unit area or the mass flux is given by

$$G = \dot{m}/A = \rho_x V_x = \rho_y V_y = \text{constant} \tag{5.1}$$

(b) Momentum equation

Since the flow across the normal shock wave is assumed to be frictionless, the pressure forces acting on the control volume of Figure 5.5 must be balanced by the rate of change of momentum across the control volume. Hence,

$$p_x A_x - p_y A_y = \rho_y A_y V_y^2 - \rho_x A_x V_x^2$$

As long as the cross-sectional area and the mass flow rate are constant across the normal shock wave, then

$$p_x - p_y = G(V_y - V_x) \tag{5.2}$$

(c) Energy equation

For an adiabatic process across a normal shock wave with no external work, the first law of thermodynamics becomes

$$h_{0x} = h_x + \frac{V_x^2}{2} = h_y + \frac{V_y^2}{2} = h_{0y} = h_0 = \text{constant} \tag{5.3}$$

(d) The second law of thermodynamics

The flow through a normal shock wave is irreversible, because of the almost discontinuous property changes across the shock wave. Thus,

$$s_y > s_x \tag{5.4}$$

(e) Equation of state

For any fluid, the equation of state may be given as

$$h = h(s, \rho) \tag{5.5a}$$

and

$$s = s(p, \rho) \tag{5.5b}$$

5.3.2 The Fanno line

Now, assume that the conditions upstream of the normal shock wave (state x) are fixed, and those downstream (state y) are to be found. Then the continuity equation (5.1), the energy equation (5.3) and the equation of state (5.5a) define a locus of states passing through state x. This locus is called the **Fanno line** on the h–s diagram, as shown in Figure 5.6. For example, if state x is completely fixed, the following may be determined for a particular value of V_y:

1 ρ_y may be calculated from the continuity equation (5.1).
2 h_y may be evaluated from the energy equation (5.3).
3 s_y may be found from the equation of state (5.5a).

Since the momentum equation (5.2) has not yet been introduced, **the Fanno line represents the states with the same mass flux and the same stagnation enthalpy, but not necessarily the same value of the impulse function**. Therefore, on the Fanno line, frictional effects are present because the energy equation does not contain the friction term but does assume adiabatic conditions.

For any state on the Fanno line, from the energy equation (5.3), $h_0 = h + V^2/2 =$ constant, and differentiating

$$\mathrm{d}h + V\,\mathrm{d}V = 0 \tag{5.6}$$

Also, from the continuity equation (5.1), $G = \rho V =$ constant across the normal shock wave. Then differentiation yields

$$\rho\,\mathrm{d}V + V\,\mathrm{d}\rho = 0 \tag{5.7}$$

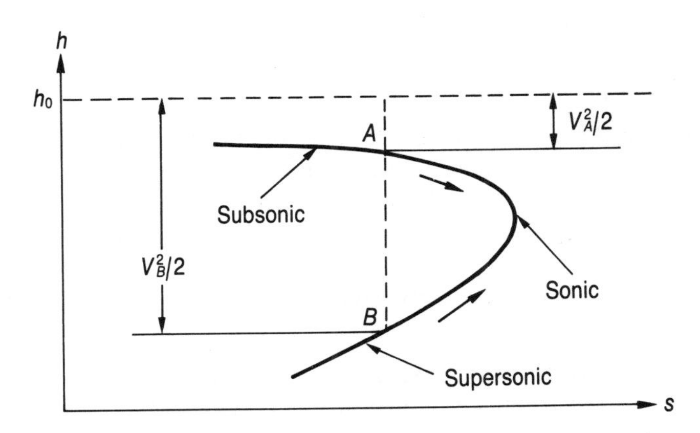

Figure 5.6 The Fanno line

Finally, from the thermodynamic relation

$$T\ \mathrm{d}s = dh - \mathrm{d}p/\rho$$

and at the point of maximum entropy, this relation reduces to

$$\mathrm{d}h = \mathrm{d}p/\rho \tag{5.8}$$

Now, in Equation (5.8), dh may be eliminated by using Equation (5.6), and ρ may be eliminated by using Equation (5.7). Thus

$$\mathrm{d}p/\mathrm{d}\rho = V^2$$

But at the point of maximum entropy d$s = 0$, that is the entropy is constant. Then

$$\frac{\mathrm{d}p}{\mathrm{d}\rho} = \left(\frac{\partial p}{\partial \rho}\right)_s$$

Hence, from Equation (3.5),

$$V^2 = \left(\frac{\partial p}{\partial \rho}\right)_s = a^2$$

Therefore, the Mach number is unity, that is $M = 1$, at the point of maximum entropy.

In Figure 5.6, states A and B are chosen such that they are on the upper and lower branches of the same Fanno line, respectively, and have the same entropy. Then both points will have the same stagnation state, and the vertical distances between these states and the stagnation state represent the kinetic energies per unit mass of states A and B. Hence, it may be observed from Figure 5.6 that the kinetic energy of state B is greater than that of state A. Therefore the upper branch of the Fanno line represents subsonic flow, while the lower branch represents supersonic flow. For any physically possible flow, the entropy must increase in the direction of flow. For this reason, the subsonic and supersonic flows, which are on the upper and lower branches of the Fanno line, respectively, reach a sonic velocity at the point of maximum entropy.

5.3.3 The Rayleigh line

Again, assume that the conditions upstream of the normal shock wave (state x) are fixed, and those downstream (state y) are to be found. Then the continuity equation (5.1), the momentum equation (5.2) and the equation of state (5.5b) define a locus of states passing through state x. This locus is called the **Rayleigh line** on the h–s diagram, as indicated in Figure 5.7.

For example, if state x is completely fixed, the following may be determined for a particular value of V_y:

1 ρ_y may be calculated from the continuity equation (5.1).
2 p_y may be evaluated from the momentum equation (5.2).
3 s_y may be found from the equation of state (5.5b).

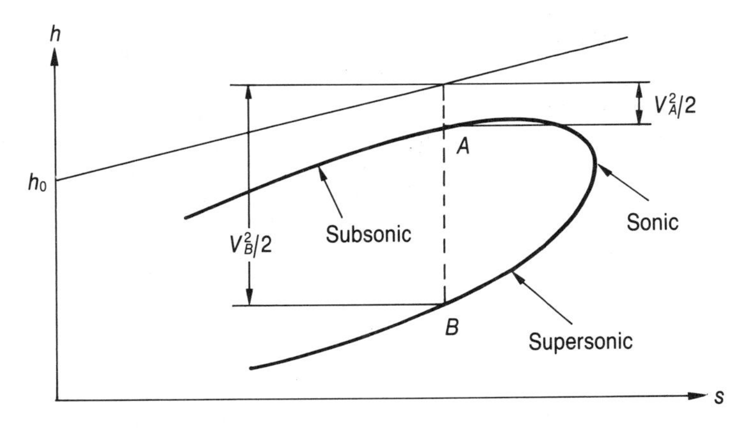

Figure 5.7 The Rayleigh line

Since the energy equation (5.3) has not yet been introduced, **the Rayleigh line represents the states with the same mass flux and the same impulse function but not necessarily the same value of the stagnation enthalpy**. Therefore, a state on the Rayleigh line is reachable from another state on the same line in the presence of heat transfer effects.

For any state on the Rayleigh line, from the momentum equation (5.2), $p + GV = \text{constant}$, and differentiating

$$\mathrm{d}p + G\,\mathrm{d}V = 0$$

Also, from the continuity equation (5.1), $G = \rho V$. Thus

$$\mathrm{d}p + \rho V\,\mathrm{d}V = 0 \tag{5.9}$$

Finally, using the differential form of the continuity equation (5.7)

$$\mathrm{d}p/\mathrm{d}\rho = V^2$$

But at the point of maximum entropy $\mathrm{d}s = 0$, that is the entropy is constant. Then,

$$\frac{\mathrm{d}p}{\mathrm{d}\rho} = \left(\frac{\partial p}{\partial \rho}\right)_s$$

Hence, from Equation (3.5),

$$V^2 = \left(\frac{\partial p}{\partial \rho}\right)_s = a^2$$

Therefore, at the point of maximum entropy the Mach number is unity, that is the flow is sonic.

In Figure 5.7, states A and B are chosen such that they are on the upper and lower branches of the same Rayleigh line, respectively, and have the same entropy. Then both points will have the same stagnation state, and the vertical distances between these states and the stagnation state represent the kinetic energies per unit mass of states A and B. Hence, it may be observed from Figure 5.7 that the kinetic energy of state B is greater than that of state A. Therefore, the upper branch of the Rayleigh line represents subsonic flow, while the lower branch represents supersonic flow. For any physically possible flow, the entropy must increase in the direction of flow. For this reason, the subsonic and supersonic flows, which are on the upper and lower branches of the Rayleigh line, respectively, reach a sonic velocity at the point of maximum entropy.

5.3.4 Mathematical description of the normal shock wave

The normal shock wave must satisfy the continuity equation (5.1), the momentum equation (5.2), the energy equation (5.3), the equation of state (5.5) and the second law of thermodynamics (5.4). The Fanno line satisfies Equations (5.1), (5.3) and (5.5), while the Rayleigh line satisfies Equations (5.1), (5.2) and (5.5). Therefore, for a given state x, the end state, y, must then lie on both the Fanno line and the Rayleigh line passing through state x. Hence, the intersection of the two lines at state y represents the conditions at the downstream side of the normal shock wave corresponding to the prescribed conditions at state x, as shown in Figure 5.8.

The analysis, thus far, has no restriction on the direction of the process, that is whether from state x to state y or from state y to state x. For all known fluids, state y lies to the right of state x (Figure 5.8). Thus far, the second law of thermodynamics (5.4) has not yet been introduced, which states that the entropy cannot decrease during an adiabatic process. Therefore, with the aid of this law, the direction of the normal shock wave may be fixed. Hence, the normal shock wave can only proceed from state x to state y, that is from a supersonic flow to a subsonic one with a consequent rise in pressure.

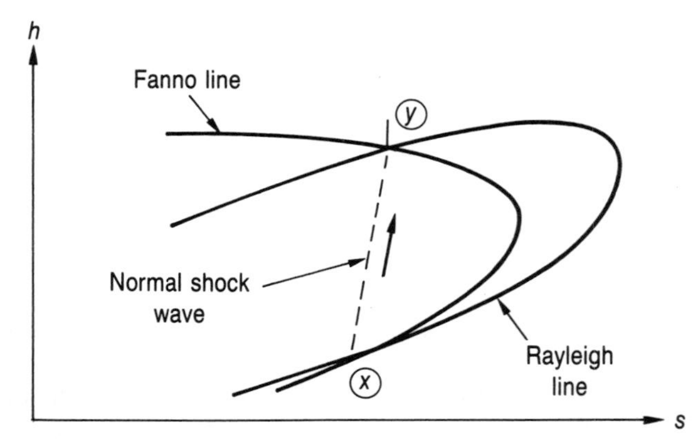

Figure 5.8 Normal shock wave on an h–s diagram

5.4 RELATIONS FOR THE FLOW OF A PERFECT GAS ACROSS A NORMAL SHOCK WAVE

The equations governing one-dimensional flow across a normal shock wave were given in Section 5.3. Although any problem related to the flow across a normal shock wave may be handled by solving these equations, the solution can be facilitated by rearranging the equations in a more suitable form. In this section, these relations will be derived.

5.4.1 Equation of the Fanno line

For a perfect gas $h = c_p T$; then Equation (5.3) may be arranged to give

$$c_p T_x + \frac{V_x^2}{2} = c_p T_y + \frac{V_y^2}{2}$$

or, using $c_p = kR/(k-1)$ and the definition of the Mach number from Equation (3.8) as $M^2 = V^2/(kRT)$,

$$\frac{T_y}{T_x} = \frac{1 + \frac{1}{2}(k-1)M_x^2}{1 + \frac{1}{2}(k-1)M_y^2} \tag{5.10}$$

Substituting the equation of state for a perfect gas, $\rho = p/(RT)$, into the continuity equation (5.1) yields

$$\frac{T_y}{T_x} = \frac{p_y}{p_x}\frac{V_y}{V_x}$$

and recalling $M = V/a$ and $a = \sqrt{(kRT)}$ gives

$$\frac{T_y}{T_x} = \left(\frac{p_y}{p_x}\right)^2 \left(\frac{M_y}{M_x}\right)^2 \tag{5.11}$$

Now, combining Equations (5.10) and (5.11) yields an equation for the Fanno line:

$$\frac{p_y}{p_x} = \frac{M_x}{M_y}\sqrt{\left(\frac{1 + \frac{1}{2}(k-1)M_x^2}{1 + \frac{1}{2}(k-1)M_y^2}\right)} \tag{5.12}$$

5.4.2 Equation of the Rayleigh line

For a perfect gas $p = \rho RT$, and the momentum equation (5.2) may be transformed to read

$$p_x + \frac{p_x}{RT_x} V_x^2 = p_y + \frac{p_y}{RT_y} V_y^2$$

Using $M^2 = V^2/(kRT)$ yields an equation for the Rayleigh line:

$$\frac{p_y}{p_x} = \frac{1 + kM_x^2}{1 + kM_y^2} \tag{5.13}$$

5.4.3 Relation between the upstream and the downstream Mach numbers

A relation between the Mach numbers upstream and downstream of a normal shock wave may be obtained by combining the equation of the Fanno line (5.12) and the equation of the Rayleigh line (5.13) as

$$\frac{M_x\sqrt{[1+\frac{1}{2}(k-1)M_x^2]}}{1+kM_x^2}=\frac{M_y\sqrt{[1+\frac{1}{2}(k-1)M_y^2]}}{1+kM_y^2} \tag{5.14}$$

This equation may be rearranged as

$$[kM_x^2-\tfrac{1}{2}(k-1)]M_y^4-(kM_x^4+1)M_y^2+[M_x^2+\tfrac{1}{2}(k-1)M_x^4]=0$$

which is a quadratic equation for M_y^2. The roots of this equation are

$$M_y=M_x \tag{5.15a}$$

and

$$M_y=\sqrt{\left(\frac{(k-1)M_x^2+2}{2kM_x^2-(k-1)}\right)} \tag{5.15b}$$

The first of these solutions is quite trivial, since it expresses the obvious fact that the conditions upstream and downstream of a normal shock wave may be identical. For this trivial solution, Equations (5.13) and (5.11) indicate that $p_y=p_x$ and $T_y=T_x$, respectively. The second solution gives the relation between the states that must exist on the two sides of the normal shock wave.

5.4.4 Relation between M_x^* and M_y^* (Prandtl's equation)

If M_x and M_y in Equation (5.15b) are replaced by using Equation (4.19b), then it is possible to obtain a relation between M_x^* and M_y^* as

$$M_x^*M_y^*=1 \tag{5.16a}$$

This relation is known as Prandtl's equation. From Equation (4.12), it is possible to note that $a^*=a_x^*=a_y^*$, since the stagnation temperature does not change across a normal shock wave. Then, using $M^*=V/a^*$, it is possible to rewrite Equation (5.16a) as

$$V_xV_y=a^{*2} \tag{5.16b}$$

From Equation (5.16a), it is obvious that if the flow at state x is supersonic, then the flow at state y must be subsonic.

5.4.5 Pressure ratio across a normal shock wave

An equation for the pressure ratio across a normal shock wave may be obtained by eliminating the downstream Mach number M_y with the aid of Equation (5.15b) either in the equation of the Fanno line (5.12) or in the equation of the Rayleigh line (5.13) as

$$\frac{p_y}{p_x}=\frac{2k}{k+1}M_x^2-\frac{k-1}{k+1} \tag{5.17}$$

5.4.6 Temperature ratio across a normal shock wave

The temperature ratio across a normal shock wave may be obtained by substituting Equation (5.15b) into Equation (5.10) as

$$\frac{T_y}{T_x}=\left(1+\frac{k-1}{2}M_x^2\right)\left(\frac{2k}{k-1}M_x^2-1\right)\Big/\left(\frac{(k+1)^2}{2(k-1)}M_x^2\right) \tag{5.18}$$

5.4.7 Density ratio across a normal shock wave

The density ratio across a normal shock wave may be calculated by using the equation of state for a perfect gas, $p=\rho RT$, as

$$\frac{\rho_y}{\rho_x}=\frac{p_y}{p_x}\frac{T_x}{T_y} \tag{5.19}$$

Using Equations (5.17) and (5.18) yields

$$\frac{\rho_y}{\rho_x}=\frac{(k+1)M_x^2}{2+(k-1)M_x^2} \tag{5.20}$$

5.4.8 Velocity ratio across a normal shock wave

The combination of the continuity equation (5.1) and the density ratio (5.20) yields

$$\frac{V_y}{V_x}=\frac{\rho_x}{\rho_y}=\frac{2+(k-1)M_x^2}{(k+1)M_x^2} \tag{5.21}$$

5.4.9 Stagnation pressure ratio across a normal shock wave

The stagnation pressure ratio across a normal shock wave may be rearranged as

$$\frac{p_{0y}}{p_{0x}}=\frac{p_{0y}}{p_y}\frac{p_y}{p_x}\frac{p_x}{p_{0x}}$$

Then, substituting for p_{0y}/p_y and p_x/p_{0x} from Equation (4.28) and for p_y/p_x from Equation (5.17) and using Equation (5.15b) yields the following equation:

$$\frac{p_{0y}}{p_{0x}}=\left(\frac{\frac{1}{2}(k+1)M_x^2}{1+\frac{1}{2}(k-1)M_x^2}\right)^{k/(k-1)}\left(\frac{2k}{k+1}M_x^2-\frac{k-1}{k+1}\right)^{1/(1-k)} \tag{5.22}$$

5.4.10 Critical area ratio across a normal shock wave

For isentropic flow, the area at which the Mach number is equal to unity was defined as A^*, with this area being used as a reference. However, a normal shock is not an isentropic process. Therefore, if a normal shock occurs in a channel (Figure 5.9), flow areas downstream of the normal shock cannot be referenced to the critical area upstream of the normal shock, that is $A_x^* \neq A_y^*$. For steady flow across a normal shock wave

$$\dot{m}_x = \dot{m}_y = \text{constant}$$

Then with the aid of Equation (4.35) it is possible to write

$$\frac{\dot{m}_x \sqrt{T_{0x}}}{p_{0x} A_x^*} = \frac{\dot{m}_y \sqrt{T_{0y}}}{p_{0y} A_y^*}$$

Since the flow is adiabatic across a normal shock wave, that is $T_{0x} = T_{0y}$, then

$$p_{0y} A_y^* = p_{0x} A_x^*$$

or with the aid of Equation (5.22)

$$\frac{A_y^*}{A_x^*} = \frac{p_{0x}}{p_{0y}} = \left(\frac{\frac{1}{2}(k+1)M_x^2}{1 + \frac{1}{2}(k-1)M_x^2} \right)^{k/(1-k)} \left(\frac{2k}{k+1} M_x^2 - \frac{k-1}{k+1} \right)^{1/(k-1)} \tag{5.23}$$

5.4.11 Entropy change across a normal shock wave

The entropy change for a perfect gas may be given as

$$s_y - s_x = c_p \ln \left(\frac{T_y}{T_x} \right) - R \ln \left(\frac{p_y}{p_x} \right)$$

Also, for a perfect gas, $R = (k-1)c_p/k$. Then

$$s_y - s_x = c_p \ln \left(\frac{T_y / T_x}{(p_y / p_x)^{(k-1)/k}} \right)$$

or

$$s_y - s_x = c_p \ln \left(\frac{(T_y / T_{0y})(T_{0y} / T_{0x})(T_{0x} / T_x)}{[(p_y / p_{0y})(p_{0y} / p_{0x})(p_{0x} / p_x)]^{(k-1)/k}} \right)$$

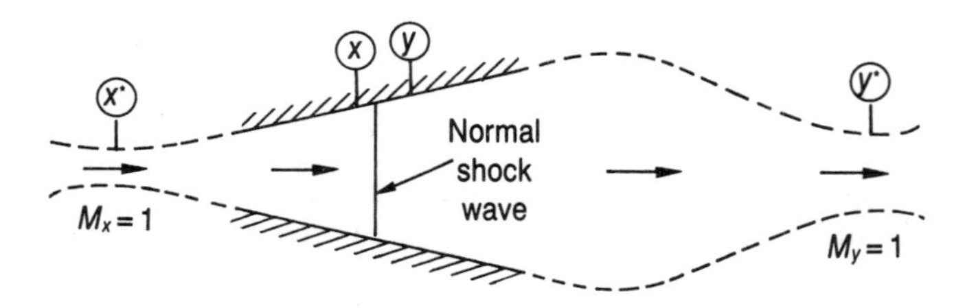

Figure 5.9 Critical areas upstream and downstream of a normal shock wave

Now, noting that $T_y/T_{0y} = (p_y/p_{0y})^{(k-1)/k}$ and $T_x/T_{0x} = (p_x/p_{0x})^{(k-1)/k}$, the above equation becomes

$$s_y - s_x = c_p \ln \left(\frac{T_{0y}/T_{0x}}{(p_{0y}/p_{0x})^{(k-1)/k}} \right)$$

But across a normal shock wave the stagnation temperature is constant, that is $T_{0x} = T_{0y}$. Also, for a perfect gas $c_p = kR/(k-1)$. Thus

$$\frac{s_y - s_x}{R} = -\ln\left(\frac{p_{0y}}{p_{0x}}\right)$$

Finally, substituting for p_{0y}/p_{0x} from Equation (5.22) yields

$$\frac{s_y - s_x}{R} = \frac{k}{k-1} \ln\left(\frac{2}{(k+1)M_x^2} + \frac{k-1}{k+1}\right) + \frac{1}{k-1} \ln\left(\frac{2k}{k+1} M_x^2 - \frac{k-1}{k+1}\right) \tag{5.24}$$

A careful study of Equation (5.24) indicates that for gases with $1 \leqslant k \leqslant 1.67$, the entropy change is always positive when M_x is greater than unity, and always negative when M_x is less than unity. The general form of Equation (5.24) is shown graphically in Figure 5.10. Therefore, for a perfect gas, the shock proceeds only from supersonic flow to a subsonic one.

5.4.12 The Rankine–Hugoniot relation

The equation relating the pressure and density ratios is known as the

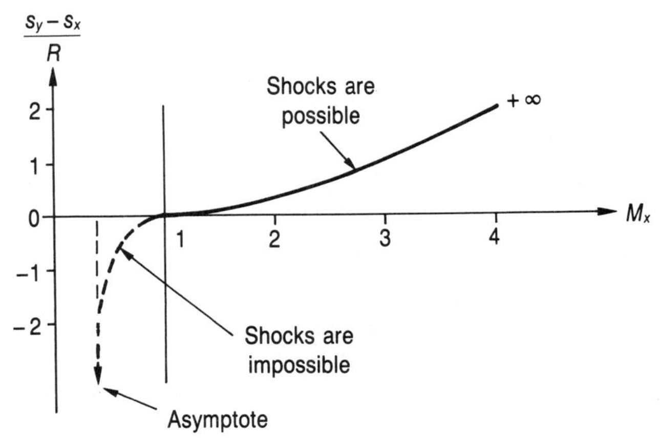

Figure 5.10 Entropy change across a normal shock wave

Rankine–Hugoniot relation. Combining Equations (5.17) and (5.20), and also eliminating M_x^2, yields

$$\frac{p_y}{p_x} = \frac{[(k+1)/(k-1)](\rho_y/\rho_x) - 1}{[(k+1)/(k-1)] - (\rho_y/\rho_x)} \tag{5.25}$$

However, the isentropic relation between the pressure and density ratios may be given as

$$\frac{p_y}{p_x} = \left(\frac{\rho_y}{\rho_x}\right)^k$$

The relations between the pressure and density ratios across a normal shock wave and for isentropic flow are shown in Figure 5.11. Across a weak shock, the pressure rise is small, and then these curves indicate that **weak shocks are nearly isentropic.**

5.4.13 Strength of a normal shock wave

The strength of a normal shock wave, $\mathbb{P}$, may be defined as the ratio of the pressure increase to the initial pressure, that is

$$\mathbb{P} = \frac{p_y - p_x}{p_x} = \frac{p_y}{p_x} - 1 \tag{5.26}$$

Now, eliminating the pressure ratio by using Equation (5.17) yields

$$\mathbb{P} = \frac{2k}{k+1}(M_x^2 - 1) \tag{5.27}$$

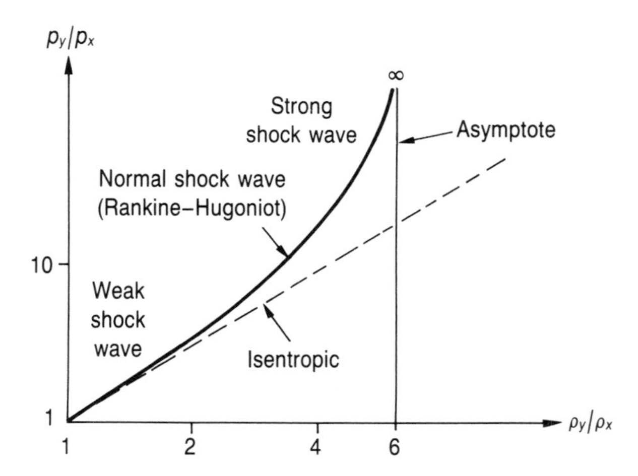

Figure 5.11 Rankine–Hugoniot curve for a perfect gas with $k = 1.4$

Also, the pressure ratio in Equation (5.26) may be eliminated by using the Rankine–Hugoniot relation (5.25). Then

$$\mathbb{P} = \frac{[2k/(k-1)](\rho_y/\rho_x - 1)}{[2/(k-1)] - (\rho_y/\rho_x - 1)} \tag{5.28}$$

If the normal shock wave is relatively weak, then the pressure rise across it is quite small. Therefore, the strength of the normal shock wave approaches zero, as indicated by Equation (5.26). Also, from Equation (5.27), it may be observed that the value of M_x approaches unity. Furthermore, from Equations (5.15b) and (5.28), M_y and the density ratio approach unity, respectively. Since M_x is unity for such a shock wave, the shock wave of infinitesimal strength travels at the speed of sound with respect to the fluid in which it is propagating. For this reason, a shock of infinitesimal strength is identical to a sound wave, and it is the limiting case of a normal shock. This same conclusion was reached in Chapter 3.

5.5 WORKING CHART AND WORKING TABLE FOR FLOW ACROSS A NORMAL SHOCK WAVE

The formulas derived thus far introduce quite extensive numerical calculations, and their solutions often involve a trial-and-error procedure. The numerical computations may be facilitated by introducing a working chart and a working table. The following equations are the property ratios for flow across a normal shock wave:

$$M_y = \sqrt{\left(\frac{(k-1)M_x^2 + 2}{2kM_x^2 - (k-1)}\right)} \tag{5.15b}$$

$$\frac{T_y}{T_x} = \left(1 + \frac{k-1}{2} M_x^2\right)\left(\frac{2k}{k-1} M_x^2 - 1\right) \Big/ \left(\frac{(k+1)^2}{2(k-1)} M_x^2\right) \tag{5.18}$$

$$\frac{p_y}{p_x} = \frac{2k}{k+1} M_x^2 - \frac{k-1}{k+1} \tag{5.17}$$

$$\frac{\rho_y}{\rho_x} = \frac{(k+1)M_x^2}{2 + (k-1)M_x^2} \tag{5.20}$$

$$\frac{V_y}{V_x} = \frac{2 + (k-1)M_x^2}{(k+1)M_x^2} \tag{5.21}$$

$$\frac{p_{0y}}{p_{0x}} = \frac{A_x^*}{A_y^*} = \left(\frac{\frac{1}{2}(k+1)M_x^2}{1 + \frac{1}{2}(k-1)M_x^2}\right)^{k/(k-1)} \left(\frac{2k}{k+1} M_x^2 - \frac{k-1}{k+1}\right)^{1/(1-k)} \tag{5.22}$$

$$\frac{s_y - s_x}{R} = \frac{k}{k-1} \ln\left(\frac{2}{(k+1)M_x^2} + \frac{k-1}{k+1}\right) + \frac{1}{k-1} \ln\left(\frac{2k}{k+1} M_x^2 - \frac{k-1}{k+1}\right) \tag{5.24}$$

5.5.1 Working chart for flow across a normal shock wave

The right-hand sides of the above equations are all functions of the upstream Mach number, M_x, and the specific heat ratio, k. The dimensionless ratios on the left-hand sides of these equations can be represented in graphical form with the upstream Mach number as the independent variable for a fixed value of the specific heat ratio. Such

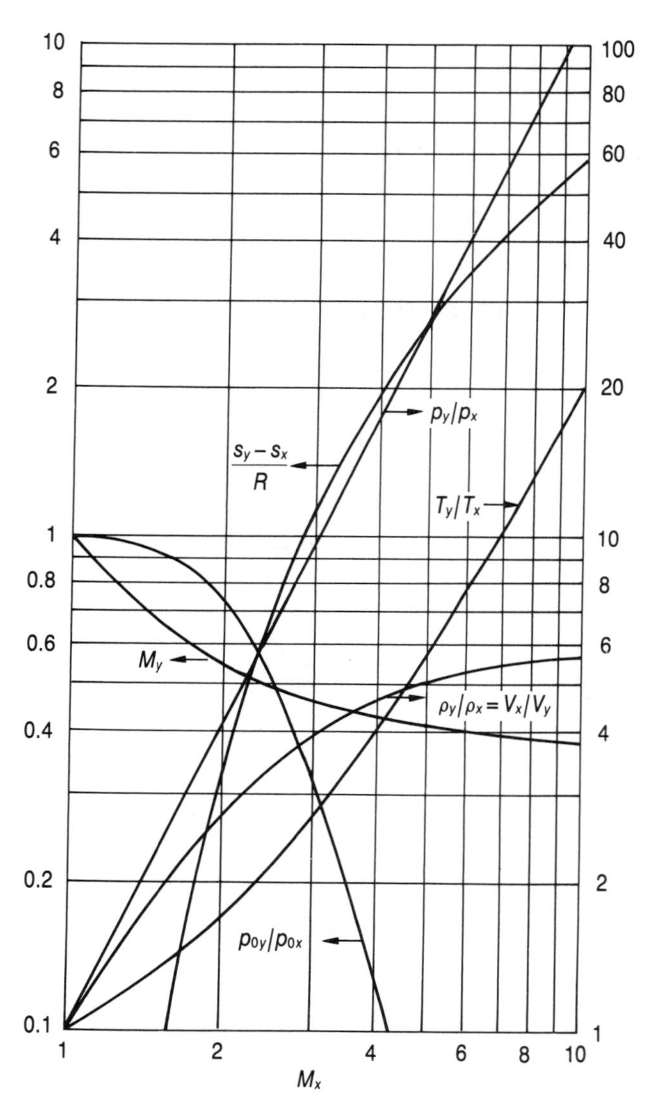

Figure 5.12 Working chart for the flow across a normal shock wave with $k = 1.4$

a working chart for the flow of a perfect gas across a normal shock wave is presented in Figure 5.12 for $k = 1.4$.

5.5.2 Working table for flow across a normal shock wave

The dimensionless ratios can also be represented in tabular form with the upstream Mach number as the independent variable. For $k = 1.4$, the values of these non-dimensional ratios are given in Appendix D as a function of the upstream Mach number.

Example 5.1

An airstream with a velocity of 500 m s^{-1}, a static pressure of 70 kPa and a static temperature of 300 K undergoes a normal shock, as shown in Figure 5.13. Determine

a the Mach number and the velocity after the normal shock wave,
b the static conditions after the normal shock wave,
c the stagnation conditions after the normal shock wave, and
d the entropy change across the normal shock wave.

Solution

(a) The speed of sound upstream of the normal shock wave, from Equation (3.7a), is

$$a_x = \sqrt{(kRT_x)} = \sqrt{[(1.4)(287.1\ \mathrm{J\,kg^{-1}\,K^{-1}})(300\ \mathrm{K})]} = 347.3\ \mathrm{m\,s^{-1}}$$

Then the upstream Mach number may be evaluated as

$$M_x = \frac{V_x}{a_x} = \frac{500\ \mathrm{m\,s^{-1}}}{347.3\ \mathrm{m\,s^{-1}}} = 1.440$$

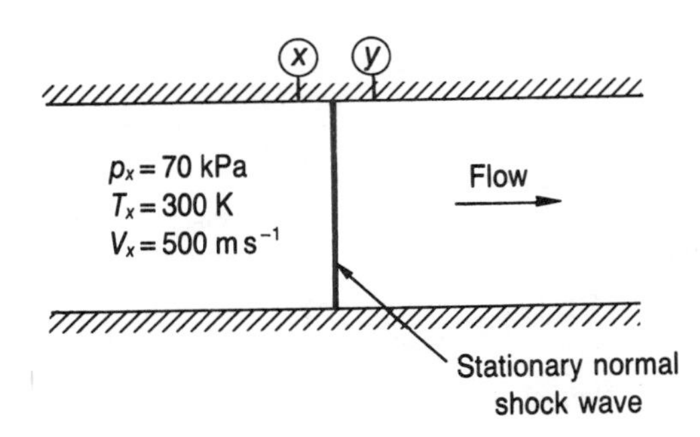

Figure 5.13 Sketch for Example 5.1

From Appendix D, corresponding to $M_x = 1.44$,

$$M_y = \mathbf{0.7235}$$

$$V_x/V_y = 1.759 \quad \text{or} \quad V_y = 500\ \text{m s}^{-1}/1.759 = \mathbf{284.3\ m\,s^{-1}}$$

(b) Again from Appendix D, corresponding to $M_x = 1.44$,

$$p_y/p_x = 2.253 \quad \text{or} \quad p_y = (2.253)(70\ \text{kPa}) = \mathbf{157.7\ kPa}$$

$$T_y/T_x = 1.281 \quad \text{or} \quad T_y = (1.281)(300\ \text{K}) = \mathbf{384.3\ K}$$

(c) The stagnation pressure and the stagnation temperature upstream of the normal shock wave may be calculated from Appendix C, corresponding to $M_x = 1.44$:

$$p_x/p_{0x} = 0.2969 \quad \text{or} \quad p_{0x} = 70\ \text{kPa}/0.2969 = 235.8\ \text{kPa}$$

$$T_x/T_{0x} = 0.7069 \quad \text{or} \quad T_{0x} = 300\ \text{K}/0.7069 = 424.4\ \text{K}$$

Then, the stagnation pressure downstream of the normal shock wave corresponding to $M_x = 1.44$ is

$$p_{0y}/p_{0x} = 0.9476 \quad \text{or} \quad p_{0y} = (235.8\ \text{kPa})(0.9476) = \mathbf{223.4\ kPa}$$

Since the flow is adiabatic across a normal shock wave, then

$$T_{0y} = T_{0x} = \mathbf{424.4\ K}$$

(d) The entropy change across the normal shock wave may be evaluated from Appendix D, corresponding to $M_x = 1.44$;

$$\frac{s_y - s_x}{R} = 0.053\,77 \text{ or } s_y - s_x = (0.053\,77)(287.1\ \text{J kg}^{-1}\,\text{K}^{-1}) = \mathbf{15.44\ J\,kg^{-1}\,K^{-1}}$$

Example 5.2

An airstream with a Mach number of 2.0, a pressure of 150 kPa and a temperature of 350 K enters a diverging channel (Figure 5.14). If the ratio of the

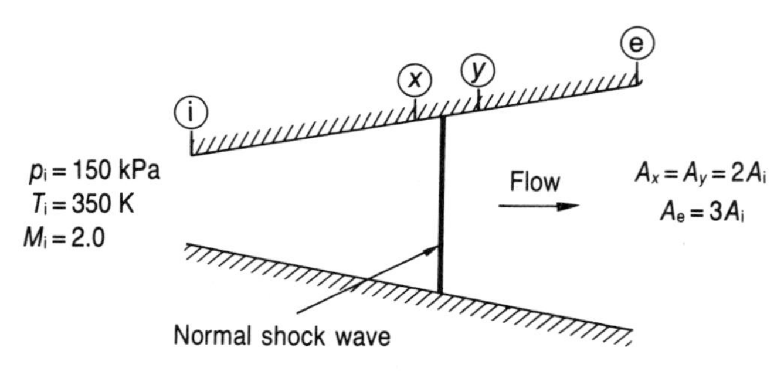

Figure 5.14 Sketch for Example 5.2

exit cross-sectional area to the inlet cross-sectional area is 3.0, determine the back pressure which is necessary to produce a normal shock wave in the channel with a cross-sectional area equal to twice the inlet cross-sectional area. Assume steady, one-dimensional isentropic flow except through the normal shock wave.

Solution

From Appendix C, at $M_i = 2.0$,

$$A_i/A_x^* = 1.687 \quad \text{and} \quad p_i/p_{0x} = 0.1278$$

Then the area ratio, A_x/A_x^*, upstream of the normal shock wave may be evaluated as

$$\frac{A_x}{A_x^*} = \frac{A_x}{A_i}\frac{A_i}{A_x^*} = (2.0)(1.687) = 3.374$$

and corresponding to this area ratio, from Appendix C,

$$M_x = 2.76$$

Now, the properties downstream of the normal shock wave may be evaluated using Appendix D with $M_x = 2.76$:

$$A_x^*/A_y^* = p_{0y}/p_{0x} = 0.4028$$

Hence

$$\frac{A_e}{A_y^*} = \frac{A_e}{A_i}\frac{A_i}{A_x^*}\frac{A_x^*}{A_y^*} = (3.0)(1.687)(0.4028) = 2.039$$

Corresponding to the area ratio of $A_e/A_y^* = 2.039$, there exist two solutions for the Mach number: one for subsonic flow with $M = 0.30$, and the other for supersonic flow with $M = 2.22$. But because the flow after the shock wave must be subsonic, then

$$M_e = 0.30$$

Now from Appendix C, corresponding to $M_e = 0.30$,

$$\frac{p_e}{p_{0y}} = 0.9395$$

Then

$$\frac{p_e}{p_i} = \frac{p_e}{p_{0y}}\frac{p_{0y}}{p_{0x}}\frac{p_{0x}}{p_i} = (0.9395)(0.4028)\frac{1}{(0.1278)} = 2.961$$

or

$$p_e = (2.961)(150 \text{ kPa}) = 444.2 \text{ kPa}$$

With subsonic flow at the channel exit, the channel back pressure is equal to the exit plane pressure. Therefore

$$p_b = p_e = \mathbf{444.2\ kPa}$$

5.6 MOVING NORMAL SHOCK WAVES

In the previous sections, stationary normal shock waves were discussed. However, many physical situations arise in which a normal shock wave is moving. When an explosion occurs, a shock wave propagates through the atmosphere from the point of the explosion. When a valve in a gas line is suddenly closed, the shock wave propagates back through the gas. To treat these cases, it is necessary to extend procedures that have already been developed for the stationary normal shock wave.

Consider a normal shock wave moving in a stationary gas at a constant velocity, V_s, with respect to a fixed observer. The conditions before and after the moving normal shock wave are denoted by subscripts b and a, respectively. Then, V_a represents the velocity of gases behind the shock wave with respect to a fixed observer, as shown in Figure 5.15(a).

Now, consider the same physical situation with respect to an observer moving at the shock wave velocity. In this case, the shock is stationary with respect to the observer, as shown in Figure 5.15(b), and this is exactly the same case that was discussed in the previous sections. To apply these results to a moving shock, consideration must be given to the effect of the observer's velocity on the static and stagnation properties.

Static properties are defined as those measured with an instrument which is moving

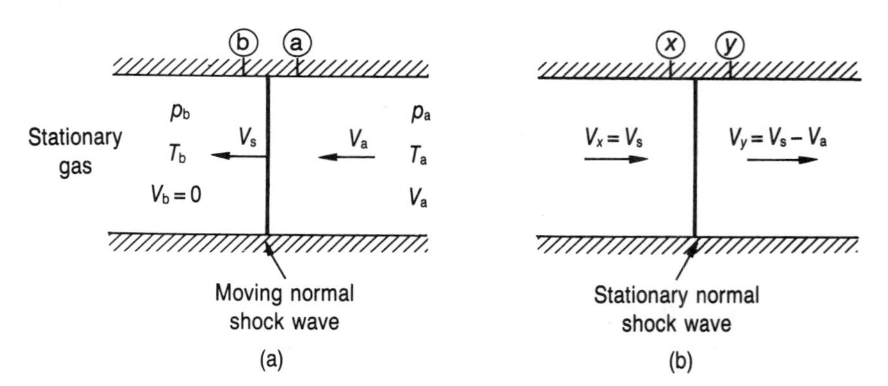

Figure 5.15 Transformation of a moving normal shock wave into a stationary normal shock wave

at the absolute flow velocity. Therefore, **static properties are independent of the observer's velocity**. Hence,

$$p_x = p_\mathrm{b} \tag{5.29a}$$

$$p_y = p_\mathrm{a} \tag{5.29b}$$

and

$$T_x = T_\mathrm{b} \tag{5.30a}$$

$$T_y = T_\mathrm{a} \tag{5.30b}$$

The velocity and the Mach number before and after the stationary normal shock wave may be given as

$$V_x = V_\mathrm{s} \tag{5.31a}$$

$$V_y = V_\mathrm{s} - V_\mathrm{a} \tag{5.31b}$$

and

$$M_x = \frac{V_x}{a_x} = \frac{V_x}{\sqrt{(kRT_x)}} = \frac{V_\mathrm{s}}{\sqrt{(kRT_\mathrm{b})}} \tag{5.32a}$$

$$M_y = \frac{V_y}{a_y} = \frac{V_y}{\sqrt{(kRT_y)}} = \frac{V_\mathrm{s} - V_\mathrm{a}}{\sqrt{(kRT_\mathrm{a})}} \tag{5.32b}$$

Hence, **stagnation properties** are measured by bringing the flow to rest with respect to the observer, and therefore they are **dependent on the observer's velocity**. For this reason, the stagnation properties in each reference frame must be evaluated by using the isentropic relations. At this point, it is worth while to note that there is a change in the stagnation temperature across a moving normal shock wave, and this change may be expressed by using the definition of the stagnation temperature (4.8a) as

$$T_{0\mathrm{a}} - T_{0\mathrm{b}} = T_\mathrm{a} - T_\mathrm{b} + \frac{V_\mathrm{a}^2}{2c_\mathrm{p}} - \frac{V_\mathrm{b}^2}{2c_\mathrm{p}}$$

Using Equations (5.30) and considering that $V_\mathrm{b} = 0$, the above expression becomes

$$T_{0\mathrm{a}} - T_{0\mathrm{b}} = T_y - T_x + \frac{V_\mathrm{a}^2}{2c_\mathrm{p}}$$

But across a stationary normal shock wave the stagnation temperature is constant, and with the aid of Equation (4.8a) it is possible to write

$$T_{0x} = T_x + \frac{V_x^2}{2c_p} = T_y + \frac{V_y^2}{2c_p} = T_{0y}$$

or using Equations (5.31)

$$T_y - T_x = \frac{V_s^2}{2c_p} - \frac{(V_s - V_a)^2}{2c_p}$$

Therefore, the change in the stagnation temperature across a moving shock wave is

$$T_{0a} - T_{0b} = \frac{V_s V_a}{c_p} \tag{5.33}$$

The following general procedure may be applied during the analysis of moving-shock-wave problems:

1 Transform the known static fluid properties and velocities into a reference frame for the equivalent stationary shock wave.
2 Solve for the unknown static fluid properties and velocities across the stationary shock wave.
3 Transform the evaluated static fluid properties and velocities back into the reference frame for the moving shock wave.

In the following examples, two cases dealing with moving shock waves are considered. In the first one, the shock velocity is known, and it is required to find the gas velocity and other properties behind the moving wave. In the second one, the gas velocity behind the wave is known, and it is required to determine the shock velocity.

Example 5.3
A normal shock wave moves at a constant velocity of 500 m s^{-1} into still air with a temperature and pressure of 300 K and 100 kPa, respectively (Figure 5.16).

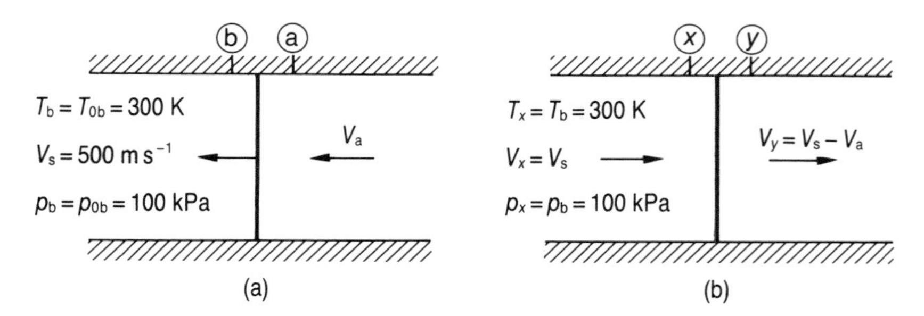

Figure 5.16 Sketch for Example 5.3: (a) for a fixed observer; (b) for an observer moving with the normal shock wave

Determine the static and stagnation conditions present in the air after the passage of the wave, as well as the gas velocity behind the wave.

Solution

For a fixed observer, the physical situation is as shown in Figure 5.16(a). To apply the stationary normal shock-wave equations, the situation transforms to that shown in Figure 5.16(b) for an observer moving with the normal shock wave. Then

$$V_x = V_s = 500 \text{ m s}^{-1}$$

Since the static properties are independent of the observer's velocity, then

$$p_x = p_b = 100 \text{ kPa}$$

and

$$T_x = T_b = 300 \text{ K}$$

The Mach number upstream of the stationary shock wave is then

$$M_x = \frac{V_x}{\sqrt{(kRT_x)}} = \frac{500 \text{ m s}^{-1}}{\sqrt{[(1.4)(287.1 \text{ J kg}^{-1}\text{ K}^{-1})(300 \text{ K})]}} = 1.440$$

The downstream conditions may now be evaluated by using Appendix D with $M_x = 1.44$:

$$M_y = 0.7235$$

$$p_y/p_x = 2.253 \quad \text{or} \quad p_y = (2.253)(100 \text{ kPa}) = 225.3 \text{ kPa}$$

$$T_y/T_x = 1.281 \quad \text{or} \quad T_y = (1.281)(300 \text{ K}) = 384.3 \text{ K}$$

The gas velocity behind the moving shock wave may be evaluated from

$$M_y = \frac{V_s - V_a}{\sqrt{(kRT_y)}} \quad \text{or} \quad V_a = V_s - M_y\sqrt{(kRT_y)}$$

as

$$V_a = 500 \text{ m s}^{-1} - (0.7235)\sqrt{[(1.4)(287.1 \text{ J kg}^{-1}\text{ K}^{-1})(384.3 \text{ K})]} = \mathbf{215.7 \text{ m s}^{-1}}$$

The static conditions behind the moving shock wave are identical to those downstream of the stationary shock wave. Therefore

$$p_a = p_y = \mathbf{225.3 \text{ kPa}}$$

and

$$T_a = T_y = \mathbf{384.3 \text{ K}}$$

To determine the stagnation conditions after the moving shock wave, the Mach number after the wave must be evaluated as

$$M_a = \frac{V_a}{\sqrt{(kRT_a)}} = \frac{215.7 \text{ m s}^{-1}}{\sqrt{[(1.4)(287.1 \text{ J kg}^{-1} \text{ K}^{-1})(384.3 \text{ K})]}} = 0.5488$$

Then, using Appendix C, for $M_a = 0.55$:

$$p_a/p_{0a} = 0.8142 \quad \text{or} \quad p_{0a} = 225.3 \text{ kPa}/0.8142 = \mathbf{276.7 \text{ kPa}}$$

$$T_a/T_{0a} = 0.9430 \quad \text{or} \quad T_{0a} = 384.3 \text{ K}/0.9430 = \mathbf{407.5 \text{ K}}$$

Note that for a fixed observer, the stagnation temperature after the passage of the shock wave is greater than that before the passage of the shock wave. However, for an observer moving with the shock wave, there is no change in the stagnation temperature across the shock wave.

Example 5.4

A piston confined in a tube is suddenly accelerated to a velocity of 50 m s^{-1}, which causes a normal shock wave to move into the air at rest in the tube. Calculate the velocity of the normal shock wave if the temperature of the stationary air is 27 °C.

Solution

The air next to the piston must move at the same velocity as the piston, as shown in Figure 5.17(a). For an observer moving with the normal shock wave, the situation is as shown in Figure 5.17(b). Since the static properties are independent of the observer's velocity, then

$$T_x = T_b = 300 \text{ K}$$

The speed of sound upstream of the stationary shock wave is then

$$a_x = \sqrt{(kRT_x)} = \sqrt{[(1.4)(287.1 \text{ J kg}^{-1} \text{ K}^{-1})(300 \text{ K})]} = 347.3 \text{ m s}^{-1}$$

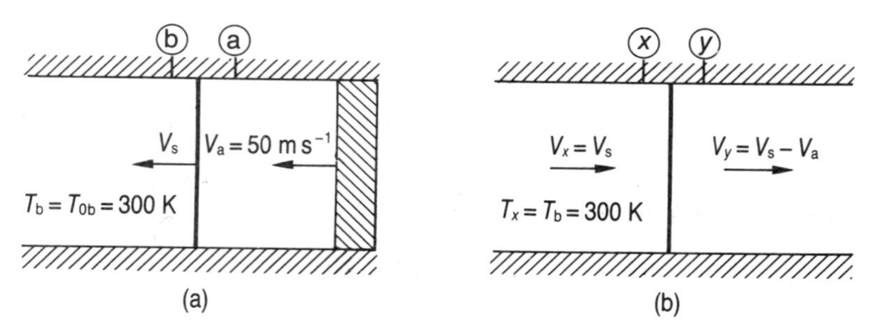

Figure 5.17 Sketch for Example 5.4: (a) for a fixed observer; (b) for an observer moving with the normal shock wave

and the upstream Mach number is

$$M_x = \frac{V_x}{a_x} = \frac{V_s}{347.3 \text{ m s}^{-1}}$$

Now, from Equation (5.21),

$$\frac{V_y}{V_x} = \frac{V_s - V_a}{V_s} = \frac{V_s - 50}{V_s} = \frac{2 + (k-1)M_x^2}{(k+1)M_x^2}$$

$$= \frac{2 + [(1.4-1)V_s^2/(347.3)^2]}{(1.4+1)V_s^2/(347.3)^2}$$

or rearranging,

$$2V_s^2 - 120V_s - 2.412 \times 10^5 = 0$$

The solution of this quadratic equation gives two solutions for V_s: 378.6 m s^{-1} and −318.6 m s^{-1}. The negative value of V_s has no physical meaning in this problem. Therefore,

$$V_s = \mathbf{378.6 \text{ m s}^{-1}}$$

5.7 REFLECTED NORMAL SHOCK WAVES

To complete the study of moving normal shock waves, consider the result of a wave impinging on the end of a tube. Two cases will be studied: a closed tube and a tube open to the atmosphere.

Consider a normal shock wave moving into a stationary gas at a constant velocity, V_{si}, in a closed tube, as shown in Figure 5.18(a). The gas behind the incident shock wave moves to the right at a velocity of V_g. When the shock wave impinges on the closed end of the tube, a wave will be reflected back into the gas at a velocity V_{sr}, as shown in Figure 5.18(b). The gas behind the reflected wave must be stationary, since the closed end of the tube is not moving. For an observer moving with the

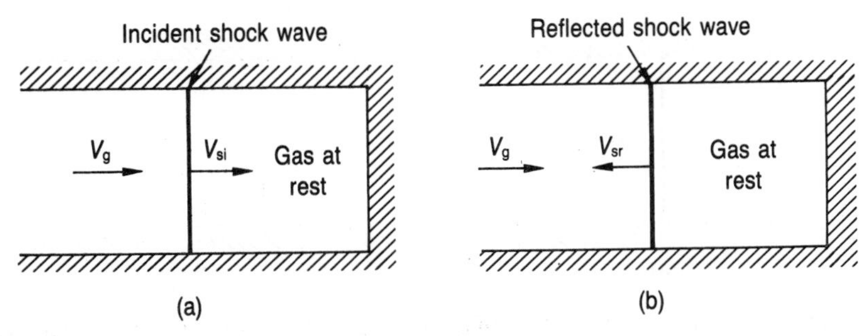

Figure 5.18 A normal shock wave reflecting from the closed end of a tube

reflected wave, the physical situation is as shown in Figure 5.19, which indicates a decrease in velocity across the reflected wave. Therefore, **a normal shock wave reflects back from the closed end of a tube as a normal shock wave.**

In Figure 5.20(a) a normal shock wave is moving into a stationary gas at a constant velocity, V_{si}, in a tube which is exposed to the atmosphere. The pressure of the gas ahead of the incident shock wave is at atmospheric pressure, while the pressure behind this wave is greater than atmospheric pressure. As the shock wave impinges on the open end of the tube, a wave will be reflected back into the gas at a velocity V_{wr}. The pressure of the gas behind the reflected wave is also atmospheric, since it is exposed to the atmospheric conditions, as shown in Figure 5.20(b). However, the pressure ahead of this reflected wave is greater than atmospheric pressure. Therefore, **a normal shock wave reflects back from the open end of a tube as a series of expansion waves.**

Example 5.5

A normal shock wave with a pressure ratio of 4.5 impinges on a plane wall, as shown in Figure 5.21. Determine the static pressure and temperature behind the

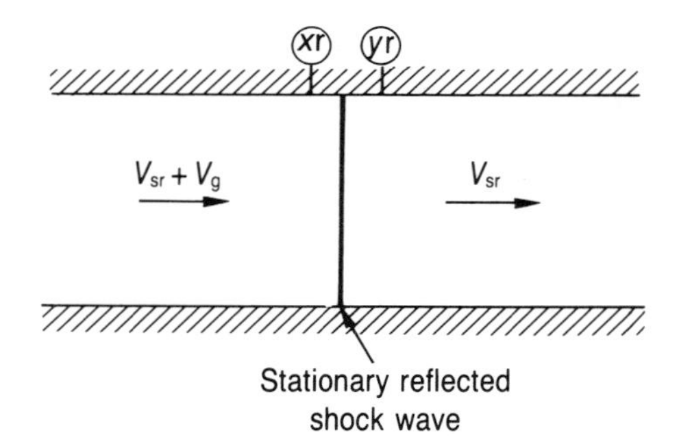

Figure 5.19 Physical situation for an observer moving with the reflected shock wave

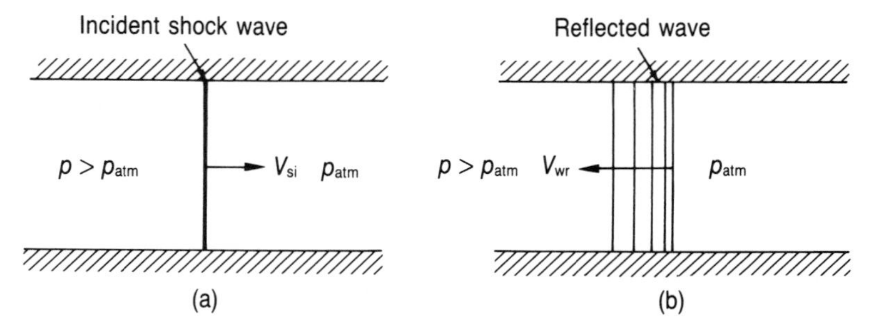

Figure 5.20 A normal shock wave reflecting from the open end of a tube which is exposed to the atmosphere

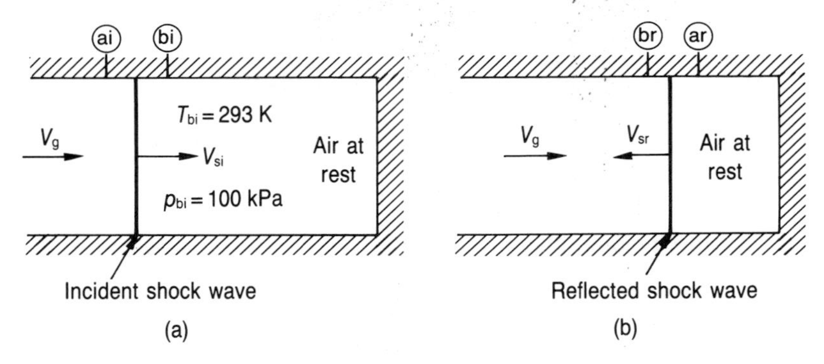

Figure 5.21 Sketch for Example 5.5

reflected normal shock wave, if the air pressure and temperature in front of the incident wave are 100 kPa and 20 °C, respectively.

Solution

To apply the stationary normal shock-wave equations, the situation is transformed to that shown in Figures 5.22(a) and 5.22(b) for observers moving with the incident and reflected shock waves, respectively. Since the static properties are independent of the observer's velocity

$$T_{xi} = T_{bi} = 293 \text{ K}$$

$$p_{xi} = p_{bi} = 100 \text{ kPa}$$

and

$$p_{yi}/p_{xi} = 4.5 \quad \text{or} \quad p_{yi} = (4.5)(100 \text{ kPa}) = 450 \text{ kPa}$$

The speed of sound before the incident shock wave is

$$a_{xi} = \sqrt{(kRT_{xi})} = \sqrt{[(1.4)(287.1 \text{ J kg}^{-1}\text{ K}^{-1})(293 \text{ K})]} = 343.2 \text{ m s}^{-1}$$

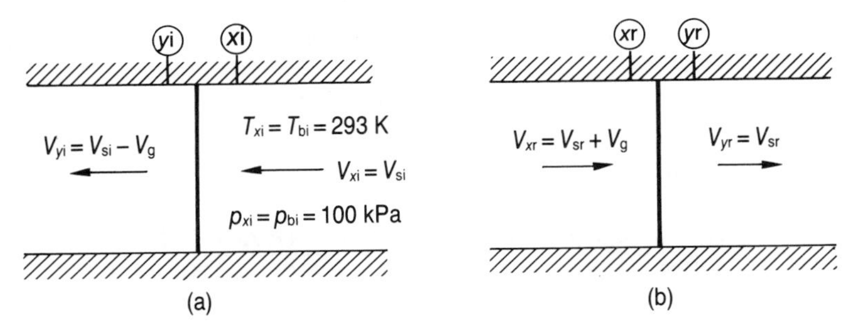

Figure 5.22 Transformation of (a) incident and (b) reflected shock waves into stationary ones, for Example 5.5

From Appendix D, corresponding to $p_{yi}/p_{xi} = 4.5$,

$$M_{xi} = V_{xi}/a_{xi} = 2.00 \quad \text{or} \quad V_{xi} = (2.00)(343.2\ \text{m s}^{-1}) = 686.4\ \text{m s}^{-1}$$

$$T_{yi}/T_{xi} = 1.687 \quad \text{or} \quad T_{yi} = (1.687)(293\ \text{K}) = 494.3\ \text{K}$$

$$V_{xi}/V_{yi} = 2.667 \quad \text{or} \quad V_{yi} = 686.4\ \text{m s}^{-1}/2.667 = 257.4\ \text{m s}^{-1}$$

The velocity of the incident shock wave is

$$V_{si} = V_{xi} = 686.4\ \text{m s}^{-1}$$

Then the velocity of the gas behind the incident shock wave is

$$V_{yi} = V_{si} - V_g \quad \text{or} \quad V_g = 686.4\ \text{m s}^{-1} - 257.4\ \text{m s}^{-1} = 429\ \text{m s}^{-1}$$

As long as the static properties are not affected by the observer's velocity, then the static properties before the reflected shock wave are

$$T_{xr} = T_{br} = T_{yi} = 494.3\ \text{K}$$

and

$$p_{xr} = p_{br} = p_{yi} = 450\ \text{kPa}$$

From Figure 5.22(b) the velocities before and after the stationary reflected shock wave are

$$V_{xr} = V_{sr} + V_g = V_{sr} + 429\ \text{m s}^{-1}$$

and

$$V_{yr} = V_{sr}$$

respectively. The speed of sound upstream of the stationary reflected wave is

$$a_{xr} = \sqrt{(kRT_{xr})} = \sqrt{[(1.4)(287.1\ \text{J kg}^{-1}\ \text{K}^{-1})(494.3\ \text{K})]} = 445.7\ \text{m s}^{-1}$$

and the Mach number before the reflected wave is

$$M_{xr} = \frac{V_{xr}}{a_{xr}} = \frac{V_{sr} + 429\ \text{m s}^{-1}}{445.7\ \text{m s}^{-1}}$$

Now, from Equation (5.21),

$$\frac{V_{yr}}{V_{xr}} = \frac{V_{sr}}{V_{sr} + 429} = \frac{2 + (k-1)M_{xr}^2}{(k+1)M_{xr}^2} = \frac{2 + [(1.4-1)(V_{sr} + 429)^2/(445.7)^2]}{(1.4+1)(V_{sr} + 429)^2/(445.7)^2}$$

or rearranging,

$$2V_{sr}^2 + 686.4V_{sr} - 4.709 \times 10^5 = 0$$

The solution of this quadratic equation gives two solutions for V_{sr}: 343.1 m s^{-1} and -686.3 m s^{-1}. The negative value of V_{sr} has no physical meaning in this problem. Therefore

$$V_{sr} = 343.1 \text{ m s}^{-1}$$

and

$$M_{xr} = \frac{V_{sr} + 429}{445.7} = \frac{343.1 \text{ m s}^{-1} + 429 \text{ m s}^{-1}}{445.7 \text{ m s}^{-1}} = 1.732$$

Now, from Appendix D, for $M_{xr} = 1.73$,

$$p_{yr}/p_{xr} = 3.325 \quad \text{or} \quad p_{yr} = (3.325)(450 \text{ kPa}) = 1496 \text{ kPa}$$

$$T_{yr}/T_{xr} = 1.480 \quad \text{or} \quad T_{yr} = (1.480)(494.3 \text{ K}) = 731.6 \text{ K}$$

Since the static properties are independent of the observer's velocity, then

$$p_{ar} = p_{yr} = \mathbf{1496 \text{ kPa}}$$

and

$$T_{ar} = T_{yr} = \mathbf{731.6 \text{ K}}$$

5.8 NON-ISENTROPIC FLOW REGIME IN CONVERGING–DIVERGING NOZZLES

After studying normal shock waves, it is possible to complete the discussion of flow in converging–diverging nozzles operating under varying back pressures, which were previously discussed in Section 4.8.2.

Flow pattern 4 of Figure 4.27(b) is a typical representation of the non-isentropic flow regime ($p_{e5y} < p_b < p_{e3}$). If the back pressure is lowered below that of flow pattern 3 of Figure 4.27(b), a normal shock wave appears downstream of the throat. The flow in the diverging portion of the nozzle is supersonic upstream of the normal shock. Since the flow is subsonic after the shock, the flow decelerates with an accompanying increase in pressure through the diverging portion of the nozzle. In this regime, the flow is choked at the throat where the Mach number is unity, and the mass flow rate is insensitive to the changes in the back pressure. As long as the flow is subsonic at the exit plane, the exit pressure is identical to the back pressure, and the jet leaves the nozzle as a cylindrical parallel stream. As the back pressure is lowered, the shock moves down the nozzle until it appears at the exit plane of the nozzle. This situation is known as the exit plane shock condition, as indicated by flow pattern 5 of Figure 4.27(b).

Example 5.6

At a point upstream of the throat in a converging–diverging nozzle, the velocity, pressure and temperature of the air are 172 m s^{-1}, 200 kPa and 22 °C, respectively. If the throat and exit cross-sectional areas of the nozzle are 0.01 m^2 and 0.020 37 m^2, respectively, determine

a the maximum back pressure to choke the nozzle,
b the back pressure for the nozzle to be perfectly expanded to the design Mach number,
c the back pressure such that there is a normal shock wave at the exit plane, and
d the range of back pressures for subsonic, non-isentropic, overexpansion and underexpansion flow regimes.

Solution

For the converging–diverging nozzle, which is shown in Figure 5.23(a), the speed of sound at the given point is

$$a = \sqrt{(kRT)} = \sqrt{[(1.4)(287.1\ \text{J kg}^{-1}\,\text{K}^{-1})(295\ \text{K})]} = 344.3\ \text{m s}^{-1}$$

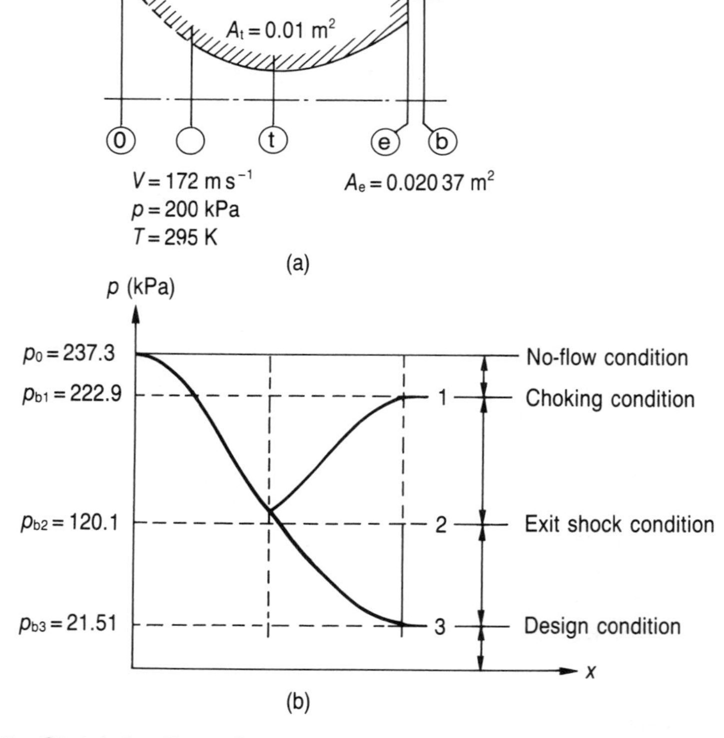

Figure 5.23 Sketch for Example 5.6

and the Mach number is

$$M = \frac{V}{a} = \frac{172 \text{ m s}^{-1}}{344.3 \text{ m s}^{-1}} = 0.4996$$

Then the stagnation pressure may be evaluated using Appendix C corresponding to $M = 0.50$ as

$$p/p_0 = 0.8430 \quad \text{or} \quad p_0 = 200 \text{ kPa}/0.8430 = 237.3 \text{ kPa}$$

(a) The maximum back pressure to choke the nozzle is obtained whenever the Mach number is unity at the throat and the flow is entirely subsonic in the diverging section. In this case,

$$A_e/A^* = A_e/A_t = 0.020\,37 \text{ m}^2/0.01 \text{ m}^2 = 2.037$$

Corresponding to the area ratio of $A_e/A^* = 2.037$, there exist two solutions for the Mach number: one for subsonic flow with $M_{e1} = 0.30$ and the other for supersonic flow with $M_{e3} = 2.22$ from Appendix C. But for the maximum back pressure to choke the nozzle, the exit Mach number must be subsonic. Therefore, $M_{e1} = 0.30$. From Appendix C, corresponding to $M_{e1} = 0.30$,

$$p_{e1}/p_0 = 0.9395 \quad \text{or} \quad p_{e1} = (0.9395)(237.3 \text{ kPa}) = 222.9 \text{ kPa}$$

But for subsonic flow at the exit area, the back pressure must be identical to the pressure at the exit plane, that is

$$p_{b1} = p_{e1} = \mathbf{222.9 \text{ kPa}}$$

Flow pattern 1, which corresponds to the choking condition, is shown in Figure 5.23(b).

(b) For the design condition, the Mach number is again unity at the throat and the flow is entirely supersonic in the diverging portion of the nozzle. Then the exit Mach number is $M_{e3} = 2.2$. From Appendix C, corresponding to $M_{e3} = 2.22$,

$$p_{e3}/p_0 = 0.090\,64 \quad \text{or} \quad p_{e3} = (0.090\,64)(237.3 \text{ kPa}) = 21.51 \text{ kPa}$$

For the design condition, the back pressure is equal to the exit plane pressure. Therefore

$$p_{b3} = p_{e3} = \mathbf{21.51 \text{ kPa}}$$

In Figure 5.23(b) the design condition is indicated by flow pattern 3.

(c) When there is a normal shock wave just at the exit plane of the nozzle, the Mach number and the static pressure before the shock wave are $M_{e2x} = 2.22$ and $p_{e2x} = 21.51$ kPa, respectively. Then, from Appendix D, corresponding to $M_{e2x} = 2.22$,

$$p_{e2y}/p_{e2x} = 5.583 \quad \text{or} \quad p_{e2y} = (5.583)(21.51 \text{ kPa}) = 120.1 \text{ kPa}$$

Again, the flow is subsonic at the exit plane. Therefore

$$p_{b2} = p_{e2y} = \mathbf{120.1\ kPa}$$

Flow pattern 2, which corresponds to the exit shock condition, is shown in Figure 5.23(b).

(d) Ranges of back pressure for the regimes are as follows:

1 Subsonic flow regime:

$$p_0 \geqslant p_b \geqslant p_{b1} \quad \text{or} \quad \mathbf{237.3\ kPa \geqslant p_b \geqslant 222.9\ kPa}$$

2 Non-isentropic flow regime:

$$p_{b1} \geqslant p_b \geqslant p_{b2} \quad \text{or} \quad \mathbf{222.9\ kPa \geqslant p_b \geqslant 120.1\ kPa}$$

3 Overexpansion flow regime:

$$p_{b2} \geqslant p_b \geqslant p_{b3} \quad \text{or} \quad \mathbf{120.1\ kPa \geqslant p_b \geqslant 21.51\ kPa}$$

4 Underexpansion flow regime:

$$p_{b3} \geqslant p_b \geqslant 0 \quad \text{or} \quad \mathbf{21.51\ kPa \geqslant p_b \geqslant 0\ kPa}$$

Example 5.7

If the back pressure for the converging–diverging nozzle of Example 5.6 is 170 kPa, determine the cross-sectional area of the diverging portion of the nozzle where the normal shock wave is located.

Solution

A sketch and the nomenclature for Example 5.7 are shown in Figure 5.24(a). The

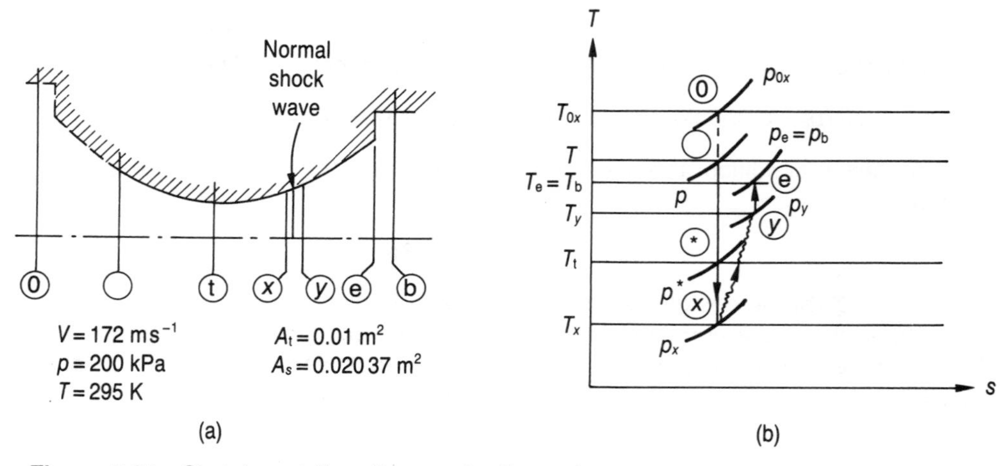

Figure 5.24 Sketch and T–s diagram for Example 5.7

given back pressure is in the non-isentropic flow regime. Therefore, there is a normal shock wave which is located at the diverging portion of the nozzle. The location of this normal shock wave can be found by an iterative process. For this reason, it is necessary to assume a value for the cross-sectional area at which the shock wave is located. As an initial trial, assume

$$A_x = A_y = 0.0115 \text{ m}^2$$

As long as the nozzle is choked, $A_t = A_x^*$. Then

$$A_x/A_x^* = 0.0115 \text{ m}^2/0.01 \text{ m}^2 = 1.15$$

Corresponding to the area ratio of $A_x/A_x^* = 1.15$, the supersonic Mach number from Appendix C is

$$M_x = 1.46$$

and

$$p_x/p_{0x} = 0.2886 \quad \text{or} \quad p_x = (0.2886)(237.3 \text{ kPa}) = 68.48 \text{ kPa}$$

Now, from Appendix D, for $M_x = 1.46$,

$$M_y = 0.7157$$

and

$$p_y/p_x = 2.320 \quad \text{or} \quad p_y = (2.320)(68.48 \text{ kPa}) = 158.9 \text{ kPa}$$

The critical area and the stagnation pressure after the shock may be evaluated by using Appendix C, with $M_y = 0.72$, as

$$A_y/A_y^* = 1.081 \quad \text{or} \quad A_y^* = 0.0115 \text{ m}^2/1.081 = 0.010\,64 \text{ m}^2$$

$$p_y/p_{0y} = 0.7080 \quad \text{or} \quad p_{0y} = 158.9 \text{ kPa}/0.7080 = 224.4 \text{ kPa}$$

Now, the critical area ratio at the exit plane of the nozzle is

$$A_e/A_y^* = 0.020\,37 \text{ m}^2/0.010\,64 \text{ m}^2 = 1.914$$

and from Appendix C for subsonic flows

$$p_e/p_{0y} = 0.9315 \quad \text{or} \quad p_e = (0.9315)(224.4 \text{ kPa}) = 209.0 \text{ kPa}$$

Since the exit plane pressure is greater than the back pressure, the normal shock wave must be located further downstream. The rest of the iterations are shown in Table 5.1. After convergence, the area of the nozzle, where the normal shock wave is located, is

$$A_x = A_y = \mathbf{0.0148 \text{ m}^2}$$

The process is shown in the T–s diagram of Figure 5.24(b). Air is first accelerated isentropically to the sonic Mach number in the converging part of the

Table 5.1 Iterations for Example 5.7

Iteration	1	2	3	4	5
$A_x = A_y$ (m^2)	0.0115	0.0185	0.0150	0.0140	0.0148
A_x/A_x^*	1.15	1.85	1.50	1.40	1.48
M_x	1.46	2.11	1.85	1.76	1.84
p_x/p_{0x}	0.2886	0.1077	0.1612	0.1850	0.1637
p_x (kPa)	68.49	25.56	38.25	43.90	38.85
M_y	0.7157	0.5598	0.6057	0.6257	0.6078
p_y/p_x	2.320	5.027	3.826	3.447	3.783
p_y (kPa)	158.9	128.5	146.4	151.3	147.0
A_y/A_y^*	1.081	1.240	1.177	1.155	1.177
A_y^* (m^2)	0.010 64	0.014 92	0.012 74	0.012 12	0.012 57
p_y/p_{0y}	0.7080	0.8082	0.7778	0.7654	0.7778
p_{0y} (kPa)	224.4	159.0	188.2	197.7	189.0
A_e/A_y^*	1.915	1.365	1.599	1.681	1.621
p_e/p_{0y}	0.9315	0.8486	0.8956	0.9098	0.9004
p_e (kPa)	209.0	134.9	168.6	179.9	170.2
Shock movement	→	←	←	→	OK

nozzle. Then it is accelerated to a supersonic Mach number of 1.84 in the diverging part. There, after an irreversible normal shock wave, the flow is subsonic, and the rest of the diverging portion of the nozzle acts as a diffuser to decelerate the flow.

5.9 NON-ISENTROPIC OPERATION OF SUPERSONIC DIFFUSERS

When the flow at the entrance of a converging–diverging duct is supersonic, the fluid will be slowed down and compressed to a Mach number of unity at the throat. At this Mach number at the throat, the fluid may either expand again to supersonic speeds or be compressed beyond the throat at subsonic speeds. In the latter case, the converging–diverging duct is often referred to as a supersonic diffuser; it is used in wind tunnels with a supersonic test section and in engine intakes. The flow in these devices will be considered by ignoring all losses except those across normal shock waves.

5.9.1 Supersonic diffusers in wind tunnels

A very simple supersonic wind tunnel may be constructed, as shown in Figure 5.25. In this wind tunnel, air is drawn through a converging–diverging nozzle by means of a fan. As the back pressure is lowered, the nozzle first passes through the subsonic and the non-isentropic flow regimes, and a supersonic flow is then obtained in the constant-area test section. Although the construction of such a wind tunnel is very simple, its operation at supersonic speeds may be either quite difficult or even impossible since the fan may not be operated at these high speeds. For this reason, it is necessary to use a supersonic diffuser after the test section in order to decelerate the air to subsonic speeds.

(a) Supersonic diffusers in fixed-geometry wind tunnels

A wind tunnel with a supersonic diffuser is shown in Figure 5.26(a). The air is drawn from a large reservoir where the stagnation temperature, T_0, and the stagnation pressure, p_{0i}, are fixed. The wind tunnel discharges into another reservoir where the back pressure, p_b, is controllable by means of a fan. The possible flow patterns from the start-up of the wind tunnel up to the operating condition are shown in Figure 5.26(b) as the effects of variations in the back pressure on the distribution of the pressure along the wind tunnel.

(i) No-flow condition ($p_b = p_{0i}$)

To begin with, suppose that the fan is not operating, as shown by flow pattern 1 of Figure 5.26(b). In this case, the pressure is constant throughout the wind tunnel such that $p_b = p_{0i}$ and there is no flow.

(ii) Subsonic flow regime ($p_{0i} > p_b > p_{e3}$)

When the fan is operated at low speeds, the flow will be subsonic throughout the wind tunnel, as indicated by flow pattern 2 of Figure 5.26(b). The air is first accelerated with expansion and then decelerated with compression in the converging and diverging parts of both variable-area ducts, respectively.

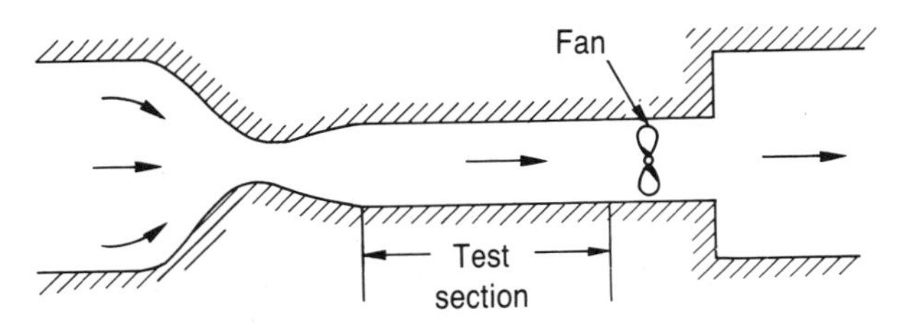

Figure 5.25 A very simple supersonic wind tunnel

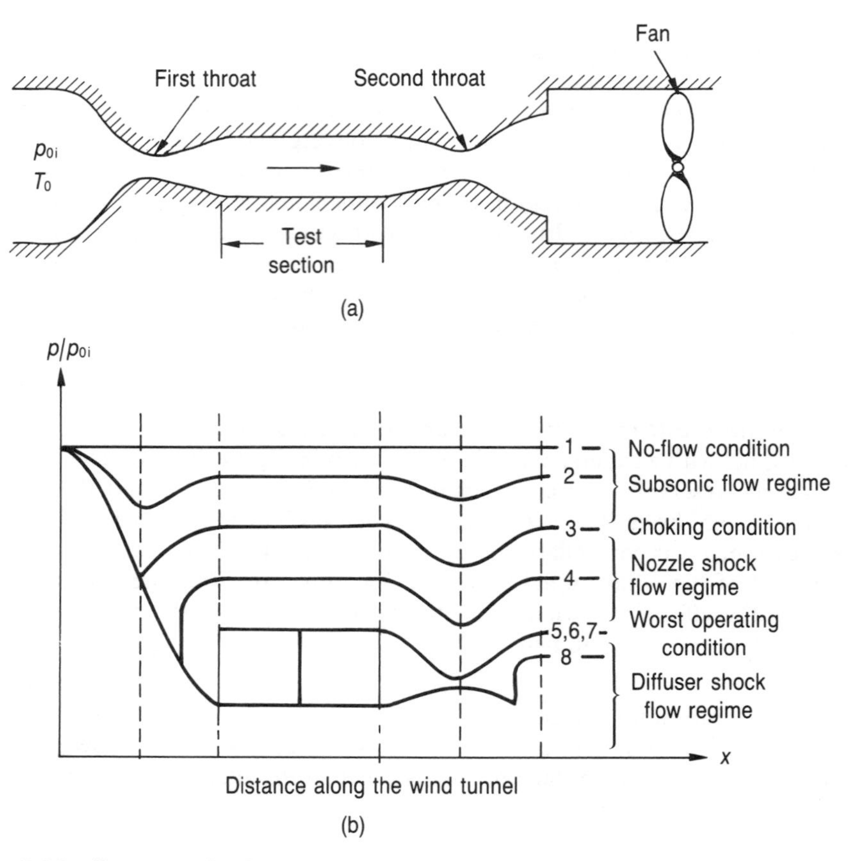

Figure 5.26 Pressure distributions during the start-up of a fixed-geometry supersonic wind tunnel

(iii) Choking condition ($p_b = p_{e3}$)

During the reduction of the back pressure, the flow will first be choked in one of the two throats having the smaller cross-sectional area.

If the cross-sectional area of the second throat is smaller than that of the first, then the flow will certainly be choked at the second throat. In this case, the second converging–diverging duct acts as a supersonic nozzle and passes through its non-isentropic, overexpansion and underexpansion flow regimes when the back pressure is further reduced. However, the flow upstream of the second throat is always subsonic, and a supersonic flow cannot be obtained in the test section.

When the cross-sectional areas of the two throats are the same, one might think that both throats will be choked at the same time. However, this is not the case. As the back pressure is reduced, disturbances proceed towards the second throat. Therefore, the second throat is choked first and the second converging–diverging duct again passes through its non-isentropic, overexpansion and underexpansion flow regimes with further reductions in the back pressure. Hence, the flow will again be subsonic in the test section.

Therefore, the only possible way of obtaining supersonic flow in the test section is to have the first throat with a cross-sectional area smaller than the second one. In this case, the first nozzle will be choked as shown by flow pattern 3 of Figure 5.26(b). However, the flow will again be subsonic in the test section and in the second converging–diverging duct.

(iv) Nozzle shock flow regime ($p_{e3} > p_b > p_{e5}$)

With further reductions in the back pressure, a normal shock wave appears in the diverging part of the first converging–diverging duct, which is a supersonic nozzle. Flow pattern 4 of Figure 5.26(b) is a typical representation of this flow regime. The flow is subsonic downstream of the normal shock wave.

(v) Worst operating condition ($p_b = p_{e5} = p_{e6} = p_{e7}$)

When the back pressure is reduced to p_{e5} the normal shock wave moves to the exit plane of the supersonic nozzle, as indicated by flow pattern 5 of Figure 5.26(b). However, as long as the test section is isentropic, then the normal shock wave can stand neither at the exit plane of the supersonic nozzle nor at the test section, but moves to the inlet of the second converging–diverging duct, as indicated by flow pattern 7 of Figure 5.26(b).

One should also notice that the greatest stagnation pressure loss occurs when the normal shock wave stands in the test section, as shown in Figure 5.27, since the Mach number is maximum at the test section. In this case, the maximum mass flow rate per unit area is given by Equation (4.35) as

$$\frac{\dot{m}}{A^*} = p_0 \sqrt{\left(\frac{k}{RT_0}\right)\left(\frac{2}{k+1}\right)^{(k+1)/[2(k-1)]}}$$

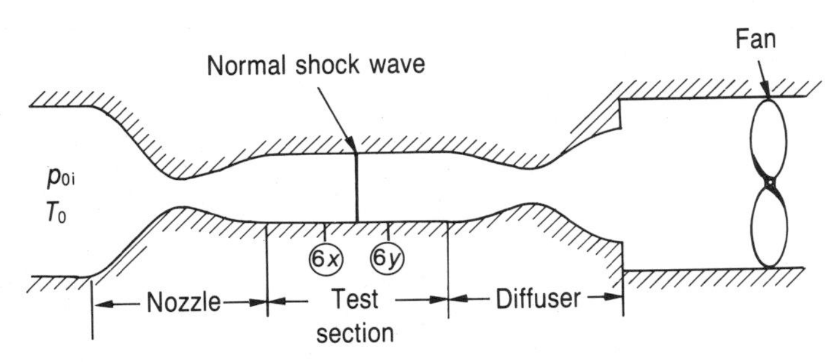

Figure 5.27 Worst operating condition of a wind tunnel with a fixed-geometry supersonic diffuser

As long as the mass flow rate is constant and the flow is adiabatic along the wind tunnel, $A^* p_0$ is constant, so that

$$A^*_{6x} p_{06x} = A^*_{6y} p_{06y} = A^* p_0 = \text{constant}$$

Hence, the minimum cross-sectional area of the diffuser throat is

$$(A_{\mathrm{dt}})_{\min} = A_{\mathrm{nt}}(p_{06x}/p_{06y}) \tag{5.34}$$

where A_{nt} and A_{dt} represent the cross-sectional areas of the nozzle throat and the diffuser throat, respectively.

(vi) Diffuser shock flow regime ($p_b < p_{e5} = p_{e6} = p_{e7}$)
When the back pressure is slightly reduced, the normal shock wave proceeds to the converging part of the second converging–diverging duct, which is now behaving as a supersonic diffuser. However, the normal shock wave cannot stand in the converging part of the diffuser. One may prove this fact by considering that $A^* p_0$ is constant along the wind tunnel. As long as the flow is decelerated to a lower supersonic Mach number in the diffuser, then it undergoes the normal shock wave at a Mach number which is less than the test section Mach number. In this case, the stagnation pressure loss is lower so that $p_{08y} > p_{06y}$. As long as $A^* p_0$ is constant along the wind tunnel, then

$$A^*_{8x} p_{08x} = A^*_{6x} p_{06x} = A^*_{8y} p_{08y} = A^*_{6y} p_{06y}$$

Hence, $A^*_{8y} < A^*_{6y} = (A_{\mathrm{dt}})_{\min}$, which is physically impossible. Therefore, the normal shock wave has to move back either to the test area or to the diverging part of the supersonic diffuser. The former case with a normal shock wave in the test section results in a subsonic flow leaving the wind tunnel at a higher pressure than the back pressure, which is physically impossible. Therefore, the normal shock wave quickly travels through the converging part of the supersonic diffuser and stabilizes in the diverging part having a cross-sectional area slightly larger than that of the test section. This process is often referred to as the **swallowing of a shock wave**. Flow pattern 8 of Figure 5.26(b) is a typical representation of this flow regime.

(vii) Best operating condition ($p_b = p_{e10}$)
The back pressure may now be increased so that the normal shock wave is forced to move upstream towards the diffuser throat where the strength of the shock is minimum, and the corresponding stagnation pressure loss is likewise a minimum. Flow pattern 10 of Figure 5.28(b) indicates the best operating condition of a fixed-geometry supersonic wind tunnel. However, in practice, the normal shock wave is positioned slightly downstream of the diffuser throat. If the normal shock wave is exactly located at the diffuser throat, then any small disturbance in the flow may cause the normal shock wave to move upstream of the diffuser throat where it is unstable. In this case, the normal shock wave will continue its movement through the test section and locate itself once again at the nozzle exit plane. This is known as the

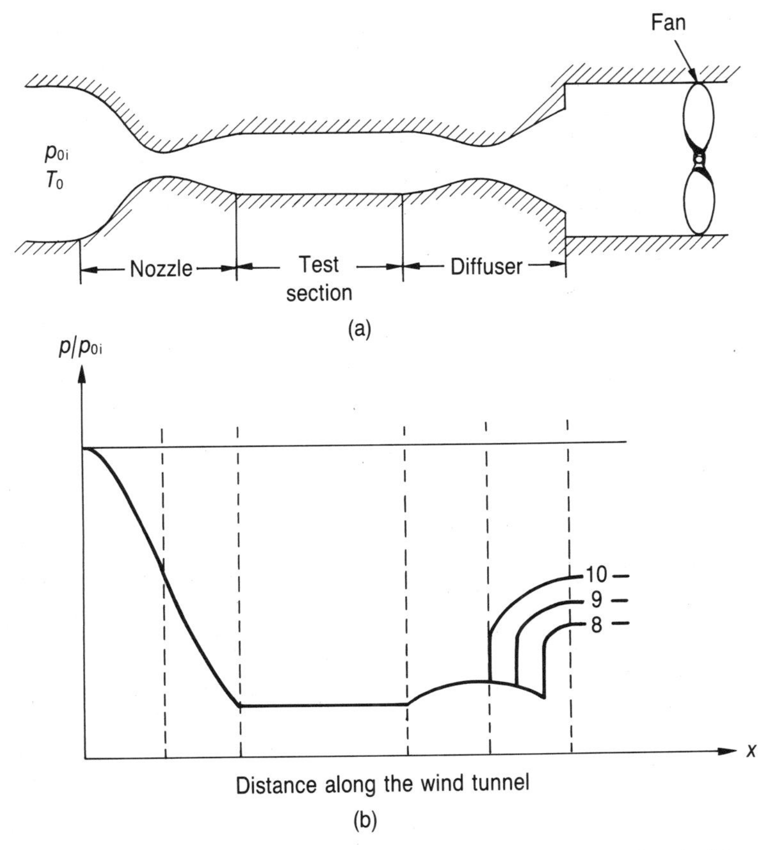

Figure 5.28 Pressure distributions in a fixed-geometry supersonic wind tunnel as the back pressure is increased

disgorging of a shock wave. Consequently, the entire procedure for swallowing the shock wave must be repeated.

Example 5.8
In a fixed-geometry wind tunnel, air is drawn in by means of a fan, as shown in Figure 5.29(a). In the supply reservoir, the stagnation pressure and the stagnation temperature are 100 kPa and 300 K, respectively. The nozzle has a throat cross-sectional area of 1 m^2 and discharges to a test section with a constant cross-sectional area of 1.687 m^2. The throat and exit cross-sectional areas of the diffuser are 1.5 m^2 and 2 m^2, respectively. Determine

a the maximum possible Mach number at the test section,
b the minimum allowable throat cross-sectional area,

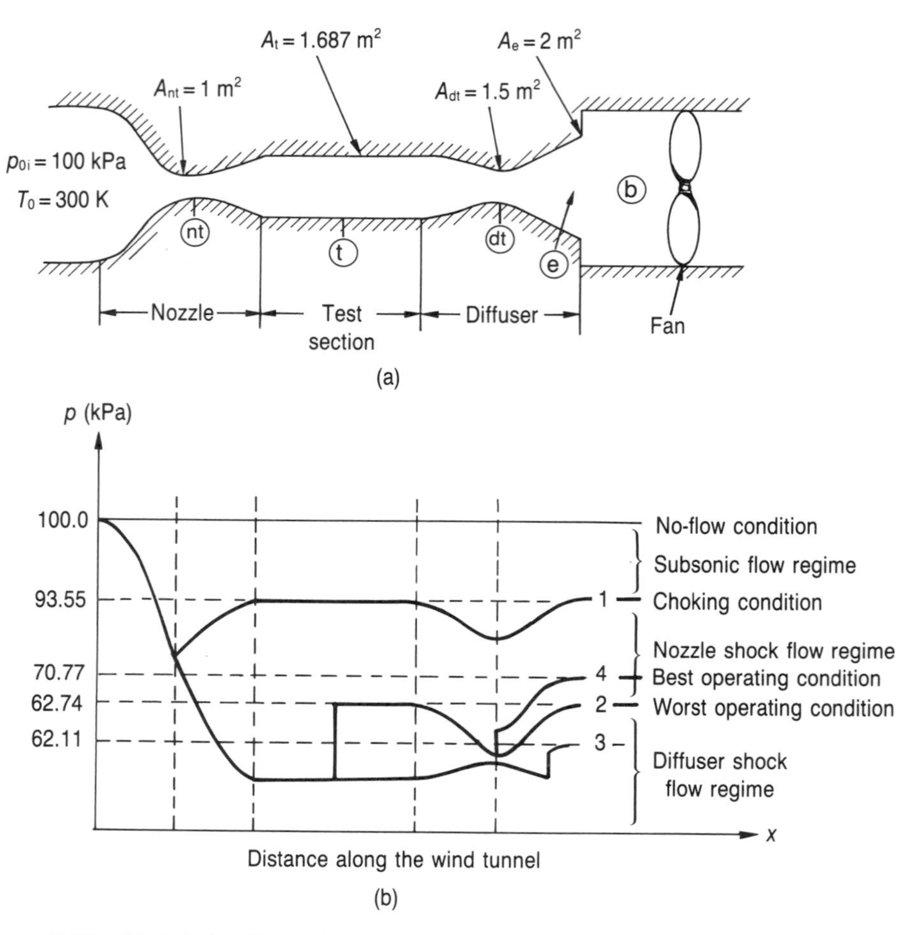

Figure 5.29 Sketch for Example 5.8

c whether or not a supersonic flow can be obtained at the test section,
d the back pressure for the worst operating condition,
e the back pressure to choke the flow at the nozzle throat,
f the location of the normal shock wave when the back pressure for the worst operating condition is reduced by 1 per cent,
g the back pressure for the best operating condition, and
h the ranges of back pressure for the subsonic, nozzle shock and diffuser shock flow regimes

Solution

(a) The maximum possible Mach number at the test section is obtained when the flow is choked at the nozzle throat and is entirely supersonic in the diverging part of the nozzle. In this case,

$$A_t/A^* = A_t/A_{nt} = 1.687\ \text{m}^2/1\ \text{m}^2 = 1.687$$

Corresponding to the area ratio of $A_t/A^* = 1.687$, there exist two solutions for the Mach number: one for subsonic flow with $M_{t1} = 0.37$, and the other for supersonic flow with $M_{t2x} = M_{t3} = M_{t4} = 2.00$ from Appendix C. Then, the maximum possible Mach number at the test section is

$$M_{t2x} = M_{t3} = M_{t4} = \mathbf{2.00}$$

(b) For the worst operating condition, the flow is choked at the nozzle throat so that $A^*_{2x} = A_{nt}$. Now, from Appendix D, corresponding to $M_{t2x} = 2.0$,

$$M_{t2y} = 0.5774$$

and

$$A^*_{2x}/A^*_{2y} = 0.7209 \quad \text{or} \quad A^*_{2y} = 1\ \text{m}^2/0.7209 = 1.387\ \text{m}^2$$

Then, the minimum allowable diffuser throat cross-sectional area for a supersonic wind tunnel is

$$(A_{dt})_{min} = A^*_{2y} = \mathbf{1.387\ m^2}$$

(c) As long as the given throat cross-sectional area of the diffuser is larger than the minimum allowable one, that is

$$A_{dt} = 1.5\ \text{m}^2 > (A_{dt})_{min} = 1.387\ \text{m}^2$$

then it is possible to obtain supersonic flow in the test section.

(d) The stagnation pressure after the normal shock wave in the worst operating condition may be obtained from Appendix D by using $M_{t2x} = 2.0$ as

$$p_{02y}/p_{02x} = 0.7209 \quad \text{or} \quad p_{02y} = (0.7209)(100\ \text{kPa}) = 72.09\ \text{kPa}$$

After the normal shock wave, the flow is entirely subsonic in the diffuser. Then from Appendix C corresponding to the area ratio of $A_e/A^*_{2y} = 2\ \text{m}^2/1.387\ \text{m}^2 = 1.442$

$$M_{e2} = 0.45$$

and

$$p_{e2}/p_{02y} = 0.8703 \quad \text{or} \quad p_{e2} = (0.8703)(72.09\ \text{kPa}) = 62.74\ \text{kPa}$$

Since the flow is subsonic at the exit plane,

$$p_{b2} = p_{e2} = \mathbf{62.74\ kPa}$$

In Figure 5.29(b), the worst operating condition is indicated by flow pattern 2.

(e) When the flow is just choked at the nozzle throat, then the flow is entirely subsonic in the wind tunnel. Hence $A_1^* = A_{nt}$ and $p_{01} = p_{0i}$. In this case, from Appendix C, corresponding to an area ratio of $A_e/A_1^* = 2\ \text{m}^2/1\ \text{m}^2 = 2.0$,

$$M_{e1} = 0.31$$

and

$$p_{e1}/p_{01} = 0.9355 \quad \text{or} \quad p_{e1} = (0.9355)(100\ \text{kPa}) = 93.55\ \text{kPa}$$

Again, the flow is subsonic at the exit plane. Therefore

$$p_{b1} = p_{e1} = \mathbf{93.55\ kPa}$$

Flow pattern 1, which corresponds to the choking condition, is shown in Figure 5.29(b).

(f) The location of the normal shock wave when $p_{b3} = 0.99p_{b2} = 62.11$ kPa may be determined by an iterative process. For this reason, it is necessary to assume a value for the cross-sectional area at which the shock wave is located in the diverging part of the diffuser. As an initial trial, assume

$$A_{3x} = A_{3y} = 1.6\ \text{m}^2$$

As long as the nozzle is choked, $A_{3x}^* = A_{nt}$, then corresponding to the area ratio of $A_{3x}/A_{3x}^* = 1.6\ \text{m}^2/1\ \text{m}^2 = 1.6$, the supersonic Mach number from Appendix C is

$$M_{3x} = 1.94$$

The Mach number, the critical area and the stagnation pressure after the normal shock wave may be evaluated by using Appendix D with $M_{3x} = 1.94$ as

$$M_{3y} = 0.5880$$

$$A_{3x}^*/A_{3y}^* = 0.7488 \quad \text{or} \quad A_{3y}^* = 1\ \text{m}^2/0.7488 = 1.336\ \text{m}^2$$

$$p_{03y}/p_{03x} = 0.7488 \quad \text{or} \quad p_{03y} = (0.7488)(100\ \text{kPa}) = 74.88\ \text{kPa}$$

Now, corresponding to the area ratio of $A_e/A_{3y}^* = 2\ \text{m}^2/1.336\ \text{m}^2 = 1.497$, the subsonic Mach number and the exit pressure from Appendix C are

$$M_{e3} = 0.43$$

and

$$p_{e3}/p_{03y} = 0.8807 \quad \text{or} \quad p_{e3} = (0.8807)(74.88\ \text{kPa}) = 65.95\ \text{kPa}$$

Since the exit plane pressure is greater than the back pressure, the normal shock wave must be located further downstream. The rest of the iterations are shown in Table 5.2. After convergence, the area of the diffuser, where the normal shock wave is located, is

$$A_{3x} = A_{3y} = \mathbf{1.7\ m^2}$$

This condition is indicated by flow pattern 3 in Figure 5.29(b).

Table 5.2 Iterations for Example 5.8

Iteration	1	2	3
$A_x = A_y$ (m^2)	1.6	1.8	1.7
A_{3x}/A^*_{3x}	1.6	1.8	1.7
M_{3x}	1.94	2.08	2.01
M_{3y}	0.5880	0.5643	0.5757
A^*_{3x}/A^*_{3y}	0.7488	0.6835	0.7162
A^*_{3y} (m^2)	1.336	1.463	1.396
p_{03y}/p_{03x}	0.7488	0.6835	0.7162
p_{03y} (kPa)	74.88	68.35	71.62
A_e/A^*_y	1.497	1.367	1.433
M_{e3}	0.43	0.49	0.46
p_{e3}/p_{03y}	0.8807	0.8486	0.8650
p_{e3} (kPa)	65.95	58.00	61.95
Shock movement	→	←	OK

(g) During the best operating condition, the flow is choked at the nozzle throat and a normal shock wave stands at the diffuser throat. Therefore, $A^*_{4x} = A_{nt}$ and $A_{4x} = A_{4y} = A_{dt}$. Then, the supersonic Mach number before the normal shock wave corresponding to an area ratio of $A_{4x}/A^*_{4x} = 1.5\ m^2/1\ m^2 = 1.5$ may be obtained from Appendix C as

$$M_{4x} = 1.85$$

The Mach number, the critical area and the stagnation pressure after the normal shock wave may now be determined from Appendix D with $M_{4x} = 1.85$ as

$$M_{4y} = 0.6057$$

$$A^*_{4x}/A^*_{4y} = 0.7902 \quad \text{or} \quad A^*_{4y} = 1\ m^2/0.7902 = 1.266\ m^2$$

$$p_{04y}/p_{04x} = 0.7902 \quad \text{or} \quad p_{04y} = (0.7902)(100\ kPa) = 79.02\ kPa$$

Finally, the subsonic Mach number and the pressure at the exit plane may be evaluated from Appendix C by using the area ratio of $A_e/A^*_{4y} = 2\ m^2/1.266\ m^2 = 1.580$ as

$$M_{e4} = 0.40$$

and

$$p_{e4}/p_{04y} = 0.8956 \quad \text{or} \quad p_{e4} = (0.8956)(79.02\ kPa) = 70.77\ kPa$$

As long as the flow leaves the wind tunnel at subsonic speeds, then

$$p_{b4} = p_{e4} = \mathbf{70.77\ kPa}$$

(h) Ranges of back pressure for the following regimes are as follows:

1 Subsonic flow regime:

$$p_0 \geqslant p_b \geqslant p_{b1} \text{ or } \mathbf{100\ kPa \geqslant p_b \geqslant 93.55\ kPa}$$

2 Nozzle shock flow regime:

$$p_{b1} \geqslant p_b \geqslant p_{b2} \text{ or } \mathbf{93.55\ kPa \geqslant p_b \geqslant 62.74\ kPa}$$

3 Diffuser shock flow regime:

$$p_{b2} \geqslant p_b \geqslant 0 \text{ or } \mathbf{62.74\ kPa \geqslant p_b \geqslant 0\ kPa}$$

(b) Supersonic diffusers in variable-geometry wind tunnels

One of the major problems in a fixed-geometry supersonic wind tunnel is the large stagnation pressure drop associated with the normal shock wave in the diffuser. This problem may be significantly reduced by employing a variable-geometry wind tunnel. In practice, either the nozzle throat area or the diffuser throat area, or both, may be made variable. In the following discussion, it is assumed that the nozzle geometry is fixed, but the diffuser throat area is variable. The results are also applicable to the case where the nozzle geometry is variable, because the significant parameter is the ratio of the diffuser throat area to the nozzle throat area A_{dt}/A_{nt}.

During start-up, the throat of the variable-geometry diffuser is opened to an area larger than that specified by Equation (5.34). As the back pressure is decreased, the sequence of events is identical to that illustrated in Figure 5.26 for a fixed-geometry diffuser. Once the normal shock wave is swallowed, as indicated by flow pattern 8 of Figure 5.26(b) the diffuser throat area is decreased while the back pressure is simultaneously increased, so that the shock wave is forced to move to the diffuser throat. The decrease in the diffuser throat area, A_{dt}, increases the area ratio A_{nt}/A_{dt}, causing the Mach number in the diffuser throat to decrease and to approach unity as A_{dt} approaches A_{nt}. In the limit, the Mach number in the diffuser throat becomes unity, so that the normal shock wave at the diffuser throat has vanishing strength, and the entire flow field becomes isentropic. The sequence of events is shown in Figure 5.30, and flow pattern 11 indicates an entirely isentropic flow.

5.9.2 Supersonic diffusers in jet engine inlets

In this section, jet engine inlets are discussed when they are operating at supersonic speeds. As long as these engines reach their operating speed by being accelerated from lower speeds, then it is necessary to investigate the operation of supersonic inlets during starting. In the following discussion all losses except those across normal shock waves will be ignored.

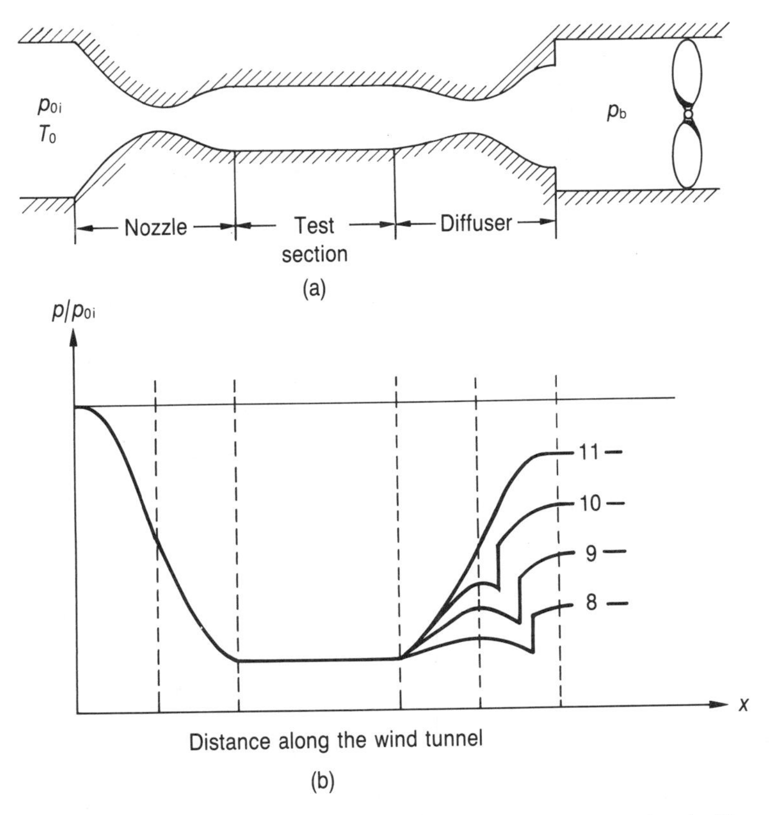

Figure 5.30 Pressure distributions in a variable-geometry supersonic wind tunnel as the diffuser throat area is decreased while the back pressure is increased

(a) Supersonic diffusers in fixed-geometry jet engine inlets

Consider a converging–diverging inlet which has an inlet area of A_i and throat area of A_t. The sequence of operating conditions, which is attained during start-up, is illustrated in Figure 5.31 for a fixed-geometry converging–diverging diffuser. At low subsonic flight speeds, the entire flow is subsonic, as illustrated in Figure 5.31(a). The **capture area**, A_c, through which the diffuser mass flow rate passes, may be greater than, equal to or less than the inlet area A_i depending on the mass flow rate requirements of the jet engine. As the subsonic flight Mach number, M_∞, is increased, the mass flow rate requirement of the diffuser increases. In this case, the Mach number reaches unity at the diffuser throat and then decreases in the diverging part of the diffuser, as shown in Figure 5.31(b). The streamline pattern upstream of the inlet is no longer influenced by the conditions downstream of the diffuser throat, but only by the flight Mach number, M_∞, and by the ratio A_i/A_t. When the flight Mach number M_∞ reaches unity, the capture area A_c decreases and becomes equal to the

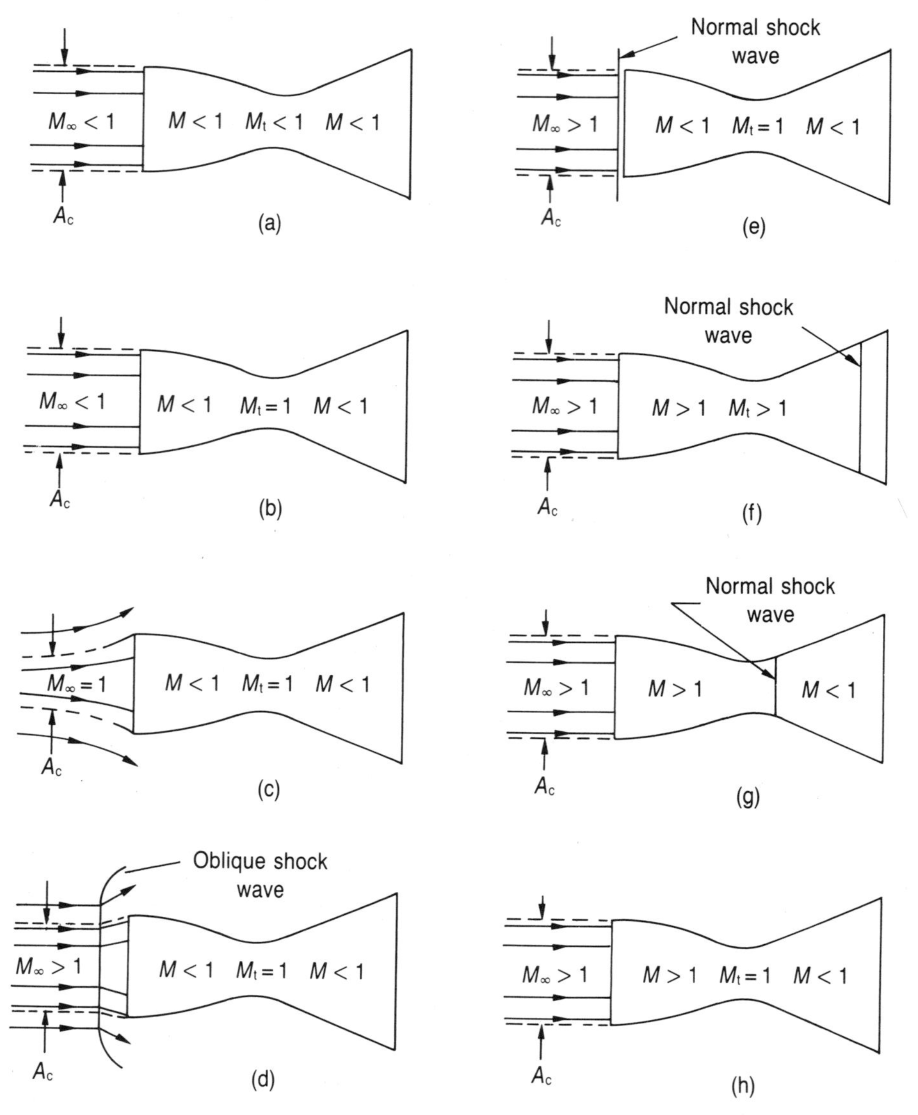

Figure 5.31 Flow conditions for a converging–diverging diffuser when the flight Mach number is increased

diffuser throat area A_t. In this case, the air spills over the lips of the inlet, as shown in Figure 5.31(c). Further increases in the flight Mach number to supersonic values cause an oblique shock wave to form upstream of the inlet, as shown in Figure 5.31(d). The capture area is less than the inlet area so that the air again spills over the lips of the inlet. When the flight Mach number reaches a sufficiently high value, a normal shock wave is located at the entrance of the inlet with the capture area being equal to the inlet area, as shown in Figure 5.31(e). A slight increase in the flight Mach number causes the normal shock wave to enter the converging part of the diffuser where its position is unstable. As a result, the normal shock wave proceeds through the diffuser throat and positions itself at the diverging part of the diffuser, as shown in Figure 5.31(f). At this point, the normal shock wave is swallowed, and the inlet is said to be started. The stagnation pressure loss may now be decreased either by increasing the back pressure or by decreasing the flight Mach number. As the flight Mach number is decreased, the normal shock wave will locate itself slightly downstream from the diffuser throat, as shown in Figure 5.31(g). The maximum stagnation pressure recovery is obtained when the normal shock wave is located exactly at the diffuser throat, as shown in Figure 5.31(h). However, the normal shock wave is not stable at this position, and it will be disgorged from the diffuser with a slight decrease in the flight Mach number. In this case, the shock has to be swallowed again. For this reason, the design point is chosen such that the normal shock wave will locate itself slightly downstream of the diffuser throat, as shown in Figure 5.31(g).

When the normal shock wave is attached to the entrance of the diffuser, as shown in Figure 5.31(e), the area ratio A_t/A_i of the diffuser corresponding to a given value of the flight Mach number M_∞ is as given in Figure 5.32. This curve may be obtained

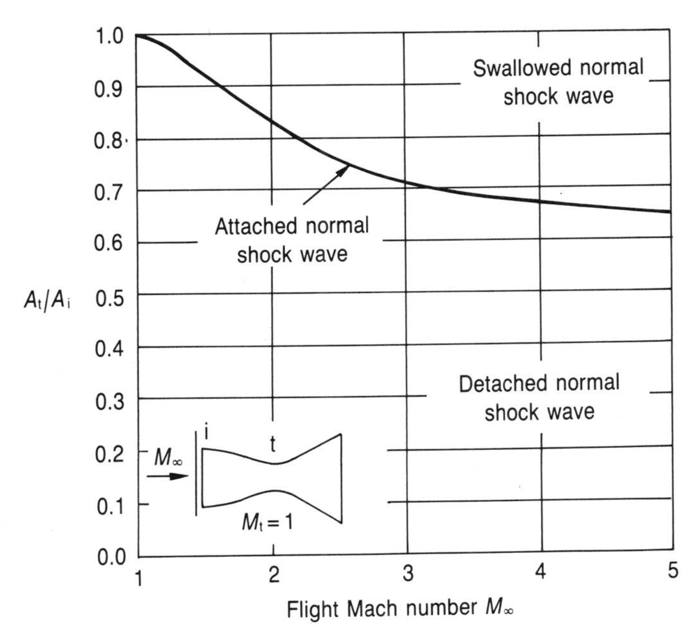

Figure 5.32 Area ratio A_t/A_i of the supersonic diffuser when a normal shock wave is attached at the entrance of the diffuser ($k = 1.4$)

by first determining the subsonic Mach number immediately downstream of the normal shock wave from Appendix D, and then determining the isentropic area ratio required for accelerating this subsonic flow to a Mach number of unity at the throat by using Appendix C for $k = 1.4$. Operation of the diffuser at any point in the region below the curve corresponds to a situation where the normal shock wave is detached in front of the entrance of the diffuser. Operation of the diffuser at any point in the region above the curve corresponds to a condition where the shock wave is swallowed.

When the isentropic supersonic flow reaches a Mach number of unity at the diffuser throat and continues subsonically thereafter, as shown in Figure 5.31(h), then the area ratio A_t/A_i of the diffuser corresponding to a given value of the flight Mach number, M_∞, can be obtained from Figure 5.33. The curve may be obtained by determining the isentropic area ratio required for decelerating the flow from the flight Mach number to a Mach number of unity by using Appendix C for $k = 1.4$. A point, located in the region below this curve, corresponds to a diffuser throat area, which is too small for passing all of the captured flow at that particular flight Mach number. In this case, a normal shock wave propagates upstream through the diffuser throat and the converging part of the diffuser, and positions itself upstream of the entrance of the diffuser. For any operating point above the curve of Figure 5.33, the flow at the converging part of the diffuser and also at the diffuser throat is supersonic. A normal

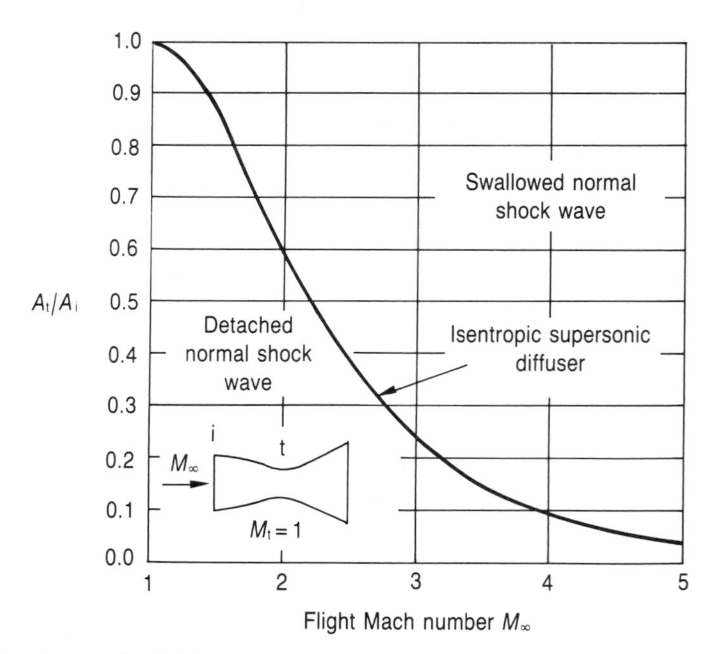

Figure 5.33 Area ratio A_t/A_i of the supersonic diffuser for isentropic supersonic flow with a sonic Mach number at the throat ($k = 1.4$)

shock wave would then position itself somewhere at the diverging part of the diffuser, whose location depends on the back pressure exerted by the remainder of the jet engine.

The curve presented in Figure 5.33 applies only after the supersonic diffuser is started. Consequently, an operating point on or above the curve of Figure 5.32 must first be obtained before Figure 5.33 becomes applicable. The two curves of Figures 5.32 and 5.33 are presented together in Figure 5.34 in order to illustrate the overall operating features of a converging–diverging supersonic diffuser. Any operating point in the region above the two curves indicates that there is a swallowed normal shock wave in the diverging part of the supersonic diffuser. However, any operating point in the region below the two curves indicates that there is a detached normal shock wave upstream of the entrance of the supersonic diffuser. Finally, any operating point in the region between these two curves indicates that there is either a swallowed or a detached normal shock wave depending on the direction of movement. If the operating point moves towards the upper curve from below, then there is a detached normal shock wave upstream of the supersonic diffuser. However, if the operating point moves towards the lower curve from above, then there is a swallowed normal shock wave at the diverging part of the diffuser.

During the start-up of a supersonic inlet, only the upper curve of Figure 5.34 is physically meaningful. The horizontal line *ab* traces the operation of the inlet as the

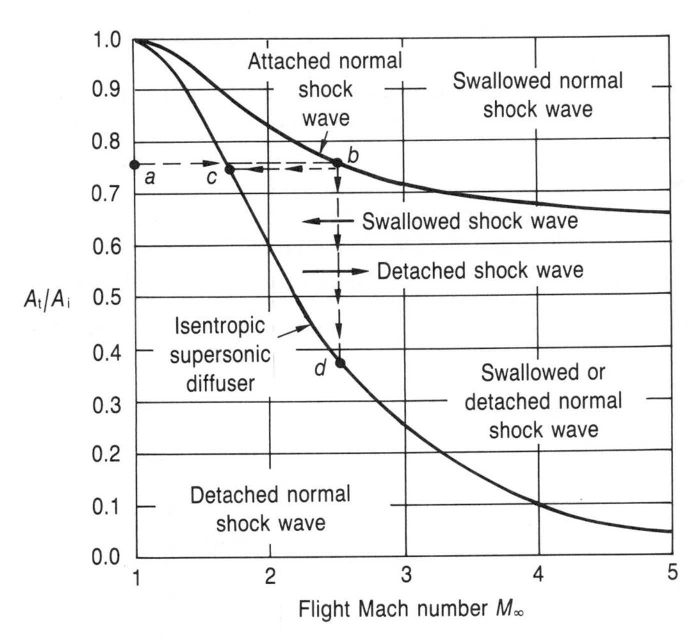

Figure 5.34 Area ratio A_t/A_i of the supersonic diffuser for detached, attached and swallowed normal shock waves as a function of the flight Mach number

flight Mach number, M_∞, is increased. At point b, the normal shock wave is swallowed, and the lower curve becomes physically meaningful. If the flight Mach number, M_∞, is now decreased, the operating point moves from point b towards point c. In the region between points b and c, the normal shock wave is swallowed and is located in the diverging part of the diffuser. At point c, the normal shock wave is located at the throat of the diffuser where the Mach number is unity and the shock strength is vanishingly small. This operating point corresponds to Figure 5.31(h). If the flight Mach number is decreased further, then the normal shock wave will be disgorged. Therefore, there is a hysteresis effect in that the normal shock wave is swallowed at a flight Mach number corresponding to point b, but the normal shock wave is not disgorged until the flight Mach number decreases to a value corresponding to point c. Increasing the flight speed above the design value in order to swallow the shock wave is known as **overspeeding**.

Overspeeding for starting the supersonic fixed-geometry inlet is only practical for very low supersonic design Mach numbers. However, for higher supersonic design Mach numbers, overspeeding is either unrealistic or impossible. As the flight Mach number in Figure 5.31(e) approaches infinity, the Mach number after the normal shock wave asymptotically approaches a Mach number of 0.3780, as indicated in Appendix D. In this case, one may determine from Appendix C that the upper curve of Figure 5.34 approaches an asymptotic value of $A_t/A_i = 0.6028$. Consequently, one may easily determine either from Appendix C or from the lower curve of Figure 5.34 that for a design flight Mach number of 1.98 corresponding to $A_i/A_t = 1.659$, the inlet would have to be overspeeded to a Mach number of infinity to start it. Therefore, the supersonic fixed-geometry inlet is impossible for design flight Mach numbers above 1.98. At design flight Mach numbers which are less than but closer to a Mach number of 1.98, overspeeding is unrealistic due to excessive power requirements.

Example 5.9

A supersonic fixed-geometry inlet is operating at sea level where the temperature and pressure are 300 K and 100 kPa, respectively (Figure 5.35). The inlet and throat cross-sectional areas are 0.1 m^2 and 0.07 m^2, respectively.

a Determine the Mach number to which the inlet must be overspeeded for starting.

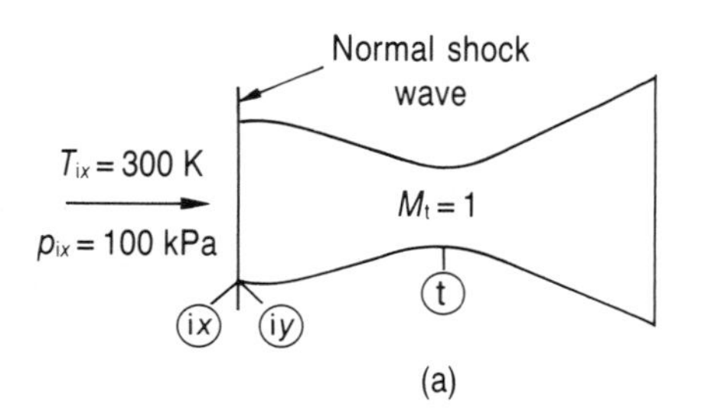

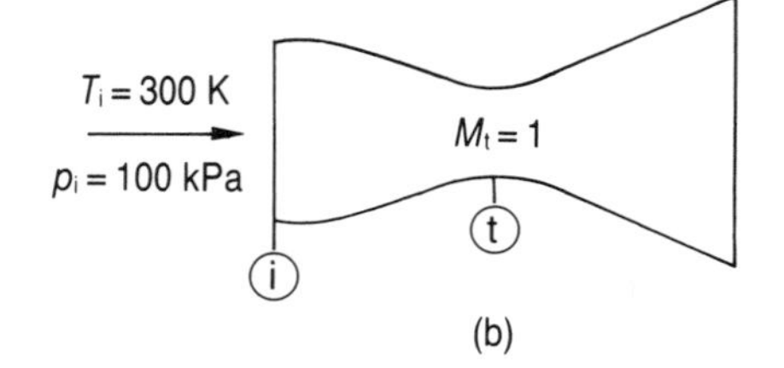

Figure 5.35 Sketch for Example 5.9

b Determine the Mach number for which the inlet is designed.
c Determine the mass flow rate when the inlet is overspeeded.
d Determine the mass flow rate at the design flight Mach number.
e Comment on whether or not it is feasible to use this inlet.

Solution
(a) During the start-up of a supersonic fixed-geometry inlet, the normal shock wave is swallowed just after the inlet is overspeeded so that there is a normal shock wave attached to the entrance of the inlet, as shown in Figure 5.35(a). The Mach number after the normal shock wave corresponding to an area ratio of $A_i/A_t = A_i/A^* = 0.1\ \text{m}^2/0.07\ \text{m}^2 = 1.429$ may be obtained from Appendix C as

$$M_{iy} = 0.46$$

The Mach number before the normal shock wave may now be determined from Appendix D as

$$M_{ix} = \mathbf{3.29}$$

(b) The inlet is designed for the isentropic deceleration of the air from supersonic to subsonic speeds, as shown in Figure 5.35(b). Then, the design flight Mach number corresponding to an area ratio of $A_i/A_t = A_i/A^* = 1.429$ may now be found from Appendix C as

$$M_i = \mathbf{1.79}$$

(c) The overspeeding of the inlet is shown in Figure 5.35(a). The stagnation properties before the normal shock wave corresponding to $M_{ix} = 3.29$ found from Appendix C are

$$p_{ix}/p_{0x} = 0.017\,73 \quad \text{or} \quad p_{0x} = 100\ \text{kPa}/0.017\,73 = 5640\ \text{kPa}$$

$$T_{ix}/T_{0x} = 0.3160 \quad \text{or} \quad T_{0x} = 300\ \text{K}/0.3160 = 949.4\ \text{K}$$

The stagnation properties after the normal shock wave corresponding to $M_{ix} = 3.29$ may be obtained from Appendix D as

$$p_{0y}/p_{0x} = 0.2555 \quad \text{or} \quad p_{0y} = (5640\ \text{kPa})(0.2555) = 1441\ \text{kPa}$$

$$T_{0y} = T_{0x} = 949.4\ \text{K}$$

Finally, one may use Appendix C with $M_{iy} = 0.46$ to determine the mass flow rate as

$$\frac{\dot{m}\sqrt{(RT_{0y})}}{A_i p_{0y}} = 0.4806 \quad \text{or} \quad \dot{m} = \frac{(0.4806)(0.1\ \text{m}^2)(1\,441\,000\ \text{N}\,\text{m}^{-2})}{\sqrt{[(287.1\ \text{J}\,\text{kg}^{-1}\,\text{K}^{-1})(949.4\ \text{K})]}}$$

$$= \mathbf{132.7\ kg\,s^{-1}}$$

(d) The mass flow rate at the design flight Mach number may be evaluated by referring to Figure 5.35(b). In this case, the stagnation properties and the mass flow rate corresponding to $M_i = 1.79$ may be evaluated by using Appendix C as

$$p_i/p_0 = 0.1767 \quad \text{or} \quad p_0 = 100\ \text{kPa}/0.1767 = 565.9\ \text{kPa}$$

$$T_i/T_0 = 0.6095 \quad \text{or} \quad T_0 = 300\ \text{K}/0.6095 = 492.2\ \text{K}$$

$$\frac{\dot{m}\sqrt{(RT_0)}}{A_i p_0} = 0.4794 \quad \text{or} \quad \dot{m} = \frac{(0.4794)(0.1\ \text{m}^2)(565\,900\ \text{N m}^{-2})}{\sqrt{[(287.1\ \text{J kg}^{-1}\,\text{K}^{-1})(492.2\ \text{K})]}}$$

$$= \mathbf{72.17\ kg\,s^{-1}}$$

(e) In order to start up the supersonic fixed-geometry inlet, it is necessary to overspeed the inlet to a Mach number of 3.29. For this reason, an excessive propulsion unit is necessary, although the inlet can operate at a design Mach number of 1.79. Therefore, **it is not feasible to use this inlet**.

(b) Supersonic diffusers in variable-geometry jet engine inlets

The shortcomings of the supersonic fixed-geometry inlet may be eliminated by using variable geometry. The sequence of operating conditions for a variable-geometry inlet during start-up is exactly the same as for a fixed-geometry inlet until the normal shock wave is attached to the entrance of the inlet, as indicated by Figures 5.31(a) to 5.31(e). During this start-up procedure of the supersonic variable-geometry inlet, only the upper curve of Figure 5.34 is physically meaningful. The horizontal line *ab* traces the operation of the inlet as the flight Mach number M_∞ is increased to its design value. At point *b*, the normal shock wave is swallowed, and the lower curve of Figure 5.34 becomes physically meaningful. If the area ratio A_t/A_i of the inlet is now decreased, then the stagnation pressure loss is reduced as the operating point moves from point *b* to point *d*. In the region between points *b* and *d*, the normal shock wave is swallowed and is located in the diverging part of the diffuser. At point *d*, the normal shock wave is located at the throat of the diffuser where the Mach number is unity and the shock strength is vanishingly small. If the area ratio is decreased further, then the normal shock wave will be disgorged. Therefore, there is a hysteresis effect in that the normal shock wave is swallowed at an area ratio, A_t/A_i, corresponding to point *b*, but the normal wave is not disgorged until the area ratio, A_t/A_i, decreases to a value corresponding to point *d*.

Example 5.10

A supersonic variable-geometry inlet is designed for a Mach number of 2.0. The cross-sectional area at the entrance of the inlet is 0.1 m^2. Determine

a the throat area when the normal shock wave is just swallowed, and
b the throat area at the design Mach number.

Solution

(a) During the start-up of a supersonic variable-geometry inlet, the normal shock

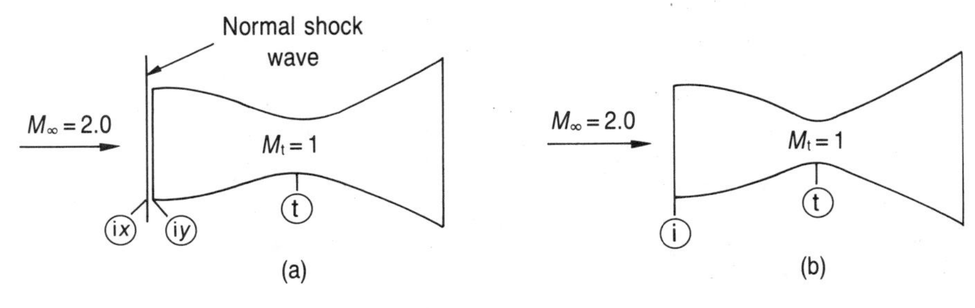

Figure 5.36 Sketch for Example 5.10

wave is swallowed just after it is attached to the entrance of the inlet, as shown in Figure 5.36(a). The critical area before the normal shock wave corresponding to a Mach number of $M_\infty = 2.0$ may be obtained from Appendix C as

$$A_i/A_x^* = 1.687 \quad \text{or} \quad A_x^* = 0.1\ \text{m}^2/1.687 = 0.059\ 28\ \text{m}^2$$

Noting that $A_t = A_y^*$, one may evaluate the throat area corresponding to $M_\infty = 2.0$ from Appendix D as

$$A_x^*/A_y^* = 0.7209 \quad \text{or} \quad A_t = A_y^* = 0.059\ 28\ \text{m}^2/0.7209 = \mathbf{0.082\ 23\ m^2}$$

(b) The inlet is designed for the isentropic deceleration of the air from supersonic to subsonic flow, as shown in Figure 5.36(b). Noting that $A_t = A^*$, the throat area corresponding to a Mach number of $M_\infty = 2.0$ may be found from Appendix C as

$$A_i/A^* = 1.687 \quad \text{or} \quad A_t = A^* = 0.1\ \text{m}^2/1.687 = \mathbf{0.059\ 28\ m^2}$$

5.10 PERFORMANCE OF REAL DIFFUSERS

Consider a diffuser which is supplied with a perfect gas having a temperature of T_i and a pressure of p_i at a velocity V_i, as shown in Figure 5.37(a). The gas decelerates adiabatically, but with increasing entropy to the exit state, e. The velocity at the exit state is practically negligible. If the gas is decelerated isentropically to the same final pressure, then the exit state would have been es, as shown in Figure 5.37(b). The diffuser efficiency, η_d, may now be defined as the ratio of the dissipated kinetic energy if the gas had been isentropically compressed to the same final pressure to the actual dissipated kinetic energy, and may be given as

$$\eta_d = \frac{h_{es} - h_i}{h_e - h_i} = \frac{h_{es} - h_i}{V_i^2/2} \tag{5.35}$$

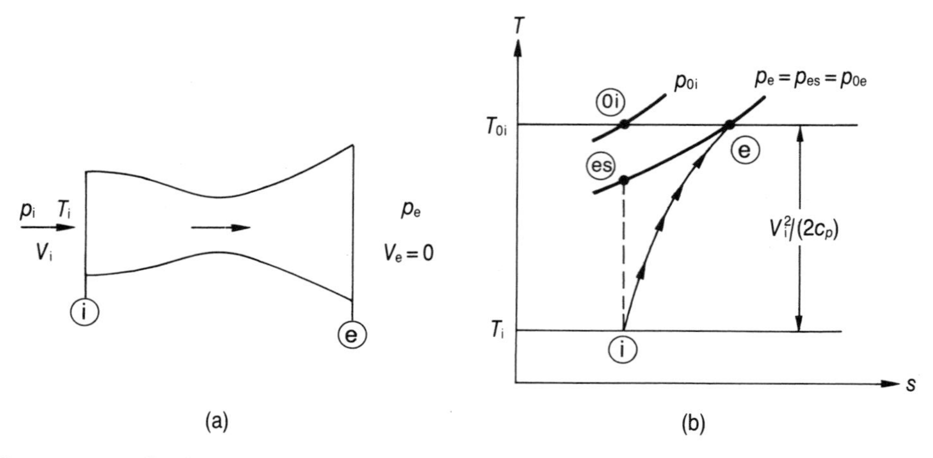

Figure 5.37 Performance of a real diffuser

However, for a perfect gas $h = c_p T$, so the above equation takes the following form:

$$\eta_d = \frac{T_{es} - T_i}{V_i^2/(2c_p)} = \frac{T_i(T_{es}/T_i - 1)}{V_i^2/(2c_p)}$$

Noting that $p_{es} = p_e$, it is possible to write the relation

$$\frac{T_{es}}{T_i} = \left(\frac{p_{es}}{p_i}\right)^{(k-1)/k} = \left(\frac{p_e}{p_i}\right)^{(k-1)/k}$$

Also, one may use the definition of the Mach number as $M_i = V_i/\sqrt{(kRT_i)}$ in order to express the diffuser efficiency as

$$\eta_d = \frac{(p_e/p_i)^{(k-1)/k} - 1}{\frac{1}{2}(k-1)M_i^2} \tag{5.36}$$

At this point, one should note that $p_e = p_{0e}$. Then

$$\frac{p_e}{p_i} = \frac{p_e}{p_{0i}}\frac{p_{0i}}{p_i} = \frac{p_{0e}}{p_{0i}}\frac{p_{0i}}{p_i}$$

Now, employing the isentropic relation (4.28) for p_{0i}/p_i and substituting into Equation (5.36), it is possible to obtain

$$\eta_d = \frac{[1 + \frac{1}{2}(k-1)M_i^2]\,(p_{0e}/p_{0i})^{(k-1)/k} - 1}{\frac{1}{2}(k-1)M_i^2} \tag{5.37}$$

Therefore the diffuser efficiency is only a function of the inlet Mach number and the stagnation pressure ratio across the diffuser.

5.11 SUPERSONIC PITOT TUBE

The Pitot tube has been used as an instrument for measuring velocities in subsonic flows for a long time. However, when a Pitot tube is used for measuring velocities in supersonic flows, modifications must be made for interpreting the measured data.

Experiments reveal that when a Pitot tube is placed in a supersonic flow, a curved shock wave stands ahead of the Pitot tube, as shown in Figure 5.38(a). The Mach number of the undisturbed flow, which is ahead of the shock wave, may be found from the measured data when the following assumptions are made:

1 The curved shock wave is normal to the stagnation streamline at the point where the stagnation streamline crosses the shock wave. This condition is only true when the Pitot tube is placed parallel to a uniform flow.
2 The fluid particles are brought to rest isentropically in the subsonic flow after the shock wave.

Considering these assumptions, the process on the T–s diagram representing the flow about the Pitot tube is as shown in Figure 5.38(b). Then, it is possible to express p_{0y}/p_x as

$$\frac{p_{0y}}{p_x} = \frac{p_{0y}}{p_y}\frac{p_y}{p_x}$$

Further, p_{0y}/p_y may be expressed in terms of M_x by using Equations (4.28) and (5.15b) and p_y/p_x is given in terms of M_x by using Equation (5.17) to yield

$$\frac{p_{0y}}{p_x} = \left(\frac{k+1}{2}M_x^2\right)^{k/(k-1)} \Bigg/ \left(\frac{2k}{k+1}M_x^2 - \frac{k-1}{k+1}\right)^{1/(k-1)} \tag{5.38}$$

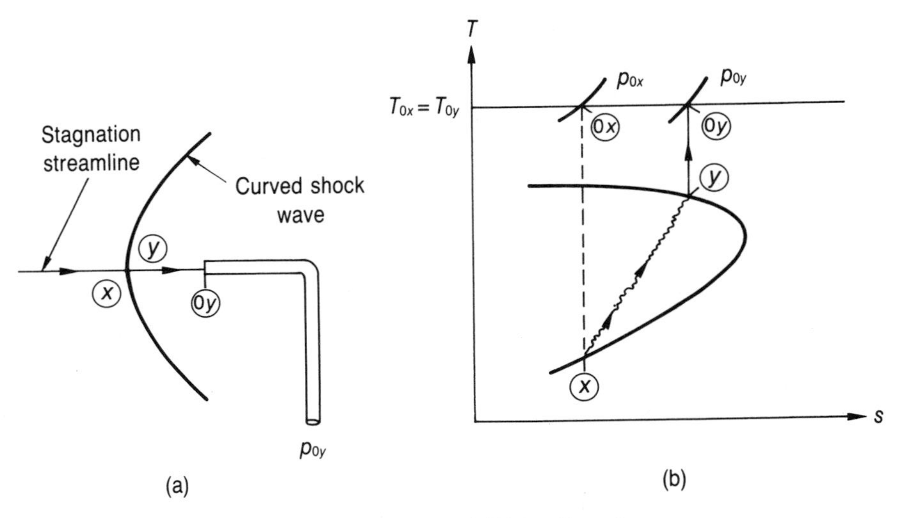

Figure 5.38 (a) A supersonic Pitot tube, and (b) its T–s diagram

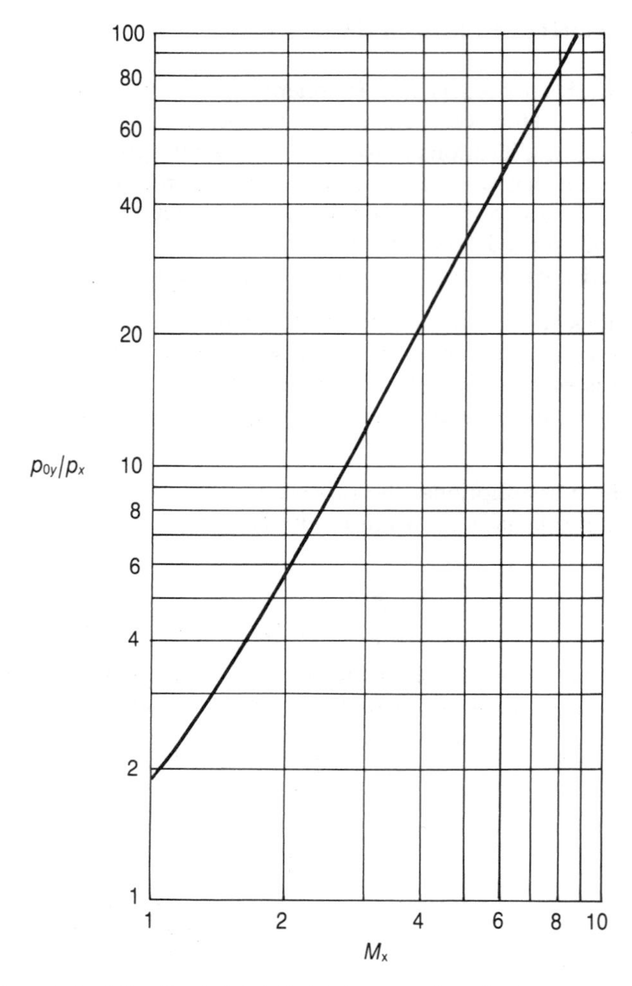

Figure 5.39 The plot of the Rayleigh Pitot tube equation for $k = 1.4$

This relation is known as the Rayleigh Pitot tube equation, and it is an implicit equation for M_x. The ratio p_{0y}/p_x is presented in Figure 5.39 for $k = 1.4$.

Example 5.11

A combined Pitot tube is used for determining the velocity of a supersonic airstream, as shown in Figure 5.40. The static pressure and the stagnation pressure are measured as 140 kPa and 200 kPa, respectively. If the temperature of the approaching air is 350 K, determine the velocity of the approaching airstream.

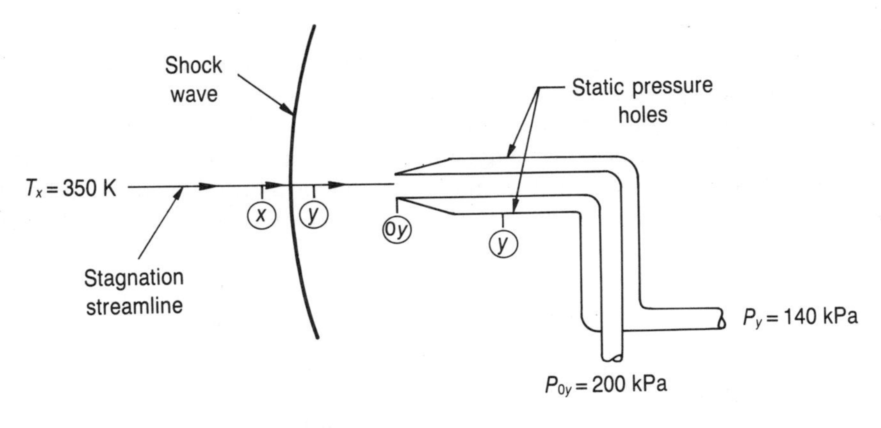

Figure 5.40 Sketch for Example 5.11

Solution

The Mach number after the shock wave corresponding to $p_y/p_{0y} =$ 140 kPa/200 kPa = 0.7 may be obtained from Appendix C as $M_y = 0.73$. Now, the Mach number before the shock wave may be determined from Appendix D by using $M_y = 0.73$ as $M_x = 1.42$. Finally, the velocity of the airstream which is approaching the Pitot tube is

$$V_x = M_x\sqrt{[(kRT_x)]} = (1.42)\sqrt{[(1.4)(287.1\ \text{J kg}^{-1}\,\text{K}^{-1})(350\ \text{K})]} = \mathbf{532.6\ m\,s^{-1}}$$

5.12 ONE-DIMENSIONAL UNSTEADY FLOW IN A SHOCK TUBE

A shock tube is a device in which a normal shock wave is generated by the rupture of a diaphragm or a quick-acting valve separating a gas at high pressure from another gas at low pressure, as shown in Figure 5.41. A shock tube is a useful tool for

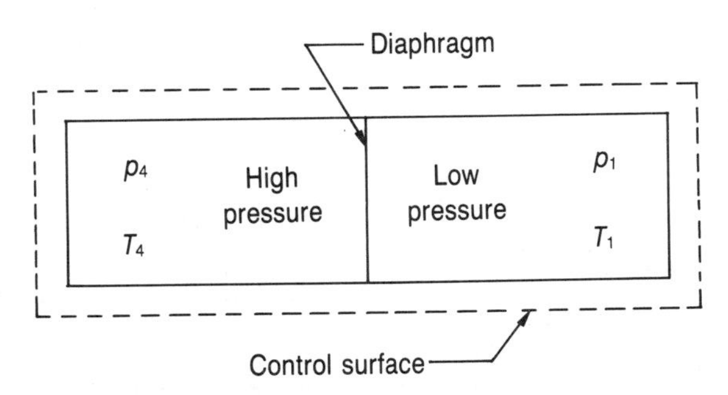

Figure 5.41 A shock tube

investigating not only shock phenomena but also the behaviour of materials and objects when they are subjected to the extreme conditions of pressure and temperature which exist in the gas flow behind a normal shock wave.

After the rupture of the diaphragm in the shock tube shown in Figure 5.41, the system eventually approaches thermodynamic equilibrium such that the final state in the closed tube can be determined from the first law of thermodynamics. When the flow is adiabatic and there is no external work, the total internal energy of the mixture

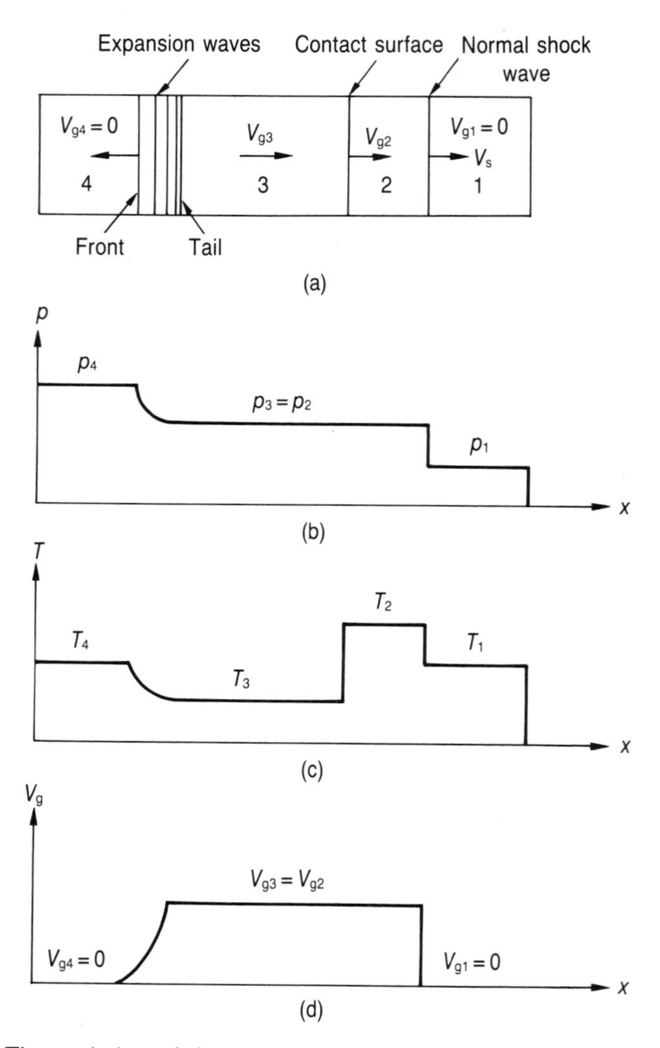

Figure 5.42 The variation of the pressure, temperature and velocity along the shock tube shortly after the rupture of the diaphragm

of gases at the final state is equal to the sum of the internal energies of the gases which are present initially on each side of the diaphragm. However, the primary interest is not the final equilibrium state of gases, but the transient shock phenomena after the rupture of the diaphragm. When the diaphragm ruptures, a normal shock wave moves into the low-pressure side to increase the pressure, while a series of expansion waves propagate into the high-pressure side to decrease the pressure. The variation of the pressure, temperature and velocity with distance, x, shortly after the rupture of the diaphragm is shown in Figure 5.42. Following the initial normal shock wave down the tube is a **contact surface**. This contact surface is a temperature discontinuity separating the gases which are compressed by the normal shock wave from those which are cooled by the expansion waves. The gases in regions 2 and 3 move at the same velocity and are at the same pressure. However, there is a discontinuity in the density and in the temperature distinguishing these regions, since there is not enough time for the required heat transfer between regions 2 and 3. The strength of the normal shock wave and the gas velocities are dependent on the initial pressure ratio across the diaphragm, the properties of the gases that are involved and the initial temperature of the gases.

The primary interest in the operation of a shock tube is to develop an expression relating the pressure ratio, p_2/p_1, across the normal shock wave to the initial pressure ratio, p_4/p_1, across the diaphragm. In the following analysis it is first necessary to develop an expression for the pressure ratio, p_4/p_3 across the expansion waves as a function of V_{g3}, and then to develop an expression for the pressure ratio, p_2/p_1, across the normal shock wave as a function of V_{g2}. Finally, these two expressions may be matched by using the fact that $V_{g2} = V_{g3}$ and $p_2 = p_3$, and a resultant equation which relates p_2/p_1 to p_4/p_1 may be obtained.

First, it is necessary to determine the variation of properties across the expansion waves. For this reason, consider the general case of an infinitesimal wave moving leftwards down a tube into a gas stream which is moving towards the infinitesimal wave, as shown in Figure 5.43(a). As long as the infinitesimal wave moves at the speed of sound relative to the gas, then the absolute velocity of the wave is $a - V_g$. The

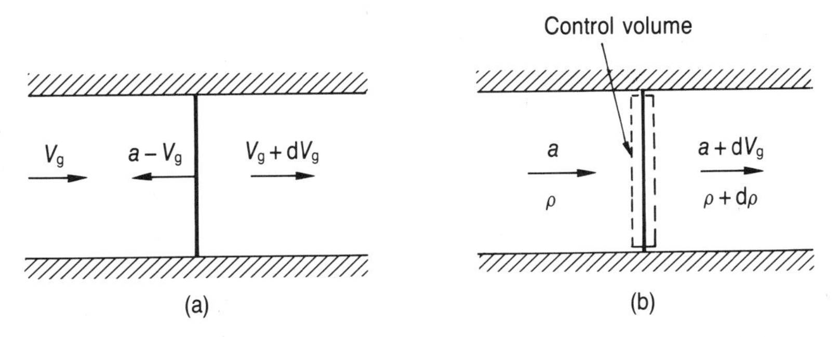

Figure 5.43 (a) A leftward-moving infinitesimal wave, and (b) when the leftward-moving infinitesimal wave is rendered stationary

leftward-moving wave may then be rendered stationary, as shown in Figure 5.43(b). By applying the continuity equation to the control volume of Figure 5.43(b), it is possible to obtain

$$\rho a = (\rho + d\rho)(a + dV_g)$$

or

$$\frac{d\rho}{\rho} = -\frac{dV_g}{a} \tag{5.39}$$

If one logarithmically differentiates the equation of state for a perfect gas $\rho = p/(RT)$, then

$$\frac{d\rho}{\rho} = \frac{dp}{p} - \frac{dT}{T} \tag{5.40}$$

Logarithmic differentiation of the speed of sound relation, $a^2 = kRT$, yields

$$\frac{dT}{T} = 2\frac{da}{a} \tag{5.41}$$

Finally, the flow across an infinitesimal wave is isentropic, so $p/\rho^k = \text{constant}$ or $p/T^{k/(k-1)} = \text{constant}$. Then, logarithmic differentiation yields

$$\frac{dp}{p} = \frac{k}{k-1}\frac{dT}{T} \tag{5.42}$$

Now, combining Equations (5.39), (5.40), (5.41) and (5.42), it is possible to obtain

$$dV_g + \frac{2}{k-1}\,da = 0$$

This equation may now be integrated to give

$$V_g + \frac{2}{k-1}\,a = \alpha = \text{constant} \tag{5.43}$$

A similar analysis may be applied to a rightward-moving infinitesimal wave, shown in Figure 5.44, to yield

$$-V_g + \frac{2}{k-1}\,a = \beta = \text{constant} \tag{5.44}$$

In Equations (5.43) and (5.44), α and β are constants of integration and are referred to as **Riemann invariants.**

Equation (5.43) may now be applied across the expansion waves in the shock tube between regions 3 and 4 in Figure 5.42(a) to yield

$$\frac{V_{g3}}{a_4} = \frac{2}{k-1}\left(1 - \frac{a_3}{a_4}\right) \tag{5.45}$$

with $V_{g4} = 0$. In region 3 to 4, expansion waves cannot catch each other, but they can

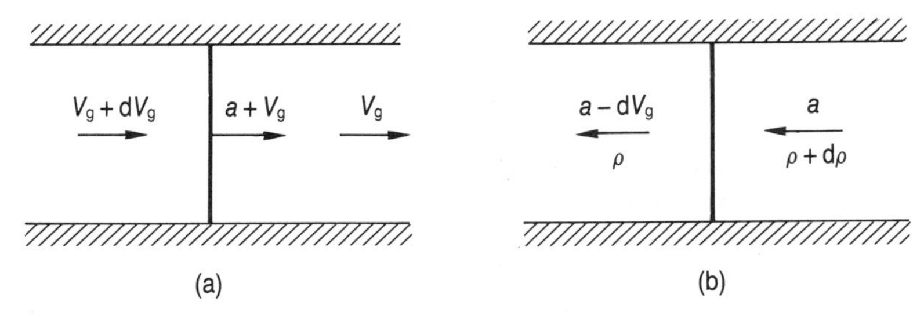

Figure 5.44 (a) A rightward-moving infinitesimal wave, and (b) when the rightward-moving infinitesimal wave is rendered stationary

get away from each other as they move down the tube. Therefore, the flow is isentropic between sections 3 and 4, so

$$\frac{p_3}{p_4} = \left(\frac{T_3}{T_4}\right)^{k/(k-1)} = \left(\frac{a_3}{a_4}\right)^{2k/(k-1)}$$

This equation may now be substituted into Equation (5.45) to yield

$$\frac{V_{g3}}{a_4} = \frac{2}{k-1}\left[1 - \left(\frac{p_3}{p_4}\right)^{(k-1)/(2k)}\right] \tag{5.46}$$

Now, one may write the governing equations for a normal shock wave moving into a gas which is at rest, as shown in Figure 5.45(a). The flow across the shock wave may be made steady by rendering the normal shock wave stationary, as shown in Figure 5.45(b). The continuity equation may now be applied to the control volume to yield

$$\rho_1 V_s = \rho_2 (V_s - V_{g2}) \tag{5.47}$$

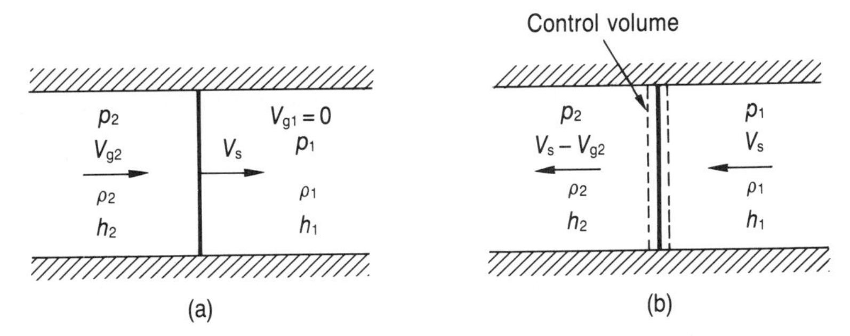

Figure 5.45 (a) A rightward-moving normal shock wave, and (b) when the rightward-moving normal shock wave is rendered stationary

For frictionless flow across the normal shock wave, the momentum equation becomes

$$p_1 - p_2 = \rho_2 (V_s - V_{g2})^2 - \rho_1 V_s^2 \tag{5.48}$$

For an adiabatic process across a stationary normal shock wave with no external work, the first law of thermodynamics becomes

$$h_1 + \frac{V_s^2}{2} = h_2 + \frac{(V_s - V_{g2})^2}{2} \tag{5.49}$$

At this stage, Equations (5.47) and (5.48) may be combined to yield

$$p_1 - p_2 = -\rho_1 V_s V_{g2} \tag{5.50}$$

However, the equation of state for a perfect gas, $p = \rho R T$, and the definition of the speed of sound, $a^2 = kRT$, may be combined to yield $p = \rho a^2/k$. In this case, Equation (5.50) becomes

$$\rho_1 a_1^2 - \rho_2 a_2^2 = -k \rho_1 V_s V_{g2}$$

where ρ_2 may be eliminated with the aid of Equation (5.47) to give

$$a_1^2 - \frac{V_s a_2^2}{V_s - V_{g2}} = -k V_s V_{g2} \tag{5.51}$$

For a perfect gas $h = c_p T$, $c_p = kR/(k-1)$ and $a^2 = kRT$. Then, Equation (5.49) becomes

$$\frac{a_2^2 - a_1^2}{k-1} = \frac{V_s^2}{2} - \frac{(V_s - V_{g2})^2}{2} = V_s V_{g2} - \frac{V_{g2}^2}{2} \tag{5.52}$$

Equations (5.51) and (5.52) may now be combined by eliminating a_2 to yield

$$V_{g2} = \frac{2}{k+1}\left(V_s - \frac{a_1^2}{V_s}\right) \tag{5.53}$$

Substituting Equation (5.53) into Equation (5.50) and noting that $\rho_1/p_1 = 1/(RT_1)$, it is possible to obtain

$$\frac{p_2}{p_1} = 1 + \frac{2k}{k+1}\left(\frac{V_s^2}{a_1^2} - 1\right)$$

In this equation, V_s may be eliminated with the aid of Equation (5.53) to yield

$$\frac{V_{g2}}{a_1} = \left(\frac{p_2}{p_1} - 1\right) \Big/ \left\{ k \sqrt{\left[1 + \frac{k+1}{2k}\left(\frac{p_2}{p_1} - 1\right)\right]} \right\} \tag{5.54}$$

When the initial temperatures of the high-pressure and low-pressure gases are the same, then $a_1 = a_4$. Finally, when Equations (5.46) and (5.54) are combined with $V_{g2} = V_{g3}$ and $p_2 = p_3$, it is possible to obtain

$$\frac{p_1}{p_4} = \frac{p_1}{p_2}\left(1 - \frac{k-1}{2k}\frac{p_2/p_1 - 1}{\sqrt{\{1 + [(k+1)/(2k)](p_2/p_1 - 1)\}}}\right)^{2k/k-1} \tag{5.55}$$

This equation is an implicit equation for the pressure ratio, p_2/p_1, across the normal shock wave as a function of the pressure ratio across the diaphragm, p_4/p_1, and it may be solved by trial and error. Figure 5.46 provides a plot of the pressure ratio, p_4/p_1, across the diaphragm versus the pressure ratio, p_2/p_1, across the normal shock wave in a shock tube with $k = 1.4$.

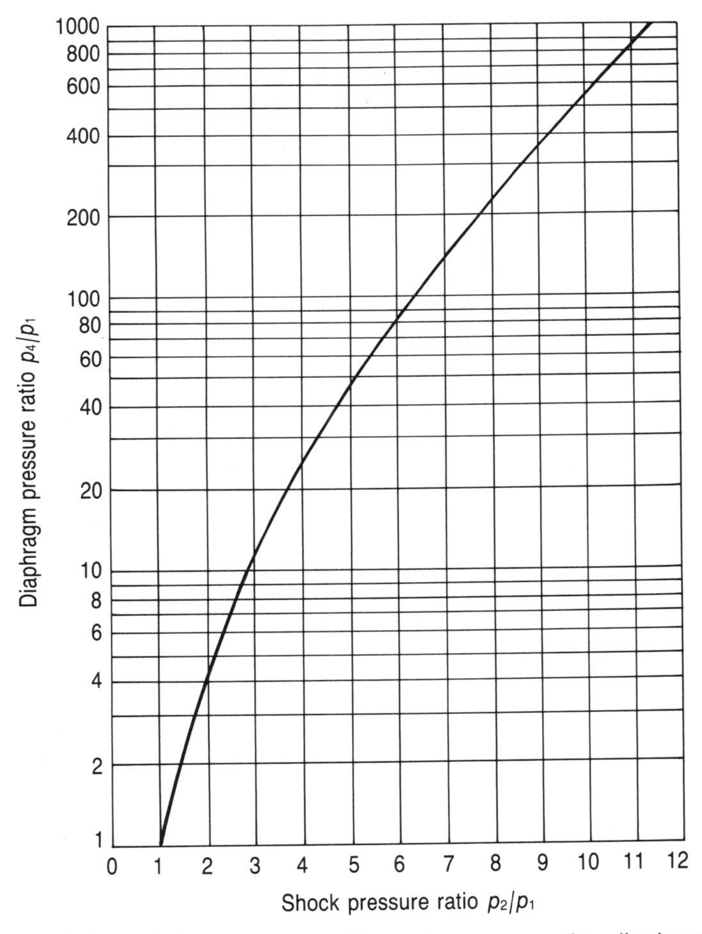

Figure 5.46 Variation of the pressure ratio, p_4/p_1, across the diaphragm with the pressure ratio, p_2/p_1, across the normal shock wave ($k = 1.4$)

Example 5.12

Air at a pressure of 500 kPa is separated from air at a pressure of 20 kPa by a diaphragm in a shock tube, as shown in Figure 5.47(a). The initial temperature of the air on both sides of the diaphragm is 300 K. The diaphragm is suddenly ruptured causing a normal shock wave to propagate rightwards into the low-pressure side and a series of expansion waves to propagate leftwards into the high-pressure side, as shown in Figure 5.47(b). Determine

a the pressure behind the normal shock wave,
b the velocity of the normal shock wave,
c the temperature and the velocity behind the normal shock wave,
d the speed of the contact surface,
e the pressure, velocity and temperature behind the contact surface,
f the speed of the expansion waves at the head and at the tail of the system of expansion waves, and
g the location of the normal shock wave, the contact surface and the system of expansion waves as a function of time.

Solution

(a) The normal shock wave moving rightwards in the shock tube, which is shown in Figure 5.48(a), may be rendered stationary, as shown in Figure 5.48(b). The

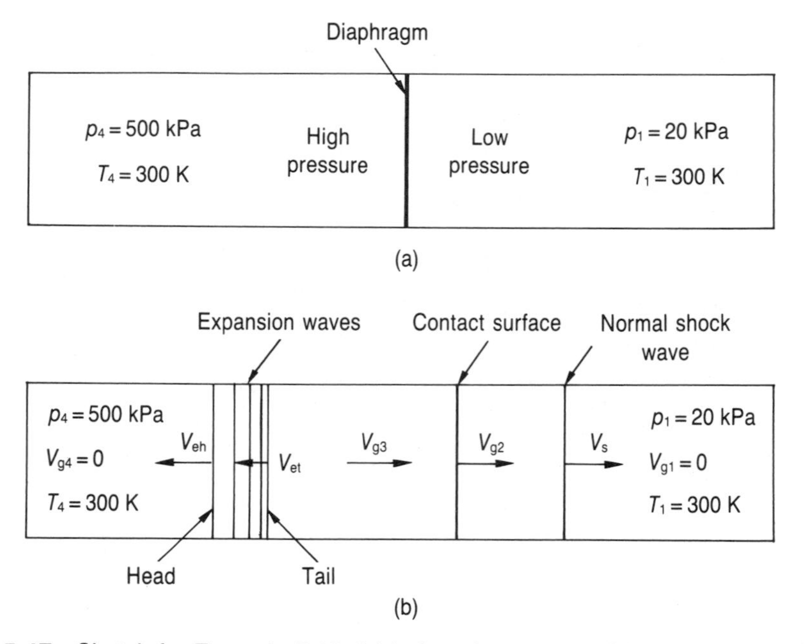

Figure 5.47 Sketch for Example 5.12: (a) before the rupture of the diaphragm; (b) after the rupture of the diaphragm

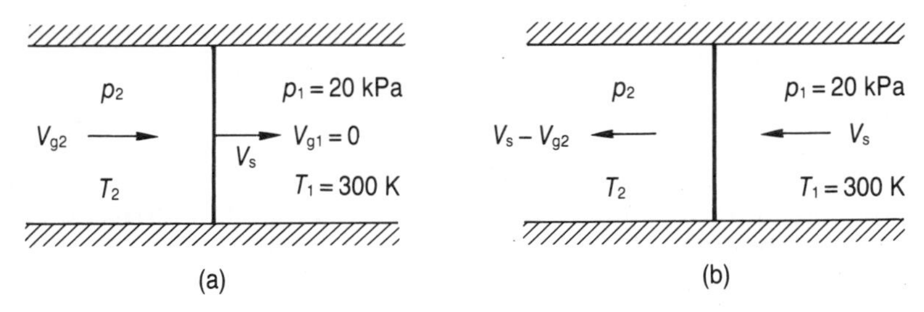

Figure 5.48 (a) Moving normal shock wave and (b) stationary normal shock wave for Example 5.12

pressure ratio p_2/p_1 across the normal shock wave may be evaluated either from Equation (5.55) or from Figure 5.46 corresponding to the pressure ratio $p_4/p_1 = 500\ \text{kPa}/20\ \text{kPa} = 25$ across the diaphragm as

$$p_2/p_1 = 4.047 \quad \text{or} \quad p_2 = (4.047)(20\ \text{kPa}) = \mathbf{80.94\ kPa}$$

(b) Now, using Appendix D with $p_2/p_1 = 4.047$, $M_s = 1.90$. The speed of the normal shock wave may now be evaluated as

$$V_s = M_s a_s = M_s\sqrt{(kRT_1)} = (1.90)\sqrt{[(1.4)(287.1\ \text{J kg}^{-1}\ \text{K}^{-1})(300\ \text{K})]} = \mathbf{659.8\ m\,s^{-1}}$$

(c) The temperature and the velocity behind the normal shock wave may now be evaluated from Appendix D, corresponding to $M_s = 1.90$, as

$$T_2/T_1 = 1.608 \qquad \text{or} \quad T_2 = (1.608)(300\ \text{K}) = \mathbf{482.4\ K}$$

$$\frac{V_s}{V_s - V_{g2}} = 2.516 \qquad \text{or} \quad V_{g2} = \frac{(1.516)(659.8\ \text{m s}^{-1})}{2.516} = \mathbf{397.6\ m\,s^{-1}}$$

(d) The contact surface moves at the same velocity as that of the air behind the normal shock wave so that

$$V_{cs} = V_{g2} = \mathbf{397.6\ m\,s^{-1}}$$

(e) The pressure and the velocity behind the contact surface are exactly the same as the pressure and the velocity behind the normal shock wave. Therefore

$$p_3 = p_2 = \mathbf{80.94\ kPa}$$

and

$$V_{g3} = V_{g2} = \mathbf{397.6\ m\,s^{-1}}$$

The temperature behind the contact surface may now be obtained by considering the isentropic process across the expansion waves between regions 3 and 4 in Figure 5.42(a) as

$$T_3 = T_4\left(\frac{p_3}{p_4}\right)^{(k-1)/k} = (300\ \text{K})\left(\frac{80.94\ \text{kPa}}{500\ \text{kPa}}\right)^{(1.4-1)/1.4} = \mathbf{178.3\ K}$$

(f) The wave at the head of the expansion wave system moves leftwards at the speed of sound relative to the gas which is at rest. Hence,

$$V_{eh} = a_4 = \sqrt{(kRT_4)} = \sqrt{[(1.4)(287.1\ \text{J kg}^{-1}\ \text{K}^{-1})(300\ \text{K})]} = \mathbf{347.3\ m\,s^{-1}}$$

However, the wave at the tail of the expansion wave system moves leftwards at the speed of sound relative to the gas which is moving rightwards at a speed of V_{g3}. Hence, the absolute velocity of the wave at the tail is

$$V_{et} = a_3 - V_{g3} = \sqrt{(kRT_3)} - V_{g3}$$

$$= \sqrt{[(1.4)(287.1\ \text{J kg}^{-1}\ \text{K}^{-1})(178.3\ \text{K})]} - 397.6\ \text{m s}^{-1}$$

$$= \mathbf{-129.9\ m\,s^{-1}}$$

The minus sign indicates that the tail of the expansion wave system is moving rightwards.

(g) The locations of the normal shock wave, the contact surface and the system of expansion waves are presented as a function of time in Figure 5.49.

In a shock tube, it is often desirable to subject a test object to uniform conditions of high pressure and high temperature for a maximum period of time. Uniform conditions behind the normal shock wave prevail until the passage of the contact

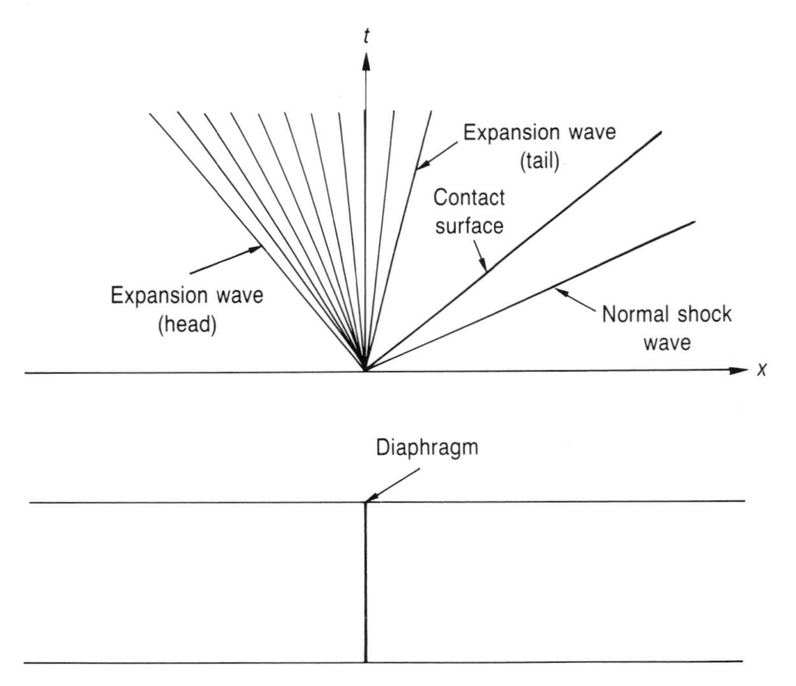

Figure 5.49 The locations of the normal shock wave, the contact surface and the system of expansion waves as a function of time

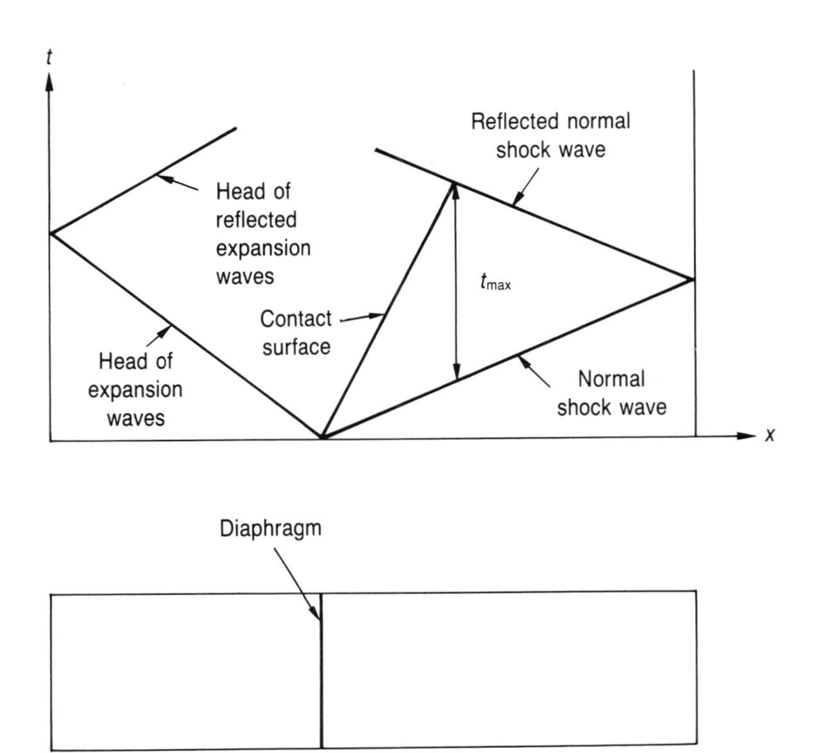

Figure 5.50 The optimum position at which the test object should be placed in order to subject it to uniform conditions of high pressure and high temperature for a maximum time interval

surface, or the shock reflected from the closed end of the tube or the front of the system of expansion waves reflected from the opposite end. Figure 5.50 shows the location of each of these waves as a function of time after the rupture of the diaphragm for a typical shock tube. The optimum position at which the test object should be placed in order to subject it to uniform flow conditions of high pressure and high temperature for a maximum time interval, t_{max}, is indicated in Figure 5.50. The test time is evidently a function of the shock-tube length, as well as the pressure ratio across the diaphragm.

Example 5.13
Air is stored in a tube of 5 m in length at a temperature of 300 K and a pressure of 200 kPa. A diaphragm at the end of the tube is suddenly ruptured causing expansion waves to propagate down the tube, as shown in Figure 5.51. If the atmospheric pressure is 100 kPa, then determine

a the velocity of the head of the expansion waves,

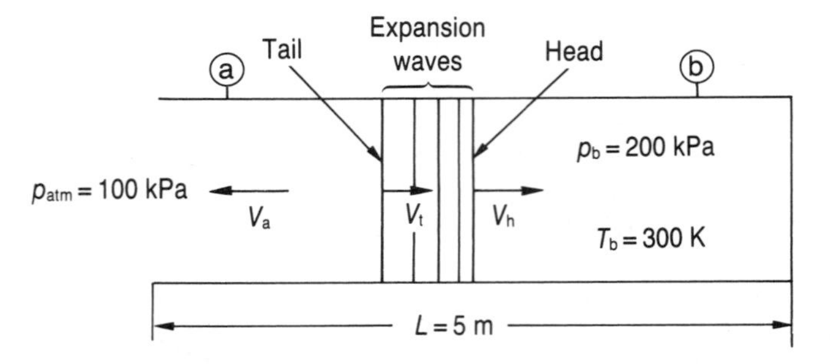

Figure 5.51 Sketch for Example 5.13

b the time for the head of expansion waves to reach the closed end of the tube,
c the pressure of the air behind the expansion waves, and
d the velocity of the tail of the expansion waves.

Solution

(a) The wave at the front of the expansion wave system moves rightwards at the speed of sound relative to the gas which is at rest. Thus

$$V_h = a_b = \sqrt{(kRT_b)} = \sqrt{[(1.4)(287.1\ \text{J kg}^{-1}\ \text{K}^{-1})(300\ \text{K})]} = \mathbf{347.3\ m\,s^{-1}}$$

(b) The time required for the first expansion wave to reach the closed end of the tube is

$$t = \frac{L}{V} = \frac{5\ \text{m}}{347.3\ \text{m s}^{-1}} = \mathbf{0.014\ 40\ s}$$

(c) As long as the open end of the tube is exposed to the atmosphere, then the pressure behind the expansion wave system must be the same as atmospheric pressure, that is

$$p_a = p_{atm} = 100\ \text{kPa}$$

(d) The velocity of the air behind the expansion wave system may then be obtained from Equation (5.46) as

$$V_a = \frac{2}{k-1}\left[1 - \left(\frac{p_a}{p_b}\right)^{(k-1)/2k}\right] a_b = \frac{2}{1.4-1}\left[1 - \left(\frac{100\ \text{kPa}}{200\ \text{kPa}}\right)^{(1.4-1)/[2(1.4)]}\right]$$

$$\times (347.3\ \text{m s}^{-1}) = \mathbf{163.7\ m\,s^{-1}}$$

Now, one might consider the isentropic process across the expansion wave system to determine the temperature behind the expansion wave as

$$T_a = T_b\left(\frac{p_a}{p_b}\right)^{(k-1)/k} = (300\ \text{K})\left(\frac{100\ \text{kPa}}{200\ \text{kPa}}\right)^{(1.4-1)/1.4} = 246.1\ \text{K}$$

The speed of sound in this region is then

$$a_a = \sqrt{(kRT_a)} = \sqrt{[(1.4)(287.1\ \mathrm{J\,kg^{-1}\,K^{-1}})(246.1\ \mathrm{K})]} = 314.5\ \mathrm{m\,s^{-1}}$$

However, the tail of the expansion wave system moves rightwards at the speed of sound relative to the gas which is moving leftwards at a speed of V_a. Hence, the absolute velocity of the tail of the expansion waves is

$$V_t = a_a - V_a = 314.5\ \mathrm{m\,s^{-1}} - 163.7\ \mathrm{m\,s^{-1}} = \mathbf{150.8\ m\,s^{-1}}$$

FURTHER READING

Cambel, A. B. and Jennings, B. H. (1958) *Gas Dynamics*, Dover: New York.

Daneshyar, H. (1976) *One-Dimensional Compressible Flow*, Pergamon: Oxford.

Fox, R. W. and McDonald, A. T. (1978) *Introduction to Fluid Mechanics*, 2nd edn, Wiley: New York.

John, J. E. A. (1984) *Gas Dynamics*, 2nd edn, Allyn & Bacon: Boston, MA.

Liepmann, H. W. and Roshko, A. (1957) *Elements of Gas Dynamics*, Wiley: New York.

Owczarek, J. A. (1964) *Fundamentals of Gas Dynamics*, International Textbook: Scranton, PA.

Shapiro, A. H. (1953) *The Dynamics and Thermodynamics of Fluid Flow*, Vol. 1, Ronald Press: New York.

Zucrow, M. J. and Hoffman, J. D. (1976) *Gas Dynamics*, Vol. 1, Wiley: New York.

PROBLEMS

5.1 A stream of air travelling at 500 m s^{-1} with a static pressure of 75 kPa and a static temperature of 15 °C undergoes a normal shock. Determine the static and the stagnation conditions and the air velocity after the shock wave.

5.2 Air approaches a normal shock wave with a Mach number of 3.0. The stagnation temperature and the stagnation pressure of the airstream are 700 K and 500 kPa, respectively. Determine the temperature and the velocity of the air leaving the normal shock wave and the entropy change across the normal shock wave.

5.3 A normal shock wave stands in a constant-area duct. Air approaches the normal shock wave with a stagnation temperature of 1000 K and a stagnation pressure of 700 kPa at a Mach number of 3.0. Determine the static pressure downstream of the shock. Compare the downstream pressure with the value that would be reached by decelerating the airstream isentropically to the same subsonic Mach number.

5.4 Air which enters a diverging duct is slowed down by a normal shock wave, as shown in the figure below. The airstream enters the duct at a Mach number of 3.0 and leaves it at a Mach number of 0.40. The exit cross-sectional area of the duct is twice the inlet cross-sectional area. Determine

a the pressure ratio across the normal shock wave, and
b the ratio of the exit pressure to the inlet pressure.

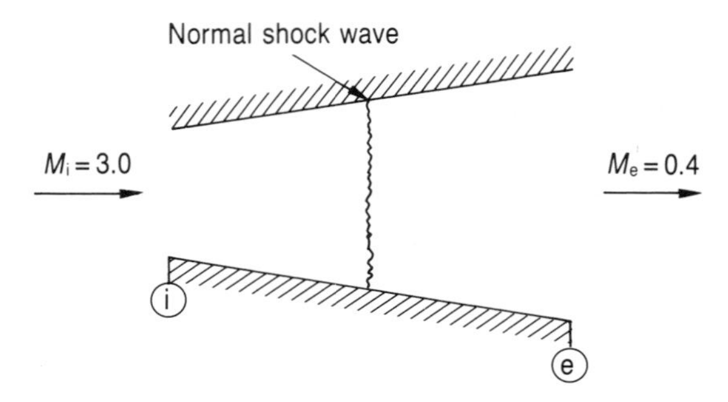

Problem 5.4

5.5 Helium enters a diverging duct with a velocity of 1800 m s^{-1}, a temperature of 400 K and a pressure of 500 kPa. The flow undergoes a normal shock wave, as shown in the figure below. The temperature at the exit of the duct is 600 K. The cross-sectional areas at the inlet and exit of the duct are 0.01 m^2 and 0.02 m^2, respectively. Assuming adiabatic flow, determine the force which is required to hold the duct in place. (METU 1988)

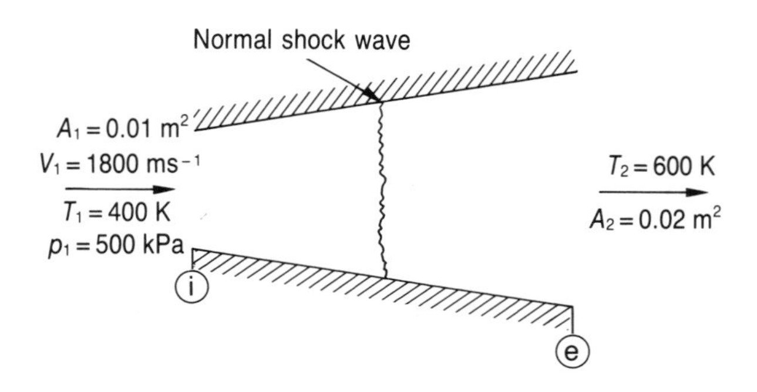

Problem 5.5

5.6 Consider a normal shock wave in a non-perfect gas, as shown in the figure below. Starting from first principles, show that

$$h_2 - h_1 = \frac{\rho_1 + \rho_2}{2\rho_1\rho_2}(p_2 - p_1)$$

(METU 1987)

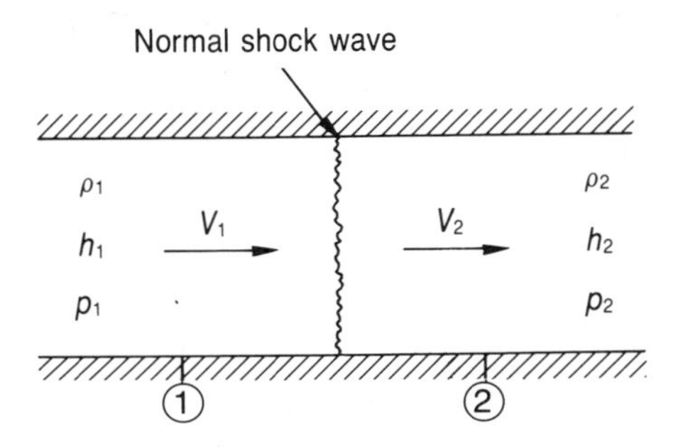

Problem 5.6

5.7 A normal shock wave moves at a constant velocity of 1000 m s^{-1} into still air with a temperature and pressure of 300 K and 100 kPa, respectively. Determine

a the velocity of the air behind the normal shock wave,
b the static pressure and temperature behind the normal shock wave, and
c the stagnation pressure and temperature behind the normal shock wave.

5.8 A normal shock wave is observed to move through a constant-area tube into still air with a temperature of 300 K. The velocity of the air behind the normal shock wave is measured to be 150 m s^{-1}. Determine the velocity of the normal shock wave.

5.9 A piston in a tube is suddenly accelerated to a velocity of 50 m s^{-1} which causes a normal shock wave to move into the stationary air in the tube. One second later, the piston is suddenly accelerated from 50 m s^{-1} to 100 m s^{-1} which causes a second normal shock wave to move down the tube. The initial temperature of the air is 300 K.

a Determine the velocity of the first and second normal shock wave.
b How much time will elapse from the initial acceleration of the piston to the intersection of the two shocks?

5.10 Air at a pressure of 100 kPa and a temperature of 290 K is flowing in a constant-area tube at a velocity of 100 m s^{-1}. Suddenly, the end of the tube is closed so that a normal shock wave propagates back through the airstream. Determine the absolute velocity of the normal shock wave.

5.11 A normal shock wave travels at a velocity of 1000 m s^{-1} into still air with a temperature of 275 K and a pressure of 100 kPa and reflects back from a plane wall. Determine

a the velocity of the reflected normal shock wave,
b the pressure ratio across the incident normal shock wave,
c the pressure ratio across the reflected normal shock wave, and
d the stagnation pressure and temperature behind the reflected normal shock wave.

5.12 The knocking in a car engine can be approximated as a normal shock wave travelling at a velocity of 800 m s^{-1} downwards into the unburned mixture of air and fuel at a pressure and temperature of 600 kPa and 450 K, respectively, as shown in the figure below. At the same time, the piston is moving upwards at a uniform speed of 8 m s^{-1}. Determine the pressure acting on the piston face after the normal shock wave reflects from its face. Assume that the air–fuel mixture behaves like a perfect gas and has the properties of air. (METU 1987)

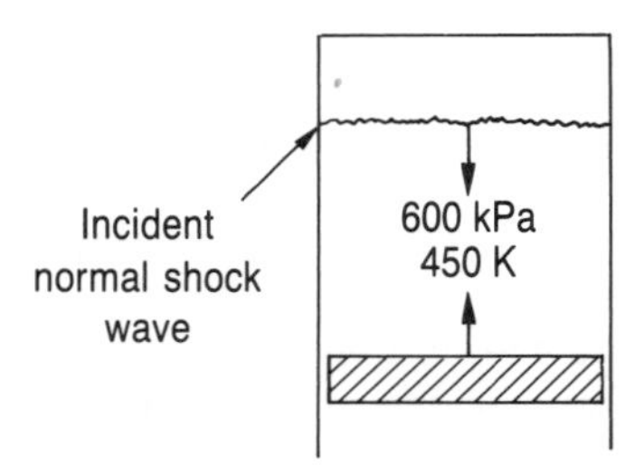

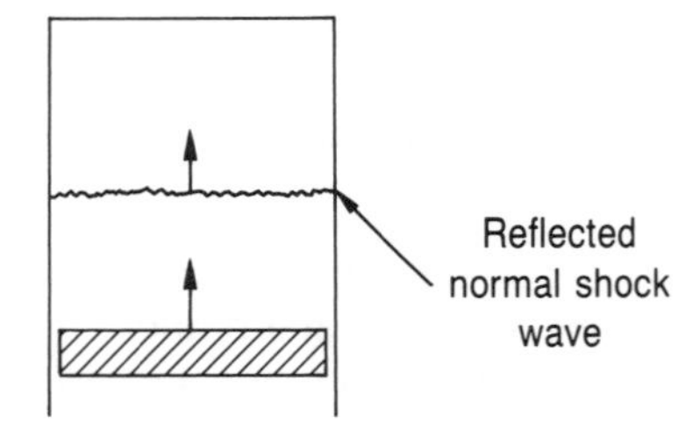

Problem 5.12

5.13 A normal shock wave moves down a shock tube at a velocity of 700 m s^{-1} into the air with a pressure of 150 kPa and a temperature of 350 K, as

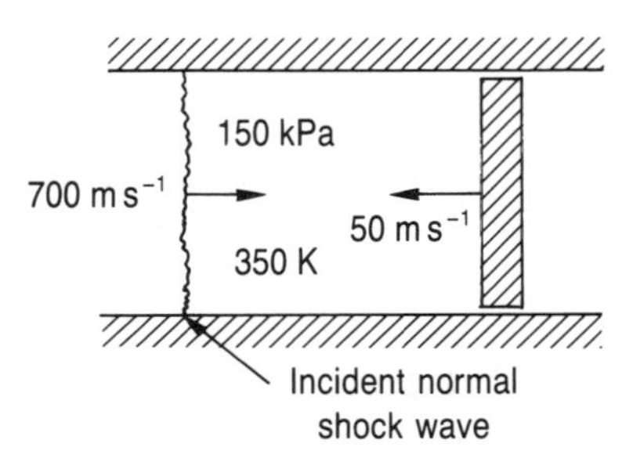

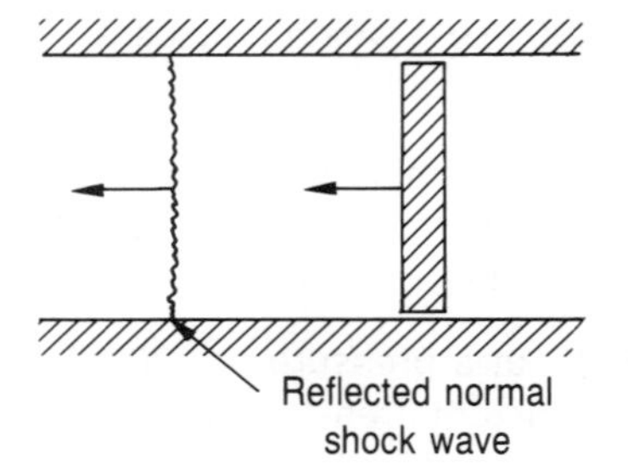

Problem 5.13

shown in the figure opposite. At the end of the tube, a piston is moving in the opposite direction at a velocity of 50 m s^{-1}. Determine

a the velocity of the reflected normal shock wave, and
b the pressure and temperature behind the reflected normal shock wave.

5.14 A converging–diverging nozzle receives air from a large reservoir where the pressure and temperature are 200 kPa and 300 K, respectively. The exit and throat cross-sectional areas of the nozzle are 0.002 m^2 and 0.001 m^2, respectively. Determine

a the maximum back pressure to choke the nozzle,
b the back pressure for the nozzle to be perfectly expanded to the design Mach number,
c the back pressure such that there is a normal shock wave at the exit plane, and
d the ranges of back pressure for subsonic, non-isentropic, overexpansion and underexpansion flow regimes.

5.15 At a section upstream of the throat in a converging–diverging nozzle, the velocity, pressure and temperature of the air are 208.4 m s^{-1}, 180 kPa and 27 °C, respectively. The ratio of the exit cross-sectional area to the throat cross-sectional area of the nozzle is 2.40.

a Determine the ranges of the back pressure for subsonic, non-isentropic, overexpansion and underexpansion flow regimes. Show the variation of the pressure with the distance along the nozzle.
b Determine the back pressure if a normal shock wave stands in the diverging section of the nozzle where the cross-sectional area is twice that of the throat of the nozzle.

(METU 1988)

5.16 A converging–diverging nozzle is supplied from a large air reservoir. The nozzle has a throat area of 10 cm^2 and an exit area of 20 cm^2. The pressure and temperature in the reservoir are 300 kPa and 30 °C, respectively.

a Determine the ranges of back pressure over which (i) a normal shock wave appears in the nozzle, (ii) oblique shock waves form at the exit, and (iii) the expansion waves form.
b Calculate the mass flow rate through the nozzle for the case where a normal shock wave appears at the exit plane.

5.17 Air flows through a converging–diverging nozzle. The flow can be considered to be frictionless and adiabatic. The pressure, temperature, velocity and cross-sectional area at a section in the diverging part of the nozzle are 45 kPa, 271 K, 528 m s^{-1} and 0.0005 m^2, respectively. The exit cross-sectional area of the nozzle is 0.0009 m^2 and the back pressure is 60 kPa.

a Determine the mass flow rate.
b Determine the area and the Mach number at the throat.
c Determine the exit Mach number and the exit pressure.
d Show the process on a T–s diagram.
(METU 1985)

5.18 A converging–diverging nozzle is fed with air from a large reservoir, where the temperature and pressure are 500 K and 450 kPa, respectively. The ratio of the exit to the throat cross-sectional areas is 1.25. The nozzle discharges to the atmosphere with a pressure of 100 kPa. Determine the Mach number, pressure and temperature at the exit plane, and also calculate the mass flux.

5.19 A large reservoir contains air at a stagnation pressure of 600 kPa and a stagnation temperature of 300 K. The air is discharged to a second reservoir through a converging–diverging nozzle, where the flow can be assumed frictionless and adiabatic. It is found that there is a normal shock wave occurring at a section where the cross-sectional area is 0.0015 m^2. At the exit plane, where the cross-sectional area is 0.0020 m^2, the Mach number is 0.40.

a Determine the stagnation pressure loss across the nozzle.
b Determine the mass flow rate.
c Determine the throat cross-sectional area.
d Determine the exit pressure.
e Determine the back pressure.
f Show the process on a T–s diagram.
(METU 1985)

5.20 Air is supplied to a converging–diverging nozzle from a large reservoir where the temperature is 400 K, as shown in the figure below. At section 1, which is upstream of the throat, the velocity is 120 m s^{-1}. A normal shock wave stands at the diverging part of the nozzle where the cross-sectional area is $A_x = A_y = 2A_1$. The exit cross-sectional area of the nozzle is $A_2 = 4A_1$. If the nozzle discharges to an ambient pressure of 100 kPa,

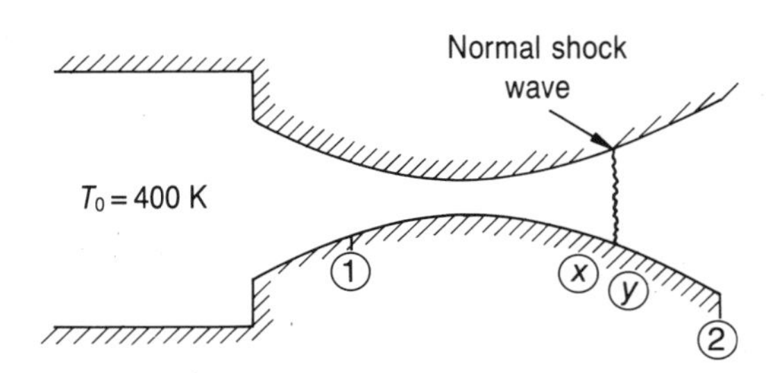

Problem 5.20

determine the pressure in the reservoir. Also, find the stagnation pressure and the stagnation temperature at the exit plane. (METU 1988)

5.21 The ratio of the exit to the throat cross-sectional areas in a converging–diverging nozzle of a rocket exhaust is 4.0. The exhaust gases are generated in a combustion chamber with a stagnation pressure of 3 MPa and a stagnation temperature of 1500 K. Assume that the exhaust gas mixture behaves like a perfect gas with a specific heat ratio of 1.3 and a molecular mass of 20 kg per kg mol. Determine the rocket exhaust velocity for the exit shock condition.

5.22 A convergent–divergent nozzle, which is shown in the figure below, is supplied with air from a large tank where the pressure is p_0 and exhausts to the atmosphere with a pressure of p_b. The area A at a distance x from the throat is given by $A = A_t(1 + 0.75x/L)$, where A_t is the throat cross-sectional area and L is the total length of the diverging part of the nozzle. At a certain pressure ratio p_0/p_b, a normal shock wave stands in the nozzle at a point where the Mach number before the normal shock wave is 1.7. Find the position of the normal shock wave from the throat and the pressure ratio p_b/p_0. (METU 1987)

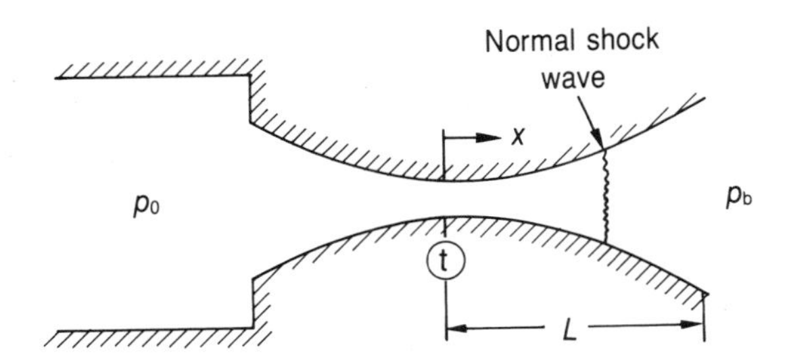

Problem 5.22

5.23 A convergent–divergent duct with a throat cross-sectional area half that of the exit cross-sectional area is supplied with air at a stagnation pressure of 140 kPa. It discharges into the atmosphere with a static pressure of 100 kPa. Show that there is a normal shock wave in the duct, and find the stagnation pressure of the air at the exit of the nozzle, the Mach numbers immediately upstream and downstream of the normal shock wave, and the ratio of the area to the throat area at the location of the normal shock wave. (METU 1987)

5.24 At a point upstream of the throat of a converging–diverging nozzle, the pressure, temperature and velocity of the air are 400 kPa, 300 K and 100 m s^{-1}, respectively. The exit and throat cross-sectional areas of the nozzle are 0.0015 m^2 and 0.001 m^2, respectively. The nozzle discharges to a back pressure of 320 kPa. Determine the pressure and the Mach number at the exit plane of the nozzle.

5.25 The duct shown in the figure below is fed with air from a large reservoir. The flow within the duct may be considered isentropic except through the normal shock waves. The cross-sectional area of the duct is constant after the normal shock wave at section 2. If the duct discharges into a region where the pressure is 100 kPa, determine

a the Mach number at the exit plane of the duct,
b the pressure at the first throat, and
c the stagnation pressure in the reservoir which feeds the duct.
(METU 1987)

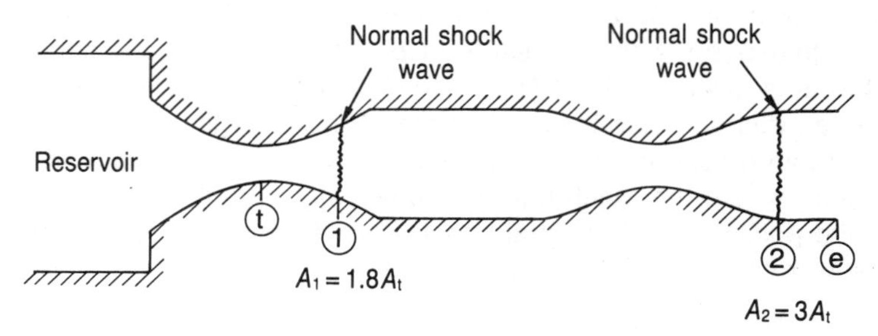

Problem 5.25

5.26 A supersonic wind tunnel is to be constructed, as shown in the figure below, with air at atmospheric pressure passing through a converging–diverging nozzle into a constant-area test section and then into a large vacuum tank. The test run is started with a pressure of 0 kPa in the tank. How long can uniform flow conditions be maintained in the test section? Assume the test section to be circular with a diameter of 0.1 m and a design Mach number of 2.4. The tank volume is 3 m^3 with atmospheric conditions of 101 kPa and 20 °C. Assume that the air is brought to rest adiabatically in the tank. (METU 1991)

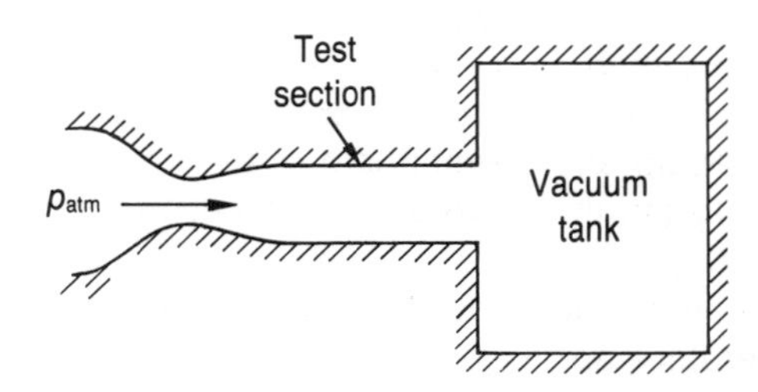

Problem 5.26

5.27 A supersonic wind tunnel, which is shown in the figure below, is fed from a large tank. The air is discharged to the atmosphere. The Mach number in the test section is 2.65 and a normal shock wave stands at the exit of the wind tunnel. The pressure at section 3 immediately after the normal shock wave is 100 kPa. The flow is isentropic except through the normal shock wave.

a Determine p_0, p_1, p_2 and M_3.
b If a Pitot tube is placed in the exit jet, then calculate the pressure measured by the Pitot tube.
(METU 1985)

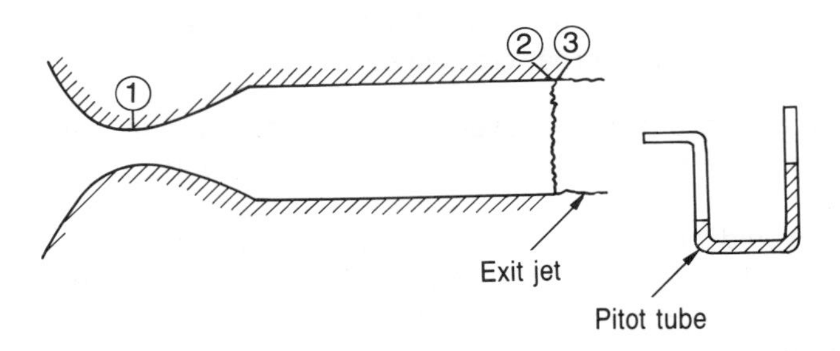

Problem 5.27

5.28 A supersonic wind tunnel is designed for obtaining supersonic flow in the test section. However, during the start-up of the wind tunnel, a normal shock wave stands in the diverging section of the nozzle. Upstream of this normal shock wave, the Mach number and pressure are 1.3 and 50 kPa, respectively. The cross-sectional area of the nozzle throat is 0.01 m^2. The temperature at the test section with a cross-sectional area of 0.02 m^2 is found to be 10 °C. Determine

a the mass flow rate, and
b the minimum possible value of the diffuser throat area for the same mass flow rate.
(METU 1986)

5.29 In a fixed-geometry wind tunnel, air is drawn in by means of a fan. The stagnation temperature and the stagnation pressure of the supply air are 300 K and 100 kPa, respectively. The nozzle has a throat cross-sectional area of 0.4 m^2 and discharges into a test section with a constant cross-sectional area of 0.8 m^2. The throat and exit cross-sectional areas of the diffuser are 0.65 m^2 and 1.0 m^2, respectively. Determine

a the maximum possible Mach number at the test section,
b the minimum allowable diffuser throat cross-sectional area,
c whether or not a supersonic flow can be obtained at the test section,

d the back pressure for the worst operating condition,
e the back pressure to choke the flow at the throat of the nozzle,
f the location of the normal shock wave when the back pressure for the worst operating condition is reduced by 4 per cent,
g the back pressure for the best operating condition, and
h the ranges of back pressure for the subsonic, nozzle shock and diffuser shock flow regimes.

5.30 In a supersonic wind tunnel, which is shown in the figure below, losses are only due to the shock waves which may be present in the tunnel. Reservoirs 1 and 2 are at the stagnation conditions. The operating Mach number in the test section is 1.7. The diffuser throat cross-sectional area is 1.25 times the nozzle throat cross-sectional area.

a Find the pressure ratio between the reservoirs for starting.
b Calculate the pressure ratio between the reservoirs at the best operating condition.
c What would be the maximum Mach number that could be obtained by changing the test section area only? What would be the ratio of the test section cross-sectional area to the nozzle throat cross-sectional area in this case? (METU 1991)

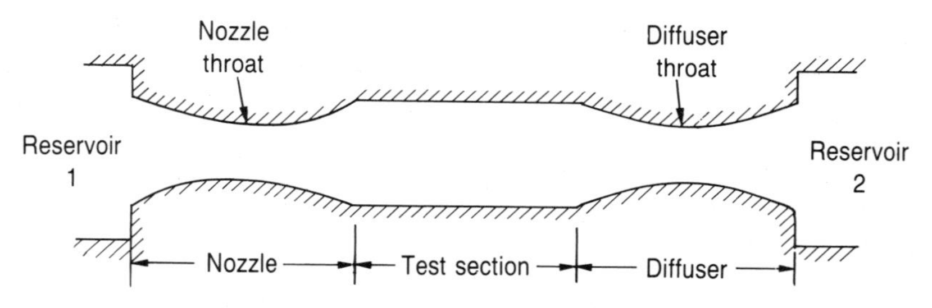

Problem 5.30

5.31 A continuous fixed-geometry supersonic wind tunnel is designed for a test section Mach number of 1.5 to simulate conditions at an altitude of 10 000 m where the pressure and temperature are 26.5 kPa and 223.3 K, respectively. The test section has a cross-sectional area of 0.05 m^2. Assume that the compressor is isentropic with a cooler located between the compressor and the nozzle, as in the figure below. In this case, the air temperature at the inlet of the compressor can be assumed equal to the stagnation temperature of the test section. Neglecting friction, determine the power requirements of the compressor during steady-state operation and during start-up of the wind tunnel. (METU 1991)

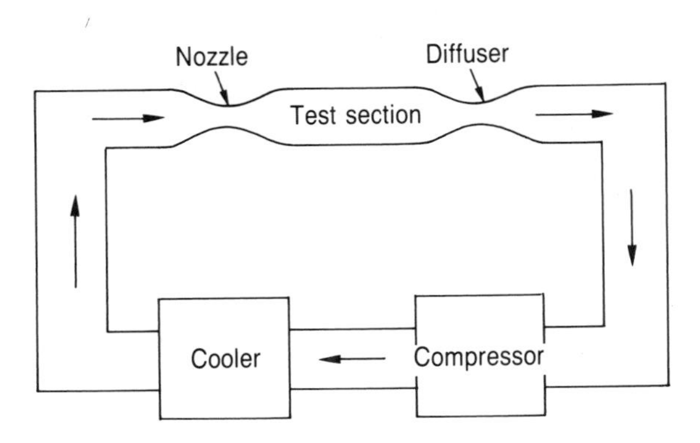

Problem 5.31

5.32 A supersonic fixed-geometry inlet is operating at sea level where the temperature and pressure are 300 K and 100 kPa, respectively. The inlet and throat cross-sectional areas are 0.08 m^2 and 0.06 m^2, respectively.

a Determine the Mach number to which the inlet must be overspeeded for starting.
b Determine the Mach number for which the inlet is designed.
c Determine the mass flow rate when the inlet is overspeeded.
d Determine the mass flow rate at the design flight Mach number.

5.33 A supersonic variable-geometry inlet is designed for a Mach number of 1.70. The cross-sectional area at the entrance of the inlet is 0.08 m^2. Determine

a the throat cross-sectional area when the normal shock wave is just swallowed, and
b the throat cross-sectional area at the design flight Mach number.

5.34 A supersonic fixed-geometry inlet, which is shown in the figure below, is designed for an aircraft velocity of 1200 $m\,s^{-1}$ at atmospheric conditions of

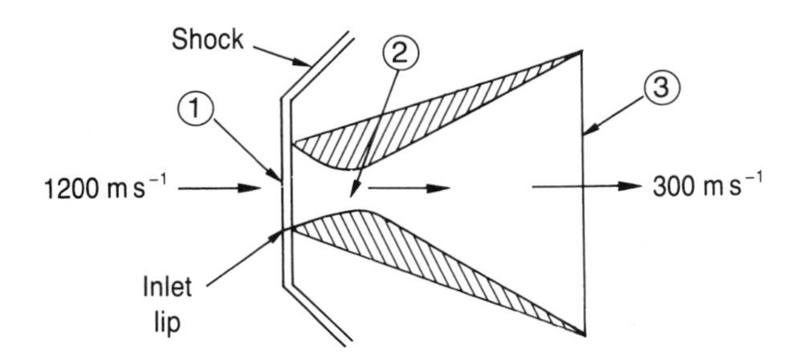

Problem 5.34

25 kPa and 224.5 K. The air velocity at cross-section 3 is going to be 300 m s^{-1}. A normal shock hangs at the inlet lip. Flow is isentropic except through the shocks.

a Calculate A_1/A_2.
b Calculate A_3/A_2.
c Calculate the static and stagnation pressures at section 3.
(METU 1991)

5.35 A combined Pitot tube is used for determining the velocity of a supersonic airstream. The static and stagnation pressures are measured to be 150 kPa and 250 kPa, respectively. If the temperature of the approaching air is 300 K, determine the velocity of the approaching airstream.

5.36 A Pitot tube on board an aircraft travelling at supersonic speed registers a 99.2 kPa stagnation pressure. It is known that atmospheric conditions are 25 kPa and 220 K, as in the figure below.

a What is the velocity of the aircraft?
b What is the absolute velocity (velocity with respect to an observer who is stationary) of the air behind the shock wave created by the aircraft? Calculate also the change in stagnation temperature of the air as the shock wave passes through the undisturbed medium.
(METU 1991)

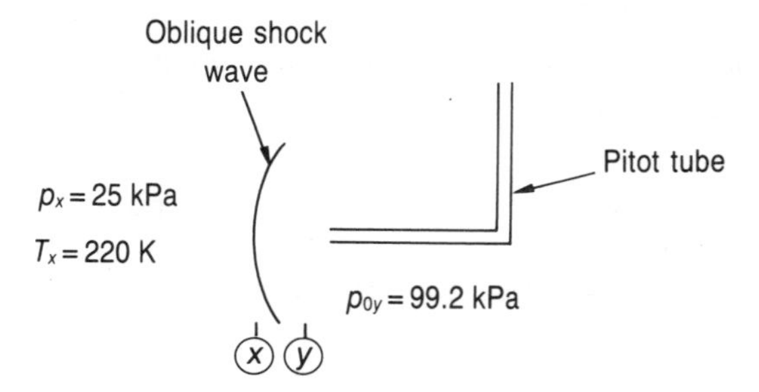

Problem 5.36

5.37 The pressure ratio across a diaphragm in a shock tube is 8. The initial pressure of the air on the low-pressure side of the diaphragm is 50 kPa, while the initial temperature on each side of the diaphragm is 300 K. Determine

a the velocity of the initial normal shock wave,
b the Mach number of the air behind the normal shock wave, and
c the static pressure and temperature behind the normal shock wave.

5.38 A tube with a length of 2 m contains air at a pressure of 5 kPa and a temperature of 300 K. When a diaphragm at the end of the tube is ruptured, a normal shock wave moves down the tube. The ambient pressure is 100 kPa. A test object is located midway along the tube. Determine

a the velocity of the initial normal shock wave,
b the velocity and Mach number of the air behind the initial normal shock wave,
c the static pressure and temperature behind the initial normal shock wave,
d the velocity of the normal shock wave which reflects from the closed end of the tube, and
e the time that the test object is subjected to the pressure and the temperature behind the initial normal shock wave before the arrival of the reflected shock wave.

5.39 A shock tube has a length of 10 m and a diaphragm located midway along the tube. The high- and low-pressure sections contain air at 200 kPa and 10 kPa, respectively. A test object is placed at the low-pressure section, 3 m from the diaphragm. The temperature of the air at both sections is 300 K. The diaphragm is suddenly ruptured causing a normal shock wave to propagate into the low-pressure region.

a Determine the shock velocity.
b Determine the velocity of the contact surface.
c Determine the Mach number behind the normal shock wave.
d Determine the time between the passage of the normal shock wave and the contact surface over the test object.
e Determine the velocity of the reflected shock wave.
f Provide an x–t diagram showing the location of the initial shock wave, the reflected shock wave and the contact surface as a function of time.
g Determine the location of the test object that would produce uniform conditions behind the initial shock wave for the maximum amount of time.

5.40 Air is stored in a tube with a length of 20 m at a temperature of 300 K and a pressure of 150 kPa. A diaphragm at the end of the tube is suddenly ruptured causing expansion waves to propagate down the tube. The ambient pressure is 100 kPa. Determine

a the velocity of the first expansion wave,
b the time required for the first expansion wave to reach the closed end of the tube,
c the velocity of air behind the expansion wave, and
d the velocity of the last expansion wave.

5.41 A normal shock wave moves in a constant-area tube at a velocity of 600 m s^{-1} and reflects from the open end of the tube as a series of expansion waves. If the ambient pressure and temperature are 100 kPa

and 300 K, respectively, determine the velocity of the first and last reflected expansion waves.
(METU 1991)

5.42 The initial pressure ratio across a diaphragm of a shock tube is 20. It is desirable to subject a test object to momentary conditions of high pressure and temperature for a period of 0.2 s between the time of passage of the initial shock wave and the time of passage of the shock wave reflected from the closed end of the tube over the test object. The test object is located 4 m from the diaphragm and the initial temperature of the air in the shock tube is 300 K. Determine a suitable length for the low-pressure side of the shock tube.
(METU 1991)

6 Frictional flow in constant-area ducts

6.1 INTRODUCTION

In Chapter 4, the compressible flow in ducts was investigated for the case in which the changes in flow properties are brought about only by changes in area. However, in a real flow situation frictional forces are present and may have an important effect on the flow. Naturally, the inclusion of frictional terms in the equations of motion makes the resultant analysis far more complex. For this reason, in order to study the effect of friction on the compressible flow in ducts certain restrictions are placed on the flow.

The first part of this chapter is concerned with compressible flow in constant-area insulated ducts which eliminates the effects of the change in area and the heat transfer. These restrictions limit the applicability of the resultant analysis to many practical and important flow problems, such as the flow in constant-area ducts, which are reasonably short. This treatment is possible because these ducts may be assumed approximately adiabatic due to negligible heat transfer to or from the fluid.

In the second part of this chapter, the other extreme case – when the constant-area ducts are extremely long – will be treated. This case approximates the flow of a natural gas through a long uninsulated pipeline where there is sufficient area for heat transfer to make the flow non-adiabatic and approximately isothermal.

These two cases cover a wide range of frictional flows which are consequently of great significance. The analysis presented in this chapter is simplified by assuming steady, one-dimensional flow of a perfect gas with constant specific heats.

6.2 GOVERNING EQUATIONS FOR ADIABATIC ONE-DIMENSIONAL FLOW OF A PERFECT GAS WITH FRICTION IN CONSTANT-AREA DUCTS

Now, consider steady, one-dimensional adiabatic flow of a perfect gas through a constant-area duct with friction, as shown in Figure 6.1. The governing equations are given below for the mentioned control volume.

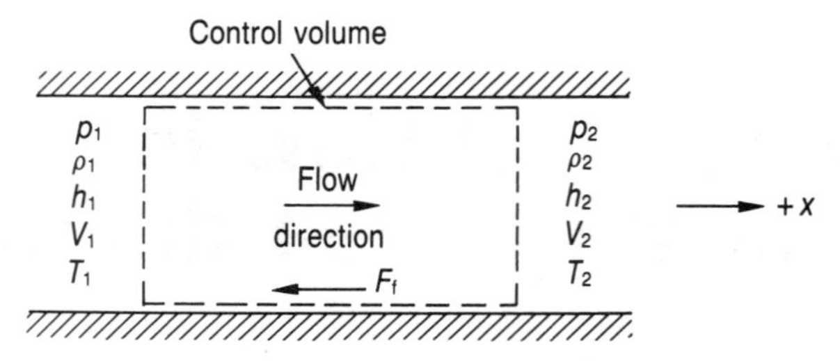

Figure 6.1 Control volume for steady, one-dimensional adiabatic flow with friction in a constant-area duct

(a) Continuity equation

For steady and uniform flow at each cross-section, the mass flow rate is constant. Then,

$$\dot{m} = \rho_1 A_1 V_1 \perp \rho_2 A_2 V_2 = \text{constant}$$

As long as the area of the duct is constant, that is $A_1 = A_2 = A$, then the mass flow rate per unit area or the mass flux, G, is given by

$$G = \dot{m}/A = \rho_1 V_1 = \rho_2 V_2 = \rho V = \text{constant} \tag{6.1}$$

(b) Momentum equation

For steady and uniform flow with friction, the sum of the pressure forces and the frictional forces acting on the control volume of Figure 6.1 must be balanced by the rate of change of linear momentum across the control volume, that is

$$-F_f + p_1 A_1 - p_2 A_2 = \rho_2 A_2 V_2^2 - \rho_1 A_1 V_1^2 \tag{6.2}$$

where F_f is the frictional force acting on the side walls of the control volume.

(c) The first law of thermodynamics

When there is no heat exchange between the control volume and its surroundings, and when there is no external shaft work on the control volume, the first law of thermodynamics with negligible elevation changes is

$$h_{01} = h_1 + \frac{V_1^2}{2} = h_2 + \frac{V_2^2}{2} = h_{02} = h + \frac{V^2}{2} \tag{6.3}$$

(d) Equation of state

For a perfect gas, the equation of state is given by

$$p = \rho R T \tag{6.4}$$

(e) The second law of thermodynamics

As long as the frictional and adiabatic flow is irreversible, then the second law of thermodynamics for the mentioned control volume may be given as

$$s_2 > s_1 \tag{6.5}$$

To calculate the entropy change, the thermodynamic relations (2.45) and (2.46)

$$T\,\mathrm{d}s = \mathrm{d}h - \frac{\mathrm{d}p}{\rho}$$

and

$$T\,\mathrm{d}s = \mathrm{d}u - RT\frac{\mathrm{d}\rho}{\rho}$$

may be used. But for a perfect gas $\mathrm{d}h = c_p\,\mathrm{d}T$, $\mathrm{d}u = c_v\,\mathrm{d}T$ and $p = \rho RT$. Then

$$\mathrm{d}s = c_p\frac{\mathrm{d}T}{T} - R\frac{\mathrm{d}p}{p} = c_v\frac{\mathrm{d}T}{T} - R\frac{\mathrm{d}\rho}{\rho}$$

If the above expression is integrated for constant specific heats, then the entropy change is

$$s_2 - s_1 = c_p \ln\left(\frac{T_2}{T_1}\right) - R\ln\left(\frac{p_2}{p_1}\right) = c_v \ln\left(\frac{T_2}{T_1}\right) - R\ln\left(\frac{\rho_2}{\rho_1}\right) \tag{6.6}$$

Example 6.1

An insulated tube with a cross-sectional area of 40 mm^2 is fed with air by an isentropic converging nozzle from a large tank where the temperature and pressure are 300 K and 100 kPa, respectively, as shown in Figure 6.2. At section 1, where the nozzle joins to the constant-area tube, the pressure is 95 kPa. At section 2, which is located some distance downstream in the tube, the air temperature is 290 K. Determine.

- **a** the mass flow rate,
- **b** the stagnation pressure at section 2, and
- **c** the frictional force on the duct wall between sections 1 and 2.

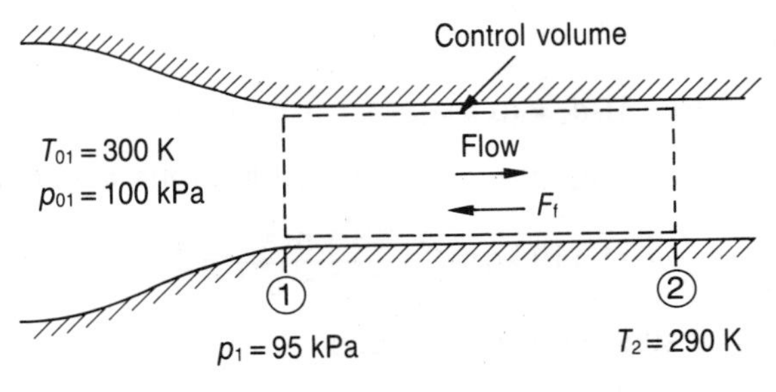

Figure 6.2 Sketch for Example 6.1

Solution

(a) For isentropic flow through the converging nozzle, the local isentropic stagnation properties are constant. Thus,

$$p_1/p_{01} = 95 \text{ kPa}/100 \text{ kPa} = 0.95$$

From Appendix C, corresponding to $p_1/p_{01} = 0.95$,

$$M_1 = 0.27$$

and

$$T_1/T_{01} = 0.9856 \quad \text{or} \quad T_1 = (0.9856)(300 \text{ K}) = 295.7 \text{ K}$$

But, for a perfect gas,

$$\rho_1 = \frac{p_1}{RT_1} = \frac{95\,000 \text{ N m}^{-2}}{(287.1 \text{ J kg}^{-1} \text{K}^{-1})(295.7 \text{ K})} = 1.119 \text{ kg m}^{-3}$$

The speed of sound at section 1 is

$$a_1 = \sqrt{(kRT_1)} = \sqrt{[(1.4)(287.1 \text{ J kg}^{-1} \text{K}^{-1})(295.7 \text{ K})]} = 344.8 \text{ m s}^{-1}$$

and the velocity is

$$V_1 = M_1 a_1 = (0.27)(344.8 \text{ m s}^{-1}) = 93.1 \text{ m s}^{-1}$$

Then, the mass flow rate from the continuity equation is

$$\dot{m} = \rho_1 V_1 A = (1.119 \text{ kg m}^{-3})(93.1 \text{ m s}^{-1})(40 \times 10^{-6} \text{ m}^2) = \mathbf{0.004\,167 \text{ kg s}^{-1}}$$

(b) As long as the flow in the duct is adiabatic, the stagnation temperature remains constant, that is $T_{02} = T_{01}$. Then,

$$T_2/T_{02} = 290 \text{ K}/300 \text{ K} = 0.9667$$

Now from Appendix C, corresponding to $T_2/T_{02} = 0.9667$,

$$M_2 = 0.41$$

The speed of sound at section 2 is

$$a_2 = \sqrt{(kRT_2)} = \sqrt{[(1.4)(287.1 \text{ J kg}^{-1} \text{K}^{-1})(290 \text{ K})]} = 341.4 \text{ m s}^{-1}$$

and the velocity is

$$V_2 = M_2 a_2 = (0.41)(341.4 \text{ m s}^{-1}) = 140 \text{ m s}^{-1}$$

The density at section 2 may now be evaluated from the continuity equation as

$$\rho_2 = \rho_1 \frac{V_1}{V_2} = (1.119 \text{ kg m}^{-3}) \frac{93.1 \text{ m s}^{-1}}{140 \text{ m s}^{-1}} = 0.7441 \text{ kg m}^{-3}$$

From the equation of state for a perfect gas

$$p_2 = \rho_2 R T_2 = (0.7441 \text{ kg m}^{-3})(287.1 \text{ J kg}^{-1} \text{K}^{-1})(290 \text{ K}) = 61.95 \text{ kPa}$$

The stagnation pressure at section 2, using Appendix C with $M_2 = 0.41$, is

$$p_2/p_{02} = 0.8907 \quad \text{or} \quad p_{02} = 61.95\ \text{kPa}/0.8907 = \mathbf{69.55\ kPa}$$

(c) The friction force may be found by applying the momentum equation to the control volume of Figure 6.2 as

$$-F_f + p_1 A - p_2 A = \rho_2 V_2^2 A - \rho_1 V_1^2 A = \dot{m}(V_2 - V_1)$$

or

$$F_f = -(0.004\,167\ \text{kg s}^{-1})(140\ \text{m s}^{-1} - 93.1\ \text{m s}^{-1})$$
$$+ (95\,000\ \text{N m}^{-2} - 61\,950\ \text{N m}^{-2})(40 \times 10^{-6}\ \text{m}^2) = 1.127\ \text{N}$$

This is the force exerted on the control volume by the duct wall. The force of the fluid on the duct is

$$F = -F_f = \mathbf{-1.127\ N}$$

6.3 THE FANNO LINE

The Fanno line was first introduced during the discussion of normal shock waves in Section 5.3.2. In this section, the equation of the Fanno line for a perfect gas will be derived.

If Equation (6.6) is applied to the control volume between any section in the duct and section 1 of Figure 6.1, then the entropy change is

$$s - s_1 = c_v \ln\left(\frac{T}{T_1}\right) - R \ln\left(\frac{\rho}{\rho_1}\right)$$

Using the continuity equation (6.1)

$$s - s_1 = c_v \ln\left(\frac{T}{T_1}\right) + R \ln\left(\frac{V}{V_1}\right)$$

From the energy equation (6.3) for a perfect gas with $h = c_p T$, $V = \surd[2c_p(T_0 - T)]$ and $V_1 = \surd[2c_p(T_0 - T_1)]$, then

$$s - s_1 = c_v \ln\left(\frac{T}{T_1}\right) + \frac{R}{2} \ln\left(\frac{T_0 - T}{T_0 - T_1}\right)$$

But as $R = c_v(k - 1)$, then

$$s = c_v\left(\ln T + \frac{k-1}{2} \ln (T_0 - T)\right) + s_1 - c_v\left(\ln T_1 + \frac{k-1}{2} \ln (T_0 - T_1)\right) \qquad (6.7)$$

which is the equation of the Fanno line for a perfect gas.

Since the momentum equation (6.2) has not yet been introduced, **the Fanno line for a perfect gas represents the states with the same mass flux and the same stagnation temperature, but not necessarily the same value of the impulse function**. Therefore, frictional effects are present on the Fanno line.

The Fanno lines corresponding to different mass fluxes are shown in Figure 6.3. To find the point of maximum entropy, Equation (6.7) may be differentiated with respect to the temperature and equated to zero, that is

$$\frac{\mathrm{d}s}{\mathrm{d}T} = c_\mathrm{v}\left(\frac{1}{T} - \frac{k-1}{2}\,\frac{1}{T_0 - T}\right) = 0$$

But, from the energy equation, $T_0 - T = V^2/2c_\mathrm{p}$. Therefore

$$V^2 = (k-1)c_\mathrm{p}T$$

Also, for a perfect gas, $c_\mathrm{p} = kR/(k-1)$. Thus

$$V^2 = kRT = a^2$$

and the Mach number is unity at the point of maximum entropy. This was the conclusion reached in Section 5.3.2. Also, in that section, it was shown that the upper branch of a Fanno line corresponds to subsonic flow, while the lower branch corresponds to supersonic flow.

Since the flow is adiabatic, according to the second law of thermodynamics, the entropy must increase. Thus, the path of states along any one of the Fanno lines must be towards the right. Consequently, if the flow at some point in the duct is subsonic, the effects of friction will tend to increase the velocity and Mach number and to decrease the temperature and pressure of the stream. If, on the other hand, the flow is initially supersonic, the effects of friction will be to decrease the velocity and Mach number and to increase the pressure and temperature of the stream. A subsonic flow may therefore never become supersonic, and a supersonic flow may never become subsonic, unless a discontinuity is present.

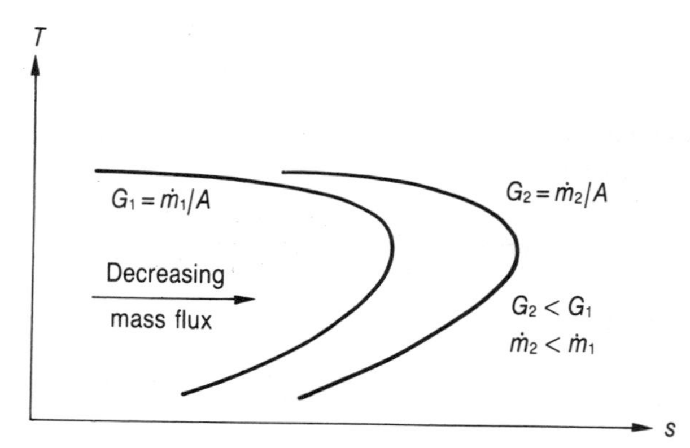

Figure 6.3 Fanno lines for different mass fluxes

6.4 EFFECT OF FRICTION ON FLOW PROPERTIES ALONG THE FANNO LINE

6.4.1 Governing equations in differential form

Consider one-dimensional adiabatic flow of a compressible fluid in a constant-area duct with friction, as shown in Figure 6.4. Friction between the fluid and the duct wall causes infinitesimal changes in the flow properties and these changes may be related by applying the basic equations to the infinitesimal control volume of Figure 6.4.

(a) Continuity equation

For steady and uniform flow at each cross-section, the mass flow rate is constant. Therefore,

$$\dot{m} = \rho VA = (\rho + d\rho)(V + dV)A$$

or

$$\rho A V = \rho A V + \rho A\ dV + AV\ d\rho + \underbrace{A\ d\rho\ dV}_{\simeq 0}$$

Neglecting the products of differentials and dividing by $\rho A V$, it is possible to obtain

$$\frac{dV}{V} + \frac{d\rho}{\rho} = \frac{d(V^2)}{2V^2} + \frac{d\rho}{\rho} = 0 \tag{6.8}$$

(b) Momentum equation

The sum of the frictional forces and the pressure forces acting on the control volume

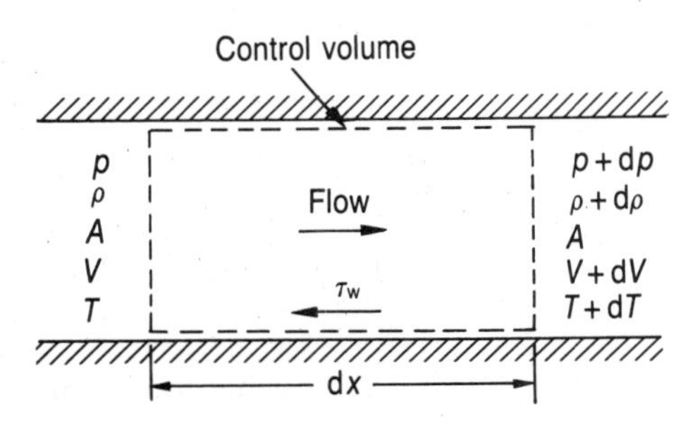

Figure 6.4 Infinitesimal control volume for the analysis of the Fanno line flow

must be balanced by the rate of change of momentum across the control volume, that is

$$pA-(p+\mathrm{d}p)A-\tau_{\mathrm{w}}\ \mathrm{d}A_{\mathrm{w}}=\dot{m}(V+\mathrm{d}V)-\dot{m}V$$

or

$$-A\ \mathrm{d}p-\tau_{\mathrm{w}}\ \mathrm{d}A_{\mathrm{w}}=\dot{m}\ \mathrm{d}V=\rho AV\ \mathrm{d}V$$

where τ_{w} is the shear stress exerted on the control volume by the duct and $\mathrm{d}A_{\mathrm{w}}$ is the wetted wall area over which the shear stress acts.

At this point, the **coefficient of friction** (Fanning's friction factor), f, which is defined as the ratio of wall shear stress to the dynamic pressure of the stream, may be introduced as

$$f=\frac{\tau_{\mathrm{w}}}{\rho V^2/2} \tag{6.9}$$

Ducts of other than circular shape can be included in this analysis by introducing the hydraulic diameter, D, as

$$D=\frac{4\times\text{cross-sectional area}}{\text{wetted perimeter}}=\frac{4A}{\mathrm{d}A_{\mathrm{w}}/\mathrm{d}x} \tag{6.10}$$

The factor of 4 is included in Equation (6.10) so that the hydraulic diameter reduces to the geometric diameter for circular ducts.

With the aid of Equations (6.9) and (6.10), the momentum equation takes the following form:

$$-\frac{4f}{D}\frac{\rho V^2}{2}\ \mathrm{d}x-\mathrm{d}p=\rho V\ \mathrm{d}V \tag{6.11}$$

(c) The first law of thermodynamics

For adiabatic flow with no external work, the first law of thermodynamics for the control volume is

$$c_{\mathrm{p}}T+\frac{V^2}{2}=c_{\mathrm{p}}(T+\mathrm{d}T)+\frac{(V+\mathrm{d}V)^2}{2}$$

or

$$\cancel{c_{\mathrm{p}}T}+\cancel{\frac{V^2}{2}}=\cancel{c_{\mathrm{p}}T}+c_{\mathrm{p}}\ \mathrm{d}T+\cancel{\frac{V^2}{2}}+V\,\mathrm{d}V+\underbrace{\cancel{\frac{(\mathrm{d}V)^2}{2}}}_{\simeq 0}$$

Neglecting second-order differentials, it is possible to obtain

$$c_{\mathrm{p}}\ \mathrm{d}T+V\ \mathrm{d}V=0$$

But, for a perfect gas, $c_p = kR/(k-1)$ and $M = V/\sqrt{(kRT)}$. Then

$$\frac{dT}{T} + \frac{k-1}{2} M^2 \frac{d(V^2)}{V^2} = 0 \tag{6.12}$$

(d) Equation of state

The equation of state for a perfect gas, $p = \rho RT$, may be logarithmically differentiated to yield

$$\frac{dp}{p} = \frac{d\rho}{\rho} + \frac{dT}{T} \tag{6.13}$$

(e) The second law of thermodynamics

For adiabatic and frictional flows

$$ds > 0 \tag{6.14}$$

6.4.2 Variation of flow properties with friction

(a) Velocity

From the definition of the Mach number, $M = V/a$, then $V^2 = M^2 a^2 = M^2 kRT$ and by logarithmic differentiation

$$\frac{d(V^2)}{V^2} = \frac{dT}{T} + \frac{d(M^2)}{M^2} \tag{6.15}$$

Dividing Equation (6.11) by p and noting that $p/\rho = RT = a^2/k$, it is possible to obtain

$$\frac{dp}{p} = -\frac{kM^2}{2} \frac{4f}{D} dx - \frac{kM^2}{2} \frac{d(V^2)}{V^2} \tag{6.16}$$

Eliminating $d\rho/\rho$ and dT/T from Equations (6.8), (6.12) and (6.13), it is possible to obtain

$$\frac{dp}{p} = -[1 + (k-1)M^2] \frac{d(V^2)}{2V^2} \tag{6.17}$$

Then Equation (6.16) becomes

$$\frac{d(V^2)}{V^2} = \frac{kM^2}{1-M^2} 4f \frac{dx}{D} \tag{6.18}$$

(b) Pressure

Combining Equations (6.17) and (6.18) yields

$$\frac{dp}{p} = -\frac{kM^2[1 + (k-1)M^2]}{2(1-M^2)} 4f \frac{dx}{D} \tag{6.19}$$

(c) Temperature

From Equations (6.12) and (6.18), it is possible to obtain

$$\frac{\mathrm{d}T}{T} = -\frac{k(k-1)M^4}{2(1-M^2)} 4f \frac{\mathrm{d}x}{D} \tag{6.20}$$

(d) Mach number

Equations (6.15), (6.18) and (6.20) may be combined to yield

$$\frac{\mathrm{d}(M^2)}{M^2} = \frac{kM^2[1+\frac{1}{2}(k-1)M^2]}{1-M^2} 4f \frac{\mathrm{d}x}{D} \tag{6.21}$$

(e) Density

If Equations (6.19) and (6.20) are substituted into Equation (6.13), then the relation for the variation of density with friction may be obtained as

$$\frac{\mathrm{d}\rho}{\rho} = -\frac{kM^2}{2(1-M^2)} 4f \frac{\mathrm{d}x}{D} \tag{6.22}$$

(f) Stagnation pressure

The previous definition of the isentropic stagnation pressure can be retained as long as the isentropic stagnation pressure corresponding to a given state is the pressure which would be attained if the stream were decelerated isentropically from the given state to the state of zero velocity. Then, from Equation (4.28),

$$p_0 = p\left(1 + \frac{k-1}{2} M^2\right)^{k/(k-1)}$$

or in differential form

$$\frac{\mathrm{d}p_0}{p_0} = \frac{\mathrm{d}p}{p} + \frac{kM^2/2}{1+\frac{1}{2}(k-1)M^2} \frac{\mathrm{d}M^2}{M^2}$$

Finally, with the aid of Equations (6.19) and (6.21), it is possible to obtain

$$\frac{\mathrm{d}p_0}{p_0} = -\frac{kM^2}{2} 4f \frac{\mathrm{d}x}{D} \tag{6.23}$$

(g) Impulse function

From the definition of the impulse function given by Equation (4.39)

$$I = pA(1+kM^2)$$

or in differential form for constant area,

$$\frac{\mathrm{d}I}{I} = \frac{\mathrm{d}p}{p} + \frac{kM^2}{1+kM^2} \frac{\mathrm{d}M^2}{M^2}$$

Now, with the aid of Equations (6.19) and (6.21), the following relation may be obtained:

$$\frac{\mathrm{d}I}{I} = -\frac{kM^2}{2(1+kM^2)}\,4f\,\frac{\mathrm{d}x}{D} \tag{6.24}$$

(h) Entropy

Considering the thermodynamic relation

$$T\,\mathrm{d}s = \mathrm{d}h - \frac{\mathrm{d}p}{\rho}$$

and noting that, $p = \rho RT$ and $c_p = kR/(k-1)$ for a perfect gas. Then

$$\frac{\mathrm{d}s}{c_p} = \frac{\mathrm{d}T}{T} - \frac{k-1}{k}\,\frac{\mathrm{d}p}{p}$$

Now, using Equations (6.19) and (6.20), it is possible to obtain

$$\frac{\mathrm{d}s}{c_p} = \frac{(k-1)M^2}{2}\,4f\,\frac{\mathrm{d}x}{D} \tag{6.25}$$

6.4.3 Summary of frictional effects

By convention, $\mathrm{d}x$ is positive in the direction of flow. The second law of thermodynamics states that the entropy cannot decrease in an adiabatic process. Therefore, it is possible to conclude that the friction coefficient must always be a positive number from Equation (6.25). In the formulation of the momentum

Table 6.1 Effect of friction on the flow properties of a perfect gas in an adiabatic constant-area duct

	Type of flow	
Flow property	Subsonic	Supersonic
Mach number, M	Increases	Decreases
Pressure, p	Decreases	Increases
Temperature, T	Decreases	Increases
Density, ρ	Decreases	Increases
Velocity, V	Increases	Decreases
Stagnation pressure, p_0	Decreases	Decreases
Impulse function, I	Decreases	Decreases
Entropy, s	Increases	Increases

equation, it is assumed that the shear stress acts in a direction opposite to the direction of the flow. Since the friction factor must be positive, the shear stress must always act in the direction assumed.

It appears from Equations (6.23) and (6.24) that both the stagnation pressure and the impulse function must decrease in the direction of flow when there is friction. This is true whether or not the flow is supersonic, and therefore the wall friction reduces the effectiveness of all types of flow.

The directions of change in the remaining flow properties, according to Equations (6.18)–(6.22) depend on whether the Mach number is greater or less than unity, since the term $(1 - M^2)$ appears in the denominator of each of these equations. These are summarized in Table 6.1.

6.5 RELATIONS FOR THE FLOW OF A PERFECT GAS ON THE FANNO LINE

The equations governing the one-dimensional adiabatic flow of a perfect gas in a constant-area duct were given in Section 6.2. Although any problem of this kind can be solved by using these equations, the solution can be simplified by rearranging them in a more suitable form. In this section, these relations are derived by choosing the Mach number as the independent variable.

6.5.1 Non-dimensional friction factor

Equation (6.21) is a differential equation which relates the changes in the Mach number to the distance along the duct, x. A general problem for the Fanno line is shown in Figure 6.5. For a given duct length, L, the integration of Equation (6.21) between states 1 and 2 would produce a complicated function of M_1 and M_2. This function would have to be evaluated numerically for each combination of M_1 and M_2 encountered in a problem. The calculations can be simplified by using reference

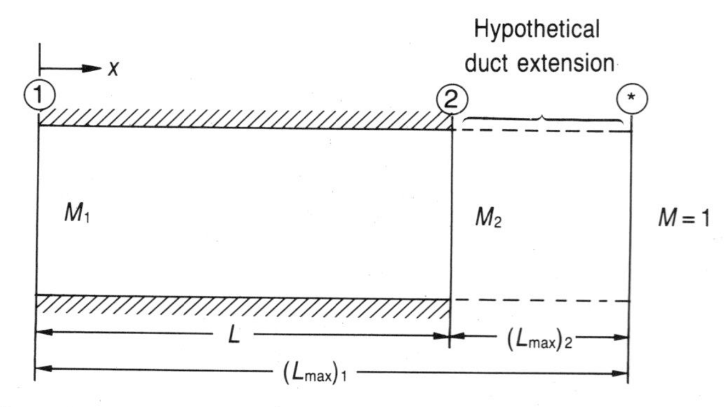

Figure 6.5 Coordinates and notation used for the analysis of the Fanno line flow

conditions. It is known that, for all subsonic and supersonic Fanno line flows, the Mach number tends towards unity. Therefore, for a given value of the inlet Mach number in the duct, there always exists a hypothetical maximum duct length, L_{max}, for which the flow is sonic at the exit of this hypothetical extension, as indicated in Figure 6.5. For this reason, reference conditions may be chosen as critical conditions corresponding to a Mach number of unity. However, it should be noted that these critical conditions are not the same as those in isentropic flow because of the different nature of these flows.

The task is to integrate Equation (6.21) between any given state and the critical state as

$$\int_0^{L_{max}} 4f\frac{dx}{D} = \int_{M^2}^1 \frac{1-M^2}{kM^4[1+\frac{1}{2}(k-1)M^2]} d(M^2) \tag{6.26}$$

The right-hand side of this equation may be evaluated using integration by parts. However, on the left-hand side, the friction factor, f, is a function of Reynolds' number, $Re = \rho VD/\mu$, where μ is the absolute viscosity of the fluid. For flows in constant-area ducts, the product, ρV, is constant along the duct as indicated by the continuity equation, and therefore the variation in Reynolds' number is caused only by variations in the fluid viscosity. Then, a mean friction factor, $\bar{f}$, over the duct length may be defined as

$$\bar{f} = \frac{1}{L_{max}} \int_0^{L_{max}} f\,dx \tag{6.27}$$

and integration of Equation (6.26) leads to

$$\frac{4\bar{f}L_{max}}{D} = \frac{1-M^2}{kM^2} + \frac{k+1}{2k} \ln\left(\frac{(k+1)M^2}{2[1+\frac{1}{2}(k-1)M^2]}\right) \tag{6.28}$$

This equation gives the maximum value of $4\bar{f}L/D$ corresponding to any given initial Mach number. Since the non-dimensional friction factor, $4\bar{f}L_{max}/D$, is a function of the Mach number, then the duct length, L, required for the flow Mach number to change from an initial value of M_1 to a final value of M_2 may be found from

$$\frac{4\bar{f}L}{D} = \left(\frac{4\bar{f}L_{max}}{D}\right)_1 - \left(\frac{4\bar{f}L_{max}}{D}\right)_2 \tag{6.29}$$

6.5.2 Non-dimensional temperature

For adiabatic flows, the stagnation temperature is constant. Thus the non-dimensional temperature, T/T^*, is

$$\frac{T}{T^*} = \frac{T/T_0}{T^*/T_0}$$

or using Equation (4.24) it is possible to obtain

$$\frac{T}{T^*} = \frac{k+1}{2[1+\frac{1}{2}(k-1)M^2]} \tag{6.30}$$

6.5.3 Non-dimensional velocity

Using the definition of Mach number as $V = M\sqrt{(kRT)}$, the non-dimensional velocity may be obtained as

$$\frac{V}{V^*} = \frac{M\sqrt{(kRT)}}{\sqrt{(kRT^*)}} = M\sqrt{\frac{T}{T^*}}$$

Then, with the aid of Equation (6.30),

$$\frac{V}{V^*} = M\sqrt{\left(\frac{k+1}{2[1+\frac{1}{2}(k-1)M^2]}\right)} \tag{6.31}$$

6.5.4 Non-dimensional density

From the continuity equation for steady flow in constant-area ducts, $\rho V = \rho^* V^*$. Then using Equation (6.31)

$$\frac{\rho}{\rho^*} = \frac{V^*}{V} = \frac{1}{M}\sqrt{\left(\frac{2[1+\frac{1}{2}(k-1)M^2]}{k+1}\right)} \tag{6.32}$$

6.5.5 Non-dimensional pressure

Using the equation of state for a perfect gas, $p = \rho RT$, the non-dimensional pressure is

$$\frac{p}{p^*} = \frac{\rho RT}{\rho^* RT^*} = \frac{\rho}{\rho^*}\frac{T}{T^*}$$

Now, use of Equations (6.30) and (6.32) leads to

$$\frac{p}{p^*} = \frac{1}{M}\sqrt{\left(\frac{k+1}{2[1+\frac{1}{2}(k-1)M^2]}\right)} \tag{6.33}$$

6.5.6 Non-dimensional stagnation pressure

The ratio of the local stagnation pressure to the reference stagnation pressure is given by

$$\frac{p_0}{p_0^*} = \frac{p_0}{p}\frac{p}{p^*}\frac{p^*}{p_0^*}$$

Then, using Equation (4.28) for p_0/p and p^*/p_0^*, and Equation (6.33) for p/p^*, it is possible to obtain

$$\frac{p_0}{p_0^*}=\frac{1}{M}\left(\frac{2[1+\frac{1}{2}(k-1)M^2]}{k+1}\right)^{(k+1)/[2(k-1)]} \tag{6.34}$$

6.5.7 Non-dimensional impulse function

From the definition of the impulse function, $I=pA(1+kM^2)$ by Equation (4.39), for constant-area ducts,

$$\frac{I}{I^*}=\frac{p}{p^*}\,\frac{1+kM^2}{1+k}$$

or with the aid of Equation (6.33)

$$\frac{I}{I^*}=\frac{1+kM^2}{M\sqrt{\{2(k+1)[1+\frac{1}{2}(k-1)M^2]\}}} \tag{6.35}$$

6.5.8 Non-dimensional entropy

The entropy change between any state and the reference state is given by Equation (6.6) as

$$s-s^*=c_p\ \ln\left(\frac{T}{T^*}\right)-R\ \ln\left(\frac{p}{p^*}\right)$$

But, for a perfect gas, $c_p=kR/(k-1)$. Thus

$$\frac{s-s^*}{R}=\ln\left(\frac{(T/T^*)^{k/(k-1)}}{p/p^*}\right)$$

Now, using Equations (6.30) and (6.33), it is possible to obtain

$$\frac{s-s^*}{R}=\ln\left\{M\left[\left(\frac{2}{k+1}\right)\left(1+\frac{k-1}{2}M^2\right)\right]^{(k+1)/[2(1-k)]}\right\} \tag{6.36}$$

6.6 WORKING CHART AND WORKING TABLE FOR THE FLOW OF A PERFECT GAS ON THE FANNO LINE

Formulas derived thus far introduce quite tedious numerical calculations and their solutions often involve a trial-and-error procedure. The numerical calculations can be facilitated by introducing a working chart and a working table. The following equations are the non-dimensional properties for the Fanno line flow of a perfect gas that were derived in the previous section:

$$\frac{p}{p^*} = \frac{1}{M}\sqrt{\left(\frac{k+1}{2[1+\frac{1}{2}(k-1)M^2]}\right)} \tag{6.33}$$

$$\frac{T}{T^*} = \frac{k+1}{2[1+\frac{1}{2}(k-1)M^2]} \tag{6.30}$$

$$\frac{\rho}{\rho^*} = \frac{1}{M}\sqrt{\left(\frac{2[1+\frac{1}{2}(k-1)M^2]}{k+1}\right)} \tag{6.32}$$

$$\frac{V}{V^*} = M\sqrt{\left(\frac{k+1}{2[1+\frac{1}{2}(k-1)M^2]}\right)} \tag{6.31}$$

$$\frac{p_0}{p_0^*} = \frac{1}{M}\left(\frac{2[1+\frac{1}{2}(k-1)M^2]}{k+1}\right)^{(k+1)/[2(k-1)]} \tag{6.34}$$

$$\frac{I}{I^*} = \frac{1+kM^2}{M\sqrt{\{2(k+1)[1+\frac{1}{2}(k-1)M^2]\}}} \tag{6.35}$$

$$\frac{s-s^*}{R} = \ln\left\{M\left[\left(\frac{2}{k+1}\right)\left(1+\frac{k-1}{2}M^2\right)\right]^{(k+1)/[2(1-k)]}\right\} \tag{6.36}$$

$$\frac{4\bar{f}L_{\text{max}}}{D} = \frac{1-M^2}{kM^2} + \frac{k+1}{2k}\ln\left(\frac{(k+1)M^2}{2[1+\frac{1}{2}(k-1)M^2]}\right) \tag{6.28}$$

6.6.1 Working chart for the Fanno line flow of a perfect gas

The right-hand sides of the above equations are all functions of the Mach number, M, and the specific heat ratio, k. The non-dimensional properties on the left-hand sides of these equations can then be represented in graphical form with the Mach number as the independent variable. For $k = 1.4$, such a working chart for the Fanno line flow of a perfect gas is given in Figure 6.6.

6.6.2 Working table for the Fanno line flow of a perfect gas

The dimensionless flow properties can also be represented in tabular form with the Mach number as the independent variable. For $k = 1.4$, the values of these non-dimensional properties are given in Appendix E as a function of the Mach number. For convenience, the limiting values of non-dimensional properties are given in Table 6.2.

Example 6.2

Air enters a constant-area insulated duct with a Mach number of 0.60, a pressure

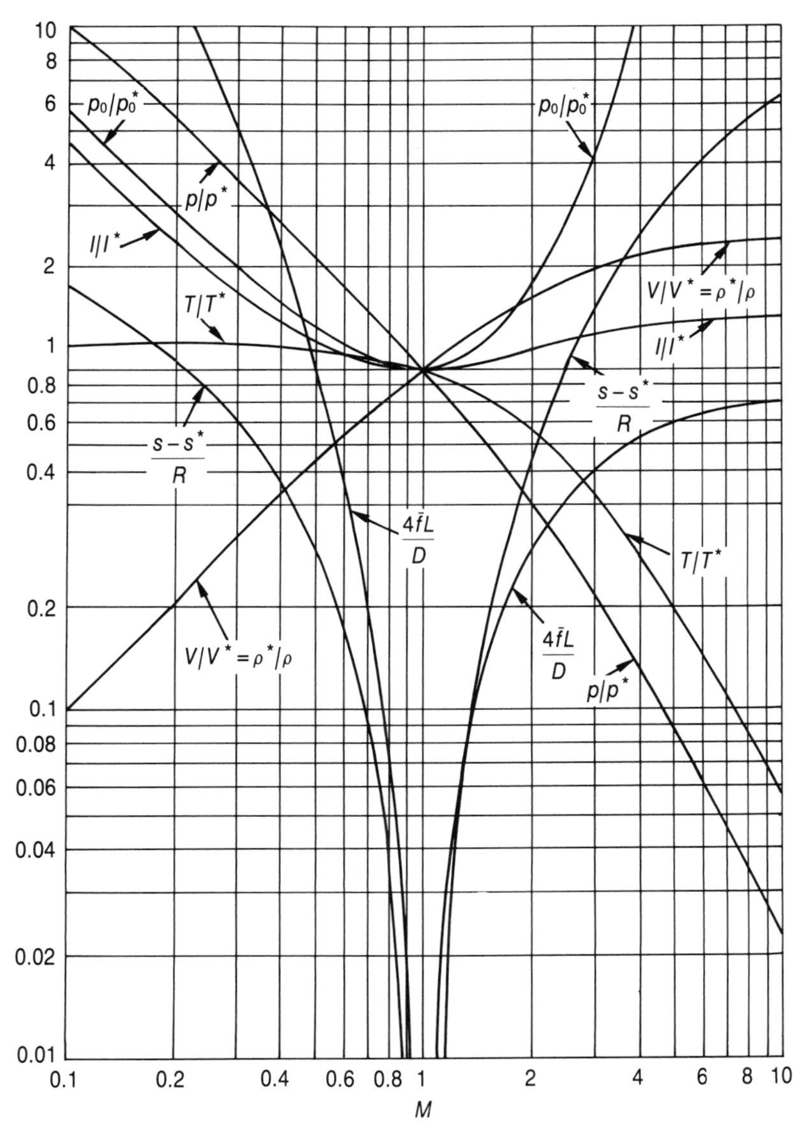

Figure 6.6 Working chart for the Fanno line flow of a perfect gas with $k = 1.4$

of 150 kPa and a temperature of 300 K, as shown in Figure 6.7. Assuming a duct length of 0.45 m, a duct diameter of 0.03 m and an average friction factor of 0.005, determine

a the Mach number, pressure and temperature at the duct exit, and

b the frictional force on the duct walls.

Table 6.2 The limiting values of the non-dimensional properties for the Fanno line flow of a perfect gas with $k = 1.4$

M	0	1	∞
p/p^*	∞	1	0.000
T/T^*	1.200	1	0.000
ρ/ρ^*	∞	1	0.4083
V/V^*	0.000	1	2.450
p_0/p_0^*	∞	1	∞
I/I^*	∞	1	1.429
$(s^* - s)/R$	∞	0.000	0.000
$4\bar{f}L_{max}/D$	∞	0.000	0.8215

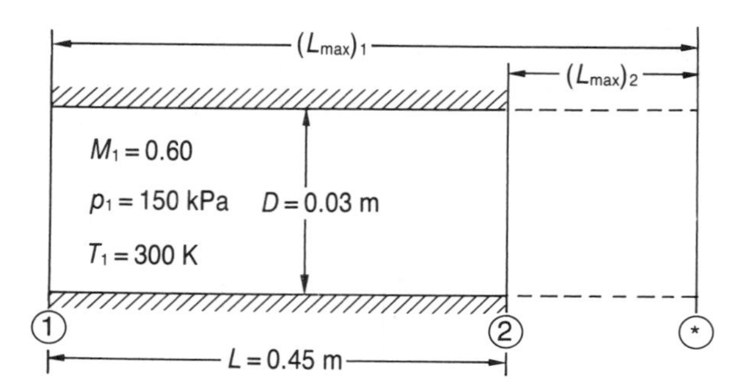

Figure 6.7 Sketch for Example 6.2

Solution

(a) From Appendix E, corresponding to $M_1 = 0.60$,

$$(4\bar{f}L_{max}/D)_1 = 0.4908$$

$$p_1/p^* = 1.763 \quad \text{or} \quad p^* = 150\ \text{kPa}/1.763 = 85.08\ \text{kPa}$$

$$T_1/T^* = 1.119 \quad \text{or} \quad T^* = 300\ \text{K}/1.119 = 268.1\ \text{K}$$

Using Figure 6.7, it is possible to write

$$(4\bar{f}L_{max}/D)_2 = (4\bar{f}L_{max}/D)_1 - 4\bar{f}L/D$$

$$= 0.4908 - \frac{(4)(0.005)(0.45\ \text{m})}{(0.03\ \text{m})} = 0.1908$$

The Mach number, pressure and temperature at section 2 may now be evaluated by using Appendix E with $(4\bar{f}L_{max}/D)_2 = 0.1908$ as

$$M_2 = \mathbf{0.71}$$

$$p_2/p^* = 1.471 \quad \text{or} \quad p_2 = (85.08 \text{ kPa})(1.471) = \mathbf{125.2\ kPa}$$

$$T_2/T^* = 1.090 \quad \text{or} \quad T_2 = (268.1)(1.090) = \mathbf{292.2\ K}$$

(b) The frictional force exerted on the fluid by the duct walls may be expressed as the difference in the impulse function between sections 1 and 2. From Equation (4.39), the impulse function at section 1 is

$$I_1 = p_1 A(1 + kM_1^2) = (150\,000 \text{ N m}^{-2}) \frac{(\pi)(0.03 \text{ m})^2}{4} [1 + (1.4)(0.6)^2]$$

$$= 159.5 \text{ N}$$

Now, from Appendix E, with $M_1 = 0.60$,

$$I_1/I^* = 1.105 \quad \text{or} \quad I^* = 159.5 \text{ N}/1.105 = 144.3 \text{ N}$$

The impulse function at section 2 may now be evaluated by using Appendix E with $M_2 = 0.71$ as

$$I_2/I^* = 1.045 \quad \text{or} \quad I_2 = (144.3 \text{ N})(1.045) = 150.8 \text{ N}$$

Then the frictional force exerted on the fluid by the duct walls is

$$F_f = I_2 - I_1 = 150.8 \text{ N} - 159.5 \text{ N} = -8.7 \text{ N}$$

The force of the fluid on the duct is

$$F = -F_f = \mathbf{8.7\ N}$$

6.7 CHOKING DUE TO FRICTION

It is known that for all subsonic and supersonic Fanno line flows, the Mach number tends towards unity. Therefore, for a specified value of the Mach number at the inlet of the duct, there always exists a maximum duct length for which the flow is sonic at the duct exit. The principal interest is to examine the variations in flow properties which are caused by increasing the duct length above its maximum value.

6.7.1 Choking due to friction in subsonic flows

Consider frictional flow through an adiabatic constant-area duct which is fed by a converging nozzle, as shown in Figure 6.8(a). The flow is subsonic at the duct inlet. Now suppose that the area A, pressure p_1, temperature T_1 and Mach number M_1 are known at the inlet section, and the duct is long enough for the initially subsonic flow to reach a Mach number of unity at the exit plane. The process, which is shown

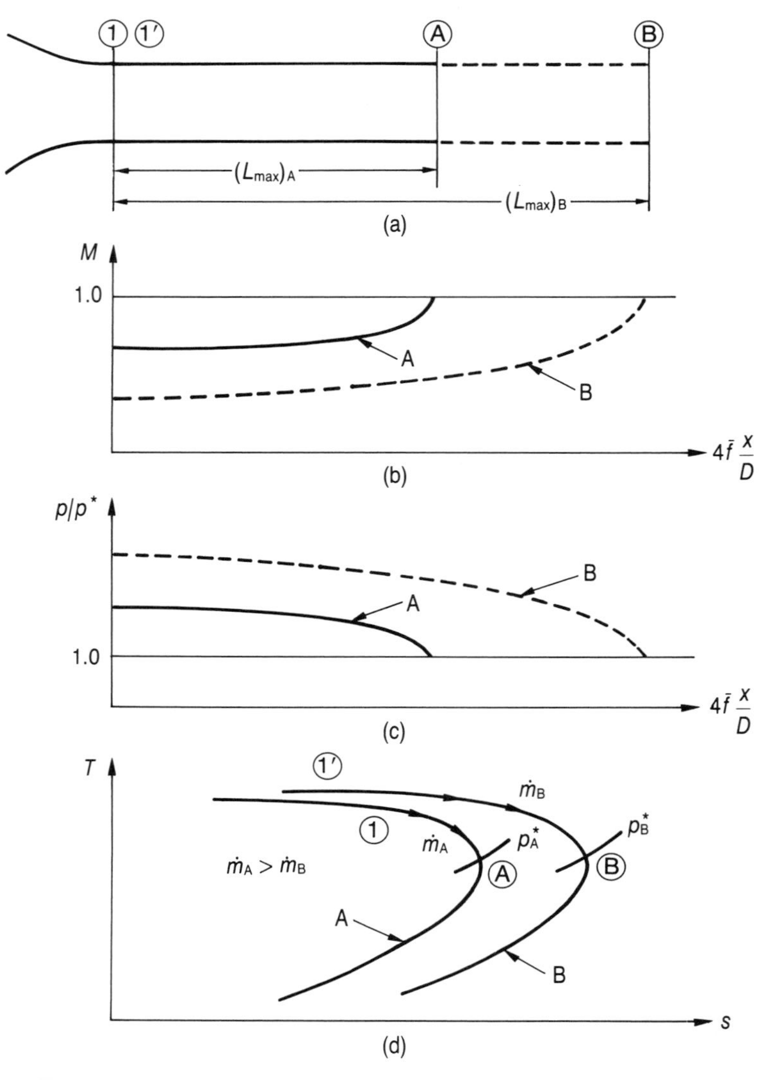

Figure 6.8 Effects of increasing the duct length above its maximum value along a subsonic Fanno line

by curve A on the T–s diagram of Figure 6.8(d), causes a certain entropy increase. Now, if the duct length is increased above its maximum value up to point B, then the resistance to the flow will also be increased. This increase in resistance will be reflected as a larger increase in the entropy. But point A is located at the point of maximum entropy on the Fanno line, where the Mach number is unity. The Mach number at the duct exit for the given mass flow cannot exceed unity without

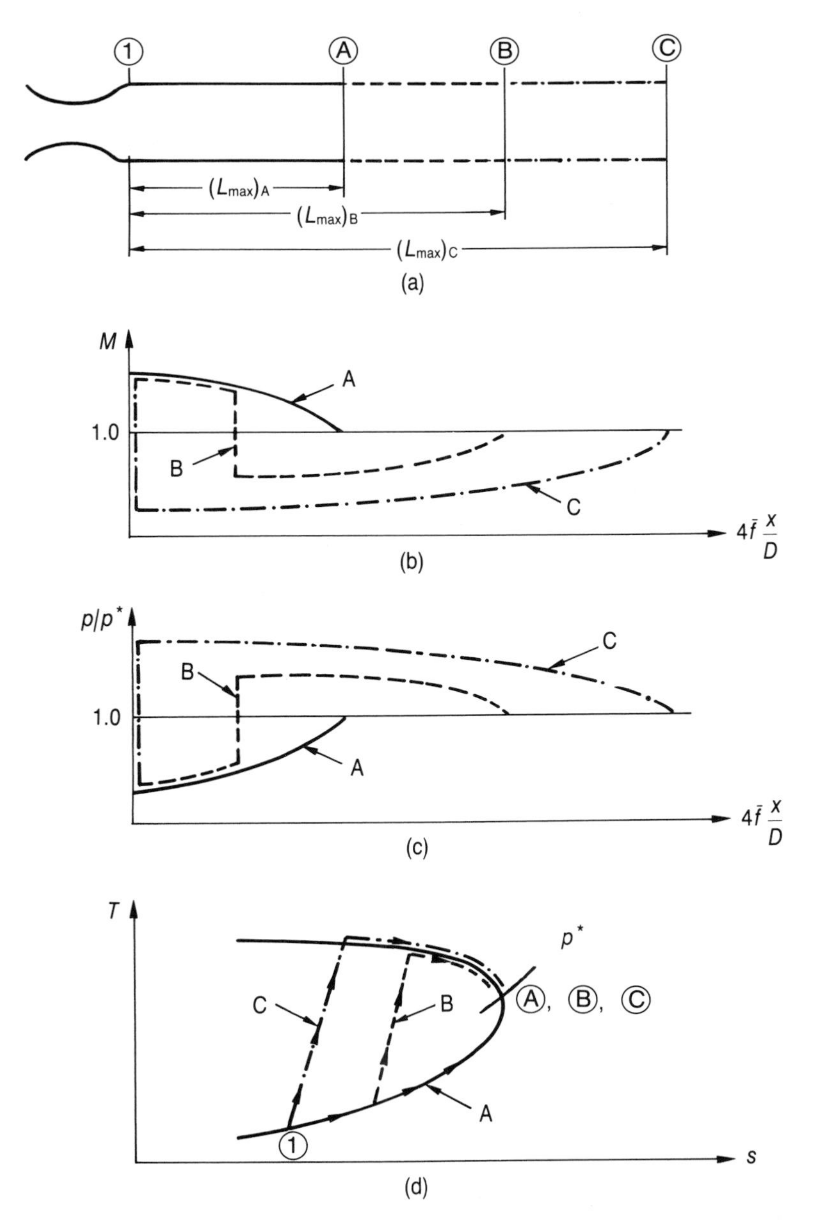

Figure 6.9 Effects of increasing the duct length above its maximum value along a supersonic Fanno line

decreasing the entropy, which will then violate the second law of thermodynamics. This is only possible by changing the steady-state conditions at the inlet of the duct. As a result, the inlet Mach number will decrease and settle at another steady-state value, after a transient period of wave propagation. Hence, the additional length brings about a reduction in the mass flow rate from $\dot{m}_A$ to $\dot{m}_B$, and the flow jumps to another Fanno line, which is known as **choking due to friction**. The new process is also shown on the *T–s* diagram of Figure 6.8(d) by curve B. In the above discussion, it must be noted that the back pressure is sufficiently low so that the changes in the flow are only due to frictional effects. The variation of the Mach number and the pressure along the duct are shown in Figures 6.8(b) and 6.8(c), respectively, for two different duct lengths.

6.7.2 Choking due to friction in supersonic flows

As a second case, consider frictional flow through an adiabatic constant-area duct which is fed by a converging–diverging nozzle, as shown in Figure 6.9(a). The duct, extending from point 1 to point A, is long enough so that the initially supersonic flow decelerates and reaches a Mach number of unity at the duct exit. The process, which is shown by curve A on the *T–s* diagram of Figure 6.9(d), causes a certain amount of entropy increase. Now, the extension of the duct up to point B will result in additional flow resistance. But point A on the *T–s* diagram of Figure 6.9(d) is located at the point of maximum entropy on the Fanno line where the Mach number is unity. The flow at the duct exit cannot go past unity without decreasing the entropy, which will then violate the second law of thermodynamics. In this case, a normal shock wave appears in the duct, as indicated by curve B of Figure 6.9(d). The normal shock wave moves upstream towards the duct inlet as the duct length is increased from point B to point C. When the duct exit is at point C, the normal shock wave is at the duct inlet. The process is shown on the *T–s* diagram of Figure 6.9(d) by the curve labelled C. The variation of the Mach number and the pressure along the duct are shown in Figures 6.9(b) and 6.9(c), respectively, for three different duct lengths. A further increase in the duct length causes the shock to move into the converging–diverging nozzle which feeds the duct. Then the Mach number at the inlet of the duct becomes subsonic. Further increases in the duct length causes the shock to move further into the nozzle, which in turn reduces the Mach number at the duct inlet. Finally, the shock vanishes in the throat, and further increases in length produce reductions in the mass flow rate.

6.8 PERFORMANCE OF ADIABATIC DUCTS AT VARIOUS PRESSURE RATIOS

To illustrate an important aspect of choking, one may investigate the variation of the flow in a constant-area duct with the back pressure. In this section, this situation will be investigated for adiabatic ducts which are fed by converging and converging–diverging nozzles.

6.8.1 Adiabatic ducts fed by converging nozzles

Consider an adiabatic duct, which is fed by an isentropic converging duct, as shown in Figure 6.10(a). Also, suppose that the supply stagnation pressure, p_{0i}, and the supply stagnation temperature, T_0, are fixed. The duct discharges into another reservoir where the back pressure, p_b, is controllable by means of a valve. Conditions at the inlet of the duct are denoted by i, while conditions at the exit of the duct are denoted by e. The effects of variations in the back pressure on the distribution of the pressure in the duct are presented in Figure 6.10(b).

(i) No-flow condition ($p_b = p_{0i}$)
To begin with, suppose that the valve is closed, as shown by flow pattern 1 of Figure 6.10(b). In this case, the pressure is constant through the converging nozzle and the duct, such that $p_b = p_e = p_i = p_{0i}$, and there is no flow.

(ii) Subcritical flow regime ($p^* < p_b < p_{0i}$)
In flow pattern 2 of Figure 6.10(b), the back pressure is slightly less than the supply stagnation pressure. There is a small pressure drop in the duct. The stream leaving the duct is subsonic, so that the exit pressure is equal to the back pressure. The process between the inlet state i2 and the exit state e2 is shown on the Fanno line of Figure 6.10(c).

A further reduction in the back pressure produces flow pattern 3 of Figure 6.10(b). In this case, the mass flow rate increases and the pressure drops further in the nozzle and in the duct. The exit Mach number is still subsonic, but it is higher than the one in flow pattern 2. As the mass flow rate increases, the process between the inlet state i3 and the exit state e3 shifts left to another Fanno line as indicated in Figure 6.10(c). Thus, flow patterns 2 and 3 are qualitatively equivalent.

(iii) Critical condition ($p_b = p^*$)
No qualitative changes are observed until the back pressure is reduced to the critical pressure. This flow pattern is indicated as 4 in Figure 6.10(b). In this case, the Mach number is unity at the exit of the duct and the flow is choked. Again, the exit pressure is equal to the back pressure. The process between the inlet state, i4, and the exit state, e4, shifts further left to another Fanno line, as shown in Figure 6.10(c) owing to the increase in the mass flow rate.

(iv) Supercritical flow regime ($p_b < p^*$)
Further reductions in the back pressure cannot produce a further increase in the mass flow rate, since the exit Mach number cannot be greater than unity. Thus, flow

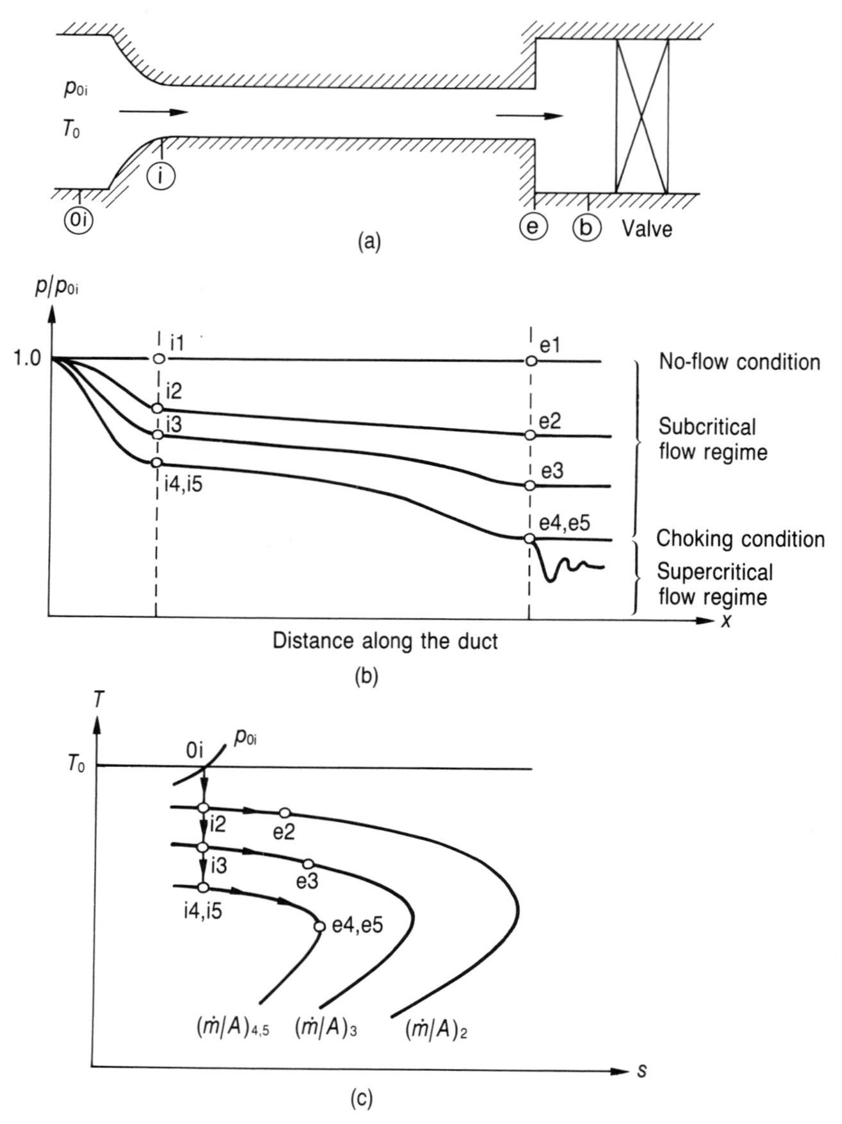

Figure 6.10 Performance of an adiabatic duct which is fed by an isentropic converging nozzle

pattern 5 within the duct is identical to flow pattern 4 of Figure 6.10(b), and the flow is choked. The expansion from the exit pressure to the back pressure involves oblique expansion waves after the stream leaves the duct. The processes between the inlet and exit states of the duct for flow patterns 4 and 5 are identical, as shown in Figure 6.10(c), as long as the flow is choked.

Example 6.3
Air at a stagnation temperature of 400 K and a stagnation pressure of 300 kPa is supplied to a constant-area insulated duct through an isentropic converging nozzle as shown in Figure 6.11(a). The duct has a diameter of 0.04 m and a length of 8.5 m, and it discharges to the atmosphere with a pressure of 100 kPa. Assume a friction factor of 0.001.

a Determine the inlet and exit Mach numbers.
b Determine the pressure and temperature at the exit of the duct.
c Determine the mass flow rate.
d Show the process on a *T–s* diagram.

Solution
(a) The first step is to check if the duct is choked. For this, one may assume that the duct is choked. Then $M_e = 1.0$ and

$$\left(\frac{4\bar{f}L_{max}}{D}\right)_i = \frac{4\bar{f}L}{D} = \frac{(4)(0.001)(8.5\text{ m})}{(0.04\text{ m})} = 0.85$$

Now, the Mach number at the inlet of the duct corresponding to $(4\bar{f}L_{max}/D)_i = 0.85$ may be obtained from Appendix E as $M_i = 0.53$. Then, using Appendix C for $M_i = 0.53$,

$$p_i/p_{0i} = 0.8259 \quad \text{or} \quad p_i = (0.8259)(300\text{ kPa}) = 247.8\text{ kPa}$$

The critical pressure may now be obtained from Appendix E corresponding to $M_i = 0.53$ as

$$p_i/p^* = 2.011 \quad \text{or} \quad p^* = 247.8\text{ kPa}/2.011 = 123.2\text{ kPa}$$

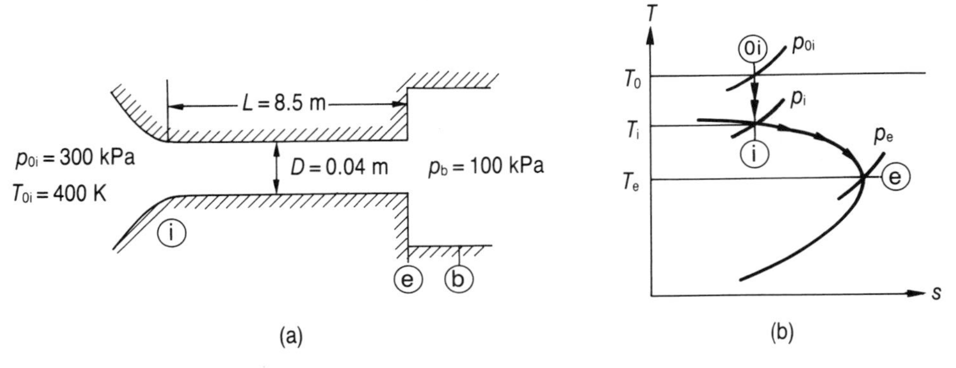

Figure 6.11 Sketch and *T–s* diagram for Example 6.3

As long as the flow is sonic at the exit of the duct, then the exit pressure is greater than the back pressure, that is

$$p_e = p^* = 123.2 \text{ kPa} > p_b = 100 \text{ kPa}$$

Hence, the flow is choked and the initial assumption is correct. Therefore, the inlet and exit Mach numbers are

$$M_i = \mathbf{0.53} \qquad M_e = \mathbf{1.0}$$

(b) The exit pressure is already calculated as

$$p_e = \mathbf{123.2\ kPa}$$

As long as the flow is adiabatic through the converging nozzle and the duct, then corresponding to $M_e = 1.0$ from Appendix C

$$T_e/T_0 = 0.8333 \quad \text{or} \quad T_e = (0.8333)(400\text{ K}) = \mathbf{333.3\ K}$$

(c) The mass flow rate through the duct may now be evaluated from Appendix C corresponding to $M_i = 0.53$ as

$$\frac{\dot{m}\sqrt{(RT_0)}}{Ap_{0i}} = 0.5323 \quad \text{or} \quad \dot{m} = \frac{(0.5323)(300\ 000\text{ N m}^{-2})(\pi)(0.04\text{ m})^2/4}{\sqrt{[(287.1\text{ J kg}^{-1}\text{ K}^{-1})(400\text{ K})]}}$$

$$= \mathbf{0.5922\ kg\ s^{-1}}$$

(d) The process is shown on the T–s diagram of Figure 6.11(b).

Example 6.4

Air at a stagnation temperature of 400 K and a stagnation pressure of 150 kPa is supplied to a constant-area insulated duct through an isentropic converging nozzle, as shown in Figure 6.12(a). The duct has a diameter of 0.04 m and a

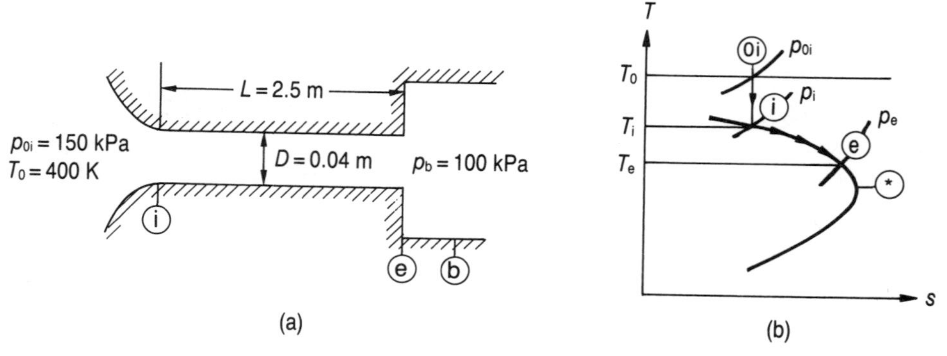

Figure 6.12 Sketch and T–s diagram for Example 6.4

length of 2.5 m, and it discharges to an atmosphere with a pressure of 100 kPa. Assume a friction factor of 0.001.

a Determine the inlet and exit Mach numbers.
b Determine the pressure and temperature at the exit of the duct.
c Determine the mass flow rate.
d Show the process on a T–s diagram.

Solution
(a) First, it is necessary to determine if the duct is choked. For this, as an initial guess one may assume that the duct is choked. Then $M_e = 1.0$ and

$$\left(\frac{4\bar{f}L_{max}}{D}\right)_i = \frac{4\bar{f}L}{D} = \frac{(4)(0.001)(2.5\text{ m})}{(0.04\text{ m})} = 0.25$$

Now, the Mach number at the inlet of the duct corresponding to $(4\bar{f}L_{max}/D) = 0.25$ may be obtained from Appendix E as $M_i = 0.68$. Thus, using Appendix C for $M_i = 0.68$,

$$p_i/p_{0i} = 0.7338 \quad \text{or} \quad p_i = (0.7338)(150\text{ kPa}) = 110.1\text{ kPa}$$

The critical pressure may now be obtained from Appendix E corresponding to $M_i = 0.68$ as

$$p_i/p^* = 1.541 \quad \text{or} \quad p^* = 110.1\text{ kPa}/1.541 = 71.45\text{ kPa}$$

According to the initial assumption, the pressure at the exit of the duct should be equal to the critical pressure, that is

$$p_e = p^* = 71.45\text{ kPa}$$

However, as long as the exit pressure is less than the back pressure, the initial assumption is incorrect, and the flow is not choked. The flow will be choked whenever the imaginary duct length is greater than the real duct length, as shown in Figure 6.13. Therefore, the determination of the imaginary duct length for which the flow is choked at the end of this imaginary duct involves a trial-and-error process. One may start the iteration by assuming that

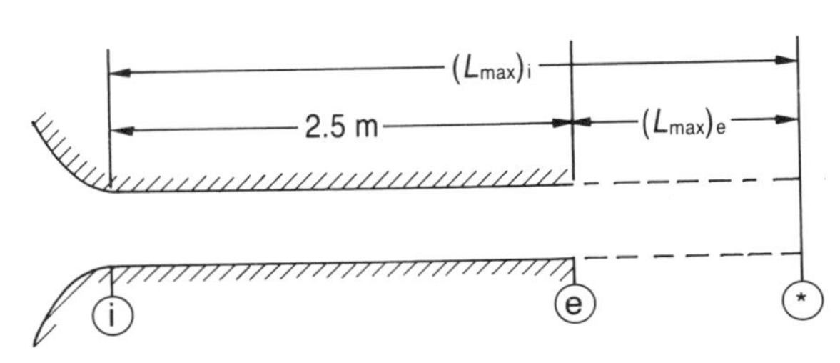

Figure 6.13 Sketch for the unchoked flow in Example 6.4

$(4\bar{f}L_{max}/D)_i = 0.3246$. In this case, the Mach number at the inlet of the duct may be determined from Appendix E as $M_i = 0.65$. Now, using Appendix C for $M_i = 0.65$

$$p_i/p_0 = 0.7528 \quad \text{or} \quad p_i = (0.7528)(150\ \text{kPa}) = 112.9\ \text{kPa}$$

The critical pressure may now be obtained from Appendix E corresponding to $M_i = 0.65$ as

$$p_i/p^* = 1.618 \quad \text{or} \quad p^* = 112.9\ \text{kPa}/1.618 = 69.78\ \text{kPa}$$

Hence,

$$(4\bar{f}L_{max}/D)_e = (4\bar{f}L_{max}/D)_i - 4\bar{f}L/D = 0.3246 - 0.25 = 0.0746$$

At this point, one may use Appendix E, corresponding to $(4\bar{f}L_{max}/D)_e$, to find $M_e = 0.80$ and

$$p_e/p^* = 1.289 \quad \text{or} \quad p_e = (1.289)(69.78\ \text{kPa}) = 89.95\ \text{kPa}$$

As long as the exit pressure is less than the back pressure, the initial assumption is incorrect, and the flow at the exit of the imaginary duct is not choked. One may then continue the trial-and-error process by extending the imaginary duct further. The iterations are summarized in Table 6.3.

According to Table 6.3, the inlet and exit Mach numbers for the duct are

$$M_i = \mathbf{0.61} \qquad M_e = \mathbf{0.70}$$

Table 6.3 Iterations for Example 6.4

Iteration	1	2	3	4
$(4\bar{f}L_{max}/D)_i$	0.25	0.3246	0.4908	0.4527
M_i	0.68	0.65	0.60	0.61
p_i/p_{0i}	0.7338	0.7528	0.7840	0.7778
p_i (kPa)	110.1	112.9	117.6	116.7
p_i/p^*	1.541	1.618	1.763	1.733
p^*(kPa)	71.45	69.78	66.71	67.34
$(4\bar{f}L_{max}/D)_e$	0	0.0746	0.2408	0.2027
M_e	1.0	0.80	0.68	0.70
p_e/p^*	1.0	1.289	1.541	1.493
p_e (kPa)	71.45	89.95	102.8	100.5

(b) As long as the flow is subsonic at the end of the duct, then

$$p_e = p_b = \mathbf{100\ kPa}$$

Also, as the flow is adiabatic through the converging nozzle and the duct, then corresponding to $M_e = 0.70$ from Appendix C

$$T_e/T_0 = 0.9107 \quad \text{or} \quad T_e = (0.9107)(400\ \text{K}) = \mathbf{364.3\ K}$$

(c) At the inlet of the duct $M_i = 0.61$

$$\frac{\dot{m}\sqrt{(RT_0)}}{Ap_{0i}} = 0.5819 \quad \text{or} \quad \dot{m} = \frac{(0.5819)(150\,000\ \text{N m}^{-2})(\pi)(0.04\ \text{m})^2/4}{\sqrt{[(287.1\ \text{J kg}^{-1}\text{K}^{-1})(400\ \text{K})]}}$$

$$= \mathbf{0.3237\ kg\ s^{-1}}$$

(d) The process is shown on the T–s diagram of Figure 6.12(b).

6.8.2 Adiabatic ducts fed by converging–diverging nozzles

Consider an adiabatic duct, which is fed by an isentropic converging–diverging nozzle, as shown in Figure 6.14. Also, suppose that the stagnation pressure p_{0i} and the stagnation temperature T_0 are fixed in the supply reservoir. The duct discharges into another reservoir, where the back pressure, p_b, is controllable by means of a valve. Conditions at the inlet of the duct are denoted by i, while conditions at the exit of the duct are denoted by e.

The discussion in this section is restricted to the cases where the flow is supersonic at the exit of the converging–diverging nozzle. Otherwise, the flow will be subsonic in the adiabatic duct, and these cases have been treated in the previous section.

As long as there is no shock wave within the converging–diverging nozzle then the pressure and Mach number at the inlet of the adiabatic duct are fixed by the area ratio of the converging–diverging nozzle. Corresponding to the inlet Mach number, M_i, of the duct, there exists a particular value of $4\bar{f}L_{max}/D$. Therefore, the types of flow may be divided into two main classes, depending on whether the actual length of the duct gives a value of $4\bar{f}L/D$ greater or less than the maximum allowable value.

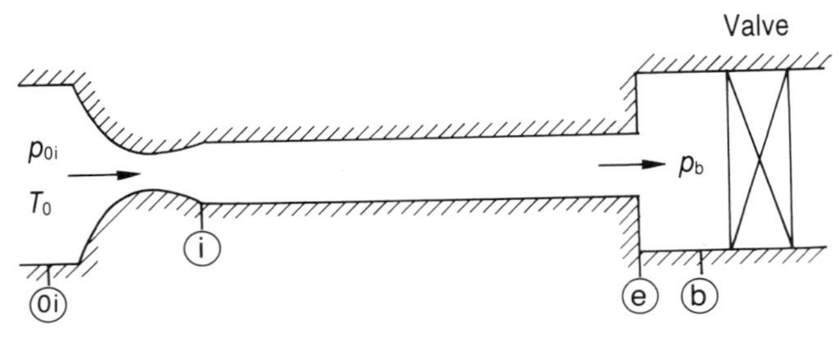

Figure 6.14 An adiabatic duct fed by an isentropic converging–diverging nozzle

(a) Adiabatic ducts with lengths less than the maximum allowable length

Consider an adiabatic duct, which is fed by an isentropic converging–diverging nozzle and whose length is less than the maximum allowable length, as shown in Figure 6.15(a). The effects of variations in the back pressure on the distribution of the pressure in the duct are presented in Figure 6.15(b).

(i) No-flow condition ($p_b = p_{0i}$)

To begin with, suppose that the valve is closed, as shown by flow pattern 1 of Figure 6.15(b). In this case, the pressure is constant throughout the converging–diverging nozzle and the duct, such that $p_b = p_e = p_i = p_{0i}$, and there is no flow.

(ii) Subsonic flow regime ($p_{0i} > p_b > p_{e2}$)

In this flow regime, the flow is always subsonic in the duct. As the back pressure is lowered, the flow in the converging–diverging nozzle passes through the subsonic flow regime, the choking condition and the non-isentropic flow regime. Since the flow in the duct is subsonic, the pressure decreases along the duct and leaves the duct such that the exit pressure is equal to the back pressure. This flow regime will not be treated in detail, since the flow patterns in a converging–diverging nozzle are discussed in Sections 4.8.2 and 5.8, and the subsonic flow patterns in a duct are discussed in Section 6.8.1.

(iii) Inlet shock condition ($p_b = p_{e2}$)

In this case, a normal shock wave appears at the inlet of the duct, and the flow pattern in the converging–diverging nozzle represents its exit shock condition. The flow throughout the duct is subsonic, and the pressure decreases along the duct, as shown in flow pattern 2 of Figure 6.15(b). Again the exit pressure is the same as the back pressure. The process 0i–i2x–i2y–e2 for this flow pattern is shown on the T–s diagram of Figure 6.15(c).

(iv) Flow regime with shocks ($p_{e2} > p_b > p_{e4y}$)

In this flow regime, a normal shock wave is always present in the duct. Flow pattern 3 of Figure 6.15(b) is a typical representation of this flow regime. The pressure increases along the duct during supersonic flow before the normal shock wave, shows an abrupt rise across the normal shock wave and decreases along the duct during subsonic flow after the normal shock wave. As long as the flow leaves the duct at subsonic speeds, the exit pressure and the back pressure are the same. The process 0i–i3–3x–3y–e3 for this flow pattern is shown on the T–s diagram of Figure 6.15(c).

(v) Exit shock condition ($p_b = p_{e4y}$)

As the back pressure is lowered, the normal shock wave moves to the exit plane of the duct, as shown by flow pattern 4 of Figure 6.15(b). In this case, the flow is supersonic within the duct and the pressure increases along the duct. After the normal shock wave, the pressure rises abruptly and the flow leaves the duct at subsonic speeds

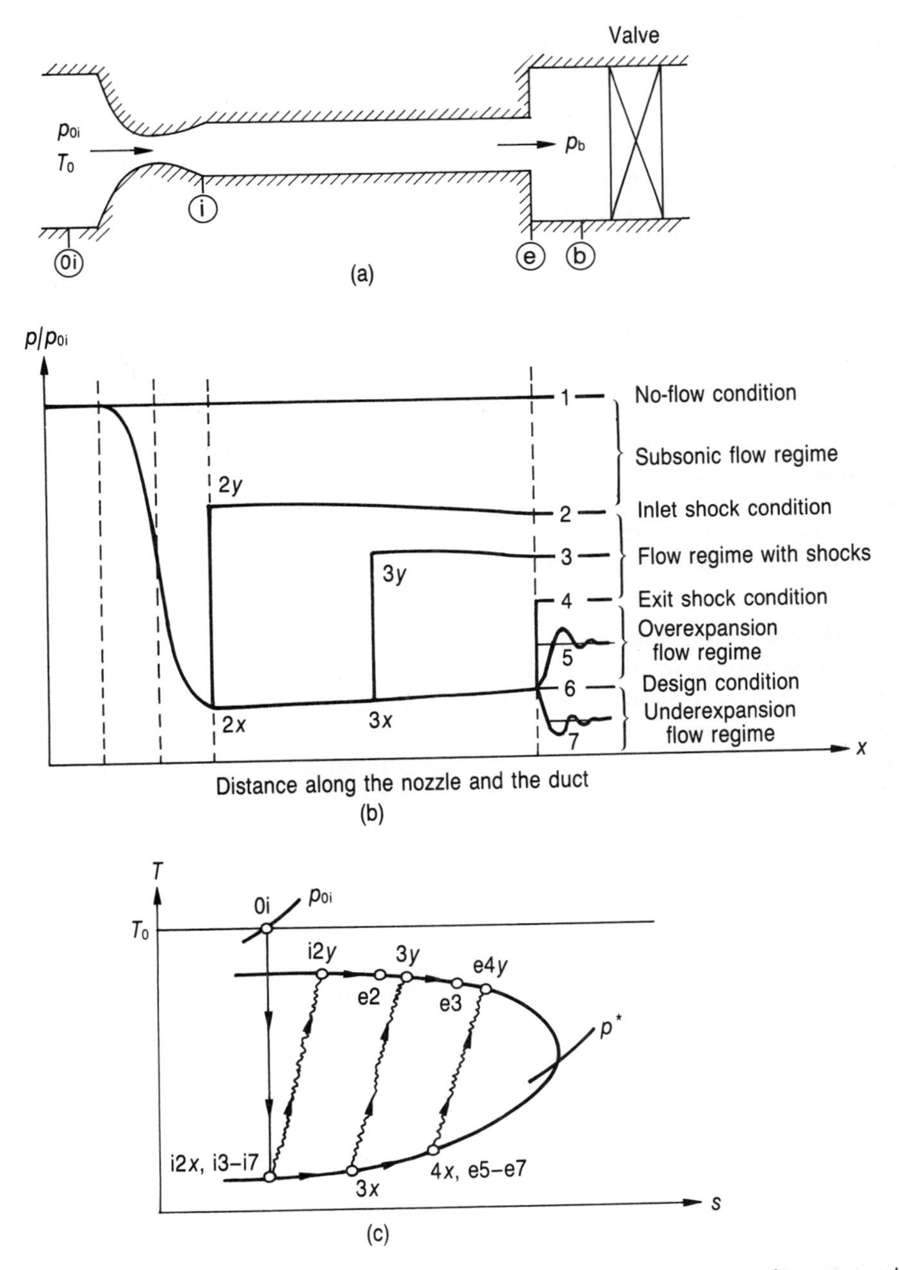

Figure 6.15 Performance of an adiabatic duct fed by an isentropic converging–diverging nozzle (the length of the duct is less than the maximum allowable length)

so that the exit pressure is equal to the back pressure. The exit shock condition is shown by the process 0i–i4–e4x–e4y on the T–s diagram of Figure 6.15(c).

(vi) Overexpansion flow regime ($p_{e4y} > p_b > p_{e6}$)
A typical example of this flow regime is shown by flow pattern 5 of Figure 6.15(b). In this flow regime, the flow is supersonic throughout the duct and the pressure increases along the duct. The gas is overexpanded at the exit of the duct, since the exit pressure is lower than the back pressure. The compression which occurs outside the duct involves non-isentropic oblique compression waves, which cannot be treated with the assumption of one-dimensional flow. The process 0i–i5–e5 for this flow pattern within the nozzle and the duct is shown in Figure 6.15(c).

(vii) Design condition ($p_b = p_{e6}$)
Flow pattern 6 of Figure 6.15(b) represents the condition for which the converging–diverging nozzle and the duct are designed. In this case, the flow is entirely supersonic within the duct and the exit pressure is identical to the back pressure. The pressure increases along the duct. The process, 0i–i6–e6, for the design condition is shown in Figure 6.15(c).

(viii) Underexpansion flow regime ($p_b < p_{e6}$)
A typical flow pattern for this flow regime is indicated by 7 in Figure 6.15(b). The flow is entirely supersonic within the duct and the pressure increases along the duct. The gas is underexpanded at the exit of the duct, since the exit pressure is higher than the back pressure. The expansion, which occurs outside the duct, involves non-isentropic oblique expansion waves, which cannot be treated with the assumption of one-dimensional flow. The process 0i–i7–e7 for this flow pattern within the nozzle and the duct is shown in Figure 6.15(c).

Example 6.5
An adiabatic duct is fed by an isentropic converging–diverging nozzle with an area ratio of 2.0, as shown in Figure 6.16(a). The stagnation pressure in the supply reservoir is 300 kPa. The length of the duct is 2 m and the diameter of the duct is 0.04 m. The average friction factor in the duct is 0.001. Determine

a the maximum allowable duct length,
b the back pressure such that there is a normal shock wave at the inlet plane of the duct,
c the back pressure for the operation of the duct at its design condition,
d the back pressure such that there is a normal shock wave at the exit plane of the duct, and
e the ranges of back pressure for the subsonic flow regime, the flow regime with shocks, the overexpansion flow regime and the underexpansion flow regime.

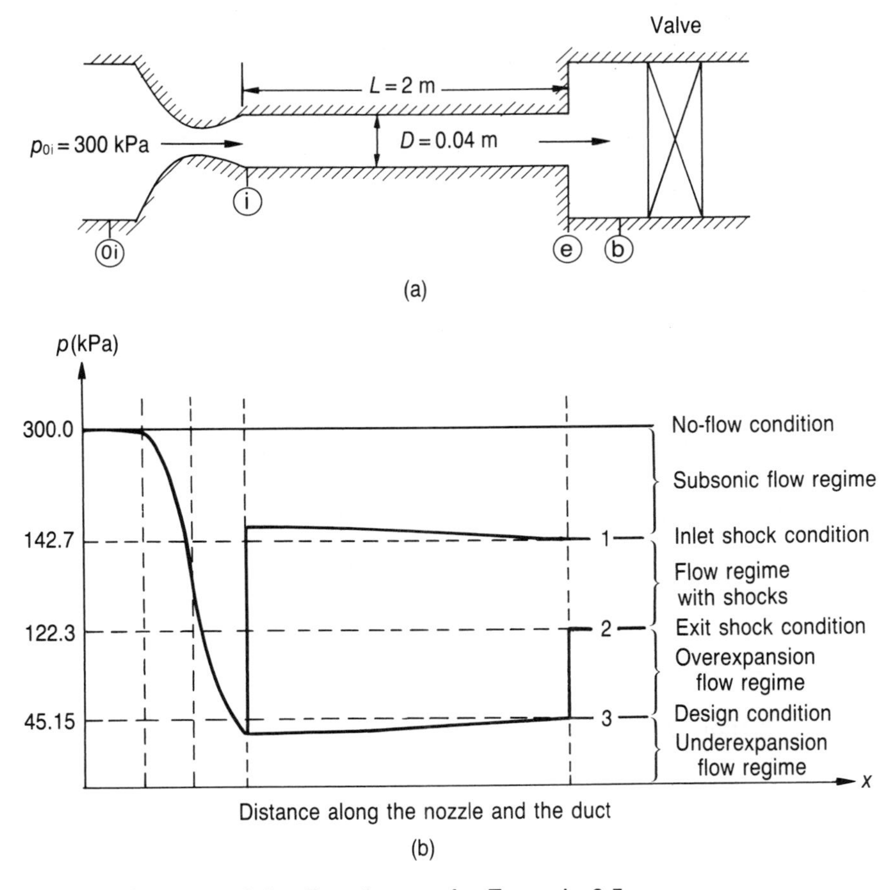

Figure 6.16 Sketch and the *T–s* diagram for Example 6.5

Solution

(a) In order to have supersonic flow at the exit of the converging–diverging nozzle, the flow must be choked at the throat. Therefore, corresponding to an area ratio of $A_i/A^* = 2.0$, one may obtain the Mach number and the pressure at the inlet of the duct from Appendix C as

$$M_i = 2.20$$

and

$$p_i/p_{0i} = 0.093\,52 \quad \text{or} \quad p_i = (0.093\,52)(300 \text{ kPa}) = 28.06 \text{ kPa}$$

The maximum allowable duct length corresponding to an inlet Mach number, $M_i = 2.20$, may be obtained from Appendix E as

$$\frac{4\bar{f}L_{max}}{D} = 0.3609 \quad \text{or} \quad L_{max} = \frac{(0.3609)(0.04\ \text{m})}{(4)(0.001)} = \mathbf{3.609\ m}$$

Hence, the given duct length is less than the maximum allowable length.

(b) When there is a normal shock wave at the inlet of the duct, then corresponding to an upstream Mach number of $M_{i1x} = 2.20$, the downstream Mach number and the pressure may be obtained from Appendix D as

$$M_{i1y} = 0.5471$$

and

$$p_{i1y}/p_{i1x} = 5.480 \quad \text{or} \quad p_{i1y} = (5.480)(28.06\ \text{kPa}) = 153.8\ \text{kPa}$$

Now, from Appendix E, corresponding to $M_{i1y} = 0.55$,

$$(4\bar{f}L_{max}/D)_{i1y} = 0.7281$$

and

$$p_{i1y}/p_1^* = 1.934 \quad \text{or} \quad p^* = 153.8\ \text{kPa}/1.934 = 79.52\ \text{kPa}$$

Using Figure 6.16(a), it is possible to write

$$(4\bar{f}L_{max}/D)_{e1} = (4\bar{f}L_{max}D)_{i1y} - 4\bar{f}L/D$$

$$= 0.7281 - \frac{(4)(0.001)(2\ \text{m})}{(0.04\ \text{m})} = 0.5281$$

The Mach number and the pressure at the exit of the duct may now be obtained from Appendix E with $(4\bar{f}L_{max}/D)_{e1} = 0.5281$ as

$$M_{e1} = 0.59$$

and

$$p_{e1}/p_1^* = 1.795 \quad \text{or} \quad p_{e1} = (1.795)(79.52\ \text{kPa}) = 142.7\ \text{kPa}$$

As long as the flow leaves the duct at subsonic speeds, the back pressure and the exit pressure must be the same, that is

$$p_{b1} = p_{e1} = \mathbf{142.7\ kPa}$$

Flow pattern 1 which corresponds to the inlet shock condition is shown in Figure 6.16(b).

(c) When the duct is operating at its design condition, there is no normal shock wave in the duct. In this case, using Appendix E with $M_{i3} = 2.20$,

$$(4\bar{f}L_{max}/D)_{i3} = 0.3609$$

and

$$p_{i3}/p_3^* = 0.3549 \quad \text{or} \quad p_3^* = 28.06\ \text{kPa}/0.3549 = 79.07\ \text{kPa}$$

Now, using Figure 6.16(a), it is possible to write

$$(4\bar{f}L_{max}/D)_{e3} = (4\bar{f}L_{max}/D)_{i3} - 4\bar{f}L/D = 0.3609 - \frac{(4)(0.001)(2\ \text{m})}{(0.04\ \text{m})}$$

$$= 0.1609$$

The Mach number and the pressure at the exit of the duct may now be obtained from Appendix E with $(4\bar{f}L_{max}/D)_{e3} = 0.1609$ as

$$M_{e3} = 1.57$$

and

$$p_{e3}/p_3^* = 0.5710 \quad \text{or} \quad p_{e3} = (0.5710)(79.07\ \text{kPa}) = 45.15\ \text{kPa}$$

For the design condition, the exit and the back pressures are the same, so that

$$p_{b3} = p_{e3} = \mathbf{45.15\ kPa}$$

In Figure 6.16(b), the design condition is indicated by flow pattern 3.

(d) When there is a normal shock wave, just at the exit plane of the duct, the Mach number and the static pressure before the shock wave are $M_{e2x} = 1.57$ and $p_{e2x} = 45.15$ kPa, respectively. Then, from Appendix D, corresponding to $M_{e2x} = 1.57$,

$$p_{e2y}/p_{e2x} = 2.709 \quad \text{or} \quad p_{e2y} = (2.709)(45.15\ \text{kPa}) = 122.3\ \text{kPa}$$

Again, the flow is subsonic at the exit plane. Therefore

$$p_{b2} = p_{e2y} = \mathbf{122.3\ kPa}$$

Flow pattern 2 which corresponds to the exit shock condition is shown in Figure 6.16(b).

(e) Ranges of back pressure for the regimes are as follows:

1 Subsonic flow regime:

$$p_0 \geqslant p_b \geqslant p_{b1} \quad \text{or} \quad \mathbf{300\ kPa} \geqslant \boldsymbol{p_b} \geqslant \mathbf{142.7\ kPa}$$

2 Flow regime with shocks:

$$p_{b1} \geqslant p_b \geqslant p_{b2} \quad \text{or} \quad \mathbf{142.7\ kPa} \geqslant \boldsymbol{p_b} \geqslant \mathbf{122.3\ kPa}$$

3 Overexpansion flow regime:

$$p_{b2} \geqslant p_b \geqslant p_{b3} \quad \text{or} \quad \mathbf{122.3\ kPa} \geqslant \boldsymbol{p_b} \geqslant \mathbf{45.15\ kPa}$$

4 Underexpansion flow regime:

$$p_{b3} \geqslant p_b \geqslant 0 \quad \text{or} \quad \mathbf{45.15\ kPa} \geqslant \boldsymbol{p_b} \geqslant \mathbf{0\ kPa}$$

Example 6.6
If the back pressure for the duct of Example 6.5 is 130 kPa, then determine the distance from the inlet of the duct at which the normal shock wave is located.

Solution
A sketch and the nomenclature for this example is shown in Figure 6.17(a). The given back pressure is in the flow regime with shocks; therefore there is a normal shock wave in the duct. The location of this normal shock wave can be found by an iterative process.

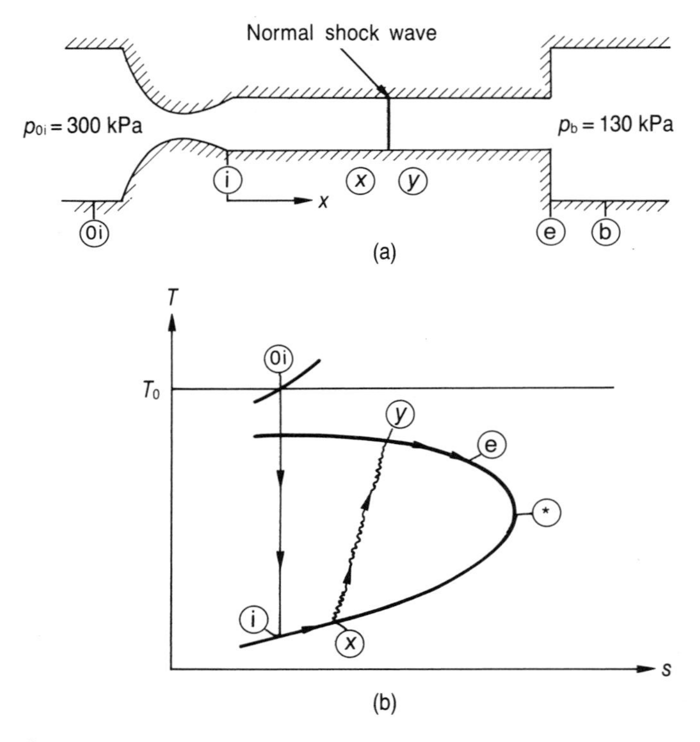

Figure 6.17 Sketch for Example 6.6

In Example 6.5, the Mach number and the pressure at the inlet of the duct were found to be $M_i = 2.20$ and $p_i = 28.06$ kPa, respectively. Now, from Appendix E, corresponding to $M_i = 2.20$,

$$(4\bar{f}L_{max}/D)_i = 0.3609$$

and

$$p_i/p^* = 0.3549 \quad \text{or} \quad p^* = 28.06 \text{ kPa}/0.3549 = 79.07 \text{ kPa}$$

At this stage, one may assume that the location of the normal shock wave is at $x = 1$ m. Then,

$$(4\bar{f}L_{max}/D)_x = (4\bar{f}L_{max}/D)_i - 4\bar{f}x/D = 0.3609 - \frac{(4)(0.001)(1 \text{ m})}{(0.04 \text{ m})} = 0.2609$$

The Mach number and the pressure upstream of the normal shock wave may now be obtained from Appendix E with $(4\bar{f}L_{max}/D)_x = 0.2609$ as

$$M_x = 1.86$$

and

$$p_x/p^* = 0.4528 \quad \text{or} \quad p_x = (0.4528)(79.07 \text{ kPa}) = 35.80 \text{ kPa}$$

Now, the Mach number and the pressure downstream of the normal shock wave corresponding to $M_x = 1.86$ may be obtained from Appendix D as

$$M_y = 0.6036$$

and

$$p_y/p_x = 3.870 \quad \text{or} \quad p_y = (3.870)(35.80 \text{ kPa}) = 138.6 \text{ kPa}$$

From Appendix E, corresponding to $M_y = 0.60$,

$$(4\bar{f}L_{max}/D)_y = 0.4908$$

However, it is possible to write

$$(4\bar{f}L_{max}/D)_e = (4\bar{f}L_{max}/D)_y - 4\bar{f}(2 - x)/D$$

$$= 0.4908 - \frac{(4)(0.001)(2 \text{ m} - 1 \text{ m})}{(0.04 \text{ m})} = 0.3908$$

The Mach number and the pressure at the exit of the duct may now be found via Appendix E with $(4\bar{f}L_{max}/D)_e = 0.3908$ as

$$M_e = 0.63$$

and

$$p_e/p^* = 1.674 \quad \text{or} \quad p_e = (1.674)(79.07 \text{ kPa}) = 132.4 \text{ kPa}$$

For subsonic flow at the exit plane

$$p_b = p_e = 132.4 \text{ kPa}$$

Table 6.4 Iterations for Example 6.6

Iteration	1	2	3
x (m)	1	1.2	1.1
$(4\bar{f}L_{max}/D)_x$	0.2609	0.2409	0.2509
M_x	1.86	1.80	1.83
p_x/p^*	0.4528	0.4741	0.4632
p_x (kPa)	35.80	37.49	36.63
M_y	0.6036	0.6165	0.6099
p_y/p_x	3.870	3.613	3.740
p_y (kPa)	138.6	135.5	137.0
$(4\bar{f}L_{max}/D)_y$	0.4908	0.4172	0.4527
$(4\bar{f}L_{max}/D)_e$	0.3908	0.3372	0.3627
M_e	0.63	0.65	0.64
p_e/p^*	1.674	1.618	1.646
p_e (kPa)	132.4	127.9	130.2
Shock movement	→	←	OK

As long as the calculated back pressure is higher than the given one, the location of the normal shock wave must move downwards in the duct. The rest of the iterations are shown in Table 6.4. After convergence, the location of the normal shock wave is found to be at

$$x = \mathbf{1.1\ m}$$

(b) Adiabatic ducts with lengths greater than the maximum allowable length

Consider an adiabatic duct, which is fed by an isentropic converging–diverging nozzle and whose length is larger than the maximum allowable length, as shown in Figure 6.18(a). The effects of variations in the back pressure on the distributions of the pressure in the duct are presented in Figure 6.18(b).

(i) No-flow condition ($p_b = p_{0i}$)

To begin with, suppose that the valve is closed, as shown by flow pattern 1 of Figure 6.18(b). In this case, the pressure is constant throughout the converging–diverging nozzle and the duct, such that $p_b = p_e = p_i = p_{0i}$ and there is no flow.

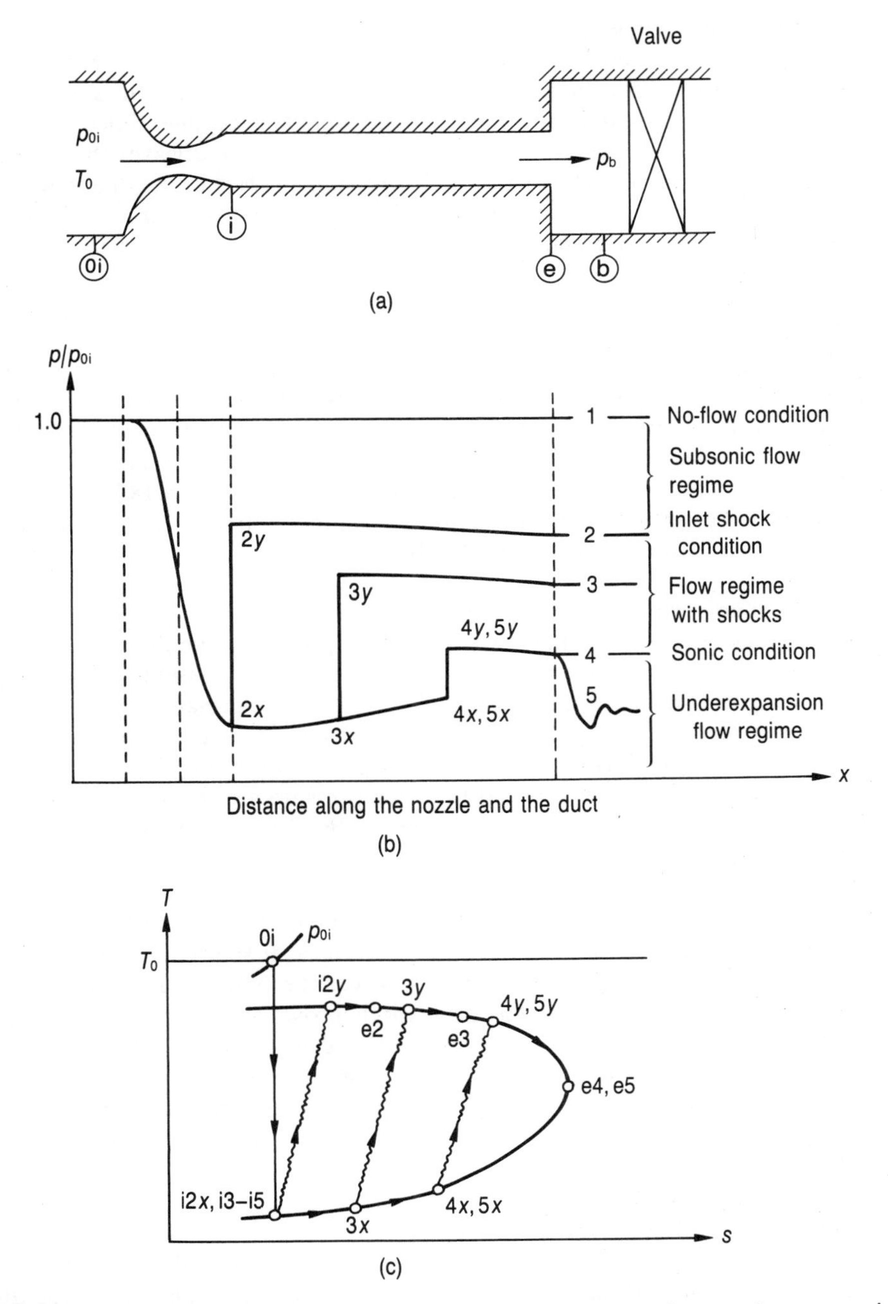

Figure 6.18 Performance of an adiabatic duct fed by an isentropic converging–diverging nozzle (the length of the duct is greater than the maximum allowable length)

(ii) Subsonic flow regime ($p_{0i} > p_b > p_{e2}$)
In this flow regime, the flow is always subsonic in the duct. As the back pressure is lowered, the flow in the converging–diverging nozzle passes through the subsonic flow regime, the choking condition and the non-isentropic flow regime. Since the flow in the duct is subsonic, the pressure decreases along the duct and leaves the duct such that the exit pressure is equal to the back pressure. This flow regime will not be treated in detail, since the flow patterns in a converging–diverging nozzle are discussed in Sections 4.8.2 and 5.8, and the subsonic flow patterns in a duct are discussed in Section 6.8.1.

(iii) Inlet shock condition ($p_b = p_{e2}$)
In this case, a normal shock wave appears at the inlet of the duct, and the flow pattern in the converging–diverging nozzle is such that there is a normal shock wave standing at its exit. Thus, the pressure increases abruptly at the inlet of the duct owing to the normal shock wave. The flow throughout the duct is subsonic and the pressure decreases along the duct, as shown by flow pattern 2 of Figure 6.18(b). Again, the exit pressure is the same as the back pressure. The process 0i–i2x–i2y–e2 for this flow pattern is shown in the T–s diagram of 6.18(c).

(iv) Flow regime with shocks ($p_{e2} > p_b > p^*$)
In this flow regime, a normal shock wave is always present in the duct. Flow pattern 3 of Figure 6.18(b) represents a typical flow pattern of this regime. The pressure increases along the duct during supersonic flow before the normal shock wave, shows an abrupt rise across the normal shock wave and decreases along the duct during subsonic flow after the normal shock wave. As long as the flow leaves the duct at subsonic speeds, the exit pressure and the back pressure are the same. The process 0i–i3–3x–3y–e3 for this flow pattern is shown on the T–s diagram of Figure 6.18(c).

(v) Sonic condition ($p_b = p^*$)
As the back pressure is lowered, the exit Mach number reaches unity, as shown by flow pattern 4 of Figure 6.18(c). In this case, the normal shock wave has a definite place in the duct. Before the normal shock wave, the flow is supersonic and the pressure increases along the duct. The pressure then rises abruptly across the normal shock wave. After the normal shock wave, the flow becomes subsonic and the pressure decreases along the duct with increasing velocity. The sonic condition is reached at the exit where the exit pressure is exactly equal to the back pressure. This condition is shown by the process 0i–i4–4x–4y–e4 on the T–s diagram of Figure 6.18(c).

(vi) Underexpansion flow regime ($p_b < p^*$)
Flow pattern 5 of this flow regime in Figure 6.18(b) follows exactly flow pattern 4 of the choking condition within the converging–diverging nozzle and within the duct. However, the gas is underexpanded at the exit of the duct, since the exit pressure is higher than the back pressure. The expansion which occurs outside the duct involves

non-isentropic oblique expansion waves, which cannot be treated with the assumption of one-dimensional flow. The process 0i–i5–5x–5y–e5 for this flow pattern is shown on the T–s diagram of Figure 6.18(c).

Example 6.7

An adiabatic duct is fed by an isentropic converging–diverging nozzle with an area ratio of 1.76, as shown in Figure 6.19(a). The stagnation pressure in the supply reservoir is 400 kPa. If the length of the duct is 5.408 m, the diameter of the duct is 0.08 m and the average friction factor in the duct is 0.002, then determine

a the maximum allowable duct length,
b the back pressure such that the flow is sonic at the end of the duct,

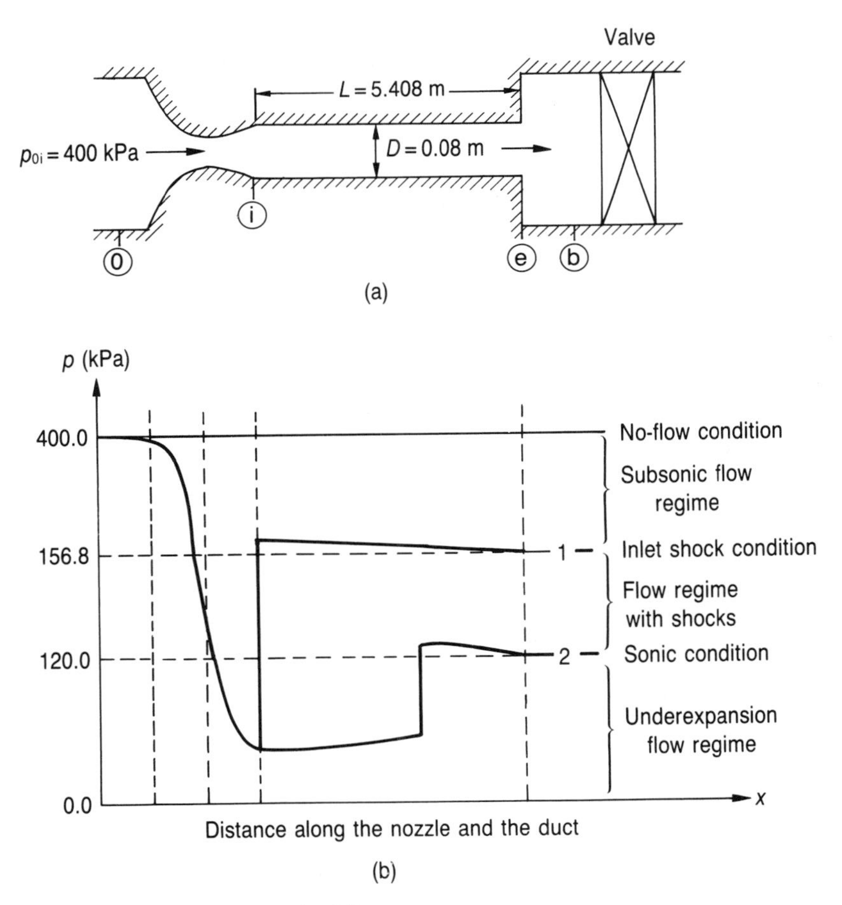

Figure 6.19 Sketch for Example 6.7

c the distance from the inlet of the duct at which the shock wave is located when the flow is sonic at the exit of the duct,

d the back pressure such that there is a normal shock wave at the inlet of the duct, and

e the ranges of back pressure for the subsonic flow regime, the flow regime with shocks and the underexpansion flow regime.

Solution

(a) In order to have supersonic flow at the exit of the converging–diverging nozzle, the flow must be choked at the throat. Therefore, corresponding to an area ratio of $A/A^* = 1.76$, one may obtain the Mach number and the pressure at the inlet of the duct from Appendix C as

$$M_i = 2.05$$

and

$$p_i/p_{0i} = 0.1182 \quad \text{or} \quad p_i = (0.1182)(400\ \text{kPa}) = 47.28\ \text{kPa}$$

The maximum allowable duct length corresponding to an inlet Mach number of $M_i = 2.05$ may be obtained from Appendix E as

$$\frac{4\bar{f}L_{max}}{D} = 0.3197 \quad \text{or} \quad L_{max} = \frac{(0.3197)(0.08\ \text{m})}{(4)(0.002)} = \mathbf{3.197\ m}$$

Therefore, the given duct length is longer than the maximum allowable length.

(b) When the flow is sonic at the end of the duct, the back pressure is equal to the critical pressure. Corresponding to an inlet Mach number of $M_i = 2.05$ from Appendix E

$$p_i/p^* = 0.3939 \quad \text{or} \quad p^* = 47.28\ \text{kPa}/0.3939 = 120.0\ \text{kPa}$$

Then, the back pressure is

$$p_{b2} = p^* = \mathbf{120.0\ kPa}$$

(c) When the flow is sonic at the end of the duct, $p_{e2} = p_{b2}$ and $M_{e2} = 1.0$. Hence

$$p_{e2} = p_{b2} = 120.0\ \text{kPa}$$

The location of the normal shock wave may now be found by an iterative solution. Initially, assuming that the normal shock wave is located at $x = 0.25$ m,

$$(4\bar{f}L_{max}/D)_{2x} = (4\bar{f}L_{max}/D)_{i2} - 4\bar{f}x/D$$

$$= 0.3197 - \frac{(4)(0.002)(0.25\ \text{m})}{(0.08\ \text{m})} = 0.2947$$

The Mach number upstream of the normal shock wave may now be obtained from Appendix E with $(4\bar{f}L_{max}/D)_{2x} = 0.2947$ as $M_{2x} = 1.97$. Now, the Mach number downstream of the normal shock wave corresponding to $M_{2x} = 1.97$ may be obtained from Appendix D as $M_{2y} = 0.5826$. From Appendix E, corresponding to $M_{2y} = 0.58$,

$$(4\bar{f}L_{max}/D)_{2y} = 0.5757$$

then

$$(4\bar{f}L_{max}/D)_{e2} = (4\bar{f}L_{max}/D)_{2y} - 4\bar{f}(5.408 \text{ m} - x)/D$$

$$= 0.5757 - \frac{(4)(0.002)(5.408 \text{ m} - 0.25 \text{ m})}{(0.08 \text{ m})} = 0.05990$$

The Mach number at the exit of the duct is then $M_{e2} = 0.81$. As long as the flow is not sonic at the duct exit, the location of the normal shock wave is incorrect and should be downstream of the assumed one. The rest of the iterations are shown in Table 6.5. After convergence, the location of the normal shock wave is found to be at

$$x = \mathbf{0.50 \ m}$$

(d) When there is a normal shock wave at the inlet of the duct, then, corresponding to an upstream Mach number of $M_{i1x} = 2.05$, the downstream Mach number and the pressure may be obtained from Appendix D as

$$M_{i1y} = 0.5691$$

and

$$p_{i1y}/p_{i1x} = 4.736 \quad \text{or} \quad p_{i1y} = (4.736)(47.28 \text{ kPa}) = 223.9 \text{ kPa}$$

Table 6.5 Iterations for Example 6.7

Iteration	1	2	3
x (m)	0.25	0.75	0.50
$(4\bar{f}L_{max}/D)_{2x}$	0.2947	0.2447	0.2697
M_{2x}	1.97	1.81	1.89
M_{2y}	0.5826	0.6143	0.5976
$(4\bar{f}L_{max}/D)_{2y}$	0.5757	0.4527	0.4908
$(4\bar{f}L_{max}/D)_{e2}$	0.059 90	−0.013 10	0.000
M_{e2}	0.81	–	1.00
Shock movement	→	←	OK

Now, from Appendix E, corresponding to $M_{i1y} = 0.57$,

$$(4\bar{f}L_{max}/D)_{i1y} = 0.6229$$

Also, it is possible to write

$$(4\bar{f}L_{max}/D)_{e1} = (4\bar{f}L_{max}/D)_{i1y} - 4\bar{f}L/D$$

$$= 0.6229 - \frac{(4)(0.002)(5.408 \text{ m})}{(0.08 \text{ m})} = 0.0821$$

The Mach number and the pressure at the duct exit may now be obtained from Appendix E with $(4\bar{f}L/D)_{e1} = 0.0821$ as

$$M_{e1} = 0.79$$

and

$$p_{e1}/p^* = 1.307 \quad \text{or} \quad p_{e1} = (1.307)(120.0 \text{ kPa}) = 156.8 \text{ kPa}$$

As long as the flow leaves the duct at subsonic speeds, then the back pressure must be the same, that is

$$p_{b1} = p_{e1} = \mathbf{156.8\ kPa}$$

(e) Ranges of back pressure for regimes are as follows:

1 Subsonic flow regime:

$$p_0 \geqslant p_b \geqslant p_{b1} \quad \text{or} \quad \mathbf{400\ kPa \geqslant p_b \geqslant 156.8\ kPa}$$

2 Flow regime with shocks:

$$p_{b1} \geqslant p_b \geqslant p_{b2} \quad \text{or} \quad \mathbf{156.8\ kPa \geqslant p_b \geqslant 120.0\ kPa}$$

3 Underexpansion flow regime:

$$p_{b2} \geqslant p_b \geqslant 0 \quad \text{or} \quad \mathbf{120.0\ kPa \geqslant p_b \geqslant 0\ kPa}$$

6.9 GOVERNING EQUATIONS FOR ISOTHERMAL ONE-DIMENSIONAL FLOW WITH FRICTION IN CONSTANT-AREA DUCTS

Now, consider steady, one-dimensional isothermal flow of a perfect gas through a constant-area duct with friction, as shown in Figure 6.20. The governing equations are given below for the mentioned control volume.

(a) Continuity equation

For steady and uniform flow at each cross-section, the mass flow rate is constant. Then,

$$\dot{m} = \rho_1 V_1 A_1 = \rho_2 V_2 A_2 = \text{constant}$$

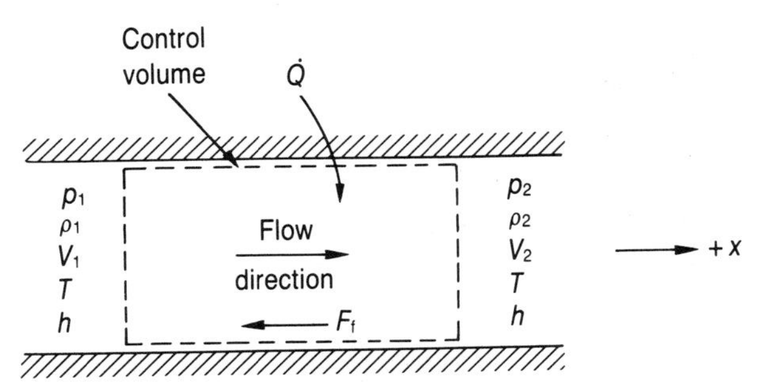

Figure 6.20 Control volume for steady, one-dimensional isothermal flow with friction in a constant-area duct

As long as the cross-sectional area of the duct is constant, that is $A_1 = A_2 = A$, then the mass flow rate per unit area or the mass flux, G, is given by

$$G = \frac{\dot{m}}{A} = \rho_1 V_1 = \rho_2 V_2 = \rho V = \text{constant} \tag{6.37}$$

(b) Momentum equation

For steady and uniform flow with friction, the sum of the pressure forces and frictional forces acting on the control volume of Figure 6.20 must be balanced by the rate of change of linear momentum across the control volume, that is

$$-F_f + p_1 A_1 - p_2 A_2 = \rho_2 A_2 V_2^2 - \rho_1 A_1 V_1^2$$

or

$$\frac{F_f}{A} + (p_2 - p_1) = \rho_1 V_1^2 - \rho_2 V_2^2 = G(V_1 - V_2) \tag{6.38}$$

where F_f is the frictional force acting on the side walls of the control volume.

(c) The first law of thermodynamics

When there is no external shaft work on the control volume, the first law of thermodynamics with negligible elevation changes is

$$\dot{Q} = \dot{m}q = \dot{m}(h_{02} - h_{01}) = \dot{m}\left(h_2 + \frac{V_2^2}{2} - h_1 - \frac{V_1^2}{2}\right)$$

However, for a perfect gas $h = c_p T$ and for an isothermal flow the static temperature is constant, that is $T_1 = T_2$. Hence, the energy equation becomes

$$\dot{Q} = \dot{m}\left(\frac{V_2^2}{2} - \frac{V_1^2}{2}\right) \tag{6.39}$$

(d) Equation of state

For a perfect gas, the equation of state is given as

$$p = \rho RT \tag{6.40}$$

(e) The second law of thermodynamics

For one-dimensional and steady flow with friction and heat transfer in a constant-area duct, the second law of thermodynamics is given by Equation (2.32) as

$$\dot{Q} < \dot{m}T(s_2 - s_1) \tag{6.41}$$

To calculate the entropy change, the thermodynamic relation

$$T\,\mathrm{d}s = \mathrm{d}h - \frac{\mathrm{d}p}{\rho}$$

may be used. But, for a perfect gas, $\mathrm{d}h = c_p\,\mathrm{d}T$ and $p = \rho RT$, and for an isothermal process $\mathrm{d}T = 0$. Then

$$\mathrm{d}s = -R\,\frac{\mathrm{d}p}{p} = -R\,\frac{\mathrm{d}\rho}{\rho}$$

which may be integrated to give the entropy change as

$$s_2 - s_1 = -R\ln\left(\frac{p_2}{p_1}\right) = -R\ln\left(\frac{\rho_2}{\rho_1}\right) \tag{6.42}$$

Example 6.8

Air flows isothermally through a duct with a cross-sectional area of 0.025 m^2, as shown in Figure 6.21. At section 1, the pressure, temperature and velocity are 200 kPa, 300 K and 100 m s^{-1}, respectively. At section 2, the pressure is 100 kPa. Determine

a the mass flow rate,
b the properties at section 2,
c the frictional force acting on the duct walls,

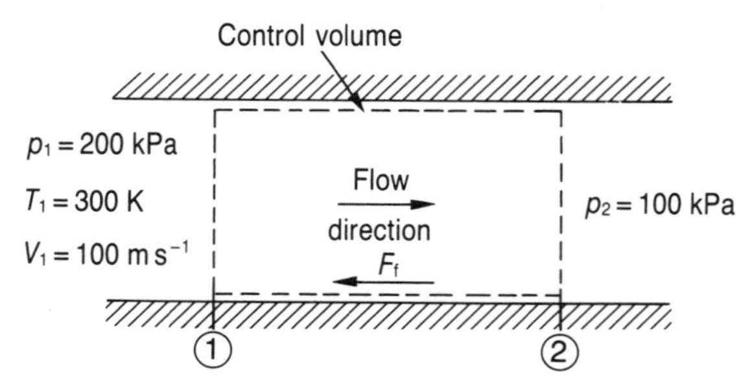

Figure 6.21 Sketch for Example 6.8

d the rate of heat transfer, and
e the entropy change.

Solution

(a) The density at section 1 may be calculated by using the equation of state as

$$\rho_1 = \frac{p_1}{RT_1} = \frac{200\,000 \text{ N m}^{-2}}{(287.1 \text{ J kg}^{-1}\text{ K}^{-1})(300 \text{ K})} = 2.322 \text{ kg m}^{-3}$$

Then, the mass flow rate from the continuity equation is

$$\dot{m} = \rho_1 A_1 V_1 = (2.322 \text{ kg m}^{-3})(0.025 \text{ m}^2)(100 \text{ m s}^{-1}) = \mathbf{5.805 \text{ kg s}^{-1}}$$

(b) As long as the flow is isothermal, then the static temperature remains constant. Hence

$$T_2 = T_1 = \mathbf{300 \text{ K}}$$

The density at section 2 may now be evaluated from the equation of state as

$$\rho_2 = \frac{p_2}{RT_2} = \frac{100\,000 \text{ N m}^{-2}}{(287.1 \text{ J kg}^{-1}\text{ K}^{-1})(300 \text{ K})} = \mathbf{1.161 \text{ kg m}^{-3}}$$

Using the continuity equation for the constant-area duct, the velocity at section 2 is

$$V_2 = V_1 \frac{\rho_1}{\rho_2} = (100 \text{ m s}^{-1}) \frac{(2.322 \text{ kg m}^{-3})}{(1.161 \text{ kg m}^{-3})} = \mathbf{200 \text{ m s}^{-1}}$$

The local isentropic stagnation temperature at section 2 may now be calculated via Equation (4.8a) as

$$T_{02} = T_2 + \frac{V_2^2}{2c_p} = 300 \text{ K} + \frac{(200 \text{ m s}^{-1})^2}{(2)(1005 \text{ J kg}^{-1}\text{ K}^{-1})} = \mathbf{319.9 \text{ K}}$$

Considering the imaginary isentropic process between state 2 and its local isentropic state, the stagnation pressure at section 2 is

$$p_{02} = p_2 \left(\frac{T_{02}}{T_2}\right)^{k/(k-1)} = (100 \text{ kPa}) \left(\frac{319.9 \text{ K}}{300 \text{ K}}\right)^{1.4/(1.4-1)} = \mathbf{125.2 \text{ kPa}}$$

(c) The frictional force, F_f, acting on the control volume may be found by applying the momentum equation to the control volume of Figure 6.21 as

$$-F_f + p_1 A_1 - p_2 A_2 = \rho_2 A_2 V_2^2 - \rho_1 A_1 V_1^2 = \dot{m}(V_2 - V_1)$$

or

$$F_f = (p_1 - p_2)A - \dot{m}(V_2 - V_1) = (200\,000 \text{ N m}^{-2} - 100\,000 \text{ N m}^{-2})(0.025 \text{ m}^2)$$
$$- (5.805 \text{ kg s}^{-1})(200 \text{ m s}^{-1} - 100 \text{ m s}^{-1}) = 1920 \text{ N}$$

This is the force exerted on the control volume by the duct walls. The force acting on the duct is then

$$F = -F_f = \mathbf{-1920\ N}$$

(d) Using Equation (4.8a), the local isentropic stagnation temperature at section 1 is

$$T_{01} = T_1 + \frac{V_1^2}{2c_p} = 300\ \text{K} + \frac{(100\ \text{m s}^{-1})^2}{(2)(1005\ \text{J kg}^{-1}\ \text{K}^{-1})} = 305.0\ \text{K}$$

Then, the amount of heat transfer may be determined from the first law of thermodynamics as

$$\dot{Q} = \dot{m}c_p(T_{02} - T_{01}) = (5.805\ \text{kg s}^{-1})(1005\ \text{J kg}^{-1}\ \text{K}^{-1})(319.9\ \text{K} - 305.0\ \text{K})$$

$$= \mathbf{86.93\ kW}$$

(e) The entropy change may now be evaluated by using Equation (6.42) as

$$s_2 - s_1 = -R\ \ln\left(\frac{p_2}{p_1}\right) = -(287.1\ \text{J kg}^{-1}\ \text{K}^{-1})\ \ln\left(\frac{100\ \text{kPa}}{200\ \text{kPa}}\right) = \mathbf{199.0\ J\,kg^{-1}\,K^{-1}}$$

6.10 EFFECT OF FRICTION ON FLOW PROPERTIES IN AN ISOTHERMAL FLOW

6.10.1 Governing equations in differential form

Consider one-dimensional isothermal flow of a compressible fluid in a constant-area duct with friction, as shown in Figure 6.22. Heat transfer and friction between the fluid and the duct wall cause infinitesimal changes in flow properties, and these changes may be related to each other by applying the basic equations to an infinitesimal control volume, as shown in Figure 6.22.

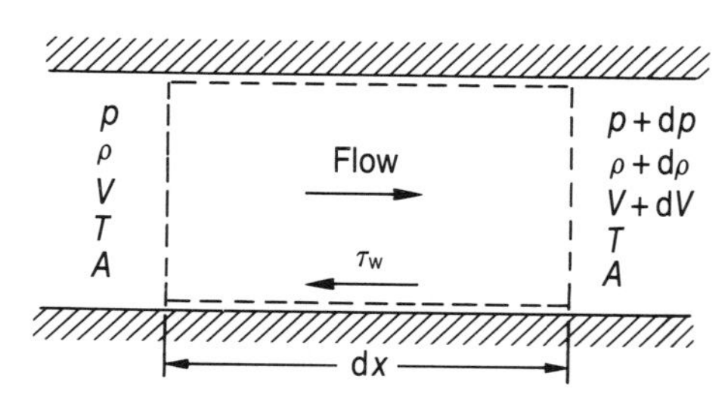

Figure 6.22 Infinitesimal control volume for the analysis of isothermal flow

(a) Continuity equation

For steady and uniform flow at each cross-section, the mass flow rate is constant. Therefore,

$$\dot{m} = \rho VA = (\rho + \mathrm{d}\rho)(V + \mathrm{d}V)A$$

or

$$\cancel{\rho A}V = \cancel{\rho A}V + \rho A\ \mathrm{d}V + AV\ \mathrm{d}\rho + A\ \overset{\simeq 0}{\cancel{\mathrm{d}\rho}}\ \mathrm{d}V$$

Neglecting the products of differentials and dividing by ρAV, it is possible to obtain

$$\frac{\mathrm{d}V}{V} + \frac{\mathrm{d}\rho}{\rho} = 0 \tag{6.43}$$

(b) Momentum equation

The sum of the frictional forces and the pressure forces acting on the control volume must be balanced by the rate of change of linear momentum across the control volume, that is

$$pA - (p + \mathrm{d}p)A - \tau_{\mathrm{w}}\ \mathrm{d}A_{\mathrm{w}} = \dot{m}(V + \mathrm{d}V) - \dot{m}V$$

or

$$-A\ \mathrm{d}p - \tau_{\mathrm{w}}\ \mathrm{d}A_{\mathrm{w}} = \dot{m}\ \mathrm{d}V = \rho AV\ \mathrm{d}V$$

Now, using the definitions of the friction factor, $\bar{f}$, and the hydraulic diameter, D, which are given by Equations (6.9) and (6.10), respectively, the momentum equation takes the following form:

$$-\frac{4\bar{f}}{D}\frac{\rho V^2}{2}\ \mathrm{d}x - \mathrm{d}p = \rho V\ \mathrm{d}V \tag{6.44}$$

(c) The first law of thermodynamics

For isothermal flow with no external shaft work, the first law of thermodynamics for the control volume is

$$c_{\mathrm{p}}T + \frac{V^2}{2} + \mathrm{d}q = c_{\mathrm{p}}T + \frac{(V + \mathrm{d}V)^2}{2}$$

or

$$\cancel{\frac{V^2}{2}} + \mathrm{d}q = \cancel{\frac{V^2}{2}} + V\ \mathrm{d}V + \overset{\simeq 0}{\cancel{\frac{(\mathrm{d}V)^2}{2}}}$$

Neglecting second-order differentials, it is possible to obtain

$$\mathrm{d}q = V\,\mathrm{d}V \tag{6.45}$$

(d) Equation of state

The equation of state for a perfect gas, $p = \rho RT$, may be logarithmically differentiated to yield

$$\frac{\mathrm{d}p}{p} = \frac{\mathrm{d}\rho}{\rho} \tag{6.46}$$

for an isothermal process.

(e) The second law of thermodynamics

For an isothermal process with friction and heat transfer

$$\mathrm{d}s > \mathrm{d}q/T \tag{6.47}$$

6.10.2 Variation of flow properties with friction in an isothermal flow

(a) Pressure

Dividing Equation (6.44) by p and noting that $p/\rho = RT = a^2/k$, it is possible to obtain

$$\frac{\mathrm{d}p}{p} = -kM^2\frac{\mathrm{d}V}{V} - \frac{kM^2}{2}\,4f\frac{\mathrm{d}x}{D}$$

However, by using Equations (6.43) and (6.46), it is possible to obtain

$$\frac{\mathrm{d}V}{V} = -\frac{\mathrm{d}\rho}{\rho} = -\frac{\mathrm{d}p}{p}$$

Therefore

$$\frac{\mathrm{d}p}{p} = -\frac{kM^2}{2(1-kM^2)}\,4f\frac{\mathrm{d}x}{D} \tag{6.48}$$

(b) Density

Combining Equations (6.46) and (6.48) yields

$$\frac{\mathrm{d}\rho}{\rho} = -\frac{kM^2}{2(1-kM^2)}\,4f\frac{\mathrm{d}x}{D} \tag{6.49}$$

(c) Velocity

From Equations (6.43) and (6.49), it is possible to obtain

$$\frac{\mathrm{d}V}{V} = \frac{kM^2}{2(1-kM^2)}\,4f\frac{\mathrm{d}x}{D} \tag{6.50}$$

(d) Mach number

From the definition of the Mach number, $M^2 = V^2/a^2 = V^2/(kRT)$, and by logarithmic differentiation with temperature as a constant

$$\frac{d(M^2)}{M^2} = 2\frac{dV}{V} = \frac{kM^2}{1-kM^2}\,4f\frac{dx}{D} \tag{6.51}$$

(e) Stagnation pressure

The previous definition of the isentropic stagnation pressure can be retained, as long as the isentropic stagnation pressure corresponding to a given local state is the pressure which would be attained if the stream were decelerated isentropically from the given state to the state of zero velocity. Then, from Equation (4.28),

$$p_0 = p\left(1 + \frac{k-1}{2}M^2\right)^{k/(k-1)}$$

or in differential form

$$\frac{dp_0}{p_0} = \frac{dp}{p} + \frac{kM^2/2}{1+\frac{1}{2}(k-1)M^2}\frac{dM^2}{M^2}$$

Finally, with the aid of Equations (6.48) and (6.51), it is possible to obtain

$$\frac{dp_0}{p_0} = -\frac{kM^2[1-\frac{1}{2}(k+1)M^2]}{2(1-kM^2)[1+\frac{1}{2}(k-1)M^2]}\,4f\frac{dx}{D} \tag{6.52}$$

(f) Stagnation temperature

The previous definition of the adiabatic stagnation temperature can be retained, since the adiabatic stagnation temperature corresponding to a given local state is the temperature which would be attained if the stream were decelerated adiabatically from the given state to the state of zero velocity. Then, from Equation (4.24),

$$T_0 = T\left(1 + \frac{k-1}{2}M^2\right)$$

Taking logarithmic differentials and noting that $dT = 0$, it is possible to obtain

$$\frac{dT_0}{T_0} = \frac{(k-1)M^2}{2[1+\frac{1}{2}(k-1)M^2]}\frac{dM^2}{M^2}$$

or using Equation (6.51) it is possible to obtain

$$\frac{dT_0}{T_0} = \frac{k(k-1)M^4}{2(1-kM^2)[1+\frac{1}{2}(k-1)M^2]}\,4f\frac{dx}{D} \tag{6.53}$$

(g) Impulse function

From the definition of the impulse function given by Equation (4.39)

$$I = pA(1 + kM^2)$$

or in differential form for constant area

$$\frac{\mathrm{d}I}{I} = \frac{\mathrm{d}p}{p} + \frac{kM^2}{1 + kM^2}\frac{\mathrm{d}M^2}{M^2}$$

Now, with the aid of Equations (6.48) and (6.51), the following relation may be obtained:

$$\frac{\mathrm{d}I}{I} = -\frac{kM^2}{2(1 + kM^2)} 4f \frac{\mathrm{d}x}{D} \tag{6.54}$$

(h) Entropy

One might recall the thermodynamic relation

$$T\,\mathrm{d}s = \mathrm{d}h - \frac{\mathrm{d}p}{\rho}$$

Table 6.6 Effect of friction on flow properties for the isothermal flow of a perfect gas in a constant-area duct

Flow property	Type of flow Subsonic $M < 1/\sqrt{k}$	Subsonic or supersonic $M > 1/\sqrt{k}$
Mach number, M	Increases	Decreases
Pressure, p	Decreases	Increases
Density, ρ	Decreases	Increases
Velocity, V	Increases	Decreases
Stagnation pressure, p_0	Decreases	Increases for $M < \sqrt{[2/(k+1)]}$ decreases for $M > \sqrt{[2/(k+1)]}$
Stagnation temperature, T_0	Increases	Decreases
Impulse function, I	Decreases	Decreases
Entropy, s	Increases	Decreases

However, for a perfect gas $dh = c_p\, dT$ and $p = \rho RT$. Also, for an isothermal process $dT = 0$. Therefore

$$\frac{ds}{R} = -\frac{dp}{p} = \frac{kM^2}{2(1 - kM^2)}\, 4f \frac{dx}{D} \tag{6.55}$$

6.10.3 Summary of frictional effects in an isothermal flow

In the formulation of the momentum equation, it is assumed that the shear stress acts in a direction opposite to the direction of flow. Since the friction factor must be positive, the shear stress must always act in a direction opposite to the direction of flow. Also, by convention, dx is positive in the direction of flow, therefore, $4f\, dx/D$ is always positive.

The directions of change in the fluid properties depend on whether the Mach number is greater or less than $1/\sqrt{k}$ according to Equations (6.48)–(6.55) (except Equation (6.54)), since the term $(1 - kM^2)$ appears in the denominator of each of these equations. One should also note that the direction of change in the stagnation pressure also depends on the term $[1 - \frac{1}{2}(k + 1)M^2]$ appearing in the numerator of Equation (6.52) when $M > 1/\sqrt{k}$. These are summarized in Table 6.6.

6.11 RELATIONS FOR THE ISOTHERMAL FLOW OF A PERFECT GAS IN A CONSTANT-AREA DUCT

The equations governing the one-dimensional isothermal flow of a perfect gas in a constant-area duct were presented in Section 6.9. Although any problem of this kind may be solved by using these equations, the solution can be simplified by rearranging them in a more suitable form. In this section, these relations are derived by choosing the Mach number as the independent variable.

6.11.1 Non-dimensional friction factor

Equation (6.51) is a differential equation which relates the changes in the Mach number to the distance along the duct, x. A general problem for the isothermal flow is shown in Figure 6.23. For a given duct length, L, the integration of Equation (6.51) between states 1 and 2 would produce a complicated function of M_1 and M_2. This function would have to be evaluated numerically for each combination of M_1 and M_2 encountered in a problem. These calculations can be simplified by using reference conditions. It is known that for subsonic flows with an initial Mach number less than $1/\sqrt{k}$ and for subsonic and supersonic flows with an initial Mach number greater than $1/\sqrt{k}$, the Mach number always tends towards $1/\sqrt{k}$. Therefore, for a given value of the inlet Mach number in the duct, there always exists a hypothetical maximum duct length, L_{max}, for which the Mach number is $1/\sqrt{k}$ at the exit of this hypothetical extension, as indicated in Figure 6.23. For this reason, reference conditions may be

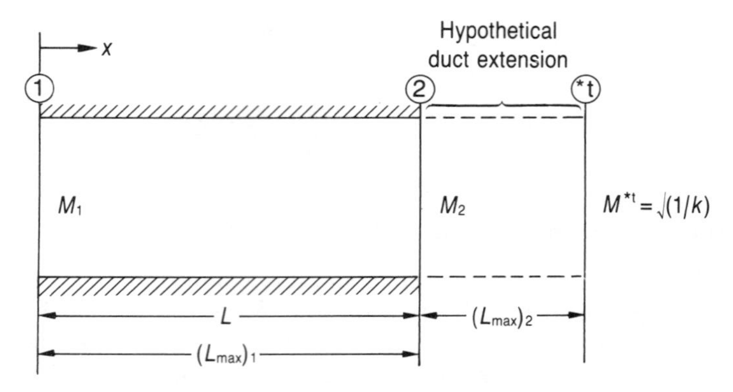

Figure 6.23 Coordinates and notation used for the analysis of isothermal flow

chosen as critical conditions corresponding to a Mach number of $1/\sqrt{k}$ and indicated by the superscript *t.

The task is to integrate Equation (6.51) between any given state and the critical state as,

$$\int_0^{L_{max}} 4f \frac{dx}{D} = \int_{M^2}^{1/k} \frac{1 - kM^2}{kM^4} \, d(M^2) \tag{6.56}$$

The right-hand side of this equation may be evaluated using integration by parts. However, on the left-hand side, the friction factor, f, is a function of Reynolds' number, $Re = \rho VD/\mu$, where μ is the absolute viscosity of the fluid. For flows in constant-area ducts, the product, ρV, is constant along the duct as indicated by the continuity equation. Therefore, the variation in Reynolds' number is caused only by variations in the fluid viscosity. Then an average friction factor, $\bar{f}$, over the duct length may be defined as

$$\bar{f} = \frac{1}{L_{max}} \int_0^{L_{max}} f \, dx \tag{6.57}$$

and integration of Equation (6.56) leads to

$$\frac{4\bar{f}L_{max}}{D} = \frac{1 - kM^2}{kM^2} + \ln (kM^2) \tag{6.58}$$

This equation gives the maximum value of $4\bar{f}L/D$ corresponding to any given initial Mach number. Since the non-dimensional friction factor, $4\bar{f}L_{max}/D$, is a function of the Mach number, then the duct length, L, required for the flow Mach number to change from an initial value of M_1 to a final value of M_2 may be found from

$$\frac{4\bar{f}L}{D} = \left(\frac{4\bar{f}L_{max}}{D}\right)_1 - \left(\frac{4\bar{f}L_{max}}{D}\right)_2 \tag{6.59}$$

6.11.2 Non-dimensional velocity

Using the definition of Mach number as $V = M\sqrt{(kRT)}$, and noting that the temperature is constant for isothermal flows, the non-dimensional velocity may be obtained as

$$\frac{V}{V^{*t}} = \frac{M\sqrt{(kRT)}}{M^{*t}\sqrt{(kRT^{*t})}} = \sqrt{(k)}M \tag{6.60}$$

6.11.3 Non-dimensional density

From the continuity equation for steady flow in constant-area ducts, $\rho V = \rho^{*t} V^{*t}$. Then using Equation (6.60)

$$\frac{\rho}{\rho^{*t}} = \frac{V^{*t}}{V} = \frac{1}{\sqrt{(k)}M} \tag{6.61}$$

6.11.4 Non-dimensional pressure

Using the equation of state for a perfect gas, $p = \rho RT$, the non-dimensional pressure may be obtained with the aid of Equation (6.61) as

$$\frac{p}{p^{*t}} = \frac{\rho}{\rho^{*t}} = \frac{1}{\sqrt{(k)}M} \tag{6.62}$$

6.11.5 Non-dimensional stagnation pressure

The ratio of the local stagnation pressure to the reference stagnation pressure is given by

$$\frac{p_0}{p_0^{*t}} = \frac{p_0}{p}\frac{p}{p^{*t}}\frac{p^{*t}}{p_0^{*t}}$$

Now, using Equation (4.28) for p_0/p and p^{*t}/p_0^{*t}, and Equation (6.62) for p/p^{*t}, it is possible to obtain

$$\frac{p_0}{p_0^{*t}} = \frac{1}{\sqrt{(k)}M}\left[\frac{2k}{3k-1}\left(1 + \frac{k-1}{2}M^2\right)\right]^{k/(k-1)} \tag{6.63}$$

6.11.6 Non-dimensional stagnation temperature

The ratio of the local stagnation temperature to the reference stagnation temperature is given by

$$\frac{T_0}{T_0^{*t}} = \frac{T_0}{T}\frac{T}{T^{*t}}\frac{T^{*t}}{T_0^{*t}}$$

Noting that $T = T^{*t}$ for an isothermal flow and using Equation (4.24), it is possible to obtain

$$\frac{T_0}{T_0^{*t}} = \frac{2k}{3k-1}\left(1 + \frac{k-1}{2} M^2\right) \tag{6.64}$$

6.11.7 Non-dimensional impulse function

From the definition of the impulse function, $I = pA(1 + kM^2)$ by Equation (4.39). Then for constant-area ducts

$$\frac{I}{I^{*t}} = \frac{p}{p^{*t}} \frac{1 + kM^2}{1 + k(M^{*t})^2}$$

Now, using Equation (6.62) and the fact that $M^{*t} = 1/\sqrt{k}$, it is possible to obtain

$$\frac{I}{I^{*t}} = \frac{1 + kM^2}{2M\sqrt{k}} \tag{6.65}$$

6.11.8 Non-dimensional entropy

The entropy change between any state and the reference state is given by using Equation (6.6) as

$$s - s^{*t} = c_p \ln\left(\frac{T}{T^{*t}}\right) - R \ln\left(\frac{p}{p^{*t}}\right)$$

However, for an isothermal flow $T = T^{*t}$ and by using Equation (6.62) it is possible to obtain

$$\frac{s - s^{*t}}{R} = \ln(\sqrt{(k)}M) \tag{6.66}$$

6.12 WORKING CHART AND WORKING TABLE FOR THE ISOTHERMAL FLOW OF A PERFECT GAS

Formulas derived thus far introduce quite tedious numerical calculations and their solutions often involve a trial-and-error procedure. The numerical calculations can be facilitated by introducing a working chart and a working table. The following equations are the non-dimensional properties for the isothermal flow of a perfect gas that were derived in the previous section:

$$\frac{p}{p^{*t}} = \frac{1}{\sqrt{(k)}M} \tag{6.62}$$

$$\frac{\rho}{\rho^{*t}} = \frac{1}{\sqrt{(k)}M} \tag{6.61}$$

$$\frac{V}{V^{*t}} = \sqrt{(k)}M \tag{6.60}$$

$$\frac{p_0}{p_0^{*t}} = \frac{1}{\sqrt{(k)}M}\left[\frac{2k}{3k-1}\left(1+\frac{k-1}{2}M^2\right)\right]^{k/(k-1)} \tag{6.63}$$

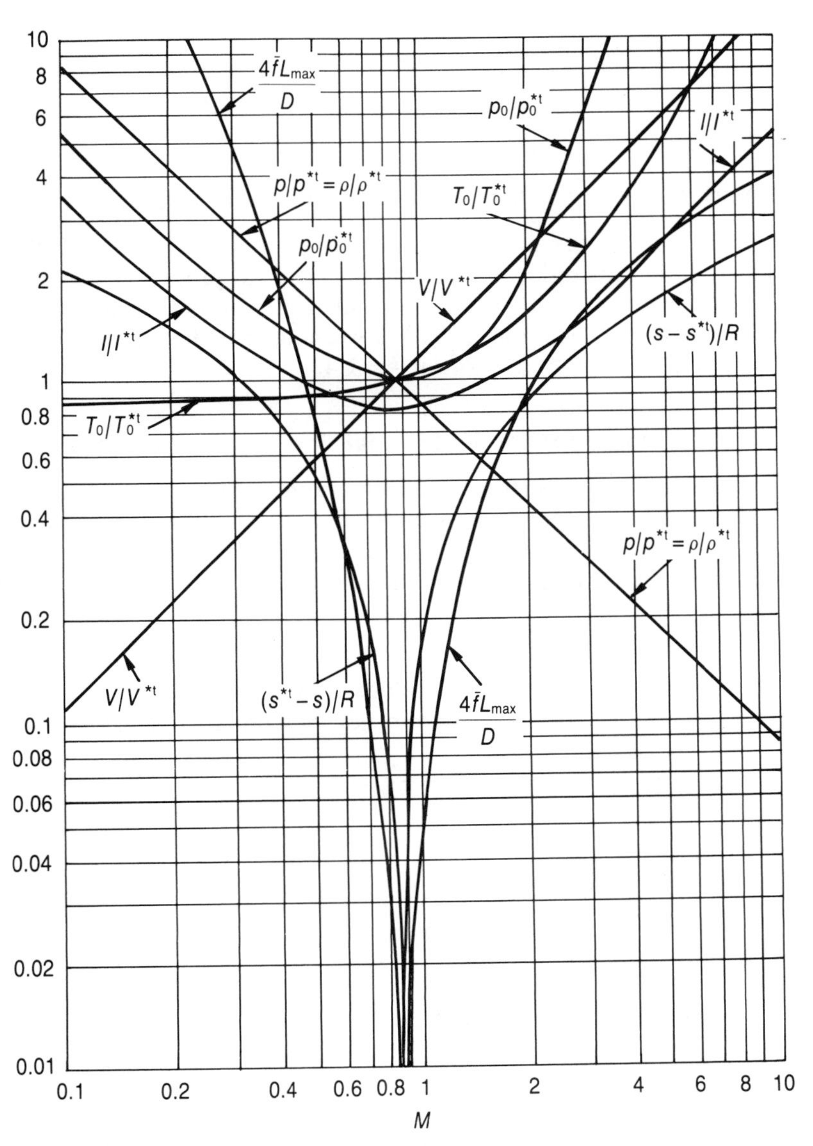

Figure 6.24 Working chart for the isothermal flow of a perfect gas with $k = 1.4$

$$\frac{T_0}{T_0^{*t}} = \frac{2k}{3k-1}\left(1 + \frac{k-1}{2} M^2\right) \tag{6.64}$$

$$\frac{I}{I^{*t}} = \frac{1 + kM^2}{2M\sqrt{k}} \tag{6.65}$$

$$\frac{s - s^{*t}}{R} = \ln\left(\sqrt{(k)}M\right) \tag{6.66}$$

$$\frac{4\bar{f}L_{max}}{D} = \frac{1 - kM^2}{kM^2} + \ln\left(kM^2\right) \tag{6.58}$$

6.12.1 Working chart for the isothermal flow of a perfect gas

The right-hand sides of the above equations are all functions of the Mach number, M, and the specific heat ratio, k. The non-dimensional properties on the left-hand sides of these equations can then be represented in graphical form with the Mach number as the independent variable. Such a working chart for the isothermal flow of a perfect gas is given in Figure 6.24 for $k = 1.4$.

6.12.2 Working table for the isothermal flow of a perfect gas

The dimensionless flow properties can also be represented in tabular form with the Mach number as the independent variable. For $k = 1.4$, the values of these non-dimensional properties as a function of the Mach number are given in Appendix F. For convenience, the limiting values of the non-dimensional properties are given in Table 6.7.

Table 6.7 The limiting values of the non-dimensional properties for the isothermal flow of a perfect gas with $k = 1.4$

M	0	0.8452	∞
p/p^{*t}	∞	1	0
ρ/ρ^{*t}	∞	1	0
V/V^{*t}	0	1	∞
p_0/p_0^{*t}	∞	1	∞
T_0/T_0^{*t}	0.8750	1	∞
I/I^{*t}	∞	1	∞
$(s^{*t} - s)/R$	∞	0	∞
$4\bar{f}L_{max}/D$	∞	0	∞

Example 6.9
Air enters an insulated constant-area duct with a Mach number of 0.50, a pressure of 200 kPa and a temperature of 300 K, as shown in Figure 6.25. Assuming a duct length of 7 m, a duct diameter of 0.04 m and an average friction factor of 0.001, determine

a the mass flow rate,
b the Mach number, pressure and temperature at the exit of the duct,
c the frictional force acting on the duct walls, and
d the rate of heat transfer.

Solution
(a) The mass flow rate may be determined by using Appendix C, with $M_1 = 0.50$, as

$$p_1/p_{01} = 0.8430 \quad \text{or} \quad p_{01} = 200\ \text{kPa}/0.8430 = 237.2\ \text{kPa}$$

$$T_1/T_{01} = 0.9524 \quad \text{or} \quad T_{01} = 300\ \text{K}/0.9524 = 315.0\ \text{K}$$

$$\frac{\dot{m}\sqrt{(RT_0)}}{Ap_0} = 0.5111 \quad \text{or} \quad \dot{m} = \frac{(0.5111)(237\,300\ \text{N m}^{-2})(\pi)(0.04\ \text{m})^2/4}{\sqrt{[(287.1\ \text{J kg}^{-1}\ \text{K}^{-1})(315.0\ \text{K})]}}$$

$$= \mathbf{0.5068\ kg\ s^{-1}}$$

(b) From Appendix F, corresponding to $M_1 = 0.50$,

$$(4\bar{f}L_{max}/D)_1 = 0.8073$$

and

$$p_1/p^{*t} = 1.690 \quad \text{or} \quad p^{*t} = 200\ \text{kPa}/1.690 = 118.3\ \text{kPa}$$

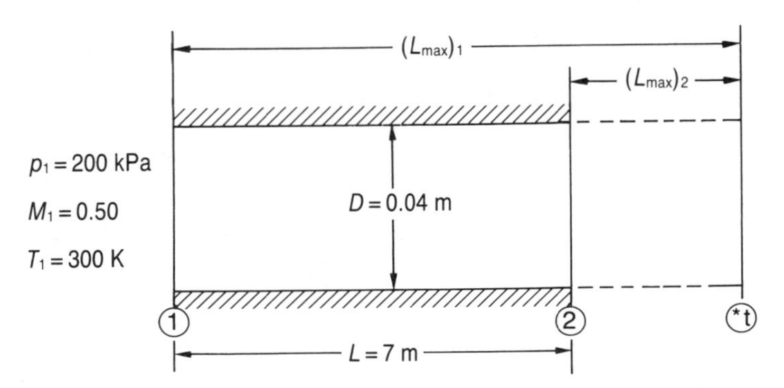

Figure 6.25 Sketch for Example 6.9

Using Figure 6.25, it is possible to obtain

$$(4\bar{f}L_{max}/D)_2 = (4\bar{f}L_{max}/D)_1 - 4\bar{f}L_{max}/D$$

$$= 0.8073 - \frac{(4)(0.001)(7\text{ m})}{(0.04\text{ m})} = 0.1073$$

The Mach number and the pressure at section 2 may now be evaluated by using Appendix F with $(4\bar{f}L_{max}/D) = 0.1073$ as

$$M_2 = \mathbf{0.68}$$

and

$$p_2/p^{*t} = 1.243 \quad \text{or} \quad p_2 = (1.243)(118.3\text{ kPa}) = \mathbf{147.0\ kPa}$$

As long as the temperature remains the same in an isothermal process, then

$$T_2 = T_1 = \mathbf{300\ K}$$

(c) The frictional force exerted on the fluid by the duct walls may be expressed as the difference between the impulse functions at sections 1 and 2. From Equation (4.39), the impulse function at section 1 is

$$I_1 = p_1A(1 + kM_1^2) = (200\ 000\text{ N m}^{-2})\,\frac{(\pi)(0.04\text{ m})^2}{4}\,[1 + (1.4)(0.5)^2]$$

$$= 339.3\text{ N}$$

Now, from Appendix F, corresponding to $M_1 = 0.50$,

$$I_1/I^{*t} = 1.141 \quad \text{or} \quad I^{*t} = 339.3\text{ N}/1.141 = 297.4\text{ N}$$

The impulse function at section 2 may now be evaluated by using Appendix F with $M_2 = 0.68$ as

$$I_2/I^{*t} = 1.024 \quad \text{or} \quad I_2 = (297.4\text{ N})(1.024) = 304.5\text{ N}$$

Then, the frictional force exerted on the fluid by the duct walls is

$$F_f = I_2 - I_1 = 304.5\text{ N} - 339.3\text{ N} = -34.8\text{ N}$$

The force exerted by the fluid on the duct is

$$F = -F_f = \mathbf{34.8\ N}$$

(d) The stagnation temperature at the exit of the duct may now be obtained from Appendix C with $M_2 = 0.68$ as

$$T_2/T_{02} = 0.9153 \quad \text{or} \quad T_{02} = 300\text{ K}/0.9153 = 327.8\text{ K}$$

Finally, the rate of heat transfer to the isothermal duct may now be evaluated as

$$\dot{Q} = \dot{m}c_p(T_{02} - T_{01}) = (0.5068\ \text{kg s}^{-1})(1005\ \text{J kg}^{-1}\ \text{K}^{-1})(327.8\ \text{K} - 315.0\ \text{K})$$

$$= \mathbf{6.52\ kW}$$

6.13 CHOKING EFFECTS IN ISOTHERMAL FLOW

For any given Mach number, there is always a maximum duct length for continuous isothermal flow. Therefore, the choking effects in isothermal pipe flow may occur in a similar fashion to those in adiabatic pipe flow.

However, it should be kept in mind that when a subsonic isothermal flow approaches the limiting Mach number, all properties change rapidly with distance. Therefore, if heat is not transferred purposefully, then the flow process is likely to be more nearly adiabatic rather than isothermal under these circumstances. Equation (6.53) indicates that at $M = 1/\sqrt{k}$, $\mathrm{d}T_0$ is infinity. For this reason, infinite heat transfer per unit length is required at this point. Therefore, this limit is artificial and not physically real.

FURTHER READING

Cambel, A. B. and Jennings, B. H. (1958) *Gas Dynamics*, Dover: New York.

Daneshyar, H. (1976) *One-Dimensional Compressible Flow*, Pergamon: Oxford.

Fox, R. W. and McDonald, A. T. (1978) *Introduction to Fluid Mechanics*, 2nd edn, Wiley: New York.

John, J. E. A. (1984) *Gas Dynamics*, 2nd edn, Allyn & Bacon: Boston, MA.

Liepmann, H. W. and Roshko, A. (1957) *Elements of Gas Dynamics*, Wiley: New York.

Owczarek, J. A. (1964) *Fundamentals of Gas Dynamics*, International Textbook: Scranton, PA.

Shapiro, A. H. (1953) *The Dynamics and Thermodynamics of Compressible Fluid Flow*, Vol. 1, Ronald Press: New York.

Zucrow, M. J. and Hoffman, J. D. (1976) *Gas Dynamics*, Vol. 1, Wiley: New York.

PROBLEMS

6.1 Air flows adiabatically in a constant-area duct. At the inlet of the duct, the Mach number, pressure and temperature are 0.60, 150 kPa and 300 K, respectively. Assuming a duct length of 0.45 m, a duct diameter of 0.03 m and a friction coefficient of 0.003, determine the Mach number, temperature and pressure at the duct outlet. Also, calculate the mass flow rate.

6.2 Air flows adiabatically in a constant-area duct. At the inlet of the duct, the velocity, pressure and temperature are 500 m s^{-1}, 200 kPa and 300 K, respectively. Assuming a duct length of 0.1 m, a duct diameter of 0.02 m and a friction factor of 0.002, determine the Mach number, temperature and pressure at the exit of the duct, and also calculate the mass flow rate.

6.3 Air is flowing through an insulated duct with a constant cross-sectional area. At the inlet of the duct, the velocity, temperature and pressure are 180 m s^{-1}, 300 K and 500 kPa, respectively. Determine the temperature and the stagnation temperature at the exit of the duct where the density is 3.5 kg m^{-3}. Also, calculate the entropy increase.

6.4 Air flows adiabatically through a duct with a constant cross-sectional area. At the inlet of the duct, the pressure, temperature and velocity are 100 kPa, 50 °C and 200 m s^{-1}, respectively. At the exit of the duct, the temperature is 20 °C. Determine the Mach number and the pressure at the exit section and the stagnation temperature midway between the inlet and the exit section.

6.5 Air flows steadily and adiabatically through a pipe with a cross-sectional area of 0.002 m^2. At the inlet of the duct, the pressure and temperature are 100 kPa and 50 °C, respectively. At the exit of the duct, the temperature and velocity are 100 °C and 600 m s^{-1}, respectively. Determine the Mach number and the velocity at the inlet section, and also find the mass flow rate.

6.6 Air flows steadily and adiabatically through a pipe with a constant cross-sectional area. Air is fed into the pipe through an isentropic converging nozzle from a large tank where the pressure and temperature are 900 kPa and 300 K, respectively. The pressure at the inlet of the duct is 750 kPa. If the temperature at the duct exit is 275 K, determine the pressure at the duct exit and the entropy increase.

6.7 An airstream flows steadily and adiabatically in a duct with a cross-sectional area of 0.1 m^2. At the inlet of the duct, the pressure, temperature and mass flux are 100 kPa, 300 K and 150 kg m^{-2} s^{-1}, respectively. The back pressure to which the tube exhausts is so low that the flow is choked at the exit. Calculate

a the Mach number at the inlet,
b the Mach number, temperature and pressure at the exit of the tube, and
c the total force in the axial direction which must be exerted in order to hold the duct stationary.

6.8 An insulated duct with a cross-sectional area of 0.001 m^2 is fed with air through an isentropic converging–diverging nozzle. At the inlet of the duct, the pressure, temperature and Mach number are 150 kPa, 300 K and 1.5, respectively. If the flow is choked at the exit of the duct, calculate the exit temperature, the net force of the fluid on the pipe and the entropy change.

6.9 In a compressed-air system, which is shown in the figure below, the pipe diameter is 2 cm and the friction factor is 0.01. The pressure drop across the valve is given as $p_2 - p_3 = 5\rho_3 V_3$. The area ratio of the frictionless nozzle is 1.136. The pipe system is insulated and there are no heat losses. The valve can be replaced by an equivalent length of pipe under these conditions. Calculate the Mach numbers at locations 4, 3, 2 and 1, if the flow is choked at the exit of the nozzle. (METU 1975)

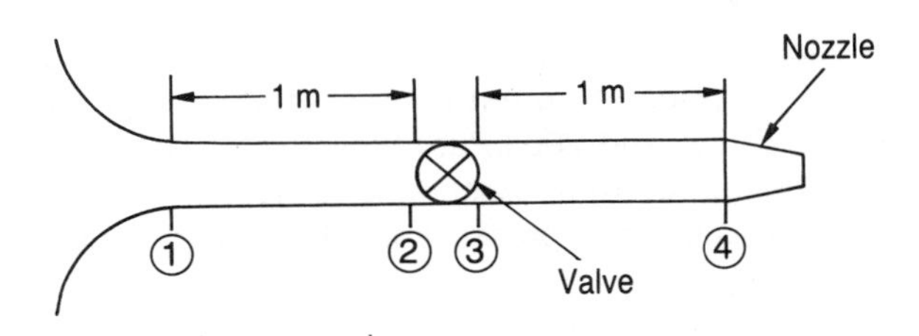

Problem 6.9

6.10 A duct with a diameter of 0.1 m is fed from a converging–diverging nozzle with a throat area of 0.003 55 m^2. The temperature in the reservoir is 300 K. The duct discharges to the atmosphere with a pressure of 100 kPa. A normal shock wave stands at the midpoint of the duct, as shown in the figure below. Assume that the flow in the nozzle is isentropic, and the flow in the duct is adiabatic. If the friction coefficient in the duct is 0.005, determine the mass flow rate. (METU 1987)

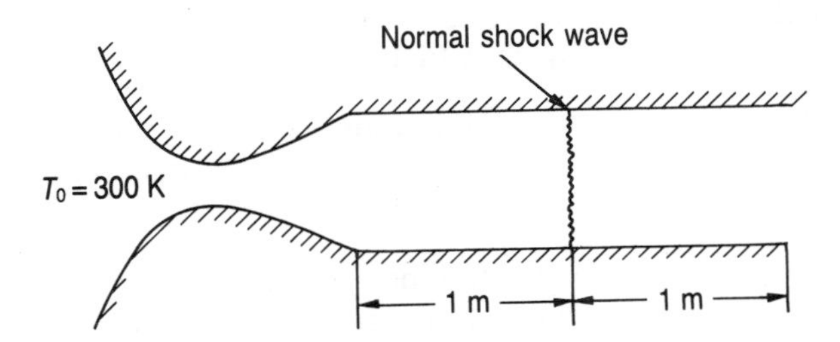

Problem 6.10

6.11 A diverging duct is fed with air from a constant-area pipe of 0.01 m^2 in cross-sectional area. At the entrance of the constant-area pipe, the pressure, temperature and velocity are 179 kPa, 39 °C and 177 m s^{-1}, respectively. The diverging duct discharges to the atmosphere where the pressure is 100 kPa. The exit cross-sectional area of the diverging duct is 0.02 m^2. Find the Mach number at the exit of the diverging duct, if the flow is adiabatic with friction. (METU 1988)

6.12 A converging nozzle feeds air to a constant-area duct of 2 m in length. The duct diameter is 0.0374 m and the friction coefficient is 0.005. The stagnation pressure and the stagnation temperature upstream of the duct are 240 kPa and 20 °C, respectively. The back pressure into which the duct discharges is 50 kPa. Find the pressure at the exit of the duct and the mass flow rate. Assume that the flow is isentropic in the nozzle and adiabatic throughout. (METU 1986)

6.13 An insulated duct with a constant cross-sectional area is fed with air by an isentropic converging nozzle from a large reservoir, where the temperature and pressure are 300 K and 200 kPa, respectively. The duct has a diameter of 0.05 m and a length of 10 m. Assuming a friction factor of 0.001, determine the ranges of back pressure for the subcritical and the supercritical flow regimes.

6.14 Air is supplied to an insulated duct with constant cross-sectional area through a converging nozzle from a large reservoir where the temperature and pressure are 350 K and 400 kPa, respectively. The duct has a length of 0.5 m and a diameter of 0.025 m, and it discharges to the atmosphere with a pressure of 100 kPa. Assuming a friction factor of 0.0025, determine

a the Mach numbers at the inlet and the exit of the duct,
b the pressure and temperature at the exit of the duct, and
c the mass flow rate.

6.15 An insulated constant-area duct is fed from a large reservoir through an isentropic converging nozzle. At the inlet of the duct, the temperature and pressure are 300 K and 400 kPa, respectively. The duct has a length of 8 m and a diameter of 0.04 m, and it discharges to the atmosphere with a pressure of 100 kPa. Assuming a friction factor of 0.002, determine

a the Mach numbers at the inlet and the exit of the duct,
b the pressure and temperature at the exit of the duct, and
c the mass flow rate.

6.16 An insulated duct with a constant cross-sectional area is fed with air by an isentropic converging nozzle from a reservoir, where the temperature and pressure are 400 K and 150 kPa, respectively. The duct has a length of 2.5 m and a diameter of 0.04 m, and it discharges to a back pressure of 119 kPa. Assuming a friction factor of 0.001, determine

a the inlet and exit Mach numbers of the duct,
b the pressure and temperature at the exit of the duct, and
c the mass flow rate.

6.17 A converging nozzle supplies air to an insulated constant-area duct having a length of 10 m and a diameter of 0.05 m. At the inlet of the duct, the temperature and pressure are 300 K and 150 kPa, respectively. The duct discharges to a back pressure of 105 kPa. Assuming a friction factor of 0.001, determine

a the Mach numbers at the inlet and the exit of the duct,
b the pressure and temperature at the exit of the duct, and
c the mass flow rate.

6.18 An adiabatic duct is fed with air by an isentropic converging–diverging nozzle with an area ratio of 1.76. The stagnation pressure in the supply reservoir is 200 kPa. The length of the duct is 2 m and the diameter of the duct is 0.04 m. The average friction factor in the duct is 0.001. Determine

a the maximum allowable duct length,
b the back pressure such that there is a normal shock wave at the inlet section of the duct,
c the back pressure for operation of the duct at its design condition,
d the back pressure such that there is a normal shock wave at the exit section of the duct, and
e the ranges of back pressure for the subsonic flow regime, flow regime with shocks, overexpansion flow regime and underexpansion flow regime.

6.19 An adiabatic duct is fed with air by an isentropic converging–diverging nozzle with an area ratio of 2.0. The stagnation temperature and the stagnation pressure are 300 K and 300 kPa, respectively. The duct has a length of 2 m and a diameter of 0.04 m, and it discharges to a back pressure of 209 kPa. If the average friction factor is 0.001, determine the Mach number, pressure and temperature at the exit of the duct, and also calculate the mass flow rate.

6.20 An adiabatic duct is fed with air by an isentropic converging–diverging nozzle with an area ratio of 1.76. The stagnation temperature and the stagnation pressure in the supply reservoir are 400 K and 200 kPa, respectively. The duct has a length of 2 m and a diameter of 0.04 m, and it discharges to a back pressure of 20 kPa. If the average friction factor is 0.001, determine the Mach number, pressure and temperature at the exit of the duct, and also calculate the mass flow rate.

6.21 An adiabatic duct is fed with air by an isentropic converging–diverging nozzle with an area ratio of 2.0. The stagnation temperature and the stagnation pressure are 350 K and 300 kPa, respectively. The duct has a length of 4 m and a diameter of 0.08 m, and it discharges to an atmosphere with a back pressure of 100 kPa. If the average friction factor is 0.001, determine the Mach number, temperature and pressure at the exit section of the duct, and also find the mass flow rate.

6.22 An adiabatic duct is fed with air by an isentropic converging–diverging nozzle with an area ratio of 2.0. The stagnation temperature and the stagnation pressure are 350 K and 600 kPa, respectively. The duct has a length of 4 m and a diameter of 0.08 m, and discharges to a back pressure of 262 kPa. If the average friction factor is 0.001, determine the Mach number, temperature and pressure at the exit section of the duct, and also find the mass flow rate.

6.23 Air is supplied to an adiabatic duct through an isentropic converging–diverging nozzle with an area ratio of 1.5. The stagnation pressure and the stagnation temperature in the supply reservoir are 300 kPa and 400 K, respectively. If the duct length is 0.7 m, the duct diameter is 0.02 m and the average friction factor is 0.002, determine

a the maximum allowable duct length,
b the back pressure such that the flow is sonic at the end of the duct,
c the distance from the inlet of the duct at which the normal shock wave is located when the flow is sonic at the exit of the duct,
d the back pressure such that there is a normal shock wave at the inlet of the duct, and
e the ranges of the back pressure for the subsonic flow regime, flow regime with shocks and underexpansion flow regime.

6.24 An adiabatic duct is fed with air by an isentropic converging–diverging nozzle with an area ratio of 1.76. The stagnation pressure and the stagnation temperature are 200 kPa and 350 K, respectively. The duct has a length of 10.81 m and a diameter of 0.08 m, and it discharges to a back pressure of 40 kPa. If the average friction factor in the duct is 0.001, determine the Mach number, temperature and pressure at the exit of the duct, and also find the mass flow rate.

6.25 Air is supplied to an adiabatic duct through an isentropic converging–diverging nozzle with an area ratio of 1.5. The stagnation pressure and the stagnation temperature in the supply reservoir are 150 kPa and 300 K, respectively. The duct has a length of 0.35 m and a diameter of 0.01 m, and discharges to a back pressure of 61 kPa. If the average friction factor in the duct is 0.002, determine the Mach number, pressure and temperature at the exit of the duct, and also find the mass flow rate.

6.26 Air is supplied to an adiabatic duct through an isentropic converging–diverging nozzle with an area ratio of 2.0. The stagnation pressure and the stagnation temperature in the supply reservoir are 250 kPa and 300 K, respectively. The duct has a length of 6 m and a diameter of 0.05 m, and discharges to a back pressure of 165 kPa. If the average friction factor in the duct is 0.001, determine the Mach number, temperature and pressure at the exit of the duct, and also calculate the mass flow rate.

6.27 Air enters a constant-area duct at a velocity of 200 m s^{-1} and is supplied isentropically from a large reservoir with a temperature of 300 K and a pressure of 100 kPa. The duct has a length of 1 m and a diameter of

0.025 m. Assuming that the flow in the duct is isothermal with an average friction factor of 0.001, determine

a the Mach number, temperature and pressure at the exit of the duct,
b the force exerted by the air on the duct,
c the mass flow rate, and
d the rate of heat transfer.

6.28 Air enters a constant-area duct at a velocity of 500 m s^{-1}, a temperature of 350 K and a pressure of 150 kPa. The duct has a length of 1 m and a diameter of 0.02 m. Assuming that the flow is isothermal with an average friction factor of 0.001, determine

a the Mach number, temperature and pressure at the exit of the duct,
b the force exerted by the air on the duct,
c the mass flow rate, and
d the rate of heat transfer.

6.29 Air enters a constant-area duct at a velocity of 150 m s^{-1}, a temperature of 300 K and a pressure of 200 kPa. The duct has a diameter of 0.02 m, and the flow is isothermal with an average friction factor of 0.001. If the flow is choked at the end of the duct, determine

a the Mach number, temperature and pressure at the exit of the duct,
b the force exerted by the air on the duct,
c the mass flow rate, and
d the rate of heat transfer.

6.30 Air enters a constant-area duct at a velocity of 600 m s^{-1}, a temperature of 400 K and a pressure of 100 kPa. The duct has a diameter of 0.03 m, and the flow is isothermal with an average friction factor of 0.002. If the flow is choked at the end of the duct, determine

a the Mach number, temperature and pressure at the exit of the duct,
b the force exerted by the air on the duct,
c the mass flow rate, and
d the rate of heat transfer.

7 Flow in constant-area ducts with heat transfer

7.1 INTRODUCTION

From a one-dimensional point of view, the three most common factors which produce continuous changes in flow properties are (i) changes in the cross-sectional area, (ii) the wall friction, and (iii) energy effects such as external heat exchange, combustion or moisture condensation. In Chapter 4, the effects of area change on flow properties were investigated in the absence of frictional and energy effects, and such flows are termed isentropic flows or simple area-change flows. However, in Chapter 6, the effects of friction on the flow were examined in the absence of area change and energy effects, and such flows are called simple friction flows.

In this chapter, attention will be focused on a process involving changes in the stagnation enthalpy or in the stagnation temperature of a gas stream which flows in a constant-area duct without frictional effects. Such flows are called **simple stagnation temperature-change flows**.

Such a process is quite difficult to achieve in practice. If the stagnation temperature is raised or lowered as a result of external heat exchange, then frictional effects will also be present owing to the connection between the mechanisms of friction and heat transfer. Similarly, if changes in the stagnation temperature are brought about by combustion, then there will be a change in chemical composition. Finally, a reduction in the stagnation temperature through evaporation of a liquid into the gas stream also causes changes in the mass flow rate and the species composition of the gas stream.

However, many important conclusions of practical significance can be achieved by analyzing simple T_0 change flows when departures from assumptions of such flows are small. For example, when heating and cooling are a result of external heat exchange, frictional effects per unit length of the duct will be relatively unimportant compared with the heating or cooling effects per unit length of the duct. In this case, the temperature difference between the duct wall and the moving gas stream is quite large. In the combustion of hydrocarbons with air, the fuel–air ratio is small, and therefore the effects due to changes in the chemical composition are quite small compared with the effects due to changes in the stagnation temperature.

7.2 GOVERNING EQUATIONS FOR ONE-DIMENSIONAL FLOW IN CONSTANT-AREA DUCTS WITH HEATING OR COOLING

Now, consider steady, one-dimensional frictionless flow of a perfect gas through a constant-area duct with heat transfer, as shown in Figure 7.1. The governing equations are given below for the mentioned control volume.

(a) Continuity equation

For steady and uniform flow at each cross-section, the mass flow rate is constant, that is

$$\dot{m} = \rho_1 A_1 V_1 = \rho_2 A_2 V_2 = \text{constant}$$

As long as the area of the duct is constant, that is $A_1 = A_2 = A$, then the mass flow rate per unit area or the mass flux, G, is given by

$$G = \dot{m}/A = \rho_1 V_1 = \rho_2 V_2 = \rho V = \text{constant} \tag{7.1}$$

(b) Momentum equation

For steady, uniform and frictionless flows, pressure forces acting on the control volume of Figure 7.1 must be balanced by the rate of change of momentum across the control volume, that is

$$p_1 A_1 - p_2 A_2 = \rho_2 A_2 V_2^2 - \rho_1 A_1 V_1^2$$

By using the definition of the impulse function given by Equation (4.37), it is possible to note that the impulse function is constant, that is

$$I = A_1(p_1 + \rho_1 V_1^2) = A_2(p_2 + \rho_2 V_2^2) = A(p + \rho V^2) = \text{constant} \tag{7.2}$$

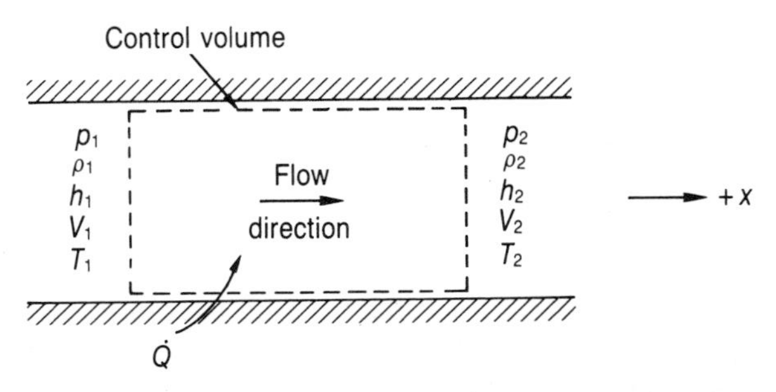

Figure 7.1 Control volume for steady, one-dimensional frictionless flow with heat transfer in a constant-area duct

(c) The first law of thermodynamics

When there is no external shaft work on the control volume and when changes in the elevation are negligible

$$\dot{Q} = \dot{m}q = \dot{m}(h_{02} - h_{01}) = \dot{m}\left(h_2 + \frac{V_2^2}{2} - h_1 - \frac{V_1^2}{2}\right) \tag{7.3}$$

where $\dot{Q}$ is the rate of heat transfer and q is the heat transfer per unit mass. The rate of heat transfer is positive during the heating of a gas, and negative during the cooling of a gas.

(d) Equation of state

For a perfect gas, the equation of state is given by

$$p = \rho RT \tag{7.4}$$

(e) The second law of thermodynamics

For a reversible heat transfer process, the second law of thermodynamics for the mentioned control volume may be given as

$$q = \int_1^2 T\,\mathrm{d}s \tag{7.5}$$

To calculate the entropy change, the thermodynamic relations

$$T\,\mathrm{d}s = \mathrm{d}h - \frac{\mathrm{d}p}{\rho}$$

or

$$T\,\mathrm{d}s = \mathrm{d}u - RT\frac{\mathrm{d}\rho}{\rho}$$

may be used. For a perfect gas, $\mathrm{d}h = c_p\,\mathrm{d}T$, $\mathrm{d}u = c_v\,\mathrm{d}T$ and $p = \rho RT$. Then

$$\mathrm{d}s = c_p\frac{\mathrm{d}T}{T} - R\frac{\mathrm{d}p}{p} = c_v\frac{\mathrm{d}T}{T} - R\frac{\mathrm{d}\rho}{\rho} \tag{7.6}$$

If the above expression is integrated for constant specific heats, then the entropy change is

$$s_2 - s_1 = c_p \ln\left(\frac{T_2}{T_1}\right) - R\ln\left(\frac{p_2}{p_1}\right) = c_v \ln\left(\frac{T_2}{T_1}\right) - R\ln\left(\frac{\rho_2}{\rho_1}\right) \tag{7.7}$$

Example 7.1

Air flows with negligible friction through a duct with a cross-sectional area of 0.025 m^2, as shown in Figure 7.2. At section 1, the pressure, temperature and velocity are 150 kPa, 350 K and 100 $m\,s^{-1}$, respectively. At section 2, the pressure is 75 kPa. If the air is heated between sections 1 and 2, determine

a the properties at section 2,

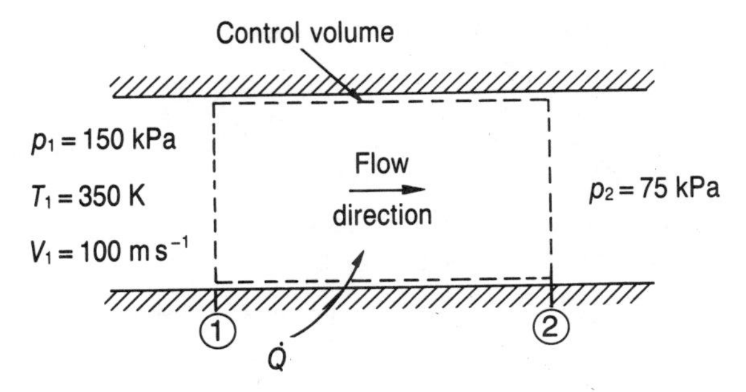

Figure 7.2 Sketch for Example 7.1

b the amount of heat transfer, and
c the entropy change.

Solution

(a) The density at section 1 may be calculated by using the equation of state as

$$\rho_1 = \frac{p_1}{RT_1} = \frac{150\,000 \text{ N m}^{-2}}{(287.1 \text{ J kg}^{-1}\text{ K}^{-1})(350 \text{ K})} = 1.493 \text{ kg m}^{-3}$$

Then, the mass flow rate from the continuity equation is

$$\dot{m} = \rho_1 V_1 A_1 = (1.493 \text{ kg m}^{-3})(100 \text{ m s}^{-1})(0.025 \text{ m}^2) = 3.733 \text{ kg s}^{-1}$$

The velocity at section 2 may be found by applying the momentum equation to the control volume of Figure 7.2 as

$$p_1 A - p_2 A = \rho_2 V_2^2 A - \rho_1 V_1^2 A = \dot{m}(V_2 - V_1)$$

or

$$V_2 = V_1 + \frac{A(p_1 - p_2)}{\dot{m}}$$

$$= 100 \text{ m s}^{-1} + \frac{(0.025 \text{ m}^2)(150\,000 \text{ N m}^{-2} - 75\,000 \text{ N m}^{-2})}{(3.733 \text{ kg s}^{-1})} = \mathbf{602.3 \text{ m s}^{-1}}$$

Using the continuity equation, the density at section 2 is

$$\rho_2 = \rho_1 \frac{V_1}{V_2} = (1.493 \text{ kg m}^{-3}) \frac{100 \text{ m s}^{-1}}{602.3 \text{ m s}^{-1}} = \mathbf{0.2479 \text{ kg m}^{-3}}$$

Then, the temperature is

$$T_2 = \frac{p_2}{RT_2} = \frac{75\,000 \text{ N m}^{-2}}{(287.1 \text{ J kg}^{-1}\text{ K}^{-1})(0.2479 \text{ kg m}^{-3})} = \mathbf{1054 \text{ K}}$$

The local isentropic stagnation temperature at section 2 may now be calculated via Equation (4.8a) as

$$T_{02} = T_2 + \frac{V_2^2}{2c_p} = 1054\ \text{K} + \frac{(602.3\ \text{m s}^{-1})^2}{(2)(1005\ \text{J kg}^{-1}\text{K}^{-1})} = \mathbf{1235\ K}$$

Considering the imaginary isentropic process between state 2 and its local isentropic state, the stagnation pressure at section 2 is

$$p_{02} = p_2\left(\frac{T_{02}}{T_2}\right)^{k/(k-1)} = (75\ \text{kPa})\left(\frac{1235\ \text{K}}{1054\ \text{K}}\right)^{1.4/(1.4-1)} = \mathbf{130.6\ kPa}$$

(b) By using Equation (4.8a), the local isentropic stagnation temperature at section 1 is

$$T_{01} = T_1 + \frac{V_1^2}{2c_p} = 350\ \text{K} + \frac{(100\ \text{m s}^{-1})^2}{(2)(1005\ \text{J kg}^{-1}\text{K}^{-1})} = 355\ \text{K}$$

Then, the amount of heat transfer may be determined from the energy equation as

$$\dot{Q} = \dot{m}c_p(T_{02} - T_{01}) = (3.733\ \text{kg s}^{-1})(1005\ \text{J kg}^{-1}\text{K}^{-1})(1235\ \text{K} - 355\ \text{K})$$
$$= \mathbf{3302\ kW}$$

(c) The entropy change may now be evaluated by using Equation (7.7)

$$s_2 - s_1 = c_p \ln\left(\frac{T_2}{T_1}\right) - R\ln\left(\frac{p_2}{p_1}\right) = (1005\ \text{J kg}^{-1}\text{K}^{-1})\ln\left(\frac{1054\ \text{K}}{350\ \text{K}}\right)$$
$$- (287.1\ \text{J kg}^{-1}\text{K}^{-1})\ln\left(\frac{75\ \text{kPa}}{150\ \text{kPa}}\right) = \mathbf{1307\ J\,kg^{-1}\,K^{-1}}$$

7.3 THE RAYLEIGH LINE

The Rayleigh line was first introduced during the discussion of normal shock waves in Section 5.3.3. If the conditions upstream of the control volume (state 1) are fixed, and the conditions are to be found at the downstream section (state 2), then for a particular value of V_2: (i) ρ_2 may be calculated from the continuity equation (7.1), (ii) p_2 may be evaluated from the momentum equation (7.2), (iii) T_2 may be found from the equation of state (7.4), and finally, (iv) s_2 may be obtained from Equation (7.7). Therefore, the continuity equation (7.1), the momentum equation (7.2) and the equation of state (7.4) define a locus of states passing through state 1, which is known as the Rayleigh line on the *T–s* diagram. Since the energy equation (7.3) has not yet been introduced, the **Rayleigh line for a perfect gas represents states with the same mass flux and the same impulse function but not necessarily the same value of the stagnation temperature.** Therefore, states on the Rayleigh line are reachable from each other in the presence of heat transfer effects. Rayleigh lines corresponding to different mass fluxes are shown in Figure 7.3.

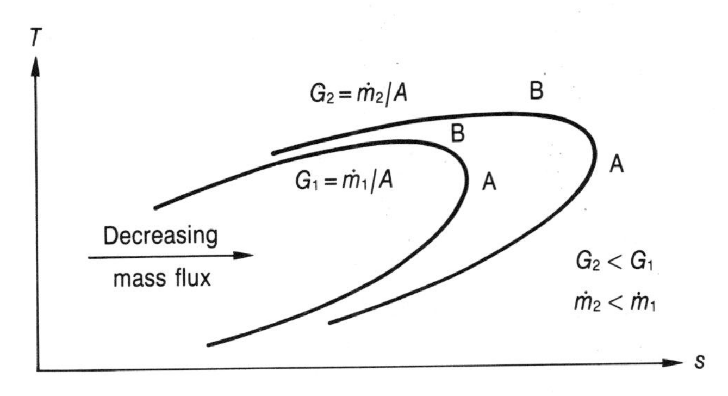

Figure 7.3 Rayleigh lines for different mass fluxes

Figure 7.3 indicates that the entropy and the temperature are maximum at points A and B, respectively. To understand the significance of these points, it is necessary to derive an equation for the slope of the Rayleigh line. For this reason, introducing the equation of state, $p=\rho RT$, and the definition of the Mach number as $V^2=M^2kRT$ into Equation (7.2) yields

$$p+\rho V^2=p(1+kM^2)=\text{constant}$$

or with logarithmic differentiation

$$\frac{\mathrm{d}p}{p}=-\frac{2kM\,\mathrm{d}M}{1+kM^2} \tag{7.8}$$

Similarly, if the equation of state, $p=\rho RT$, and the definition of the Mach number as $V=M\surd(kRT)$ are introduced into Equation (7.1)

$$\rho V=\frac{p}{RT}M\surd(kRT)=\sqrt{\left(\frac{k}{RT}\right)}pM=\text{constant}$$

then, by logarithmic differentiation,

$$\frac{\mathrm{d}T}{T}=2\left(\frac{\mathrm{d}p}{p}+\frac{\mathrm{d}M}{M}\right)$$

or eliminating $\mathrm{d}p/p$ using Equation (7.8)

$$\frac{\mathrm{d}T}{T}=2\,\frac{1-kM^2}{M(1+kM^2)}\,\mathrm{d}M \tag{7.9}$$

Now, dividing Equation (7.8) by Equation (7.9) yields

$$\frac{\mathrm{d}p}{\mathrm{d}T}=-\frac{kpM^2}{(1-kM^2)T}$$

The slope of the Rayleigh line may now be obtained with the aid of Equation (7.6) as

$$\frac{ds}{dT}=\frac{c_p}{T}-\frac{R}{p}\frac{dp}{dT}=\frac{c_p}{T}+\frac{R}{T}\frac{kM^2}{1-kM^2}$$

or for a perfect gas, $R=(k-1)c_p/k$,

$$\frac{dT}{ds}=\frac{T}{c_p}\frac{1-kM^2}{1-M^2} \tag{7.10}$$

Equation (7.10) indicates that at point A, where the slope of the Rayleigh line is infinite, the Mach number is unity. This was the conclusion reached in Section 5.3.3. Also in that section, it was shown that the lower branch of a Rayleigh line corresponds to supersonic flow. Finally, at point B, where the slope of the Rayleigh line is zero, $M=1/\sqrt{k}$.

Since the process of simple heating is thermodynamically reversible, heat addition must correspond to an entropy increase and heat rejection must correspond to an entropy decrease. Therefore, the Mach number increases with heating and decreases with cooling at subsonic speeds. However, at supersonic speeds, the Mach number decreases with heating and increases with cooling. Thus, the process of heat addition is like friction and tends to make the Mach number unity. Hence the flow cannot continue beyond $M=1$ with either heat addition or heat rejection alone. However, the combination of heating and cooling can allow a flow to cross $M=1$ without discontinuity. Finally, it is interesting to note that, during the heating of a subsonic flow between points A and B of the Rayleigh line, the static temperature decreases.

7.4 EFFECT OF HEAT TRANSFER ON FLOW PROPERTIES OF FRICTIONLESS FLOWS IN CONSTANT-AREA DUCTS

7.4.1 Governing equations in differential form

Consider one-dimensional frictionless flow of a compressible fluid in a constant-area

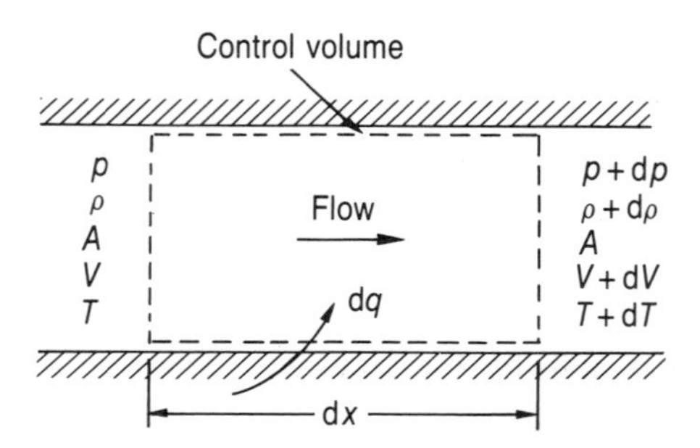

Figure 7.4 Infinitesimal control volume for the analysis of the Rayleigh line flow

duct with heat transfer, as shown in Figure 7.4. Heat transfer between the fluid and the surroundings causes infinitesimal changes in the flow properties, and these changes may be related by applying basic equations to the infinitesimal control volume of Figure 7.4.

(a) Continuity equation

For steady and uniform flow at each cross-section, the mass flow rate is constant. Therefore,

$$\dot{m} = \rho VA = (\rho + d\rho)(V + dV)A$$

or

$$\cancel{\rho} AV = \cancel{\rho} AV + \rho A\, dV + AV\, d\rho + \underbrace{A\, \cancel{d\rho}\, dV}_{\simeq 0}$$

Neglecting the product of the differentials, and dividing by ρAV, it is possible to obtain

$$\frac{dV}{V} + \frac{d\rho}{\rho} = \frac{1}{2}\frac{d(V^2)}{V^2} + \frac{d\rho}{\rho} = 0 \tag{7.11}$$

(b) Momentum equation

The sum of the pressure forces acting on the control volume must be balanced by the rate of change of momentum across the control volume, that is

$$pA - (p + dp)A = \dot{m}(V + dV) - \dot{m}V$$

or

$$dp = -\rho V\, dV$$

Now, using the equation of state $p = \rho RT$ and the definition of the Mach number $V^2 = M^2 kRT$, it is possible to obtain

$$\frac{dp}{p} = -\frac{kM^2}{2}\frac{d(V^2)}{V^2} \tag{7.12}$$

(c) The first law of thermodynamics

For frictionless flows with no external work, the energy equation for the control volume is

$$c_p T + \frac{V^2}{2} + dq = c_p(T + dT) + \frac{(V + dV)^2}{2}$$

or

$$\cancel{c_p T} + \cancel{\frac{V^2}{2}} + dq = \cancel{c_p T} + c_p\, dT + \cancel{\frac{V^2}{2}} + V\, dV + \underbrace{\cancel{\frac{(dV)^2}{2}}}_{\simeq 0}$$

Neglecting the second-order differentials, it is possible to obtain

$$c_p\,dT + V\,dV = dq$$

But for a perfect gas $c_p = kR/(k-1)$ and $M = V/\sqrt{(kRT)}$. Then

$$\frac{dT}{T} + \frac{k-1}{2} M^2 \frac{d(V^2)}{V^2} = \frac{k-1}{kR}\frac{dq}{T} \tag{7.13}$$

(d) Equation of state

Also, the equation of state $p = \rho RT$ may be logarithmically differentiated to yield

$$\frac{dp}{p} = \frac{d\rho}{\rho} + \frac{dT}{T} \tag{7.14}$$

(e) The second law of thermodynamics

For reversible heat transfer processes

$$ds = dq/T \tag{7.15}$$

7.4.2 Variation of flow properties with heat transfer

(a) Velocity

Eliminating $d\rho/\rho$ and dp/p in Equation (7.14) by using Equations (7.11) and (7.12) yields

$$\frac{dT}{T} = \frac{1-kM^2}{2}\frac{d(V^2)}{V^2} \tag{7.16}$$

Now, substituting Equation (7.16) into Equation (7.13), it is possible to obtain

$$\frac{d(V^2)}{V^2} = \frac{k-1}{kRT}\frac{2}{1-M^2}\,dq \tag{7.17}$$

(b) Temperature

Combining Equations (7.16) and (7.17) yields

$$\frac{dT}{T} = \frac{k-1}{kRT}\frac{1-kM^2}{1-M^2}\,dq \tag{7.18}$$

(c) Pressure

Using Equations (7.12) and (7.17), it is possible to obtain

$$\frac{dp}{p} = -\frac{k-1}{kRT}\frac{kM^2}{1-M^2}\,dq \tag{7.19}$$

(d) Mach number

Using the definition of the Mach number, $M = V/a$, as $V^2 = M^2 kRT$, and logarithmically differentiating, it is possible to obtain

$$\frac{\mathrm{d}(V^2)}{V^2} = \frac{\mathrm{d}T}{T} + \frac{\mathrm{d}(M^2)}{M^2}$$

Then, with the aid of Equations (7.17) and (7.18),

$$\frac{\mathrm{d}(M^2)}{M^2} = \frac{k-1}{kRT}\,\frac{1+kM^2}{1-M^2}\,\mathrm{d}q \tag{7.20}$$

(e) Density

Equations (7.11) and (7.17) may be combined to yield

$$\frac{\mathrm{d}\rho}{\rho} = -\frac{k-1}{kRT}\,\frac{1}{1-M^2}\,\mathrm{d}q \tag{7.21}$$

(f) Stagnation pressure

The previous definition of the isentropic stagnation pressure can be retained, as long as the isentropic stagnation pressure corresponding to a given state is the pressure which would be attained if the stream were decelerated isentropically from the given state to the state of zero velocity. Then, from Equation (4.28),

$$p_0 = p\left(1 + \frac{k-1}{2}\,M^2\right)^{k/(k-1)}$$

or in differential form

$$\frac{\mathrm{d}p_0}{p_0} = \frac{\mathrm{d}p}{p} + \frac{kM^2/2}{1+\frac{1}{2}(k-1)\,M^2}\,\frac{\mathrm{d}(M^2)}{M^2}$$

Finally, with the aid of Equations (7.19) and (7.20), it is possible to obtain

$$\frac{\mathrm{d}p_0}{p_0} = -\frac{k-1}{kRT}\,\frac{kM^2/2}{1+\frac{1}{2}(k-1)\,M^2}\,\mathrm{d}q \tag{7.22}$$

(g) Stagnation temperature

The variation of the stagnation temperature may be obtained by applying the energy equation to the infinitesimal control volume directly as

$$\mathrm{d}T_0 = \mathrm{d}q/c_p \tag{7.23}$$

(h) Entropy

The variation of the entropy is obtained from the second law of thermodynamics as

$$\mathrm{d}s = \mathrm{d}q/T \tag{7.24}$$

7.4.3 Summary of heat transfer effects

By convention, dq is positive during the heating of a fluid. It appears from Equations (7.23) and (7.24) that the stagnation temperature and the entropy increase, respectively and from Equation (7.22) that the stagnation pressure decreases for both subsonic and supersonic flows.

The variations in the remaining fluid properties excluding the temperature according to Equations (7.17), (7.19), (7.20) and (7.21) depend on whether the Mach number is greater or less than unity, since the term $(1-M^2)$ appears in the denominator of each of these equations.

Investigation of Equation (7.18) indicates that the temperature increases for supersonic flows. However, for subsonic flows, the temperature increases when $0 < M < 1/\sqrt{k}$ and decreases when $1/\sqrt{k} < M < 1$.

Conversely, during the cooling of a liquid, dq is negative and all the effects are as summarized in Table 7.1.

Table 7.1 Effect of heat transfer on the flow properties of a perfect gas in a frictionless constant-area duct

	Flow type			
Flow property	Subsonic flow with heating	Supersonic flow with heating	Subsonic flow with cooling	Supersonic flow with cooling
Mach number, M	Increases	Decreases	Decreases	Increases
Pressure, p	Decreases	Increases	Increases	Decreases
Temperature, T	Increases for $M < 1/\sqrt{k}$, decreases for $M > 1/\sqrt{k}$	Increases	Decreases for $M < 1/\sqrt{k}$, increases for $M > 1/\sqrt{k}$	Decreases
Density, ρ	Decreases	Increases	Increases	Decreases
Velocity, V	Increases	Decreases	Decreases	Increases
Stagnation pressure, p_0	Decreases	Decreases	Increases	Increases
Stagnation temperature, T_0	Increases	Increases	Decreases	Decreases
Entropy, s	Increases	Increases	Decreases	Decreases

7.5 RELATIONS FOR THE FLOW OF A PERFECT GAS ON THE RAYLEIGH LINE

The equations governing one-dimensional frictionless flow of a perfect gas in a constant-area duct with heat transfer were derived in Section 7.2. Although any problem of this kind can be solved using these equations, the solution can be simplified by rearranging them in a more suitable form. In this section, flow properties at the upstream and downstream sections of the control volume of Figure 7.1 will be related in terms of the Mach numbers at these sections.

7.5.1 Pressure ratio

Introducing the equation of state, $p = \rho RT$, and the definition of the Mach number as $V^2 = M^2 kRT$ into the momentum equation (7.2) yields

$$\frac{p_1}{p_2} = \frac{1 + kM_2^2}{1 + kM_1^2} \tag{7.25}$$

7.5.2 Temperature ratio

Using the equation of state for a perfect gas, $p = \rho RT$, the temperature ratio is

$$\frac{T_1}{T_2} = \frac{p_1}{p_2}\frac{\rho_2}{\rho_1}$$

Now, ρ_2/ρ_1 may be eliminated with the aid of the continuity equation (7.1) to give

$$\frac{T_1}{T_2} = \frac{p_1}{p_2}\frac{V_1}{V_2} \tag{7.26}$$

Introducing the definition of Mach number as $V = M\sqrt{(kRT)}$ yields

$$\frac{T_1}{T_2} = \left(\frac{p_1}{p_2}\frac{M_1}{M_2}\right)^2$$

or, using Equation (7.25), it is possible to obtain

$$\frac{T_1}{T_2} = \frac{M_1^2}{M_2^2}\frac{(1 + kM_2^2)^2}{(1 + kM_1^2)^2} \tag{7.27}$$

7.5.3 Velocity ratio

Substituting Equations (7.25) and (7.27) into Equation (7.26) yields

$$\frac{V_1}{V_2} = \frac{M_1^2}{M_2^2}\frac{1 + kM_2^2}{1 + kM_1^2} \tag{7.28}$$

7.5.4 Density ratio

Combining the continuity equation (7.1) with Equation (7.28), it is possible to obtain

$$\frac{\rho_1}{\rho_2}=\frac{M_2^2}{M_1^2}\frac{1+kM_1^2}{1+kM_2^2} \tag{7.29}$$

7.5.5 Stagnation temperature ratio

Using equation (4.24), it is possible to write the following equation for the stagnation temperature ratio:

$$\frac{T_{01}}{T_{02}}=\frac{T_1}{T_2}\frac{1+\frac{1}{2}(k-1)M_1^2}{1+\frac{1}{2}(k-1)M_2^2}$$

But T_1/T_2 may be eliminated by using Equation (7.27) to yield

$$\frac{T_{01}}{T_{02}}=\frac{M_1^2}{M_2^2}\frac{(1+kM_2^2)^2}{(1+kM_1^2)^2}\frac{1+\frac{1}{2}(k-1)M_1^2}{1+\frac{1}{2}(k-1)M_2^2} \tag{7.30}$$

7.5.6 Stagnation pressure ratio

The ratio of the stagnation pressure may be given as

$$\frac{p_{01}}{p_{02}}=\frac{p_1}{p_2}\left(\frac{1+\frac{1}{2}(k-1)M_1^2}{1+\frac{1}{2}(k-1)M_2^2}\right)^{k/(k-1)}$$

with the aid of Equation (4.28). Then, it is possible to eliminate p_1/p_2 by using Equation (7.25) to obtain

$$\frac{p_{01}}{p_{02}}=\frac{1+kM_2^2}{1+kM_1^2}\left(\frac{1+\frac{1}{2}(k-1)M_1^2}{1+\frac{1}{2}(k-1)M_2^2}\right)^{k/(k-1)} \tag{7.31}$$

7.5.7 Non-dimensional entropy change

It is possible to rearrange Equation (7.7) in the following form by noting that $c_p=kR/(k-1)$:

$$\frac{s_2-s_1}{R}=\ln\left(\frac{(T_2/T_1)^{k/(k-1)}}{p_2/p_1}\right)$$

Now, substituting T_2/T_1 and p_2/p_1 from Equations (7.27) and (7.25) yields

$$\frac{s_2-s_1}{R}=\ln\left[\left(\frac{1+kM_1^2}{1+kM_2^2}\right)^{(k+1)/(k-1)}\left(\frac{M_2}{M_1}\right)^{2k/(k-1)}\right] \tag{7.32}$$

7.6 WORKING CHART AND WORKING TABLE FOR THE FLOW OF A PERFECT GAS ON THE RAYLEIGH LINE

The formulas that were derived in the previous section introduce tedious numerical calculations, and their solution often involves a trial-and-error procedure. The numerical calculations can be facilitated by introducing a working chart and a working table. During the preparation of this working chart and working table, it is convenient to normalize these equations by setting the Mach number equal to unity at one of the sections. For this reason, the Mach number may be chosen to be unity at section 2, and then section 1 may be considered to be any other section of the same duct. If the properties at section 2 are denoted by a superscript asterisk, and those at section 1 are without a subscript, then the following equations are the non-dimensional properties for the Rayleigh line flow of a perfect gas that were derived in the previous section:

$$\frac{p}{p^*} = \frac{1+k}{1+kM^2} \tag{7.33}$$

$$\frac{T}{T^*} = \frac{M^2(1+k)^2}{(1+kM^2)^2} \tag{7.34}$$

$$\frac{\rho}{\rho^*} = \frac{1+kM^2}{(1+k)M^2} \tag{7.35}$$

$$\frac{V}{V^*} = \frac{(1+k)M^2}{1+kM^2} \tag{7.36}$$

$$\frac{p_0}{p_0{}^*} = \frac{1+k}{1+kM^2}\left(\frac{2+(k-1)M^2}{k+1}\right)^{k/(k-1)} \tag{7.37}$$

$$\frac{T_0}{T_0{}^*} = \frac{M^2(k+1)\,[2+(k-1)M^2]}{(1+kM^2)^2} \tag{7.38}$$

$$\frac{s^* - s}{R} = \ln\left[\left(\frac{1+kM^2}{1+k}\right)^{(k+1)/(k-1)}\left(\frac{1}{M}\right)^{2k/(k-1)}\right] \tag{7.39}$$

7.6.1 Working chart for the flow of a perfect gas on the Rayleigh line

The right-hand sides of the above equations are all functions of the Mach number, M, and the specific heat ratio, k. The non-dimensional properties on the left-hand sides of these equations can then be represented in graphical form with the Mach number as the independent variable. Such a working chart for the Rayleigh line flow of a perfect gas with $k = 1.4$ is given in Figure 7.5.

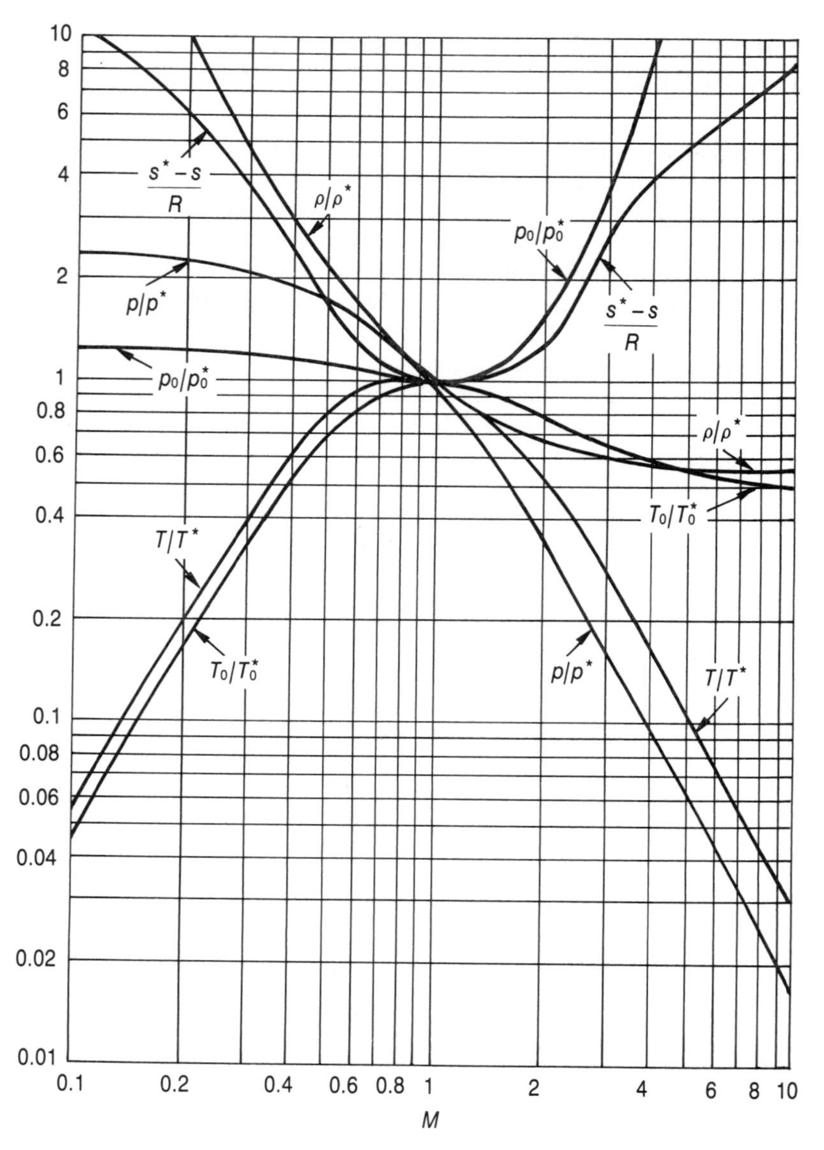

Figure 7.5 Working chart for the flow of a perfect gas with $k = 1.4$ on the Rayleigh line

7.6.2 Working table for the flow of a perfect gas on the Rayleigh line

The dimensionless flow properties can also be represented in tabular form with the Mach number as the independent variable. For $k = 1.4$, the values of these non-dimensional properties are given in Appendix G as a function of the Mach number.

For convenience, the limiting values of non-dimensional properties are given in Table 7.2.

Example 7.2

Air flows with negligible friction in a constant-area duct. At the inlet section, the temperature is 60 °C, the pressure is 135 kPa and the velocity is 732 m s^{-1}. Heat is added to the flow between the inlet section and the exit such that the Mach number is 1.2 at the exit section. Determine

a the flow properties at the exit section,
b the amount of heat transfer per unit mass, and
c the entropy change.
d Also, sketch the process on a T–s diagram.

Solution

(a) A sketch for the given problem is shown in Figure 7.6(a). To evaluate the speed of sound at the inlet section, the definition of the speed of sound may be used as

$$a_1 = \sqrt{(kRT_1)} = \sqrt{[(1.4)(287.1 \text{ J kg}^{-1} \text{ K}^{-1})(333 \text{ K})]} = 365.9 \text{ m s}^{-1}$$

Then the Mach number at the inlet section is

$$M_1 = \frac{V_1}{a_1} = \frac{732 \text{ m s}^{-1}}{365.9 \text{ m s}^{-1}} = 2.001$$

The critical properties may now be obtained from Appendix G, corresponding to $M_1 = 2.00$, as

$$p_1/p^* = 0.3636 \quad \text{or} \quad p^* = 135 \text{ kPa}/0.3636 = 371.3 \text{ kPa}$$

$$T_1/T^* = 0.5289 \quad \text{or} \quad T^* = 333 \text{ K}/0.5289 = 629.6 \text{ K}$$

Table 7.2 The limiting values of non-dimensional properties for the flow of a perfect gas with $k = 1.4$ on the Rayleigh line

M	0	1	∞
p/p^*	2.400	1	0
T/T^*	0	1	0
ρ/ρ^*	∞	1	0.5833
V/V^*	0	1	1.714
p_0/p_0^*	1.268	1	∞
T_0/T_0^*	0	1	0.4898
$(s^* - s)/R$	∞	0	∞

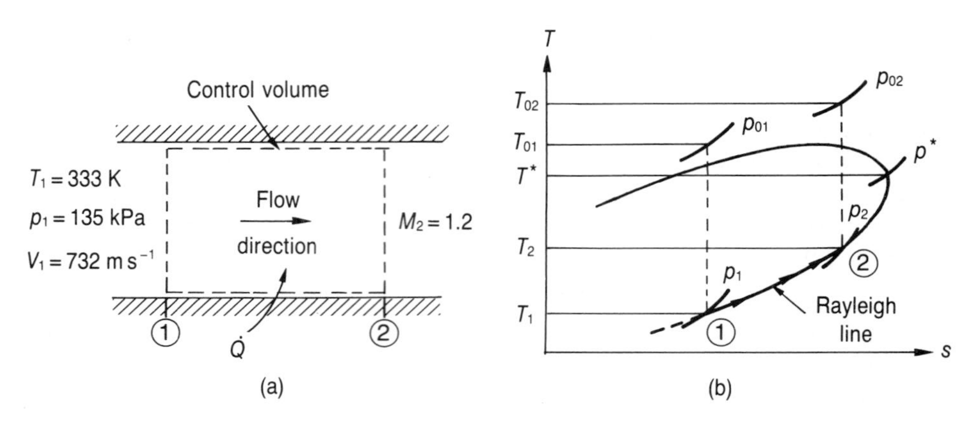

Figure 7.6 Sketch and T–s diagram for Example 7.2

Now, the pressure and temperature at the exit section of the duct may be found from Appendix G, corresponding to $M_2 = 1.2$, as

$$p_2/p^* = 0.7958 \quad \text{or} \quad p_2 = (0.7958)(371.3 \text{ kPa}) = \mathbf{295.5\ kPa}$$

$$T_2/T^* = 0.9118 \quad \text{or} \quad T_2 = (0.9118)(629.6 \text{ K}) = \mathbf{574.1\ K}$$

The density at the exit section is

$$\rho_2 = \frac{p_2}{RT_2} = \frac{295\,500 \text{ N m}^{-2}}{(287.1 \text{ J kg}^{-1}\text{ K}^{-1})(574.1 \text{ K})} = \mathbf{1.793\ kg\ m^{-3}}$$

The speed of sound at the exit section may now be calculated as

$$a_2 = \sqrt{(kRT_2)} = \sqrt{[(1.4)(287.1 \text{ J kg}^{-1}\text{ K}^{-1})(574.1 \text{ K})]} = 480.4 \text{ m s}^{-1}$$

and the velocity at the exit section is

$$V_2 = M_2 a_2 = (1.2)(480.4 \text{ m s}^{-1}) = \mathbf{576.5\ m\ s^{-1}}$$

Now, the stagnation properties at the exit section corresponding to $M_2 = 1.20$ may be found using Appendix C as

$$p_2/p_{02} = 0.4124 \quad \text{or} \quad p_{02} = 295.5 \text{ kPa}/0.4124 = \mathbf{716.5\ kPa}$$

$$T_2/T_{02} = 0.7764 \quad \text{or} \quad T_{02} = 574.1 \text{ K}/0.7764 = \mathbf{739.4\ K}$$

(b) To find the amount of heat transfer per unit mass, the stagnation temperature at the inlet section must be known. Therefore, using Appendix C corresponding to $M_1 = 2.00$

$$T_1/T_{01} = 0.5556 \quad \text{or} \quad T_{01} = 333 \text{ K}/0.5556 = 599.4 \text{ K}$$

Then the amount of heat transfer per unit mass is

$$q = c_p(T_{02} - T_{01}) = (1005 \text{ J kg}^{-1}\text{ K}^{-1})(739.4 \text{ K} - 599.4 \text{ K}) = \mathbf{140.7\ kJ\ kg^{-1}}$$

(c) To calculate the entropy change between the exit and inlet sections of the duct, Appendix G may be used for $M_1 = 2.00$ and $M_2 = 1.20$ as

$$(s^* - s_1)/R = 1.218$$

and

$$(s^* - s_2)/R = 0.094\,53$$

Therefore,

$$s_2 - s_1 = (1.218 - 0.094\,53)(287.1 \text{ J kg}^{-1} \text{K}^{-1}) = \mathbf{322.6 \text{ J kg}^{-1} \text{K}^{-1}}$$

(d) The process, which is sketched on the T–s diagram of Figure 7.6(b), follows the supersonic branch of the Rayleigh line.

7.7 CHOKING DUE TO HEAT TRANSFER

A horizontal line connecting the two branches of the curve of the stagnation temperature ratio, T_0/T_0^*, versus the Mach number, M, as shown in Figure 7.7 defines a pair of states which have the same mass flux, the same impulse function and the same stagnation temperature. Therefore, these two states correspond to states on the two sides of a normal shock wave.

When the initial state is subsonic, as shown by point 1 of Figure 7.7(a), the final state may only be subsonic, as indicated by point 2′. Supersonic speeds could only be attained if the second law of thermodynamics is violated in the form of a normal expansion shock wave from state 2′ to state 2″. However, one should note that this conclusion is valid only when the heat transfer is in one direction. Otherwise, a subsonic flow could be accelerated to a Mach number of unity by heating from state 1 to state 3 of Figure 7.7(a), after which the flow could be accelerated to supersonic speeds by cooling from state 3 to state 2″.

When the initial state is supersonic, as indicated by point 1 of Figure 7.7(b), the final state may be either supersonic or subsonic, as shown by states 2′ and 2″ in Figure 7.7(b), respectively. The subsonic speeds can be attained by a combination of simple heating and a normal compression shock wave. This may be achieved by (i) a simple heating from state 1 to state 2′ which is followed by a normal shock wave to state 2″, (ii) a normal shock wave from state 1 to state 4 followed by simple heating at subsonic speeds from state 4 to state 2″, or (iii) simple heating at supersonic speeds from state 1 to a state between states 1 and 2′ followed by a normal shock wave and concluded with simple heating at subsonic speeds to state 2″.

For fixed values of T_{02}/T_{01}, variations of the final Mach number, M_2, versus the initial Mach number, M_1, are given in Figure 7.8. When $T_{02}/T_{01} = 1.0$, the solid curve is for the trivial case in which there is no change in the state, and the dashed one represents the flow with a normal shock wave. For a typical value of $T_{02}/T_{01} = 1.1$, corresponding to heat addition, there are two branches, one corresponding to an initially subsonic flow and the other one corresponding to an initially supersonic flow. For each of these two branches, there are two values for the final Mach number M_2

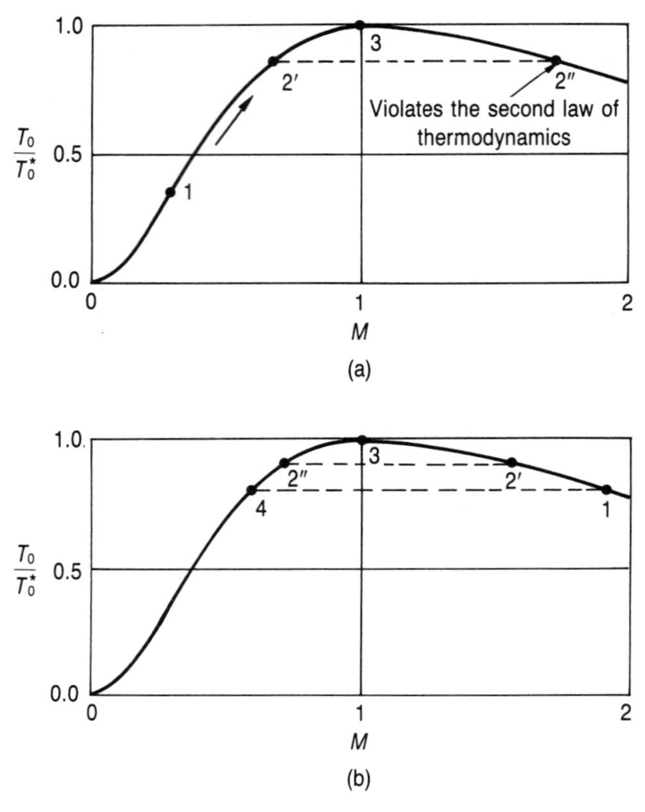

Figure 7.7 The possible end states for simple heat transfer: (a) when the flow is initially subsonic; (b) when the flow is initially supersonic

corresponding to each value of the initial Mach number, M_1. When the flow is initially supersonic, the entire curve of M_2 versus M_1 represents physically possible solutions. However, when the flow is initially subsonic, only the part of the curve, for which the final Mach number is less than unity, is valid. The remaining part of the curve implies the existence of a rarefaction normal shock wave which is physically impossible. For a typical value of $T_{02}/T_{01} = 0.9$, corresponding to heat rejection, the same discussion is also valid.

From Figures 7.7 and 7.8, it is evident that if the process involves a rise in the stagnation temperature, then for each initial Mach number there exists a maximum possible ratio of stagnation temperatures for which a solution is possible. In the limiting case, where the stagnation temperature ratio of the process is at its maximum allowable value, the Mach number reaches unity at the end of the duct. From a different point of view, if the initial stagnation temperature as well as the stagnation temperature rise are fixed, then there is a maximum allowable initial Mach number

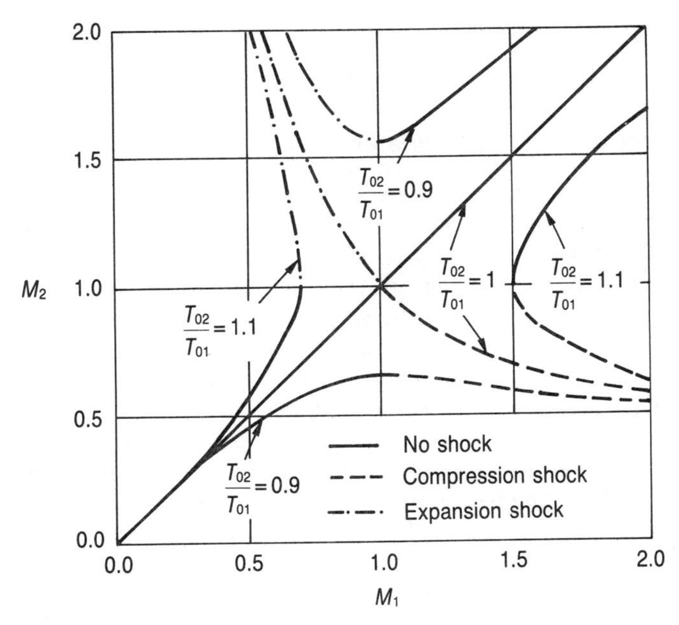

Figure 7.8 Curves of M_2 versus M_1 for fixed values of T_{02}/T_{01}

in subsonic flows, while there is a minimum allowable initial Mach number in supersonic flows for which a steady flow is possible.

When the process involves heat rejection, there are no limitations on the stagnation temperature ratio for initially subsonic flows. However, for initially supersonic flows without any normal shock wave, there is a maximum allowable reduction in the stagnation temperature ratio for any given initial Mach number. In this case, the final Mach number is infinite and corresponds to a stagnation temperature of zero at the exit of the duct, and all of the thermal energy is converted to the directed kinetic energy.

As long as the problems involving heating are more common than those involving cooling, then attention will be focused on the heating process.

7.7.1 Choking due to heat transfer in subsonic flows

Consider a subsonic flow with an initial Mach number of M_1 such that the imposed increase in the stagnation temperature yields a value of the stagnation temperature ratio T_{02}/T_{01} which is greater than the maximum allowable one. If the specified stagnation temperature rise is to occur, then some sort of readjustment is necessary. This readjustment is carried out by a series of transient effects which are characterized by the pressure waves propagating in both directions along the duct. When the flow reaches a steady condition it is said to be choked, and the Mach number at the exit of the duct is unity. In this case, the initial Mach number is reduced sufficiently to permit a specified amount of heating to occur.

7.7.2 Choking due to heat transfer in supersonic flows

Now, consider heat transfer to a supersonic stream flowing in a constant-area duct which is fed by a converging–diverging nozzle. If the stagnation temperature rise is too great to allow a solution with a specified initial supersonic Mach number, then a normal shock wave appearing in the duct does not provide the necessary readjustment, since there is no change in the stagnation temperature across an adiabatic shock wave. For example, if the state at the inlet of the duct is represented by state 1 of Figure 7.7(b), then the maximum allowable ratio of the final stagnation temperature to the initial one is T_{03}/T_{01}. If there is a shock wave within the duct where there is heating, say from state $2'$ to state $2''$, the maximum allowable stagnation temperature ratio is not altered. Therefore, the adjustment occurs with a shock wave in the diverging part of the converging–diverging nozzle. This shock wave reduces the Mach number at the inlet of the duct below the value corresponding to state 4 of Figure 7.7(b) so that a greater stagnation temperature ratio for the heating process is possible without a reduction in the mass flow rate. For very large increases in the stagnation temperature, the flow in the nozzle becomes entirely subsonic with a sonic flow at the exit of the duct.

7.8 COMBUSTION WAVES: DETONATION AND DEFLAGRATION

The phenomenon of chemical reaction behind or within a moving wave propagating in a premixed combustible mixture is called the **combustion wave** and is very important in combustion science. Two different types of combustion wave can be differentiated considering their propagation speed with respect to the reactants. These are **detonations**, which propagate at supersonic speeds, and **deflagrations**, which propagate at subsonic speeds.

If the premixed combustible gas in a long tube is ignited at the left end of the tube as shown in Figure 7.9(a), a planar combustion wave will propagate to the right into the unburned reactants, leaving burned products behind it. Across the combustion wave, the gas passes through a transition. The gas properties such as temperature, pressure and velocity are different on each side of the wave. The unsteady process of the moving wave may be rendered steady by analyzing the process with respect to an observer moving with the wave, as shown in Figure 7.9(b).

There, the combustion wave, which happens to be the **flame front** or the **explosion wave**, appears to be stationary and can be treated as a transition region in the steady flow. For further simplicity, the following assumptions will be made:

1 The flow takes place in a tube or duct of constant cross-sectional area.

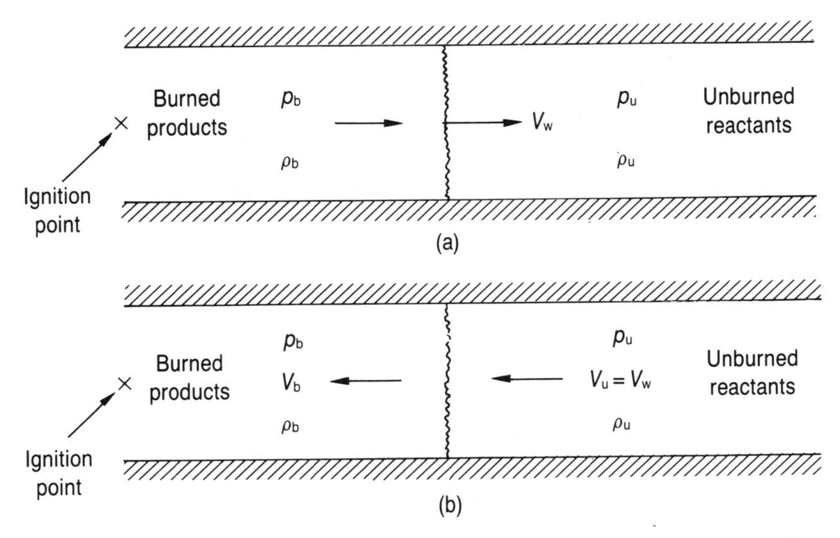

Figure 7.9 Combustion waves produced in a long tube: (a) combustion wave propagating with respect to a stationary observer; (b) combustion wave which is stationary with respect to an observer moving with the wave

2 The gases on each side of the wave can be considered uniform and inviscid. They obey the perfect gas law, and their properties such as the specific heats do not change. Moreover, no change in the molecular weight and the composition is considered to occur.
3 The chemical reaction rate is large, so that the chemistry and the affiliated molecular process are of no concern. Here, the addition of energy is instantaneous and only across the wave, which is considered to be stable.
4 The wave is considered to be infinitesimally thin.
5 As far as heat transfer is concerned, the rate of heat conduction is very much smaller than the rate of wave propagation so that negligible heat conduction can take place in the short times under consideration.

Then, across the steady combustion wave produced, as shown in Figure 7.9(b), the relations derived in Chapter 5 for normal shock waves are applicable. Therefore, the continuity, momentum and energy equations may be written across the wave between the unburned reactants and the burned products as

$$G = \rho_u V_u = \rho_b V_b \tag{7.40}$$

$$p_u + \rho_u V_u^2 = p_b + \rho_b V_b^2 \tag{7.41}$$

and

$$h_{0u} + \Delta u^0 = h_{0b} \tag{7.42a}$$

respectively, or noting that $h_0 = u + p/\rho + V^2/2$, with u being the internal energy per unit mass,

$$u_u + \frac{p_u}{\rho_u} + \frac{V_u^2}{2} + \Delta u^0 = u_b + \frac{p_b}{\rho_b} + \frac{V_b^2}{2} \tag{7.42b}$$

In these equations, the subscripts u and b refer to the unburned reactants and the burned products, respectively. Also Δu^0 is the energy per unit mass released due to the chemical reaction, and is known as the **energy of combustion.**

Combining these equations, the following equation, called the **Hugoniot equation,** can be obtained:

$$u_b - (u_u + \Delta u^0) = \tfrac{1}{2}(p_u + p_b)\left(\frac{1}{\rho_u} - \frac{1}{\rho_b}\right) \tag{7.43}$$

This equation together with the equation of state, $u = u(p, \rho)$, can be shown on a p–v diagram with $v = 1/\rho$ being the volume per unit mass, and is known as the Hugoniot curve, as shown in Figure 7.10. The lowest Hugoniot curve for $\Delta u^0 = 0$ is commonly called the **Rankine–Hugoniot curve**, which is the locus of the end points for an adiabatic normal shock wave. From the configuration of these curves, two different types of combustion waves are possible. These are **the deflagration waves** across which the pressure and density decrease and **the detonation waves** across which the pressure and density increase. Detonation waves propagate at supersonic velocity, whereas deflagration waves propagate at subsonic velocity.

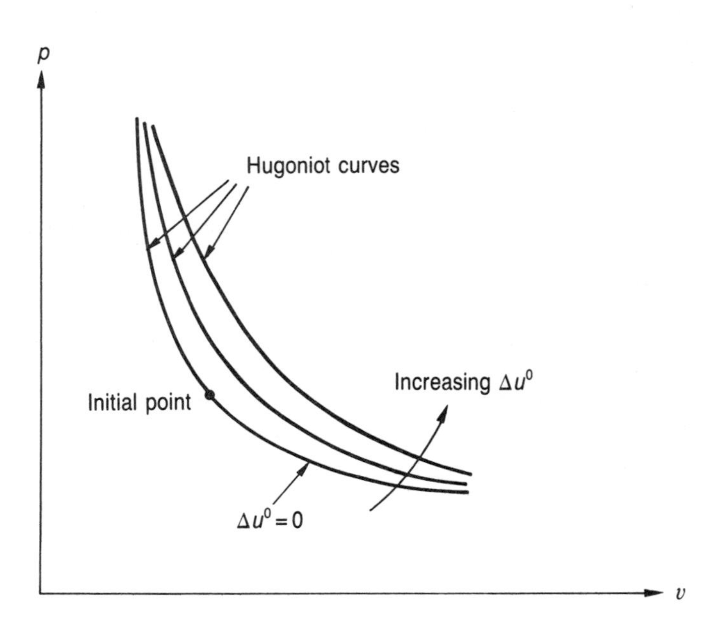

Figure 7.10 Hugoniot curves

One can also obtain the equation of the Rayleigh line by combining the continuity equation (7.40) with the momentum equation (7.41) as

$$G^2 = -\frac{p_b - p_u}{v_b - v_u} = -\left(\frac{\partial p}{\partial v}\right)_R \tag{7.44}$$

where the subscript R denotes differentiation on the Rayleigh line. As long as the mass flux is constant from Equation (7.40), then the Rayleigh lines appear to be straight lines on a p–v diagram. For this reason, the products of combustion must then satisfy the equations for both the Rayleigh line and the Hugoniot curve.

Figure 7.11 schematically illustrates a Hugoniot curve which is divided into five regions. At points J and K, the Rayleigh lines from point u having the maximum and minimum slopes, respectively, are tangents to the Hugoniot curve. At point B, the volume per unit mass is the same as that of the unburned products, while at point C the pressure is the same as that of the unburned products.

Now, denoting the slope of the Rayleigh line on a p–v diagram by α, as shown in Figure 7.12, Equation (7.44) becomes

$$\tan\alpha = \left(\frac{\partial p}{\partial v}\right)_R = -G^2 = -\rho^2 V^2$$

or

$$V = \sqrt{(-\tan\alpha)}/\rho \tag{7.45}$$

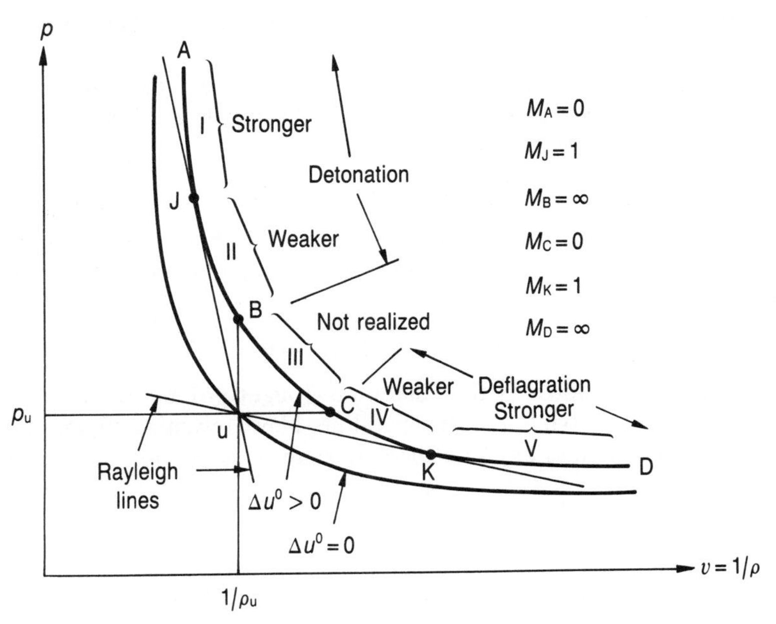

Figure 7.11 The Hugoniot curve and the Chapman–Jouget conditions

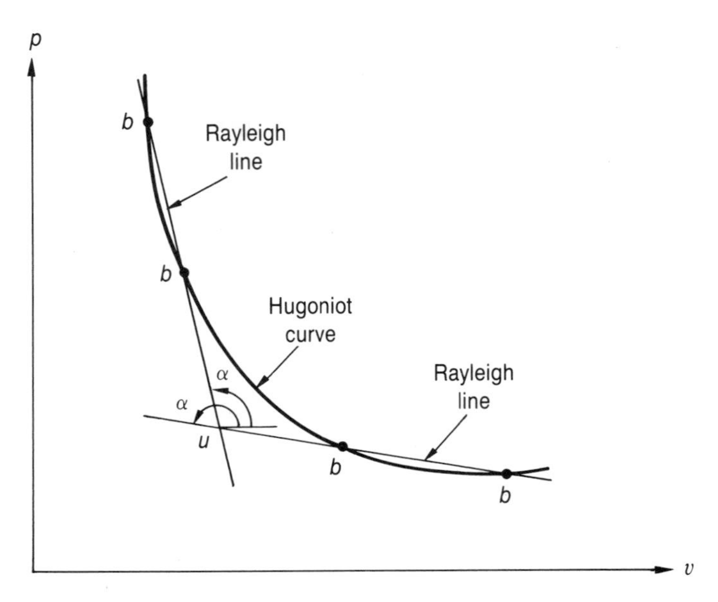

Figure 7.12 Definition of the angle α

In regions I, II, IV and V of Figure 7.11, $\tan\alpha$ is negative and physically possible values of velocity may be obtained from Equation (7.45). However, $\tan\alpha$ is positive in region III and the velocity is imaginary. As a result, the mathematical solution in region III has no physical meaning.

When one approaches point B from region II, then α decreases towards the limit of 90° and $\tan\alpha$ increases in a negative direction towards the limit of $\tan\alpha = -\infty$. Hence, at point B, Equation (7.45) becomes

$$V_B = \sqrt{[-(-\infty)]}/\rho_B = \infty \tag{7.46}$$

Meanwhile, if one approaches point C from region IV, then α increases towards the limit of 180°, while $\tan\alpha$ decreases towards the limit of $\tan\alpha = -0$. Therefore, Equation (7.45) becomes

$$V_C = \sqrt{[-(-0)]}/\rho_C = 0 \tag{7.47}$$

Points J and K are known as the **Chapman–Jouget points**. At these points, the Hugoniot curve is tangential to the Rayleigh line. Now, rewriting Equation (7.43) for the unburned reactants and any intermediate state, it is possible to obtain

$$u - (u_u + \Delta u^0) = \tfrac{1}{2}(p_u + p)(v_u - v)$$

and differentiating

$$du_H = -\tfrac{1}{2}(p_u + p)\ dv_H + \tfrac{1}{2}(v_u - v)\ dp_H \tag{7.48}$$

where the subscript H shows that the differentiation is along the Hugoniot curve.

At this point, recalling the thermodynamic relation

$$T\,\mathrm{d}s = \mathrm{d}u + p\,\mathrm{d}v$$

Equation (7.48) becomes

$$2T\left(\frac{\partial s}{\partial p}\right)_{\mathrm{H}} = (v_{\mathrm{u}} - v)\left(1 - \frac{(p_{\mathrm{u}} - p)}{(v_{\mathrm{u}} - v)}\frac{1}{(\partial p/\partial v)_{\mathrm{H}}}\right)$$

Now, using Equation (7.44) on any arbitrary Hugoniot curve, this equation becomes

$$2T\left(\frac{\partial s}{\partial p}\right)_{\mathrm{H}} = (v_{\mathrm{u}} - v)\left(1 + \frac{G^2}{(\partial p/\partial v)_{\mathrm{H}}}\right) \tag{7.49}$$

However, at points J and K, the Rayleigh lines and the Hugoniot curve are tangential so that

$$\left(\frac{\partial p}{\partial v}\right)_{\mathrm{H}} = \left(\frac{\partial p}{\partial v}\right)_{\mathrm{R}} = -G^2$$

Now, Equation (7.49) reduces to

$$\left(\frac{\partial s}{\partial p}\right)_{\mathrm{H}} = 0$$

As a result, at these points $\mathrm{d}s = 0$ and the entropy is constant, so that

$$\left(\frac{\partial p}{\partial v}\right)_{\mathrm{R}} = -\rho^2\left(\frac{\partial p}{\partial \rho}\right)_{\mathrm{s}} = -\rho^2 a^2 = -G^2 = -\rho^2 V^2$$

and thus

$$V = a \tag{7.50}$$

at points J and K.

In region I, as the final state moves away from point J, the pressure and density increase towards infinity. As a result, one may easily determine from the continuity equation (7.40) that the velocity decreases towards zero. However, in region IV, the pressure and density decrease towards zero, while the velocity increases towards infinity.

The results for possible states are summarized in Table 7.3. In region III, no real processes can occur. Also, in region V, no physical processes are possible. In this region, a subsonic flow which is entering a combustion wave should be accelerated to supersonic speeds by the addition of heat due to combustion. However, one should note that a subsonic flow cannot be accelerated to supersonic speeds by only the addition of heat, as discussed in Section 7.7. Therefore, the states in region V are not physically realizable. Weak detonation waves in region II are also impossible. A detonation wave may be considered as an adiabatic shock wave followed by the addition of heat due to combustion. Point E represents the final state corresponding to initial state, u, immediately after passing the shock wave. One should note that states u and E are on the same Hugoniot curve with $\Delta u^0 = 0$, as shown in Figure 7.13.

Table 7.3 Possible states and the relative magnitude of properties across a combustion wave

Region	Combustion wave	Realizability	M_u	M_b	$\frac{p_b}{p_u}$	$\frac{V_b}{V_u} = \frac{\rho_u}{\rho_b}$	$\frac{T_b}{T_u}$
I (A–J)	Strong detonation	Possible	>1	<1	>1	<1	>1
II (J–B)	Weak detonation	Not possible	>1	>1	>1	<1	>1
III (B–C)		Not possible	–	–	–	–	–
IV (C–K)	Weak deflagration	Possible	<1	<1	<1	>1	<1
V (K–D)	Strong deflagration	Not possible	<1	>1	<1	>1	<1

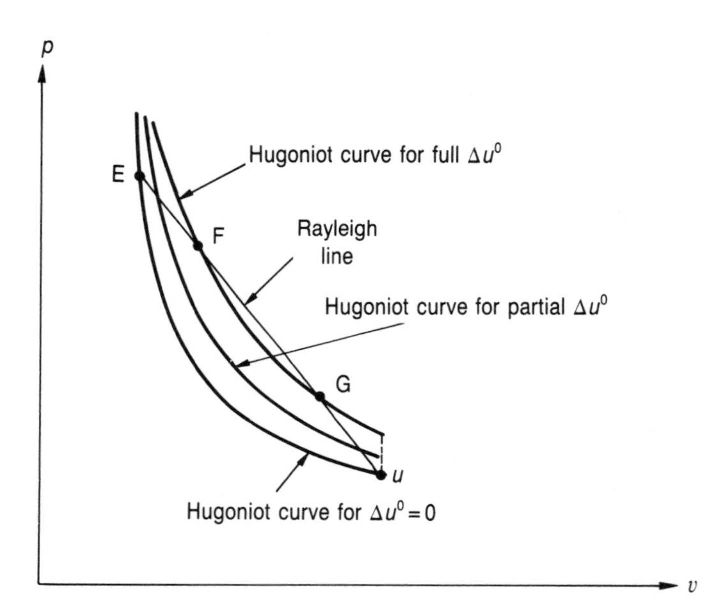

Figure 7.13 The impossibility of a weak detonation wave

Now, the Rayleigh line intersects the Hugoniot curve with full energy of formation, Δu^0 at two points, F and G. Point F corresponds to strong detonation with a final subsonic flow, while point G is for a weak detonation with a final supersonic flow. As a result, weak detonation is not possible, since subsonic flow at point E cannot be accelerated to supersonic flow at point G by only the simple addition of heat on the Rayleigh line.

Therefore, only strong detonation and weak deflagration waves are possible, as indicated in Figure 7.14. The flame speed of a deflagration wave is controlled by a

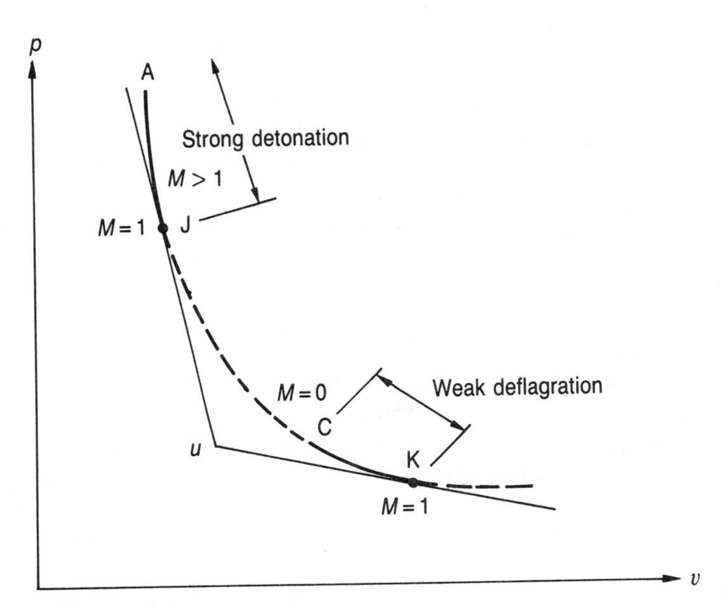

Figure 7.14 Physically attainable regions of the Hugoniot curve

combination of heat conduction and mass diffusion, whereas that of a detonation wave is controlled by the abrupt pressure, density and temperature increases associated with the normal shock wave.

FURTHER READING

Anderson, J. D. Jr (1982) *Modern Compressible Flow with Historical Perspective*, McGraw-Hill: New York.

Cambel, A. B. and Jennings, B. H. (1958) *Gas Dynamics*, Dover: New York.

Daneshyar, H. (1976) *One-Dimensional Compressible Flow*, Pergamon: Oxford.

Fox, R. W. and McDonald, A. T. (1985) *Introduction to Fluid Mechanics*, 3rd edn, Wiley: New York.

John, J. E. A. (1984) *Gas Dynamics*, 2nd edn, Allyn & Bacon: Boston, MA.

Liepmann, H. W. and Roshko, A. (1957) *Elements of Gas Dynamics*, Wiley: New York.

Owczarek, J. A. (1964) *Fundamentals of Gas Dynamics*, International Textbook: Scranton, PA.

Shapiro, A. H. (1953) *The Dynamics and Thermodynamics of Fluid Flow*, Vol. 1, Ronald Press: New York.

Zucrow, M. J. and Hoffman, J. D. (1976) *Gas Dynamics*, Vol. 1, Wiley: New York.

PROBLEMS

7.1 Air flows in a constant-area duct without friction. At the inlet of the duct, the Mach number, pressure and stagnation temperature are 0.40, 800 kPa and 300 K, respectively. As a result of heat transfer, the Mach number at the exit section is 0.80. Determine the amount of heat transfer per unit mass and the pressure difference between the inlet and exit sections of the duct.

7.2 Air flows steadily through a duct of 0.02 m^2 in cross-sectional area without friction at a mass flow rate of 2 $kg\,s^{-1}$. At the inlet of the duct, the pressure and temperature are 125 kPa and 500 K, respectively. The air leaves the duct subsonically at a pressure of 100 kPa. Determine the Mach number, temperature and stagnation temperature at the exit of the duct. Also calculate the rate of heat transfer.

7.3 A constant-area duct is fed with air from a converging–diverging nozzle. At the entrance of the duct, the stagnation pressure, stagnation temperature and Mach number are 750 kPa, 500 K and 2.0, respectively. At the exit of the duct, the density is 1.35 $kg\,m^{-3}$. Assuming frictionless flow, determine the Mach number, the velocity and pressure at the exit of the duct. Also determine the amount of heat transfer per unit mass between the inlet and exit of the duct.

7.4 Air flows without friction in a short section of a constant-area duct. At the duct inlet, the Mach number, temperature and density are 0.40, 300 K and 1.5 $kg\,m^{-3}$, respectively. At the exit of the duct, the Mach number is 0.80. Determine the heat transfer per unit mass, the entropy change and the change in the stagnation pressure for the process.

7.5 Air flows at a rate of 10 $kg\,s^{-1}$ through a duct with a cross-sectional area of 0.03 m^2. At the inlet section, the temperature and pressure are 350 K and 150 kPa, respectively. At the exit section, the flow is choked. Determine the rate of heat transfer, the entropy change and the change in the stagnation pressure. Assume frictionless flow.

7.6 Air flows without friction through a duct of constant diameter. The inlet Mach number, stagnation temperature and stagnation pressure are 0.30, 350 K and 400 kPa, respectively. Heat is added at a rate of 300 $kJ\,kg^{-1}$ to the flowing air. Determine the stagnation temperature and the Mach number at the duct outlet. Also, calculate the changes in the entropy and the stagnation pressure for the process.

7.7 A constant-area duct is connected to a high-pressure air reservoir through a converging nozzle. The walls of the duct are heated in order to supply 250 000 $J\,kg^{-1}$ to the air passing through the duct. The pressure and temperature in the reservoir are 750 kPa and 300 K, respectively. If the duct discharges to an atmosphere with a pressure of 100 kPa, then determine the mass flow rate of the air. The duct has a diameter of 0.05 m and a length of 1.2 m. Assume isentropic flow in the nozzle and frictionless flow in the duct. (METU 1987).

7.8 Air is heated as it flows through a duct 15 m in length and 0.1 m in diameter. The coefficient of friction in the duct is 0.01. The stagnation temperature and the stagnation pressure at the inlet of the duct are 500 K and 200 kPa, respectively. The total force acting on the duct is 50 N. It is found that a normal shock wave stands at a location 5 m from the entrance of the duct. The mass flow rate through the duct is 2 kg s^{-1}. If the exit pressure is 100 kPa, find the exit Mach number and the rate of heat transfer. (METU 1985)

7.9 Air is heated as it flows through a diverging duct 15 m in length. The coefficient of friction in the duct is 0.02. The stagnation temperature and the stagnation pressure at the inlet of the duct are 500 K and 200 kPa, respectively. The inlet and exit diameters of the duct are 0.1 m and 0.15 m, respectively. It is found that a normal shock wave stands at a location 10 m from the entrance of the duct. The mass flow rate through the duct is 2 kg s^{-1}. The duct discharges into a tank where the pressure is 100 kPa. If the rate of heat transfer to the duct is 70 kW, find the exit Mach number and the total force acting on the duct. (METU 1986)

7.10 A supersonic test aircraft propelled by ram jets, shown in the figure below, flies at an altitude where the ambient temperature is $-55.3\,^{\circ}C$ with a velocity of 592 m s^{-1}. The inlet is shockless and the supersonic nozzle is designed for maximum thrust. The ratio of the nozzle exit area to the throat area is 1.633. The amount of fuel burned in the combustion chamber is negligible compared with the air flow through the engine. The combustion chamber may be considered as a constant-area duct, while the flow through the duct can be approximated as a simple T_0 change flow. The specific thrust of the engine is 243 N s kg^{-1}. Calculate the heat of reaction in kilojoules per kilogram necessary under these conditions in the combustion chamber. It is assumed that the properties of both reactants and products are equivalent to air in respect to the molecular weight and specific heat. The flow at the inlet and in the nozzle is isentropic.

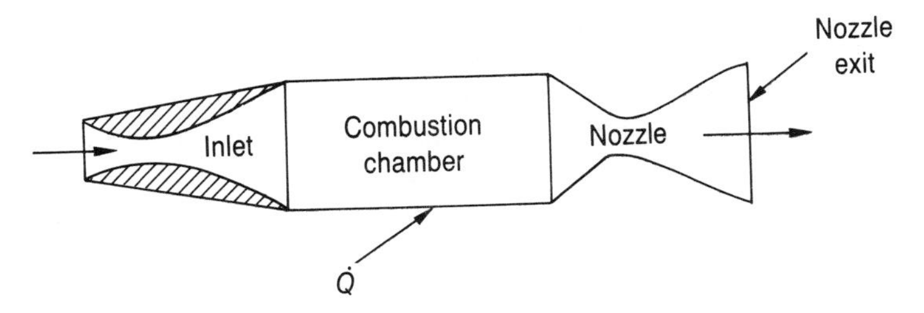

Problem 7.10

8 Steady and two-dimensional supersonic flows

8.1 INTRODUCTION

Consider a supersonic stream of gas flowing on both sides of a wall with zero thickness which has a curved section, as shown in Figure 8.1. At the point where the wall starts to turn, the streamlines adjacent to the wall begin to deflect. As a result, a pressure disturbance is created whose head represents a Mach wave. The angle of inclination of this wave to the streamlines is the Mach angle, μ_1, which is related to the Mach number by Equation (3.10).

As the gas follows the curved part of the wall, an infinite number of Mach waves are generated owing to the continuous change in the direction of streamlines. The gas properties and the flow velocity adjust to the corresponding variation at the wall through these Mach waves. Therefore, two continuous waves – one on the upper and another on the lower side of the wall – are formed. The flow is decelerated by continuous compression waves on the lower side of the wall, while it is accelerated by continuous expansion waves on the upper side. The curvature of the thin wall begins at the head of these waves and ends at the tail. The state of the gas does not change along the Mach lines.

As the Mach number of the flow increases, the angle of inclination of the Mach waves to the streamlines, that is the Mach angle, decreases, as indicated by Equation (3.10). As long as the Mach number increases across an expansion wave, then the Mach angle decreases from μ_1 to μ_2 causing the Mach waves to diverge. However, across a compression wave, the Mach number decreases and the Mach angle increases from μ_1 to μ_3 causing the Mach waves to converge and form an oblique shock wave, which is inclined to the flow direction.

If the direction of the wall changes abruptly, then the expansion waves become a centred expansion wave, while the compression waves change into an oblique shock wave, as shown in Figure 8.2. The deflection angle of the stream in Figure 8.2 is the same as that in Figure 8.1.

Supersonic flows around a bend or corner in which continuous expansion waves or continuous compression waves occur are often called Prandtl–Meyer flows.

In this chapter, oblique shock waves, Prandtl–Meyer flows and their applications are considered.

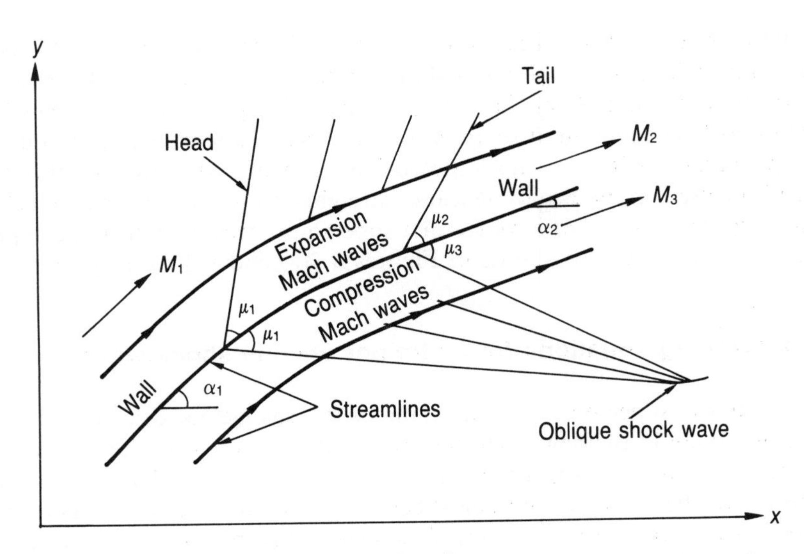

Figure 8.1 Supersonic flow over a curved wall

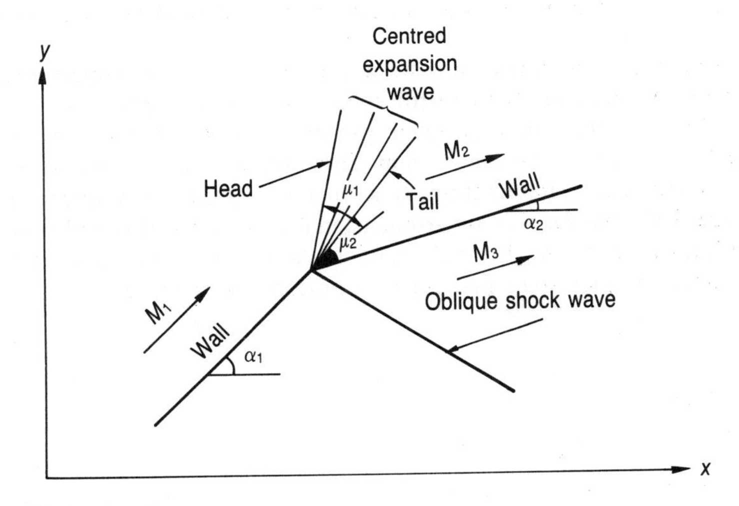

Figure 8.2 Supersonic flow over a corner

8.2 OBLIQUE SHOCK WAVES

A normal shock wave perpendicular to the flow direction was studied in Chapter 5. However, in many physical situations, an oblique shock wave, which is inclined to the flow direction, occurs.

Although the study of two-dimensional oblique shock waves represents a departure from one-dimensional flow, the method of handling the oblique shock waves is very similar to those of the normal shock waves. Even though an oblique shock wave is inclined to the flow direction, it represents a sudden and almost discontinuous change in fluid properties, with the shock process itself being adiabatic. In this section, attention is focused on the two-dimensional straight oblique shock wave, which might occur during the presence of a wedge in a supersonic stream or during a supersonic compression in a corner, as shown in Figures 8.3(a) and 8.3(b), respectively.

8.2.1 Governing equations for a straight oblique shock wave

When a uniform supersonic stream is forced to change its direction due to the presence of a body in the flow, then the stream cannot adjust to the presence of the body gradually. Such an adjustment is only possible by a sudden change in flow properties with the aid of an oblique shock wave. A simple case is supersonic flow at a two-dimensional concave corner, as shown in Figure 8.4. For small deflection angles, the flow adjusts to the presence of such a corner by means of an oblique shock wave, which is attached to the corner. The flow after the shock wave is uniform and parallel to the wall, with the entire flow having been turned through the deflection angle, δ.

The governing equations will now be written for a uniform supersonic flow at a concave corner by selecting the control volume shown in Figure 8.4.

One should note that an additional variable, the tangential flow velocity, is introduced because of the change in flow direction across the oblique shock wave. However, as the linear momentum is a vector quantity two linear momentum equations are derivable for this two-dimensional flow. With this additional variable and the additional equation, the analysis of the flow across an oblique shock wave is somewhat more complex than that across a normal shock wave.

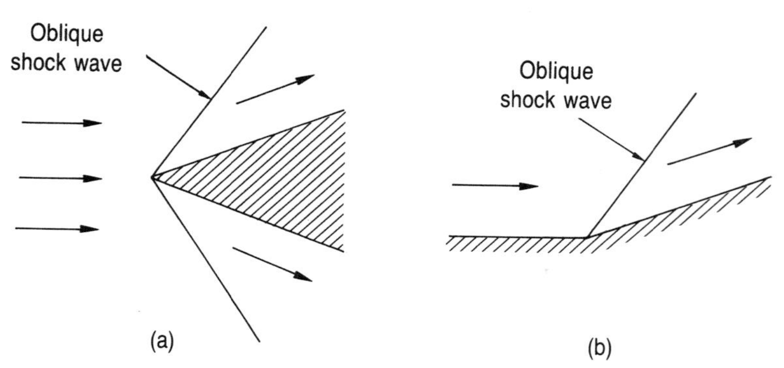

Figure 8.3 (a) Supersonic flow about a wedge; (b) supersonic compression in a corner

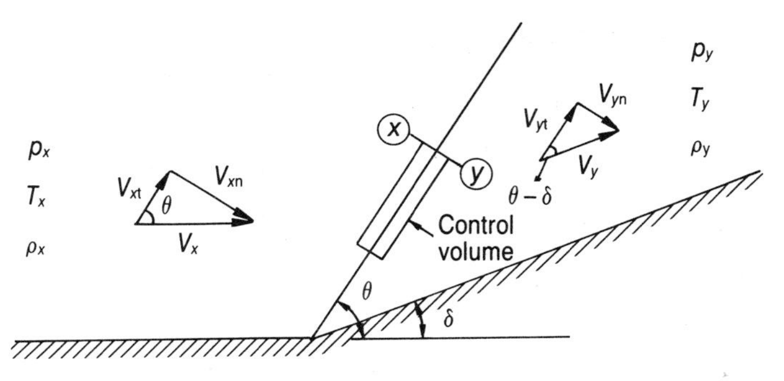

Figure 8.4 Supersonic flow at a two-dimensional concave corner

(a) Continuity equation

For steady and uniform flow upstream and downstream of the oblique shock wave, the mass flow rate is constant, so that

$$\dot{m} = \rho_x A_x V_{xn} = \rho_y A_y A_{yn}$$

where V_{xn} and V_{yn} are the velocity components normal to the shock wave. The shock wave is very thin so that $A_x = A_y = A$. Thus,

$$G = \dot{m}/A = \rho_x V_{xn} = \rho_y V_{yn} \tag{8.1}$$

(b) Momentum equation

Linear momentum is a vector quantity, so the conservation of linear momentum can be written in a direction both normal and tangential to the shock wave. Then, in a direction normal to the shock wave

$$p_x A_x - p_y A_y = \rho_y A_y V_{yn}^2 - \rho_x A_x V_{xn}^2$$

As long as the cross-sectional area and the mass flow rate are constant across a shock wave, then

$$p_x - p_y = \rho_y V_{yn}^2 - \rho_x V_{xn}^2 = G(V_{yn} - V_{xn}) \tag{8.2}$$

In the tangential direction, there is no change in the pressure so that

$$\dot{m} V_{xt} = V_{xt}(\rho_x A_x V_{xn}) = V_{yt}(\rho_y A_y V_{yn}) = \dot{m} V_{yt}$$

or cancelling

$$V_{xt} = V_{yt} \tag{8.3}$$

where V_{xt} and V_{yt} are the velocity components tangential to the oblique shock wave.

(c) The first law of thermodynamics

For adiabatic and steady flow, the first law of thermodynamics reduces to

$$h_x + \frac{V_x^2}{2} = h_y + \frac{V_y^2}{2}$$

or expanding

$$h_x + \frac{V_{xn}^2 + V_{xt}^2}{2} = h_y + \frac{V_{yn}^2 + V_{yt}^2}{2}$$

However, $V_{xt} = V_{yt}$ so that

$$h_x + \frac{V_{xn}^2}{2} = h_y + \frac{V_{yn}^2}{2} \tag{8.4}$$

(d) The second law of thermodynamics

The flow through an oblique shock wave is irreversible because of the almost discontinuous property changes across the shock wave. Then,

$$s_y > s_x \tag{8.5}$$

(e) Equation of state

For any fluid, the equation of state may be given as

$$h = h(s, \rho) \tag{8.6a}$$

or

$$s = s(p, \rho) \tag{8.6b}$$

At this point, one may observe that Equations (8.1), (8.2) and (8.4) contain only the normal velocity components and are very similar to Equations (5.1), (5.2) and (5.3) for the normal shock wave. In other words, an oblique shock wave acts as a normal shock wave for the velocity component normal to the flow direction, while the tangential velocity component remains unchanged.

8.2.2 Relations for the flow of a perfect gas across an oblique shock wave

The equations governing the flow across an oblique shock wave were given in the previous section. Although any problem related to the flow across an oblique shock wave may be handled by solving these equations, the solution can be facilitated by rearranging these equations in more suitable forms. In this section these relations will be derived.

(a) Relation between the upstream and the downstream Mach numbers

A relation between the Mach numbers upstream and downstream of an oblique shock wave may be obtained by using Equation (5.15b) in a direction normal to the oblique shock wave as

$$M_y = \frac{1}{\sin(\theta - \delta)} \sqrt{\left(\frac{(k-1)M_x^2 \sin^2\theta + 2}{2kM_x^2 \sin^2\theta - (k-1)}\right)} \tag{8.7}$$

One should note that the Mach number after an oblique shock wave can be greater than unity without violating the second law of thermodynamics. However, the normal component of the Mach number after the oblique shock wave must be less than unity.

(b) Relation between the shock-wave angle and the deflection angle

Equation (5.21) for the velocity ratio across a normal shock wave may now be applied to the flow normal to the oblique shock wave as

$$\frac{V_{yn}}{V_{xn}} = \frac{2 + (k-1)M_{xn}^2}{(k+1)M_{xn}^2} = \frac{2 + (k-1)M_x^2 \sin^2\theta}{(k+1)M_x^2 \sin^2\theta}$$

From Figure 8.4, one should note that $V_{xn} = V_{xt}\tan\theta$ and $V_{yn} = V_{yt}\tan(\theta - \delta)$. Hence

$$\frac{\tan(\theta - \delta)}{\tan\theta} = \frac{2 + (k-1)M_x^2 \sin^2\theta}{(k+1)M_x^2 \sin^2\theta}$$

since $V_{xt} = V_{yt}$. The above equation may now be solved for the deflection angle as

$$\cot\delta = \left(\frac{k+1}{2} \frac{M_x^2}{M_x^2 \sin^2\theta - 1} - 1\right)\tan\theta \tag{8.8}$$

For a given deflection angle, this equation may be solved for the shock-wave angle by trial and error. To simplify the calculations, the shock-wave angle, θ, versus the deflection angle δ, is plotted for different values of the upstream Mach number, M_x, in Figure 8.5.

For fixed values of M_x and δ, two different sets of θ and M_y are possible. These are for two oblique shock waves with different strengths. The larger of the two values of θ is for **a strong oblique shock wave**, while the smaller value is for **a weak oblique shock wave**. In practice, the weak shock wave occurs in external aerodynamic flows. In internal flows, when the downstream pressure is sufficiently high, a strong shock wave occurs, as in the case of flows in wind tunnels or in engine inlets.

One should note that the deflection angle is zero in Equation (8.8) for two shock-wave angles when M_x is known. In the first case, the flow is not deflected when $\theta = \mu$, since a Mach wave results from an infinitesimal disturbance of zero shock strength. In the second case, the flow, again, is not deflected as it passes through a normal shock wave with $\theta = \pi/2$. Between these two limiting cases, the deflection angle is always positive with a maximum value of δ_{max}. The maximum shock-wave angle, θ_{max}, corresponding to maximum flow deflection angle may be obtained as

$$\sin^2\theta_{max} = \frac{1}{kM_x^2}\left\{\frac{k+1}{4} M_x^2 - 1 + \sqrt{\left[(k+1)\left(1 + \frac{k-1}{2} M_x^2 + \frac{k+1}{16} M_x^4\right)\right]}\right\} \tag{8.9}$$

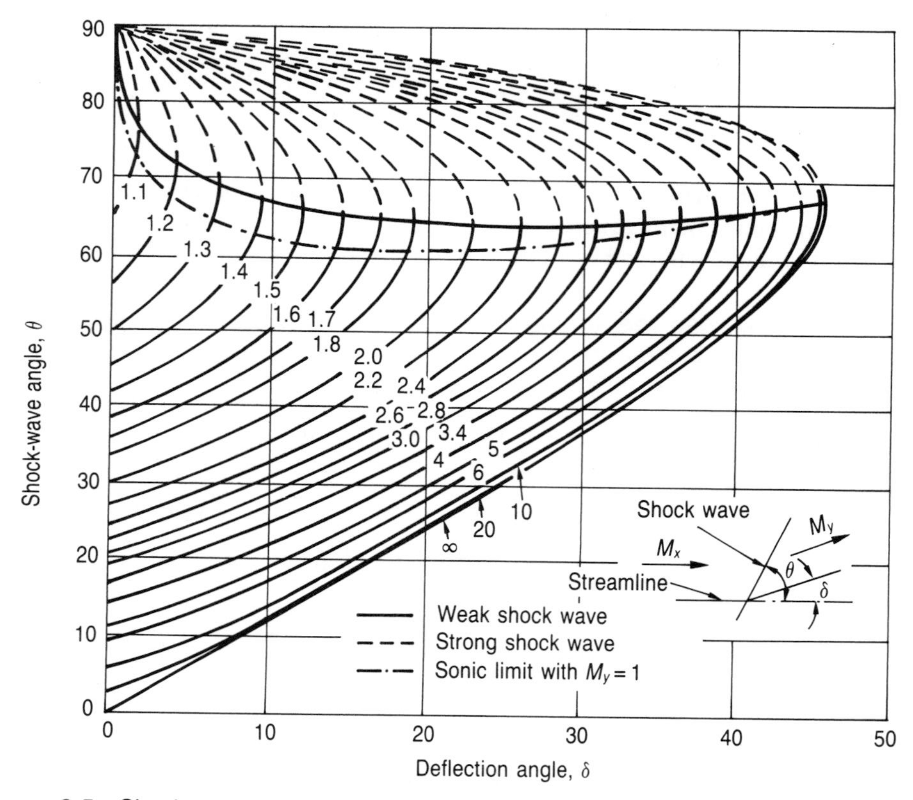

Figure 8.5 Shock-wave angle as a function of the deflection angle for different values of the upstream Mach number with $k = 1.4$

If the flow deflection angle, δ, is larger than the maximum flow deflection angle, $\delta_{\max}$, then there is no solution, as one may observe from Figure 8.5. Under these conditions, the oblique shock wave is no longer attached to the wedge but stands detached in front of the body, as shown in Figure 8.6. The detached shock is curved, and its strength decreases progressively from that of a normal shock at the apex of the wedge to that of a Mach wave far from the body. Therefore, the entire range of oblique shock solutions is obtained with a detached shock for a given upstream Mach number, M_x. The shape of the wave and the shock detachment distance are dependent on the Mach number and the shape of the body. Flow over the body is subsonic in the vicinity of the apex of the wedge where the strong oblique shock wave occurs, and supersonic farther back along the wedge where the weak oblique shock is present.

A detached oblique shock can also occur with supersonic flow in a concave corner, as shown in Figure 8.7. The characteristics of this oblique shock wave are exactly the same as those of the upper half of the detached shock in Figure 8.6. This can be seen if one replaces the centre streamline of Figure 8.6 with a plane wall. In this case, the

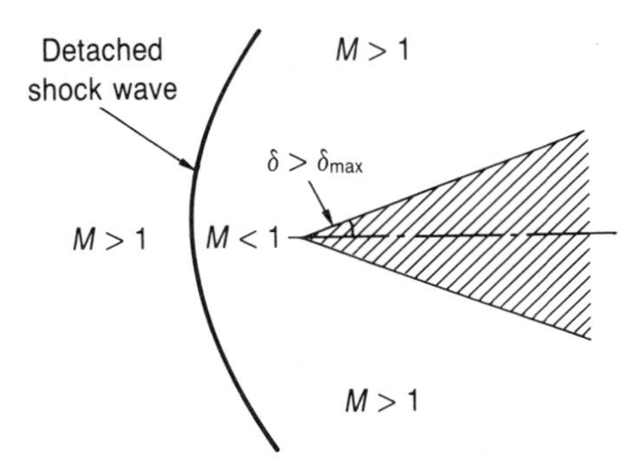

Figure 8.6 A detached oblique shock wave in front of a wedge

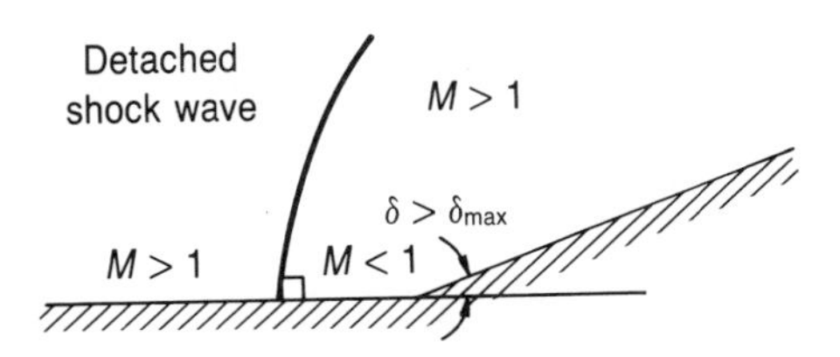

Figure 8.7 A detached shock wave in front of a corner

boundary conditions of the flow are not altered, since there can be no flow across a streamline. Thus, the flow after the oblique shock wave is subsonic near the wall and supersonic farther out in the flow.

There is also a particular value of the shock-wave angle, which is denoted by θ_s, for which the Mach number after the oblique shock wave is unity, and may be given as

$$\sin^2\theta_s = \frac{1}{kM_x^2}\left\{\frac{k+1}{4}M_x^2 - \frac{3-k}{4} + \sqrt{\left[(k+1)\left(\frac{9+k}{16} + \frac{3-k}{8}M_x^2 + \frac{k+1}{16}M_x^4\right)\right]}\right\} \quad (8.10)$$

(c) Pressure ratio across an oblique shock wave

An equation for the pressure ratio across an oblique shock wave may be obtained by using Equation (5.17) for the flow normal to the oblique shock wave as

$$\frac{p_y}{p_x} = \frac{2k}{k+1}M_x^2\sin^2\theta - \frac{k-1}{k+1} \quad (8.11)$$

(d) Temperature ratio across an oblique shock wave

The temperature ratio across an oblique shock wave may be obtained by using Equation (5.18) for the flow normal to the oblique shock wave as

$$\frac{T_y}{T_x} = \frac{[1+\frac{1}{2}(k-1)M_x^2\sin^2\theta]\,[2kM_x^2\sin^2\theta/(k-1)-1]}{[(k+1)^2/2(k-1)]\,M_x^2\sin^2\theta} \quad (8.12)$$

(e) Density ratio across an oblique shock wave

If the density ratio across a normal shock wave given by Equation (5.20) is applied to the flow normal to the oblique shock wave, then

$$\frac{\rho_y}{\rho_x} = \frac{(k+1)M_x^2 \sin^2\theta}{2 + (k-1)M_x^2 \sin^2\theta} \tag{8.13}$$

(f) Velocity ratio across an oblique shock wave

Combining the continuity equation (8.1) and the density ratio (8.13) yields

$$\frac{V_y}{V_x} = \frac{\rho_x}{\rho_y} = \frac{2 + (k-1)M_x^2 \sin^2\theta}{(k+1)M_x^2 \sin^2\theta} \tag{8.14}$$

(g) Stagnation pressure ratio across an oblique shock wave

The stagnation pressure ratio across a normal shock wave given by Equation (5.22) may now be applied to the flow normal to the oblique shock wave as

$$\frac{p_{0y}}{p_{0x}} = \left(\frac{\frac{1}{2}(k+1)M_x^2 \sin^2\theta}{1 + \frac{1}{2}(k-1)\ M_x^2 \sin^2\theta}\right)^{k/(k-1)} \times \left(\frac{2k}{k+1}\ M_x^2 \sin^2\theta - \frac{k-1}{k+1}\right)^{1/(1-k)} \tag{8.15}$$

(h) Entropy change across an oblique shock wave

An equation for the entropy change across an oblique shock wave may be obtained by applying Equation (5.24) to the flow which is perpendicular to the oblique shock wave as

$$\frac{s_y - s_x}{R} = \frac{k}{k-1} \ln\left(\frac{2}{(k+1)M_x^2 \sin^2\theta} + \frac{k-1}{k+1}\right) + \frac{1}{k-1} \ln\left(\frac{2k}{k+1}\ M_x^2 \sin^2\theta - \frac{k-1}{k+1}\right) \tag{8.16}$$

Example 8.1

A uniform supersonic air flow travelling at a Mach number of 3.0 passes over a concave corner, as shown in Figure 8.8. An oblique shock wave, which makes an

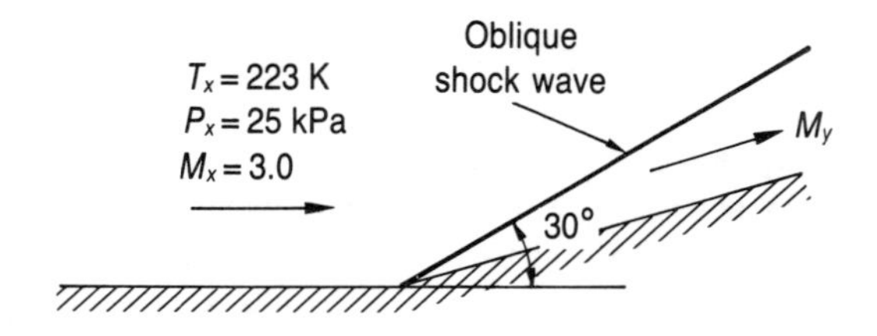

Figure 8.8 Sketch for Example 8.1

angle of 30° with the flow direction, is attached to the corner under the given conditions. If the pressure and temperature of the uniform flow are 25 kPa and −50 °C, respectively, determine the pressure and temperature behind the wave, the downstream Mach number and the deflection angle.

Solution
The normal component of the upstream Mach number is

$$M_{xn} = M_x \sin\theta = (3.0)(\sin 30^\circ) = 1.50$$

Then the pressure, temperature and normal component of the Mach number after the oblique shock wave corresponding to $M_{xn} = 1.50$ may be evaluated from Appendix D as

$$p_y/p_x = 2.458 \quad \text{or} \quad p_y = (2.458)(25\ \text{kPa}) = \mathbf{61.45\ kPa}$$

$$T_y/T_x = 1.320 \quad \text{or} \quad T_y = (1.320)(223\ \text{K}) = \mathbf{294.4\ K}$$

and

$$M_{yn} = 0.7011$$

For an adiabatic process, $T_{0x} = T_{0y}$; then from Appendix C, corresponding to $M_x = 3.0$,

$$T_x/T_{0x} = 0.3571 \quad \text{or} \quad T_{0x} = T_{0y} = 223\ \text{K}/0.3571 = 624.5\ \text{K}$$

Now, from Appendix C, corresponding to $T_y/T_{0y} = 294.4\ \text{K}/624.5\ \text{K} = 0.4714$, the downstream Mach number is

$$M_y = \mathbf{2.37}$$

At this point, one should note that the Mach number after an oblique shock wave can be greater than unity without violating the second law of thermodynamics. However, the normal component of the Mach number after the oblique shock wave must still be less than unity. From Figure 8.8

$$\sin(\theta - \delta) = \frac{V_{yn}}{V_y} = \frac{M_{yn}}{M_y} = \frac{0.7011}{2.37} = 0.2958 \quad \text{or} \quad \theta - \delta = 17.21^\circ$$

that is

$$\delta = 30^\circ - 17.21^\circ = \mathbf{12.79^\circ}$$

Example 8.2
A uniform supersonic air flow travelling at a Mach number of 3.0 passes over a concave corner which deflects the flow by 20°, as shown in Figure 8.9. If the pressure and temperature of the uniform flow are 60 kPa and −10°C, respectively, determine the Mach number, pressure and temperature downstream of the oblique shock wave and also the deflection angle above which the oblique shock wave will become detached.

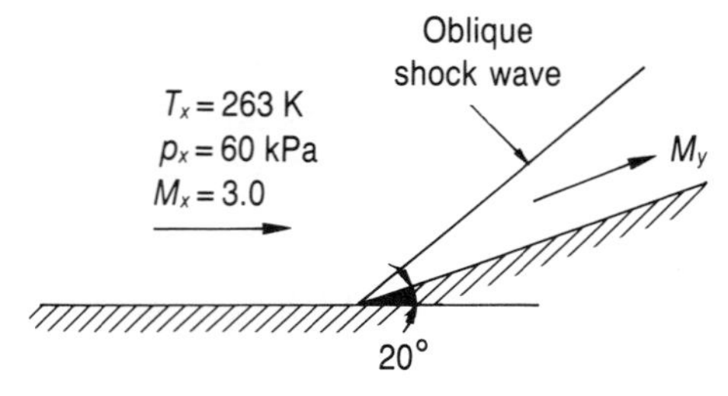

Figure 8.9 Sketch for Example 8.2

Solution

For an upstream Mach number of $M_x = 3.0$ and for a deflection angle of $\delta = 20\ ^\circ$, one may determine the shock-wave angle from Equation (8.8) as

$$\cot \delta = \left(\frac{k+1}{2}\,\frac{M_x^2}{M_x^2 \sin^2\theta - 1} - 1\right)\tan\theta$$

or

$$\cot 20^\circ = \left(\frac{1.4+1}{2}\,\frac{(3)^2}{(3)^2 \sin^2\theta - 1} - 1\right)\tan\theta$$

which may then be solved for the shock-wave angle as $\theta = 37.76\ ^\circ$ in the weak shock region, since for supersonic flow over a concave corner, the weak shock solution is found to occur. One should note that a rough estimate of the shock-wave angle may also be obtained from Figure 8.5. Now, the normal component of the upstream Mach number is

$$M_{xn} = M_x \sin\theta = (3.0)(\sin 37.76^\circ) = 1.837$$

Now, one may use Appendix D with $M_{xn} = 1.84$ to determine

$$M_{yn} = 0.6078 \quad \text{or} \quad M_y = \frac{M_{yn}}{\sin(\theta - \delta)} = \frac{0.6078}{\sin(37.76^\circ - 20^\circ)} = \mathbf{1.993}$$

$$P_y/P_x = 3.783 \quad \text{or} \quad p_y = (3.783)(60\text{ kPa}) = \mathbf{227.0\ kPa}$$

$$T_y/T_x = 1.562 \quad \text{or} \quad T_y = (1.562)(263\text{ K}) = \mathbf{410.8\ K}$$

Finally, the maximum deflection angle above which the oblique shock wave becomes detached may be obtained from Figure 8.5 as

$$\delta_{max} = \mathbf{34^\circ}$$

8.2.3 Reflection and interaction of oblique shock waves

When a supersonic flow passes over a concave corner, an oblique shock wave is generated at the corner. This shock wave deflects the flow towards itself. In this section, the rules for the reflection and interaction of oblique shock waves are discussed.

(a) Reflection of an oblique shock wave from a wall

When a two-dimensional, weak oblique shock wave impinges on a plane wall, the flow after the incident wave is deflected towards the shock wave as shown in Figure 8.10. Therefore, a reflected oblique shock wave must be present to deflect the flow back through the same angle and restore the flow in a direction parallel to the wall. Since M_{iy} is smaller than M_{ix}, then the reflected shock wave is weaker than the incident shock wave.

When the reflected shock wave impinges on the lower wall, it will again be reflected. The reflections can continue as long as the upstream Mach number and the required deflection angle permit an attached oblique shock wave. Such reflections are known as **regular reflections.**

If the Mach number after the incident shock wave is small, then the required turning angle will be greater than the maximum allowable one so that a simple shock reflection is impossible. In this case, a **Mach reflection** occurs, as shown in Figure 8.11. A curved, strong oblique shock wave stands between points O and W, and is known as the **Mach shock wave**. At point W, the shock wave must be perpendicular to the wall in order to satisfy the boundary condition that there cannot be any flow across a solid boundary. The flow downstream of the strong oblique shock wave OW is subsonic, while it is supersonic downstream of the reflected weak oblique shock wave, OR. A **slip line**, OS, which is a line across which some of the flow properties are discontinuous, is then formed. Actually, the analysis of Mach reflection is quite difficult.

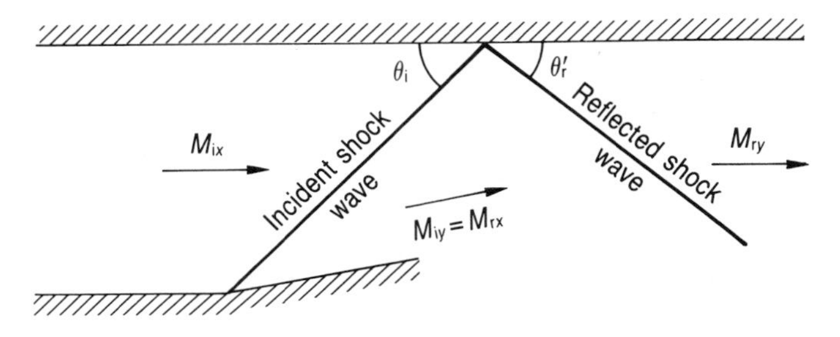

Figure 8.10 Reflection of an oblique shock wave

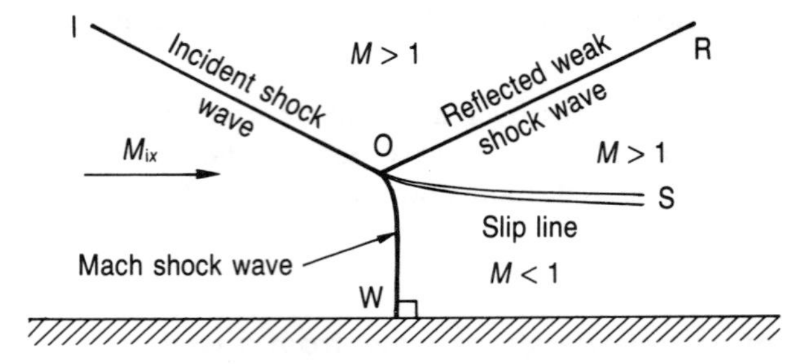

Figure 8.11 A Mach reflection

Example 8.3
Air at a Mach number of 3.0 flows through an oblique shock wave which is inclined 30° to the flow direction, as shown in Figure 8.12. Determine

a the Mach number after the incident shock wave,
b the Mach number after the reflected shock wave, and
c the angle at which the reflected shock wave is inclined to the flow direction.

Solution
(a) The deflection angle corresponding to $M_{ix} = 3.0$ and $\theta_i = 30°$ may be determined from Equation (8.8) as

$$\delta_i = \cot^{-1}\left[\left(\frac{k+1}{2}\,\frac{M_{ix}^2}{M_{ix}^2 \sin^2\theta_i - 1} - 1\right)\tan\,\theta_i\right]$$

$$= \cot^{-1}\left[\left(\frac{1.4+1}{2}\,\frac{(3)^2}{(3)^2(\sin\,30°)^2 - 1} - 1\right)\tan\,30°\right] = 12.77°$$

This corresponds to the angle through which the flow is turned after the incident shock wave and also the angle through which the flow is turned back after the

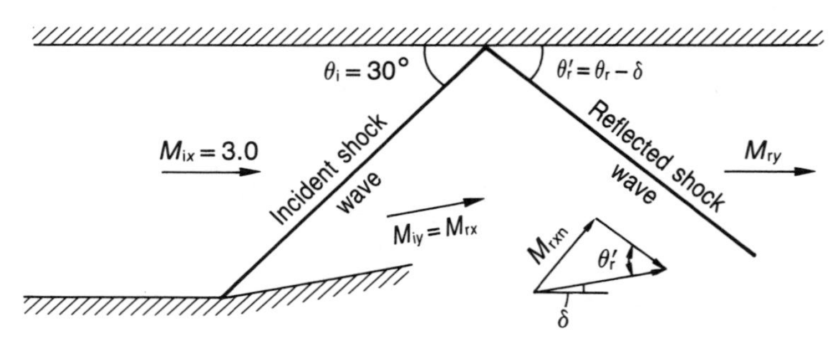

Figure 8.12 Sketch for Example 8.3

reflected wave. Now, from Appendix D, corresponding to $M_{ixn} = M_{ix} \sin\theta_i = (3)(\sin 30^\circ) = 1.5$, $M_{iyn} = 0.7011$. Hence the Mach number after the incident shock wave is

$$M_{iy} = \frac{M_{iyn}}{\sin(\theta_i - \delta_i)} = \frac{0.7011}{\sin(30^\circ - 12.77^\circ)} = \mathbf{2.367}$$

(b) Next, one may determine the wave angle after the reflected shock wave corresponding to $M_{rx} = M_{iy} = 2.367$ and $\delta_r = \delta_i = 12.77^\circ$ from Equation (8.8) as

$$\cot \delta_r = \left(\frac{k+1}{2} \frac{M_{rx}^2}{M_{rx}^2 \sin^2\theta_r - 1} - 1\right) \tan \theta_r$$

or

$$\cot 12.77^\circ = \left\{\frac{1.4+1}{2} \frac{(2.367)^2}{(2.367)^2 \sin^2\theta_r - 1} - 1\right\} \tan \theta_r$$

which may then be solved for the shock-wave angle by trial and error as $\theta_r = 36.24^\circ$ in the weak shock region, since for supersonic flow over a concave corner, the weak shock solution is found to occur. Now, from Appendix D, corresponding to $M_{rxn} = M_{rx} \sin\theta_r = (2.367)(\sin 36.24^\circ) = 1.399$,

$$M_{ryn} = 0.7397 \quad \text{or} \quad M_{ry} = \frac{M_{ryn}}{\sin(\theta_r - \delta)} = \frac{0.7397}{\sin(36.24^\circ - 12.77^\circ)} = \mathbf{1.857}$$

(c) As long as the shock-wave angle $\theta_r = 36.24^\circ$, is the angle between the flow direction in the intermediate region and the reflected shock wave, then from geometric considerations

$$\theta_r' = \theta_r - \delta = 36.24^\circ - 12.77^\circ = \mathbf{23.47^\circ}$$

(b) Reflection of an oblique shock wave from a free pressure boundary

Consider a jet of gas with atmospheric pressure p_{atm} at the boundaries, as shown in Figure 8.13. The incident oblique shock wave deflects the flow by an angle of δ towards the boundary and at the same time raises the pressure of the gas flowing through it. As a result, the pressure after the incident shock wave is greater than the ambient pressure, that is $p_{iy} > p_{ix} = p_{atm}$. However, the physical boundary condition requires that the pressure at point O must be atmospheric. Therefore, an oblique shock wave must reflect as a set of expansion waves in order to decrease the pressure back to atmospheric pressure. (Expansion waves are discussed in Section 8.3.) The expansion waves deflect the flow further towards the boundary.

(c) Neutralization of oblique shock waves

In Figure 8.14, the incident oblique shock wave deflects the flow towards itself by an angle of δ and raises the pressure of the gas. The physical boundary condition after the incident shock wave requires that the flow must be parallel to the wall. Now, if the wall after point O is turned by the same angle, δ, in the same direction, then the

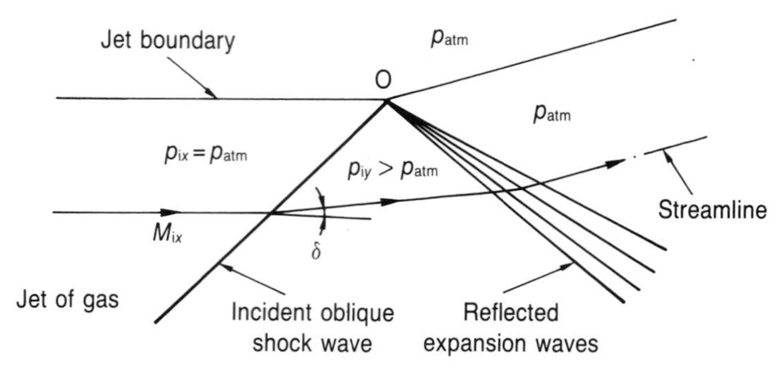

Figure 8.13 Reflection of an oblique shock wave from a free pressure boundary

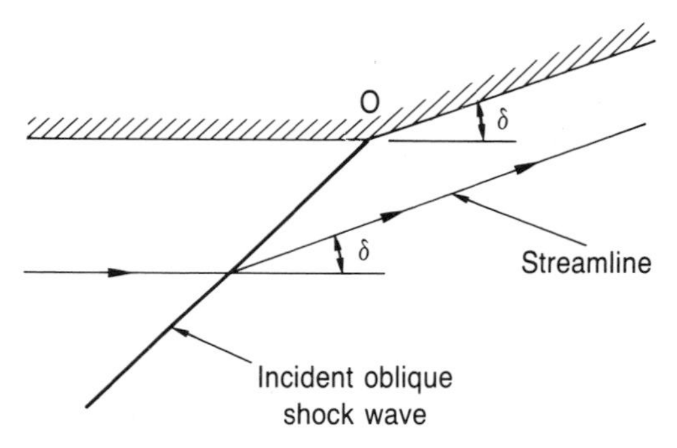

Figure 8.14 Neutralization of an oblique shock wave

wall boundary condition is satisfied and there is no reflected oblique shock wave. Therefore, by turning the wall through an appropriate angle, the reflected oblique shock wave may be **cancelled** or **neutralized**.

(d) Intersection of oblique shock waves

In Figure 8.15, two oblique shock waves OA and OB intersect and turn the uniform upstream flow of region 1 inwards. When the strengths of the two shock waves are unequal, the fluid properties in regions 2 and 3 are also different. The physical boundary condition at point O requires that the static pressures in regions 4 and 5 must be the same. However, other properties may be different. Therefore, a streamline passing through point O divides regions 4 and 5 and is known as the **slip line**. The streamlines in regions 4 and 5 are then parallel to the slip line. The transmitted oblique shock waves adjust themselves such that the pressures in regions 4 and 5 are identical. The intersection of two oblique shock waves is known as the **regular intersection**.

When there is no solution for the transmitted oblique shock waves satisfying all the flow conditions, then a more complex flow pattern, known as the **Mach intersection**,

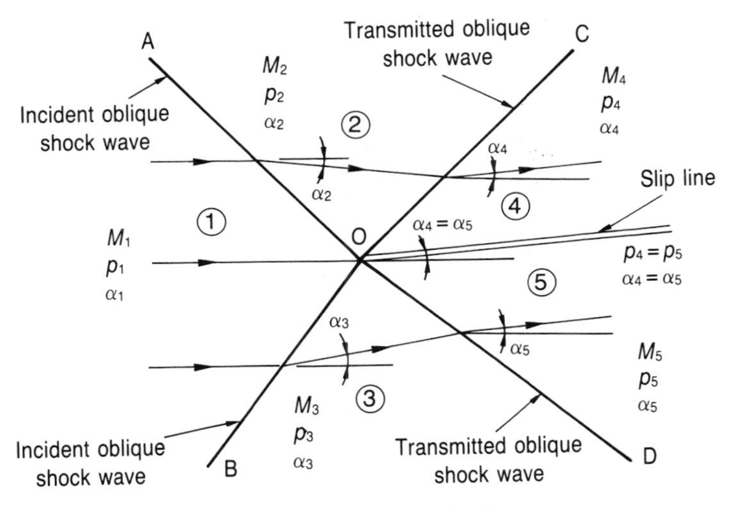

Figure 8.15 Regular intersection of two oblique shock waves

occurs as shown in Figure 8.16. This flow pattern involves a normal shock wave called a **Mach shock wave** and the flow after this shock wave is subsonic. However, the flows after the reflected oblique shock waves are supersonic and there are two slip lines.

Example 8.4

Air at a Mach number of 3.0 and a pressure of 200 kPa flows between two horizontal parallel plates, as shown in Figure 8.17. At a certain cross-section, the upper wall turns downwards through an angle of 10°, while the lower one turns upwards through an angle of 20°. At these concave corners, two oblique shock waves are generated. These

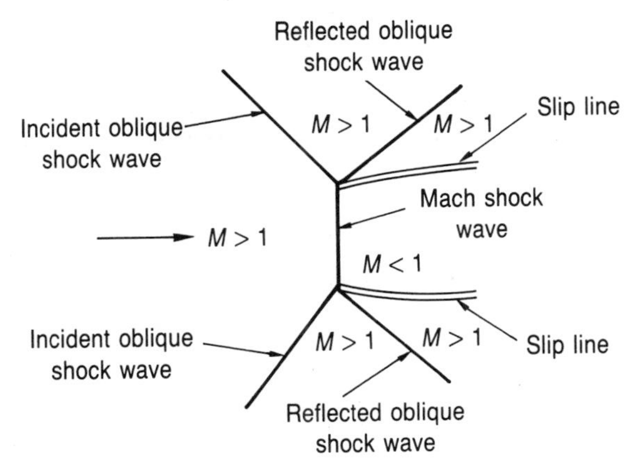

Figure 8.16 Mach intersection of two oblique shock waves

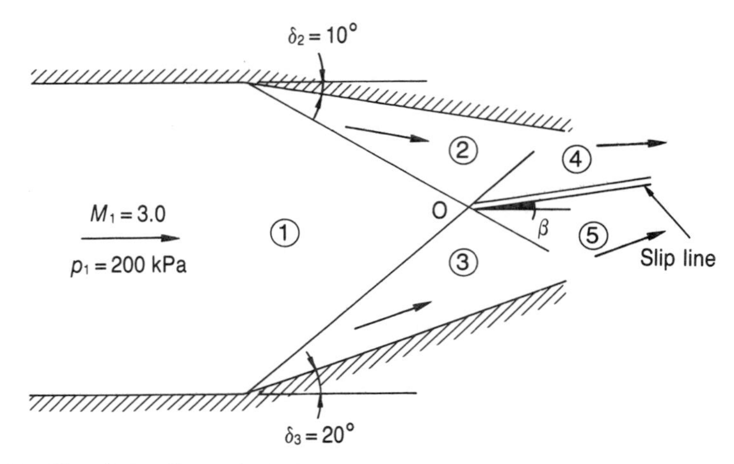

Figure 8.17 Sketch for Example 8.4

two shock waves then intersect at point O. Determine the pressure and Mach number in regions 4 and 5, and also find the net deflection of the flow.

Solution

The shock-wave angle of the upper oblique shock wave corresponding to $M_1 = 3.0$ and $\delta_2 = 10°$ may be evaluated from Equation (8.8) as

$$\cot \delta_2 = \left(\frac{k+1}{2} \frac{M_1^2}{M_1^2 \sin^2 \theta_{12} - 1} - 1\right) \tan \theta_{12}$$

or

$$\cot 10° = \left(\frac{1.4+1}{2} \frac{(3)^2}{(3)^2 \sin^2 \theta_{12} - 1} - 1\right) \tan \theta_{12}$$

which may then be solved for the weak shock-wave angle by trial and error as $\theta_{12} = 27.39°$. Then the Mach number of the flow normal to the upper oblique shock wave is

$$M_{1n} = M_1 \sin \theta_{12} = (3)(\sin 27.39°) = 1.380$$

Now, the normal component of the Mach number and the pressure in region 2 may be determined from Appendix D with $M_{1n} = 1.38$ as

$$M_{2n} = 0.7483$$

and

$$p_2/p_1 = 2.055 \quad \text{or} \quad p_2 = (2.055)(200 \text{ kPa}) = 411.0 \text{ kPa}$$

respectively. Finally, the Mach number at section 2 is

$$M_2 = \frac{M_{2n}}{\sin(\theta_{12} - \delta_2)} = \frac{0.7483}{\sin(27.39^\circ - 10^\circ)} = 2.504$$

Similarly, the shock-wave angle of the lower oblique shock wave corresponding to $M_1 = 3.0$ and $\delta_3 = 20^\circ$ may be evaluated from Equation (8.8) as

$$\cot \delta_3 = \left(\frac{k+1}{2} \frac{M_1^2}{M_1^2 \sin^2 \theta_{13} - 1} - 1\right) \tan \theta_{13}$$

or

$$\cot 20^\circ = \left(\frac{1.4+1}{2} \cdot \frac{(3)^2}{(3)^2 \sin^2 \theta_{13} - 1} - 1\right) \tan \theta_{13}$$

which may then be solved for the weak shock-wave angle by trial and error as $\theta_{13} = 37.76^\circ$. Then the Mach number of the flow normal to the lower shock wave is

$$M_{1n} = M_1 \sin \theta_{13} = (3)(\sin 37.76^\circ) = 1.837$$

Now, the normal component of the Mach number and the pressure in region 3 may be determined from Appendix D with $M_{1n} = 1.84$ as

$$M_{3n} = 0.6078$$

and

$$p_3/p_1 = 3.783 \quad \text{or} \quad p_3 = (3.783)(200 \text{ kPa}) = 756.6 \text{ kPa}$$

respectively. Finally, the Mach number at section 3 is

$$M_3 = \frac{M_{3n}}{\sin(\theta_{13} - \delta_3)} = \frac{0.6078}{\sin(37.76^\circ - 20^\circ)} = 1.993$$

In order to continue the solution of the problem, one condition in either region 4 or 5 must be known. The most convenient parameter for iteration is the net flow deflection angle, β. One may assume that the net flow deflection angle is in the upward direction as long as the deflection angle of the lower wall is greater than the upper one. Hence, an iterative solution will be necessary and to start one may assume β to be 4°. From the geometry

$$\delta_4 = 10^\circ + 4^\circ = 14^\circ$$

The shock-wave angle of the upper reflected oblique shock wave corresponding to $M_2 = 2.504$ and $\delta_4 = 14°$ may be evaluated from Equation (8.8) as

$$\cot \delta_4 = \left(\frac{k+1}{2} \frac{M_2^2}{M_2^2 \sin^2 \theta_{24} - 1} - 1 \right) \tan \theta_{24}$$

or

$$\cot 14° = \left(\frac{1.4+1}{2} \frac{(2.504)^2}{(2.504)^2 \sin^2 \theta_{24} - 1} - 1 \right) \tan \theta_{24}$$

which may then be solved for the weak shock-wave angle by trial and error as $\theta_{24} = 35.82°$. Then the Mach number of the flow normal to the upper reflected oblique shock wave is

$$M_{2n} = M_2 \sin \theta_{24} = (2.504)(\sin 35.82°) = 1.465$$

Now, the pressure in region 4 may be determined from Appendix D with $M_{2n} = 1.47$ as

$$p_4/p_2 = 2.354 \quad \text{or} \quad p_4 = (2.354)(411.0 \text{ kPa}) = 967.5 \text{ kPa}$$

From the geometry in Figure 8.17, it is possible to note that

$$\delta_5 = 20° - 4° = 16°$$

Then the shock-wave angle of the lower reflected oblique shock wave corresponding to $M_3 = 1.993$ and $\delta_5 = 16°$ may be evaluated from Equation (8.8) as

$$\cot \delta_5 = \left(\frac{k+1}{2} \frac{M_3^2}{\text{M}_3^2 \sin^2 \theta_{35} - 1} - 1 \right) \tan \theta_{35}$$

or

$$\cot 16° = \left(\frac{1.4+1}{2} \frac{(1.993)^2}{(1.993)^2 \sin^2 \theta_{35} - 1} - 1 \right) \tan \theta_{35}$$

which may then be solved for the weak shock-wave angle by trial and error as $\theta_{35} = 46.91°$. Then the Mach number of the flow normal to the lower reflected shock wave is

$$M_{3n} = M_3 \sin \theta_{35} = (1.993)(\sin 46.91°) = 1.456$$

Now, the pressure in region 5 may be determined from Appendix D with $M_{3n} = 1.46$ as

$$p_5/p_3 = 2.320 \quad \text{or} \quad p_5 = (2.320)(756.6 \text{ kPa}) = 1755 \text{ kPa}$$

At this point, one should note that $p_4 = 967.5$ kPa $\neq p_5 = 1755$ kPa. As a result, this trial is unacceptable. Since $p_5 > p_4$, the flow must turn more towards the upper wall in order to increase the pressure in region 4 and decrease it in region 5. Hence, as the next trial, one may choose a larger net flow deflection angle

Table 8.1 Iterations for Example 8.4

Iteration	1	2	3
β (°)	4	8	9.8
δ_4 (°)	14	18	19.8
θ_{24} (°)	35.82	40.34	42.58
M_{2n}	1.465	1.621	1.694
p_2/p_4	2.354	2.895	3.165
p_4 (kPa)	967.5	1190	1301
δ_5 (°)	16	12	10.2
θ_{35} (°)	46.91	41.73	39.68
M_{3n}	1.456	1.327	1.273
p_3/p_5	2.320	1.897	1.715
p_5 (kPa)	1755	1435	1298

such as $\beta = 8°$. The rest of the iterations are summarized in Table 8.1. After the third iteration

$$\beta = \mathbf{9.8°} \quad p_4 = \mathbf{1301\ kPa} \quad \text{and} \quad p_5 = \mathbf{1298\ kPa}$$

which are quite acceptable. Finally, the Mach numbers at sections 4 and 5 normal to the reflected shock waves corresponding to Mach numbers of $M_{2n} = 1.694$ and $M_{3n} = 1.273$ are

$$M_{4n} = 0.6431 \quad \text{and} \quad M_{5n} = 0.8016$$

from Appendix D, respectively. Then the Mach numbers at sections 4 and 5 are

$$M_4 = \frac{M_{4n}}{\sin(\theta_{24} - \delta_4)} = \frac{0.6431}{\sin(42.58° - 19.8°)} = \mathbf{1.661}$$

and

$$M_5 = \frac{M_{5n}}{\sin(\theta_{35} - \delta_5)} = \frac{0.8016}{\sin(39.68° - 10.2°)} = \mathbf{1.629}$$

respectively.

8.3 PRANDTL–MEYER FLOW

For a supersonic expansion flow around a convex corner, the flow properties gradually adjust to the presence of this corner. The Prandtl–Meyer fan consists of a series of

Mach waves which are centred at the convex corner, as shown in Figure 8.2. The head of this expansion fan is inclined to the approaching flow at an angle of μ_1, while the tail is inclined to the downstream flow at an angle of μ_2. The normal velocity component at each point is equal to the local speed of sound. Flow properties along each Mach wave are uniform so that they are only a function of angular position.

8.3.1 Governing equation for the Prandtl–Meyer flow

First, consider a single Mach wave, which expands a supersonic flow through an infinitesimal Prandtl–Meyer angle of $d\nu$, as shown in Figure 8.18. As long as there is no pressure gradient in the tangential direction, the tangential velocity component across the Mach wave is constant. Equating the tangential velocity components upstream and downstream of the Mach wave, then

$$V_t = V \cos \mu = (V + dV)\cos(\mu + d\nu)$$

or

$$V \cos \mu = (V + dV)(\cos \mu \cos d\nu - \sin \mu \sin d\nu)$$

Since $d\nu$ is infinitesimal, $\cos d\nu \simeq 1$ and $\sin d\nu \simeq d\nu$. Hence,

$$V \cos \mu = (V + dV)(\cos \mu - \sin \mu \, d\nu)$$

Now, expansion yields

$$V \cos \mu = V \cos \mu + \cos \mu \, dV - V \sin \mu \, d\nu - \sin \mu \, dV \, d\nu$$

Neglecting the product of the two differentials and simplifying

$$dV/V = \tan \mu \, d\nu$$

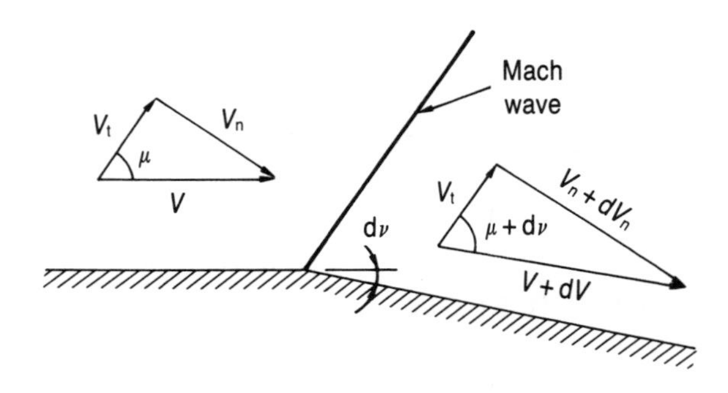

Figure 8.18 Infinitesimal expansion of a supersonic stream

Using the definition of the Mach angle which is given by Equation (3.10) as $\sin \mu = 1/M$,

$$\tan \mu = \frac{\sin \mu}{\cos \mu} = \frac{\sin \mu}{\sqrt{(1 - \sin^2 \mu)}} = \frac{1}{\sqrt{(M^2 - 1)}}$$

so that

$$\frac{dV}{V} = \frac{1}{\sqrt{(M^2 - 1)}} d\nu \tag{8.17}$$

At this point, if one uses the definition of the Mach number as $V = M\sqrt{(kRT)}$ and differentiates logarithmically

$$\frac{dV}{V} = \frac{dM}{M} + \frac{1}{2}\frac{dT}{T} \tag{8.18}$$

For adiabatic flow, there is no change in the stagnation temperature. Then, from Equation (4.24),

$$T_0 = T\left(1 + \frac{k-1}{2} M^2\right) = \text{constant}$$

Taking logarithms of both sides and differentiating

$$\frac{dT}{T} + \frac{(k-1)M\, dM}{1 + \frac{1}{2}(k-1)M^2} = 0 \tag{8.19}$$

Now, combining Equations (8.17), (8.18) and (8.19), it is possible to obtain

$$d\nu = \frac{\sqrt{(M^2 - 1)}}{1 + \frac{1}{2}(k-1)M^2} \frac{dM}{M} \tag{8.20}$$

For a finite turning from ν_1 to ν_2 as the Mach number changes from M_1 to M_2, the above expression may be integrated to yield

$$\Delta\nu = \nu_2 - \nu_1 = \int_{M_1}^{M_2} \frac{\sqrt{(M^2 - 1)}}{M} \frac{dM}{1 + \frac{1}{2}(k-1)M^2}$$

$$= \left[\sqrt{\left(\frac{k+1}{k-1}\right)} \tan^{-1} \sqrt{\left(\frac{k-1}{k+1}(M^2 - 1)\right)} - \tan^{-1}\sqrt{(M^2 - 1)}\right]_{M_1}^{M_2} \tag{8.21}$$

The results may be tabulated by defining a reference state. Choosing state 1 as the reference state with $M_1 = 1.0$ and $\nu_1 = 0$ and denoting state 2 without a subscript, Equation (8.21) becomes

$$\nu = \sqrt{\left(\frac{k+1}{k-1}\right)} \tan^{-1} \sqrt{\left(\frac{k-1}{k+1}(M^2 - 1)\right)} - \tan^{-1}\sqrt{(M^2 - 1)} \tag{8.22}$$

Hence, ν represents the angle through which a stream at a Mach number of unity must be expanded to reach a supersonic Mach number of M. The Prandtl–Meyer angle and the Mach angle are presented in tabular form in Appendix H for $k = 1.4$ with the Mach number being the independent variable.

The angle through which a flow must be turned during the expansion from a Mach number of M_1 to M_2 may easily be obtained by subtracting ν_1 from ν_2, as shown in Figure 8.19.

The variation of the thermodynamic properties through an expansion fan may be found by using the isentropic tables.

Example 8.5

A uniform airstream at a Mach number of 2.0 expands through a 10° convex corner, as shown in Figure 8.20. If the pressure and temperature of this airstream are 25 kPa and −50 °C, respectively, determine the Mach number, pressure and temperature after the expansion and also determine the fan angle.

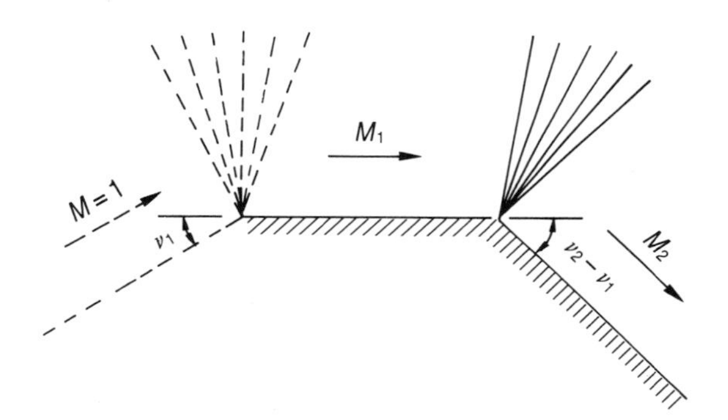

Figure 8.19 Determination of the Prandtl–Meyer angle

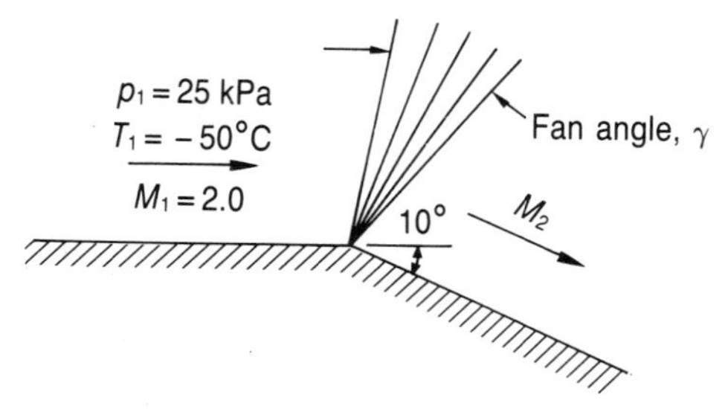

Figure 8.20 Sketch for Example 8.5

Solution

The stagnation conditions before the expansion may be determined by using Appendix C with $M_1 = 2.0$ as

$$p_1/p_{01} = 0.1278 \quad \text{or} \quad p_{01} = 25\ \text{kPa}/0.1278 = 195.6\ \text{kPa}$$

$$T_1/T_{01} = 0.5556 \quad \text{or} \quad T_{01} = 223\ \text{K}/0.5556 = 401.4\ \text{K}$$

Now, the Mach angle and the Prandtl–Meyer angle corresponding to $M_1 = 2.0$ may be obtained from Appendix H as

$$\mu_1 = 30^\circ \quad \text{and} \quad \nu_1 = 26.38^\circ$$

Then the Prandtl–Meyer angle after the expansion is

$$\nu_2 = \nu_1 + 10^\circ = 26.38^\circ + 10^\circ = 36.38^\circ$$

Again, from Appendix H, corresponding to $\nu_2 = 36.38^\circ$

$$M_2 = \mathbf{2.38} \quad \text{and} \quad \mu_2 = 24.85^\circ$$

The stagnation temperature and pressure across an expansion fan are constant so that

$$T_{02} = T_{01} = 401.4\ \text{K}$$

$$p_{02} = p_{01} = 195.6\ \text{kPa}$$

Now, using Appendix C with $M_2 = 2.38$,

$$p_2/p_{02} = 0.070\,57 \quad \text{or} \quad p_2 = (0.070\,57)(195.6\ \text{kPa}) = \mathbf{13.80\ kPa}$$

$$T_2/T_{02} = 0.4688 \quad \text{or} \quad T_2 = (0.4688)(401.4\ \text{K}) = \mathbf{188.2\ K}$$

Finally, the fan angle, γ, may be evaluated as

$$\gamma = \mu_1 - [\mu_2 - (\nu_2 - \nu_1)] = 30^\circ - [24.85^\circ - 10^\circ] = \mathbf{15.15^\circ}$$

In supersonic flow, Mach waves are generated at a smooth concave corner, as shown in Figure 8.1. These Mach waves then join further out in the stream to form an oblique shock wave. In the region between the wall and the oblique shock wave, the flow is isentropic and possesses the same characteristics as the isentropic expansion fan. Therefore, the equations derived in this section are also applicable to the isentropic flow region at a concave corner even though a compression takes place at that corner. The turning angle, $\Delta\nu$, will certainly be negative, since the flow decelerates. The presence of Prandtl–Meyer flow at a concave corner depends on the curvature of the wall. For a sharp turn, the Prandtl–Meyer flow near the corner is negligible, while a larger region has the characteristics of Prandtl–Meyer flow for a gradual turn with a large radius of curvature.

8.3.2 Maximum turning angle for the Prandtl–Meyer expansion

When a sonic flow is expanded through a convex corner to a Mach number of infinity, the pressure after the expansion fan reaches zero, as shown in Figure 8.21. In this case, the turning angle, which is given by Equation (8.22), approaches a finite value of 130.4°. The maximum turning angle is quite important for determining the shape of the exhaust gases in an underexpanded nozzle discharging into a vacuum or space. The geometrical design of a spacecraft may require modification in order to prevent possible damage from the hot exhaust gases.

8.3.3 Reflection of the Prandtl–Meyer expansion flow

When a supersonic flow passes over a convex corner, an expansion fan is generated at the corner. This expansion fan deflects the flow away from itself. In this section, the rules for the reflection of expansion waves are discussed.

(a) Reflection of expansion waves from a wall

When the expansion waves impinge on a plane wall, the flow after the incident expansion waves is deflected away from the wall, as shown in Figure 8.22. Therefore, a set of reflected expansion waves must be present to deflect the flow back through the same angle and restore the flow in a direction parallel to the wall. Therefore, each Mach wave of the initial expansion fan reflects as an expansion Mach wave to maintain the boundary condition at the wall. The interaction of the initial and reflected Mach waves makes an exact analysis of the flow within the interaction region and after the reflected expansion waves quite difficult.

(b) Reflection of expansion waves from a free pressure boundary

Consider a jet of gas with an atmospheric pressure of p_{atm} at the boundaries, as shown in Figure 8.23. The incident expansion waves deflect the flow away from the boundary and at the same time decrease the pressure of the gas flowing through them. As a

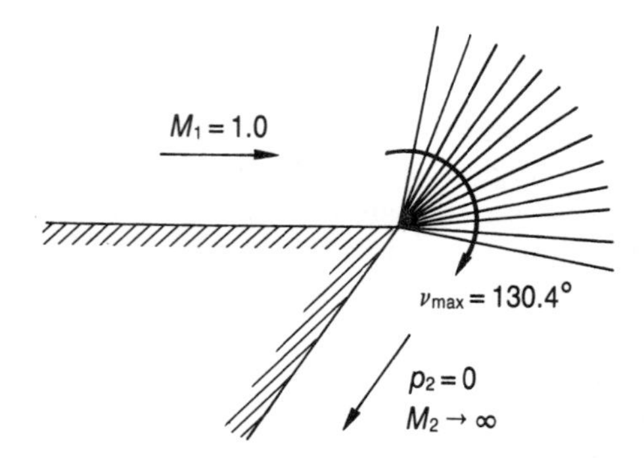

Figure 8.21 Maximum turning angle for the Prandtl–Meyer expansion

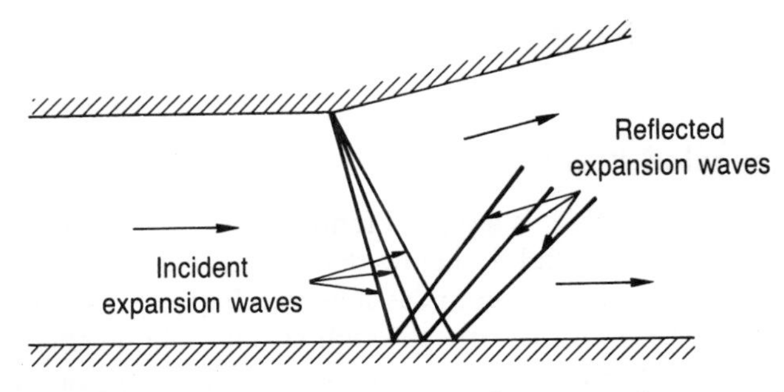

Figure 8.22 Reflection of Prandtl–Meyer expansion waves from a wall

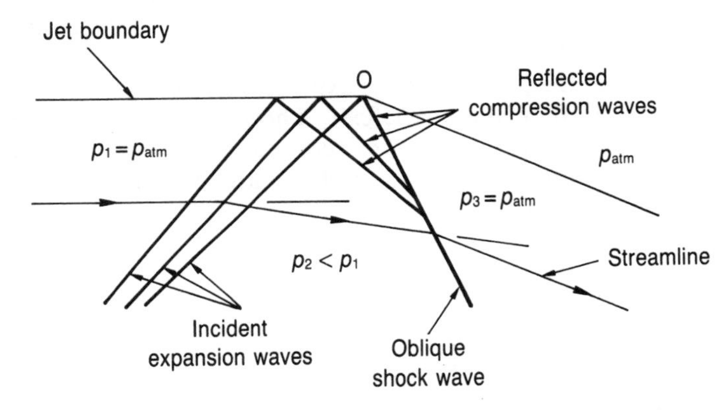

Figure 8.23 Reflection of expansion waves from a free pressure boundary

result, the pressure after the incident expansion waves is smaller than the ambient pressure, that is $p_2 < p_1 = p_{atm}$. However, the physical boundary condition requires that the pressure at point O must be atmospheric. Therefore, the expansion waves must reflect as a set of compression waves in order to increase the pressure back to atmospheric pressure. As long as the compression waves converge, then they usually unite and form an oblique shock wave. This oblique shock wave deflects the flow further away from the boundary.

8.4 OVEREXPANSION AND UNDEREXPANSION FLOW REGIMES IN CONVERGING–DIVERGING NOZZLES

Overexpansion and underexpansion flow regimes in converging–diverging nozzles were previously discussed in Section 4.8.2. The flow in both of these regimes is isentropic within the nozzle and identical to the design condition. However, the flow outside the nozzle is not isentropic and involves oblique shock waves and Prandtl–Meyer flow.

8.4.1 Overexpansion flow regime

Flow pattern 6 of Figure 4.27(b) is a typical representation of this flow regime. When the back pressure is reduced slightly below that of the exit shock condition, the jet detaches itself from the divergent section, as shown in Figure 8.24. The overexpanded gas is compressed to the back pressure through strong oblique shock waves. When the back pressure is reduced further, the jet separation disappears and the flow is compressed by weak oblique shock waves attached to the exit plane, as shown in Figure 8.25. The back pressure at which the jet separation diminishes is not known exactly. Actually, the analysis of flow regimes with jet separation is difficult and beyond the scope of this text. Therefore, in this section only flows without jet separation are considered.

In Figure 8.25, since the pressure at the exit plane is less than the back pressure, oblique shock waves form at the nozzle exit. The flow at the exit of the nozzle is assumed to be uniform and parallel. Since there can be no flow across the centreline of the jet, the velocity component normal to the centreline must be zero. Hence, the

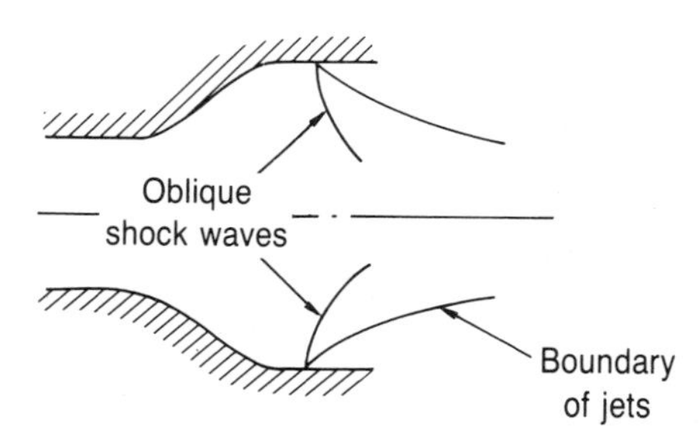

Figure 8.24 Exit flow in an overexpanded supersonic nozzle with jet separation

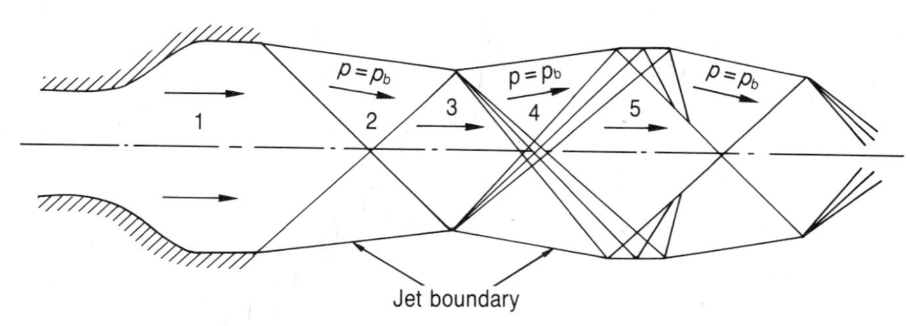

Figure 8.25 Exit flow in an overexpanded supersonic nozzle

boundary condition along the centreline is exactly the same as that on a plane wall in inviscid flow. The pressure in region 1 is increased to the specified back pressure in region 2 by means of an oblique shock wave. However, the flow is now turned towards the centreline. To turn the flow back towards the horizontal, the oblique shock wave reflects as in the case of reflection from a plane wall. In region 3, the flow is now parallel to the centreline but the pressure is greater than the back pressure owing to the reflected oblique shock wave. For this reason, the second oblique shock wave reflects from the external air as a centred expansion fan. Although the pressure is equal to the back pressure in region 4, the flow is now turned away from the centreline. To turn the flow back towards the horizontal, the expansion waves reflect from the centreline of the jet as expansion waves. Therefore, the flow is parallel to the centreline in region 5, but the pressure is now less than the back pressure. At this point, one should note that the process goes through a complete cycle and continues to repeat itself. At the exit of the supersonic rocket nozzles, the flow pattern appears as a series of diamonds. Although this pattern should extend to infinity, mixing of the jet with the ambient air along the jet boundaries causes the wave pattern to die out.

Example 8.6
An overexpanded converging–diverging nozzle designed for an exit Mach number of 2.2 is shown in Figure 8.26. The pressure and temperature of the air at the exit plane of the nozzle are 20 kPa and 150 K, respectively. The nozzle discharges to a back pressure of 40 kPa. Determine the Mach number, pressure, temperature and flow direction in regions 2, 3 and 4.

Solution
The stagnation pressure and temperature at section 1 for $M_1 = 2.2$ may be obtained from Appendix C as

$$p_1/p_{01} = 0.093\,52 \quad \text{or} \quad p_{01} = 20\text{ kPa}/0.093\,52 = 213.9\text{ kPa}$$

$$T_1 T_{01} = 0.5081 \quad \text{or} \quad T_{01} = 150\text{ K}/0.5081 = 295.2\text{ K}$$

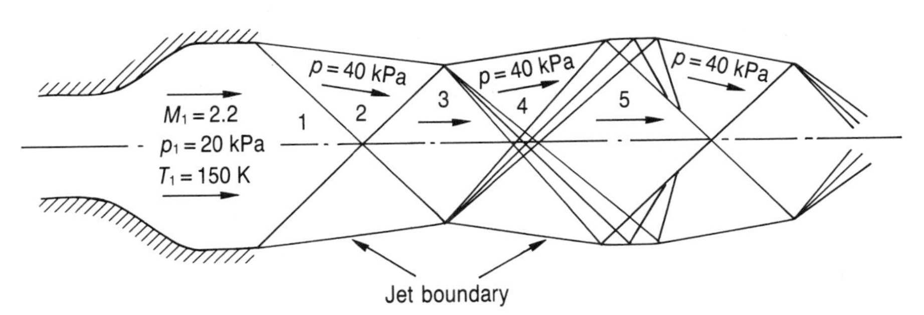

Figure 8.26 Sketch for Example 8.6

As long as the pressure in region 2 is equal to the back pressure, then

$$p_2 = p_b = \mathbf{40\ kPa}$$

Now, corresponding to the pressure ratio $p_2/p_1 = 40\ \text{kPa}/20\ \text{kPa} = 2$, one may obtain the upstream and downstream Mach numbers normal to the oblique shock wave and the stagnation pressure at section 2 from Appendix D as

$$M_{1n} = 1.36 \qquad M_{2n} = 0.7572$$

and

$$p_{02}/p_{01} = 0.9676 \quad \text{or} \quad p_{02} = (0.9676)(213.9\ \text{kPa}) = 207\ \text{kPa}$$

From the velocity triangle in region 1 of Figure 8.26

$$\sin\theta_{12} = \frac{M_{1n}}{M_1} = \frac{1.36}{2.20} = 0.6182 \quad \text{or} \quad \theta_{12} = 38.19^\circ$$

Now, the deflection angle corresponding to $M_1 = 2.20$ and $\theta_1 = 38.19^\circ$ may be obtained from Equation (8.8) as

$$\delta_2 = \cot^{-1}\left[\left(\frac{k+1}{2}\frac{M_1^2}{M_1^2\sin^2\theta_{12} - 1} - 1\right)\tan\theta_{12}\right]$$

$$= \cot^{-1}\left[\left(\frac{1.4+1}{2}\frac{(2.20)^2}{(2.20)^2(\sin 38.19^\circ)^2 - 1} - 1\right)\tan 38.19^\circ\right] = 12.30^\circ$$

The downstream Mach number may now be obtained as

$$M_2 = \frac{M_{2n}}{\sin(\theta_{12} - \delta_2)} = \frac{0.7572}{\sin(38.19^\circ - 12.30^\circ)} = \mathbf{1.734}$$

Across an oblique shock wave, the flow is adiabatic, so

$$T_{02} = T_{01} = 295.2\ \text{K}$$

The temperature at section 2 for $M_2 = 1.734$ is

$$T_2/T_{02} = 0.6256 \quad \text{or} \quad T_2 = (0.6256)(295.2\ \text{K}) = \mathbf{184.7\ K}$$

from Appendix C. The flow is deflected towards the centreline by an angle

$$\alpha_2 = \delta_2 = \mathbf{12.30^\circ}$$

The flow must now be turned back to the horizontal by $\delta_3 = \delta_2 = 12.30°$. In this case, one may determine the shock-wave angle of the reflected shock corresponding to $M_2 = 1.734$ and $\delta_3 = 12.30°$ from Equation (8.8) as

$$\cot \delta_3 = \left(\frac{k+1}{2} \frac{M_2^2}{M_2^2 \sin^2 \theta_{23} - 1} - 1\right) \tan \theta_{23}$$

or

$$\cot 12.30° = \left(\frac{1.4+1}{2} \frac{(1.734)^2}{(1.734)^2 \sin^2 \theta_{23} - 1} - 1\right) \tan \theta_{23}$$

which may then be solved for the weak shock-wave angle by trial and error as $\theta_{23} = 49.35°$. Now, one may obtain the normal component of the Mach number before the reflected normal shock wave as

$$M'_{2n} = M_2 \sin \theta_{23} = (1.734)(\sin 49.35°) = 1.316$$

so that from Appendix D

$$M_{3n} = 0.7760$$

$p_3/p_2 = 1.866$ or $p_3 = (40 \text{ kPa})(1.866) =$ **74.64 kPa**

$T_3/T_2 = 1.204$ or $T_3 = (184.7 \text{ K})(1.204) =$ **222.4 K**

$p_{03}/p_{02} = 0.9758$ or $p_{03} = (0.9758)(207 \text{ kPa}) = 202 \text{ kPa}$

Then the downstream Mach number is

$$M_3 = \frac{M_{3n}}{\sin(\theta_{23} - \delta_3)} = \frac{0.7760}{\sin(49.35° - 12.30°)} = 1.288$$

In region 3, the flow is parallel to the centreline, so

$$\alpha_3 = \mathbf{0°}$$

As long as the flow across an expansion fan is isentropic and that across the oblique shock waves is adiabatic, then

$$p_{04} = p_{03} = 202 \text{ kPa}$$

$$T_{04} = T_{03} = T_{02} = T_{01} = 295.2 \text{ K}$$

Also, the pressure in region 4 is equal to the back pressure so that

$$p_4 = p_b = \mathbf{40 \text{ kPa}}$$

Now, corresponding to a pressure ratio $p_4/p_{04} = 40 \text{ kPa}/202 \text{ kPa} = 0.1980$, one may obtain the Mach number and temperature after the expansion fan from Appendix C as

$$M_4 = \mathbf{1.72}$$

$$T_4/T_{04} = 0.6283 \quad \text{or} \quad T_4 = (0.6283)(295.2 \text{ K}) = \mathbf{185.5\ K}$$

respectively. Finally, the Prandtl–Meyer angles for Mach numbers $M_3 = 1.29$ and $M_4 = 1.72$ may be obtained from Appendix H as

$$\nu_3 = 5.898^\circ \quad \text{and} \quad \nu_4 = 18.40^\circ$$

respectively, so that the flow is turned away from the axis by an angle

$$\alpha_4 = \nu_4 - \nu_3 = 18.40^\circ - 5.898^\circ = \mathbf{12.50^\circ}$$

8.4.2 Underexpansion flow regime

Flow pattern 8 of Figure 4.27(b) is a typical representation of the underexpansion flow regime. Flow at the exit of an underexpanded nozzle is shown in Figure 8.27. As long as the exit pressure is greater than the back pressure, then a centred expansion fan forms at the nozzle exit plane to reduce the pressure. The flow at the exit of the nozzle is assumed to be uniform and parallel. The pressure is reduced to the back pressure in region 2 by means of an expansion fan. At the same time, the flow in region 2 is turned away from the jet centreline. To turn the flow back towards the horizontal direction, the expansion fan reflects from the jet centreline as another expansion fan reducing the pressure in region 3 below the back pressure. For this reason, the reflected system of expansion waves reflects from the external air as a system of compression waves. This system of compression waves will then join to form an oblique shock wave which again increases the pressure in region 4 to the back pressure. However, the flow is now turned towards the centreline. To align the flow with the horizontal direction, the oblique shock wave reflects as another oblique shock

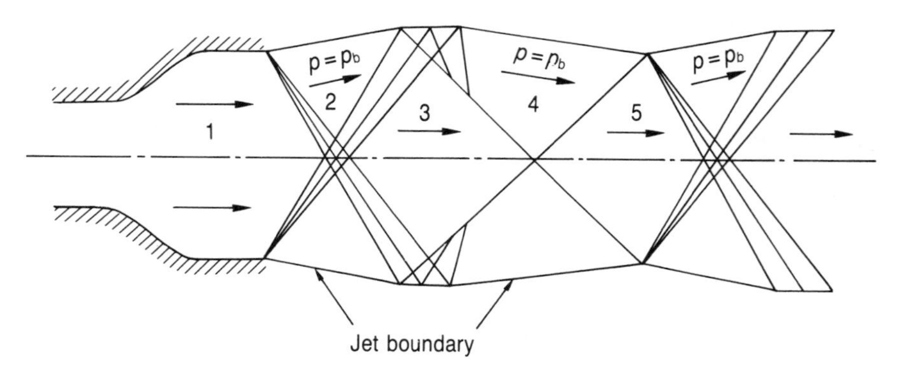

Figure 8.27 Exit flow in an underexpanded supersonic nozzle

wave. In this case, the pressure in region 5 is again higher than the back pressure. At this point, one should note that the process goes through a complete cycle and continues to repeat itself. At the exit of the supersonic rocket nozzles, the flow pattern appears as a series of diamonds. Although this pattern should extend to infinity, mixing of the jet with the ambient air along the jet boundaries causes the wave pattern to die out.

Example 8.7

An underexpanded converging–diverging nozzle designed for an exit Mach number of 2.0 is shown in Figure 8.28. The pressure and temperature of the air at the exit plane of the nozzle are 20 kPa and 200 K, respectively. If the back pressure is 10 kPa, determine the Mach number, temperature and direction of the flow downstream of the expansion fan.

Solution

The stagnation pressure and temperature before the expansion waves corresponding to $M_1 = 2.0$ may be obtained from Appendix C as

$$p_1/p_{01} = 0.1278 \quad \text{or} \quad p_{01} = 20\ \text{kPa}/0.1278 = 156.5\ \text{kPa}$$

$$T_1/T_{01} = 0.5556 \quad \text{or} \quad T_{01} = 200\ \text{K}/0.5556 = 360.0\ \text{K}$$

respectively. As long as the expansion process is isentropic, then

$$p_{02} = p_{01} = 156.5\ \text{kPa}$$

$$T_{02} = T_{01} = 360.0\ \text{K}$$

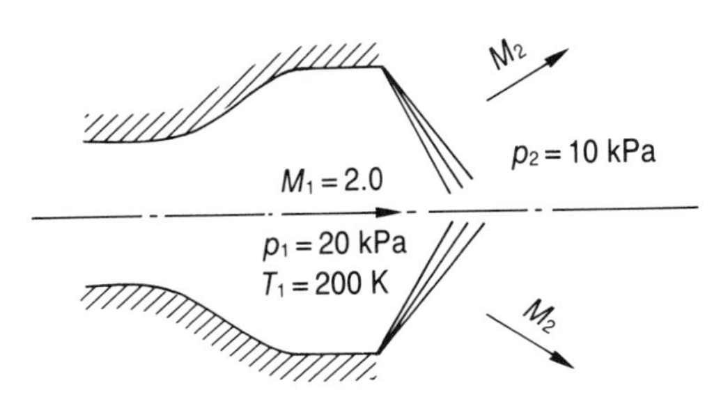

Figure 8.28 Sketch for Example 8.7

Now, corresponding to the pressure ratio $p_2/p_{02} = 10\ \text{kPa}/156.5\ \text{kPa} = 0.063\ 90$, the downstream Mach number and temperature may be obtained from Appendix C as

$$M_2 = \mathbf{2.44}$$

$$T_2/T_{02} = 0.4565 \quad \text{or} \quad T_2 = (0.4565)(360\ \text{K}) = \mathbf{164.3\ K}$$

respectively. From Appendix H, the Prandtl–Meyer angles for Mach numbers $M_1 = 2.0$ and $M_2 = 2.44$ are

$$\nu_1 = 26.38^\circ \quad \text{and} \quad \nu_2 = 37.71^\circ$$

respectively. Then, the flow has been turned through an angle

$$\Delta\nu = \nu_2 - \nu_1 = 37.71^\circ - 26.38^\circ = \mathbf{11.33^\circ}$$

8.5 SUPERSONIC OBLIQUE SHOCK DIFFUSER

In ram jets and turbojets travelling at high speeds, it is necessary to slow down the incoming air by a supersonic diffuser with a minimum loss in stagnation pressure. In Section 5.9, converging–diverging diffusers were used as inlets which can only operate isentropically at the design speed. Therefore, they are not very practical during start-up or off-design conditions. Also, one should note that the design condition cannot be reached without changing the throat area or overspeeding.

In order to eliminate the problem of starting a converging–diverging diffuser, the throat must be removed. This can be done by using a normal shock diffuser, as shown in Figure 8.29. The supersonic flow is decelerated to subsonic speeds through a normal shock wave with further deceleration in the diverging passage. The disadvantage of this configuration is the high stagnation pressure loss across the normal shock wave. For this reason, it can be used at Mach numbers close to unity.

This disadvantage may be eliminated by using a spike-type oblique shock diffuser, as shown in Figure 8.30. In this case, the supersonic flow decelerates lower supersonic speeds through a weak oblique shock wave. The decelerated supersonic flow then goes through a normal shock at the cowl inlet. The resulting subsonic flow is then further

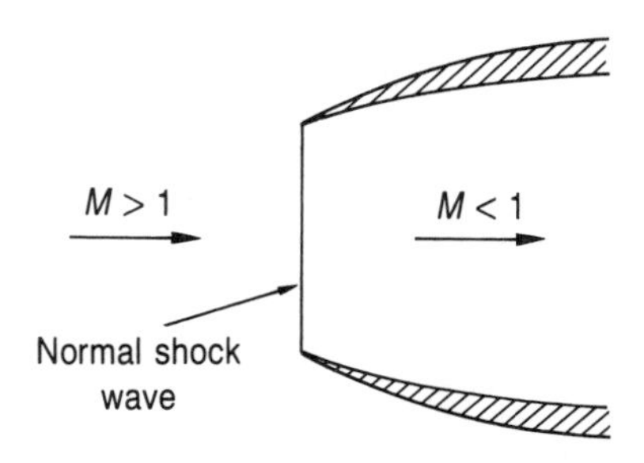

Figure 8.29 A normal shock diffuser

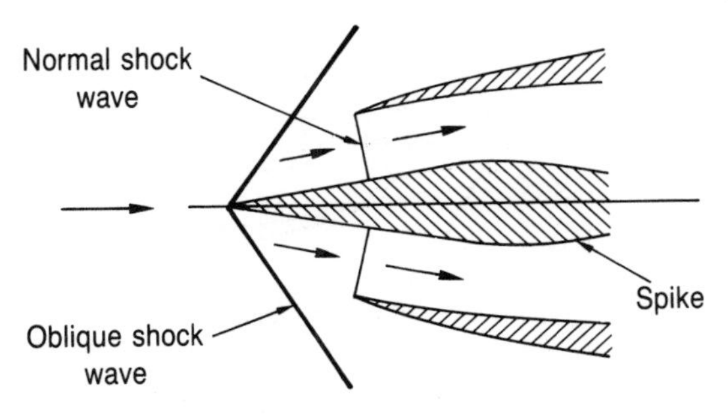

Figure 8.30 A spike-type oblique shock diffuser with one shock

decelerated in the diverging passage. As long as the flight Mach number is decreased through the oblique shock wave, the strength of the normal shock wave and the overall stagnation pressure loss are reduced.

As the number of oblique shocks is increased, the overall stagnation pressure loss decreases. A spike-type oblique shock diffuser with two oblique shocks is shown in Figure 8.31. One should note that the thickness of the boundary layer increases over the spike. The second normal shock wave creates an adverse pressure gradient which may cause flow separation. As the number of oblique shock waves is increased, the tendency for flow separation increases. Therefore, during the design of a spike-type oblique shock diffuser, the designer should make a compromise between the total stagnation pressure recovery and the increased tendency for flow separation by increasing the number of oblique shock waves.

Although the flow pattern in a converging–diverging inlet is only dependent on the inlet geometry, the flow pattern in the spike diffuser is affected by the downstream conditions. There are three modes of operation. At the critical operating condition, which is shown in Figure 8.32(a), the normal shock wave is at the cowl inlet and the engine is operating at its design speed. When the resistance to the flow at the downstream end increases, the normal shock wave moves ahead of the inlet, as shown

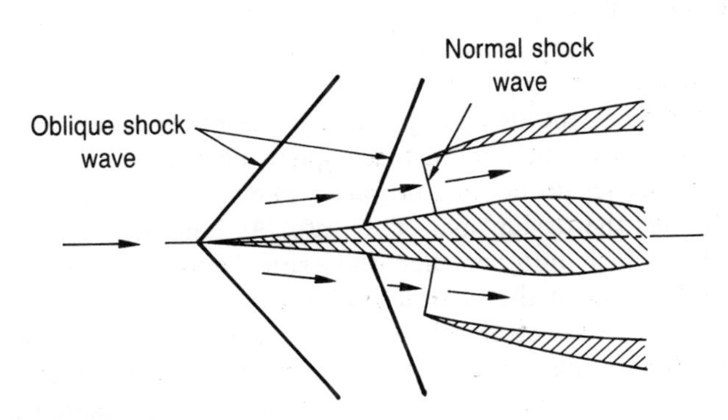

Figure 8.31 A spike-type oblique shock diffuser with two shocks

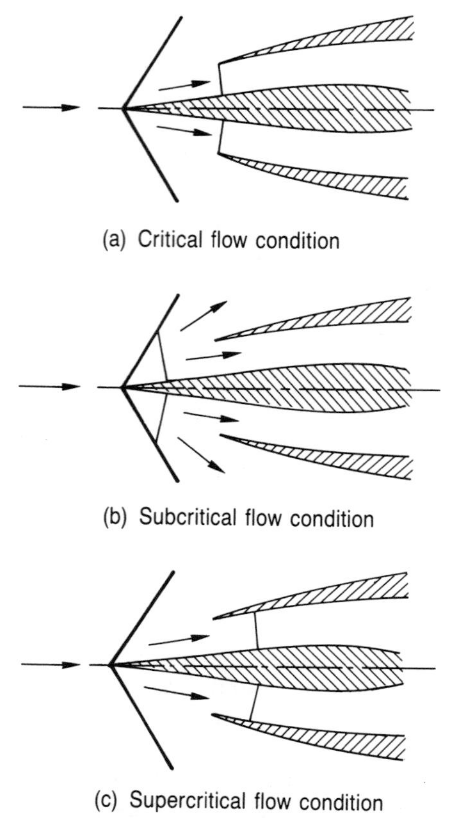

(a) Critical flow condition

(b) Subcritical flow condition

(c) Supercritical flow condition

Figure 8.32 Operation modes of a spike-type oblique shock diffuser

in Figure 8.32(b), and the subcritical operating condition is obtained. In this case, some of the subsonic flow spills over and bypasses the inlet. As a result, the inlet cannot handle the maximum mass flow rate. In addition to this disadvantage, the stagnation pressure recovery is not very good since at least some of the flow undergoes a normal shock wave at the design speed. On the other hand, if the downstream resistance decreases, then the supercritical operating condition is obtained and the normal shock wave settles inside the diffuser, as shown in Figure 8.32(c). In this case, the inlet still handles the design mass flow rate. As long as the normal shock wave occurs at a higher Mach number than the critical mode, the stagnation pressure recovery is less than that of the critical condition.

A jet engine must operate efficiently at off-design speeds and at different angles of attack. An engine operating at its critical mode may be pushed into the undesirable subcritical mode where the mass flow rate decreases and the stagnation pressure loss is high. For this reason, most inlets are designed to operate at the supercritical mode so that they are not pushed into the subcritical mode by a small decrease in engine speed. Although the supercritical mode does not provide a stagnation pressure recovery as well as the critical mode, it gives the same mass flow rate.

Example 8.8
For supersonic flow at a Mach number of 2.4, determine the stagnation pressure loss in

a a normal shock diffuser,
b a spike-type oblique shock diffuser with one shock, and
c a spike-type oblique shock diffuser with two shocks.

Assume that each oblique shock wave turns the flow through an angle of 10°, as shown in Figure 8.33.

Solution
(a) For a normal shock diffuser, the stagnation pressure ratio across the normal shock wave for $M_1 = 2.4$ may be obtained from Appendix D as

$$p_{02}/p_{01} = \mathbf{0.5401}$$

(b) The shock-wave angle of the oblique shock, corresponding to $M_1 = 2.4$ and $\delta_a = 10°$, may be obtained by using Equation (8.8) as

$$\cot \delta_a = \left(\frac{k+1}{2} \frac{M_1^2}{M_1^2 \sin^2 \theta_{1a} - 1} - 1\right) \tan \theta_{1a}$$

or

$$\cot 10° = \left(\frac{1.4+1}{2} \frac{(2.4)^2}{(2.4)^2 \sin^2 \theta_{1a} - 1} - 1\right) \tan \theta_{1a}$$

which may then be solved for the shock-wave angle by trial and error as

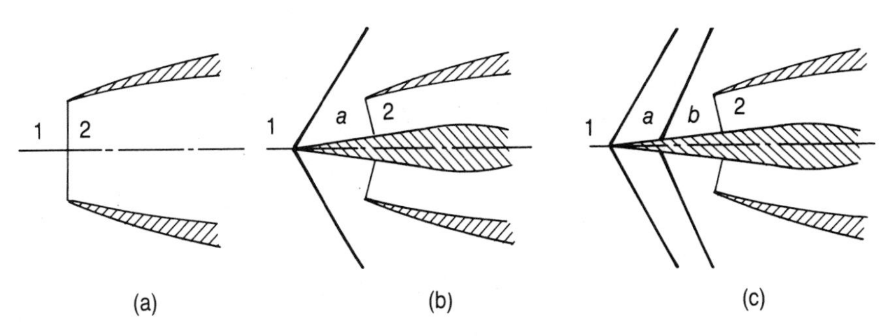

Figure 8.33 Sketch for Example 8.8

$\theta_{1a} = 33.02°$. Then the Mach number perpendicular to the oblique shock wave is

$$M_{1n} = M_1 \sin \theta_{1a} = (2.4)(\sin 33.02°) = 1.308$$

so that the normal component of the upstream Mach number and the stagnation pressure ratio across the oblique shock wave, corresponding to $M_{1n} = 1.31$, are

$$M_{an} = 0.7809$$

and

$$p_{0a}/p_{01} = 0.9776$$

from Appendix D, respectively. The Mach number downstream of the oblique shock wave is

$$M_a = \frac{M_{an}}{\sin(\theta_{1a} - \delta_a)} = \frac{0.7809}{\sin(33.02° - 10°)} = 2.083$$

Then the stagnation pressure ratio across the normal shock wave for $M_a = 2.08$ is

$$p_{02}/p_{0a} = 0.6835$$

from Appendix D. Finally, the overall stagnation pressure ratio is

$$\frac{p_{02}}{p_{01}} = \frac{p_{02}}{p_{0a}} \frac{p_{0a}}{p_{01}} = (0.6835)(0.9776) = \mathbf{0.6682}$$

(c) The shock-wave angle of the second oblique shock, corresponding to $M_a = 2.083$ and $\delta_b = 10°$, may be obtained by using Equation (8.8) as

$$\cot \delta_b = \left(\frac{k+1}{2} \frac{M_a^2}{M_a^2 \sin^2 \theta_{ab} - 1} - 1\right) \tan \theta_{ab}$$

or

$$\cot 10° = \left(\frac{1.4+1}{2} \frac{(2.083)^2}{(2.083)^2 \sin^2 \theta_{ab} - 1} - 1\right) \tan \theta_{ab}$$

which may then be solved for the shock-wave angle by trial and error as

$\theta_{ab} = 37.74°$. Then the Mach number perpendicular to the second oblique shock wave is

$$M_{an} = M_a \sin \theta_{ab} = (2.083)(\sin 37.74°) = 1.275$$

so that the normal component of the upstream Mach number and the stagnation pressure ratio across the second oblique shock wave, corresponding to $M_{an} = 1.28$, are

$$M_{bn} = 0.7963$$

and

$$p_{0b}/p_{0a} = 0.9827$$

from Appendix D, respectively. The Mach number downstream of the second oblique shock wave is

$$M_b = \frac{M_{bn}}{\sin(\theta_{ab} - \delta_b)} = \frac{0.7963}{\sin(37.74° - 10°)} = 1.711$$

Then the stagnation pressure ratio across the normal shock wave for $M_b = 1.71$ is

$$p_{02}/p_{0b} = 0.8516$$

from Appendix D. Finally, the overall stagnation pressure ratio is

$$\frac{p_{02}}{p_{01}} = \frac{p_{02}}{p_{0b}} \frac{p_{0b}}{p_{0a}} \frac{p_{0a}}{p_{01}} = (0.8516)(0.9827)(0.9776) = \mathbf{0.8181}$$

Example 8.9

A spike-type oblique shock diffuser with one shock is operating at a Mach number of 3.0 (Figure 8.34). The temperature and pressure of the approaching air are 250 K and 47 kPa, respectively. The cross-sectional area of the cowl inlet is 0.075 m². The cross-sectional area where a normal shock wave occurs is 0.1 m². The half angle of the spike is 10°. Determine the mass flow rate and the stagnation pressure loss for

a the critical operation mode, and
b the supercritical operation mode.

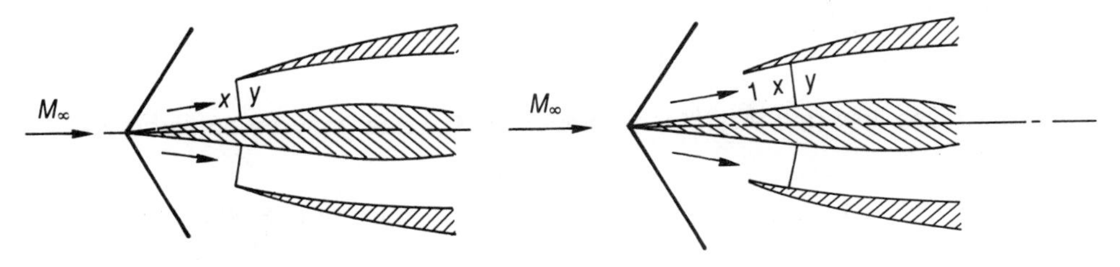

Figure 8.34 Sketch for Example 8.9

Solution

(a) The stagnation pressure and temperature at a Mach number of 3.0 may be obtained from Appendix C as

$$p_\infty/p_{0\infty} = 0.02722 \quad \text{or} \quad p_{0\infty} = 47\ \text{kPa}/0.02722 = 1727\ \text{kPa}$$

$$T_\infty/T_{0\infty} = 0.3571 \quad \text{or} \quad T_{0\infty} = 250\ \text{K}/0.3571 = 700.1\ \text{K}$$

The shock-wave angle of the oblique shock, corresponding to $M_\infty = 3.0$ and $\delta_x = 10^\circ$, may be obtained by using Equation (8.8) as

$$\cot\,\delta_x = \left(\frac{k+1}{2}\,\frac{M_\infty^2}{M_\infty^2\,\sin^2\,\theta_{\infty x} - 1} - 1\right)\tan\,\theta_{\infty x}$$

or

$$\cot\,10^\circ = \left(\frac{1.4+1}{2}\,\frac{(3)^2}{(3)^2\sin^2\,\theta_{\infty x} - 1} - 1\right)\tan\,\theta_{\infty x}$$

which may then be solved for the shock-wave angle by trial and error as $\theta_{\infty x} = 27.39^\circ$. Then the Mach number perpendicular to the oblique shock wave is

$$M_{\infty n} = M_\infty\,\sin\,\theta_{\infty x} = (3)(\sin\,27.39^\circ) = 1.380$$

so that the normal component of the downstream Mach number and the stagnation pressure after the oblique shock wave, corresponding to $M_{1n} = 1.38$, are

$$M_{xn} = 0.7483$$

and

$$p_{0x}/p_{0\infty} = 0.9630 \quad \text{or} \quad p_{0x} = (0.9630)(1727\ \text{kPa}) = 1663\ \text{kPa}$$

from Appendix D, respectively. The Mach number downstream of the oblique shock wave is

$$M_x = \frac{M_{xn}}{\sin\,(\theta_{\infty x} - \delta_x)} = \frac{0.7483}{\sin\,(27.39^\circ - 10^\circ)} = 2.504$$

One should also note that the stagnation temperature is constant across an adiabatic oblique shock wave, that is

$$T_{0x} = T_{0\infty} = 700.1\ \text{K}$$

Then the mass flow rate at a Mach number of $M_x = 2.50$ may be obtained via Appendix C as

$$\frac{\dot{m}\sqrt{(RT_{0x})}}{A_x p_{0x}} = 0.2597 \quad \text{or} \quad \dot{m} = \frac{(0.2597)(0.075\ \text{m}^2)(1\,663\,000\ \text{kPa})}{\sqrt{[(287.1\ \text{J kg}^{-1}\,\text{K}^{-1})(700.1\ \text{K})]}}$$

$$= \mathbf{72.25\ kg\ s^{-1}}$$

Then the stagnation pressure after the normal shock wave with $M_x = 2.50$ is

$$p_{0y}p_{0x} = 0.4990 \quad \text{or} \quad p_{0y} = (0.4990)(1663\ \text{kPa}) = 829.8\ \text{kPa}$$

from Appendix D. Finally, the stagnation pressure loss is

$$p_{0\infty} - p_{0y} = 1727\ \text{kPa} - 829.8\ \text{kPa} = \mathbf{897.2\ kPa}$$

(b) The state at section 1 of the supercritical flow condition is exactly the same as that at state at section x of the critical flow condition. Therefore,

$$M_1 = 2.504$$

$$T_{01} = 700.1\ \text{K}$$

$$p_{01} = 1663\ \text{kPa}$$

Also, one can easily note that the mass flow rates for both operating conditions are the same. The critical area before the normal shock wave corresponding to a Mach number of $M_1 = 2.50$ may then be obtained from Appendix C as

$$A_1/A_x^* = 2.637 \quad \text{or} \quad A_x^* = 0.075\ \text{m}^2/2.637 = 0.028\,44\ \text{m}^2$$

Hence, the Mach number corresponding to a critical area ratio of $A_x/A_x^* = 0.1\ \text{m}^2/0.028\,44\ \text{m}^2 = 3.516$ is $M_x = 2.80$ from Appendix C. Now, the stagnation pressure after the normal shock wave is

$$p_{0y}/p_{0x} = 0.3895 \quad \text{or} \quad p_{0y} = (0.3895)(1663\ \text{kPa}) = 647.7\ \text{kPa}$$

for $M_x = 2.80$ from Appendix D. Therefore, the stagnation pressure loss is

$$p_{0\infty} - p_{0y} = 1663\ \text{kPa} - 647.7\ \text{kPa} = \mathbf{1015\ kPa}$$

8.6 SUPERSONIC AEROFOILS

An aerofoil must be designed to give a lift force which is perpendicular to the free-stream direction. In addition to this lift force, there exists a drag force which is opposite to the initial flow direction as a result of frictional effects. For those cases where a compressible flow can be treated as incompressible, that is when the free-stream Mach number is less than 0.3, the shape of the aerofoil is the classical streamlined profile like a raindrop. However, when this profile is used in a supersonic flow, then either there will be an attached oblique shock wave of relatively high strength at the nose of the aerofoil (Figure 8.35(a)), or there will be a detached shock wave if the deflection angle is large enough (Figure 8.35(b)). In both of these cases, the pressure after the shock wave is quite high, producing excessive wave drag due to the presence of the oblique shock wave. For this reason, a supersonic aerofoil must possess a very sharp nose and be as thin as possible. The ideal case is a flat plate of zero thickness.

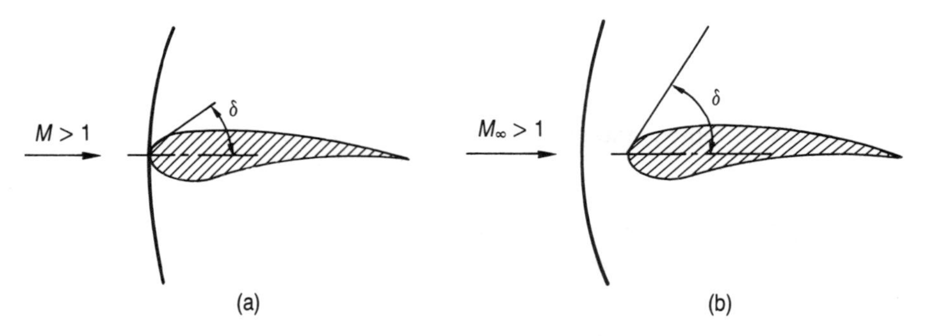

Figure 8.35 A streamlined body in a supersonic flow: (a) with an attached shock; (b) with a detached shock

In Figure 8.36, a two-dimensional flat plate is inclined at an angle of attack α to a supersonic free stream with a Mach number of M_∞. The flow on the lower surface goes through an oblique shock wave which increases the pressure on that surface. However, the flow on the upper surface goes through an expansion fan which is centred at the nose of the aerofoil decreasing the pressure on the upper side. Therefore, a positive net lift force, L, is produced acting vertically upwards, and may be given as

$$L = (p_{\mathrm{L}} - p_{\mathrm{U}})c \cos \alpha \tag{8.23}$$

where c is the chord length and p_{U} and p_{L} are the pressures on the upper and lower sides of the aerofoil respectively. This lift force is then accompanied by a drag force, D, which is opposite to the flow direction and may be given as

$$D = (p_{\mathrm{L}} - p_{\mathrm{U}})c \sin \alpha \tag{8.24}$$

Usually, it is customary to express the lift and drag forces in non-dimensional form as the lift coefficient, C_L, and the drag coefficient C_D,

$$C_L = \frac{L}{\frac{1}{2}\rho_\infty V_\infty^2 c} \tag{8.25}$$

and

$$C_D = \frac{D}{\frac{1}{2}\rho_\infty V_\infty^2 c} \tag{8.26}$$

respectively.

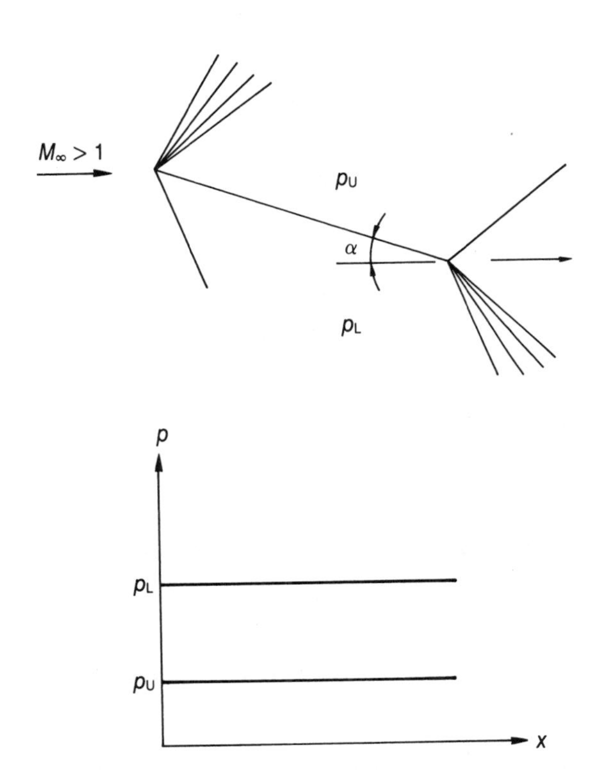

Figure 8.36 Pressure distribution on a flat plate in a supersonic stream at an angle of attack α

Example 8.10

A flat plate is placed in a supersonic flow with a Mach number of 2.0 at an angle of attack of 10°, as shown in Figure 8.37. If the chord length is 1 m, determine the lift and drag coefficients.

Solution

The stagnation pressure of the free stream corresponding to $M_\infty = 2.0$ may be obtained from Appendix C as

$$p_\infty / p_{0\infty} = 0.1278 \quad \text{or} \quad p_{0\infty} = p_\infty / 0.1278 = 7.825 p_\infty$$

For the Prandtl–Meyer expansion fan centred at the nose of the flat plate, the upstream Prandtl-Meyer angle for $M_\infty = 2.0$ may be obtained from Appendix H as $\nu_\infty = 26.38°$. Now, the Prandtl–Meyer angle after the expansion fan is

$$\nu_U = \nu_\infty + 10° = 26.38° + 10° = 36.38°$$

which corresponds to a Mach number of $M_U = 2.38$ from Appendix H. One should

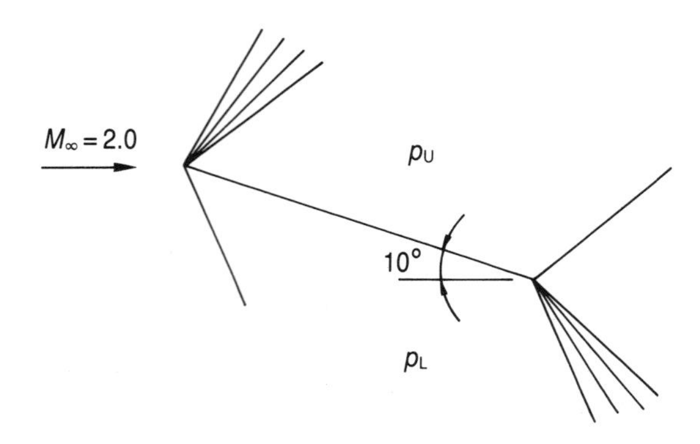

Figure 8.37 Sketch for Example 8.10

note that the flow through an expansion fan is isentropic, so that

$$p_{0U} = p_{0\infty} = 7.825 p_\infty$$

The pressure corresponding to a Mach number of $M_U = 2.38$ may now be obtained from Appendix C as

$$p_U / p_{0U} = 0.070\,57 \quad \text{or} \quad p_U = (0.070\,57)(7.825 p_\infty) = 0.5522 p_\infty$$

The shock-wave angle of the oblique shock, corresponding to $M_\infty = 2.0$ and $\delta_L = 10°$, may be obtained by using Equation (8.8) as

$$\cot \delta_L = \left(\frac{k+1}{2} \frac{M_\infty^2}{M_\infty^2 \sin^2 \theta_{\infty L} - 1} - 1 \right) \tan \theta_{\infty L}$$

or

$$\cot 10° = \left(\frac{1.4+1}{2} \frac{(2)^2}{(2)^2 \sin^2 \theta_{\infty L} - 1} - 1 \right) \tan \theta_{\infty L}$$

which may then be solved for the shock-wave angle by trial and error as $\theta_{\infty L} = 39.31°$. Then the Mach number perpendicular to the oblique shock wave is

$$M_{\infty n} = M_\infty \sin \theta_{\infty L} = (2)(\sin 39.31°) = 1.267$$

and the corresponding pressure ratio from Appendix D is

$$p_L / p_\infty = 1.715 \quad \text{or} \quad p_L = 1.715 p_\infty$$

The lift and drag forces may now be evaluated as

$$L = (p_L - p_U) \cos \alpha \, c = (1.715p_\infty - 0.5522p_\infty)(\cos 10^\circ)(1 \text{ m}) = 1.145p_\infty$$

and

$$D = (p_L - p_U) \sin \alpha \, c = (1.715p_\infty - 0.5522p_\infty)(\sin 10^\circ)(1 \text{ m}) = 0.2019p_\infty$$

respectively. Finally, the lift and drag coefficients are

$$C_L = \frac{L}{\frac{1}{2}\rho_\infty V_\infty^2 c} = \frac{L}{\frac{1}{2}kp_\infty M_\infty^2 c} = \frac{1.145p_\infty}{(0.5)(1.4)(2.0)^2(1 \text{ m})\, p_\infty} = \mathbf{0.4089}$$

and

$$C_D = \frac{D}{\frac{1}{2}\rho_\infty V_\infty^2 c} = \frac{D}{\frac{1}{2}kp_\infty M_\infty^2 c} = \frac{0.2019p_\infty}{(0.5)(1.4)(2.0)^2(1 \text{ m})\, p_\infty} = \mathbf{0.072\ 11}$$

respectively.

Physically, a flat plate aerofoil with zero thickness is not possible. Another possibility is a curved symmetrical aerofoil which satisfies the requirement that a

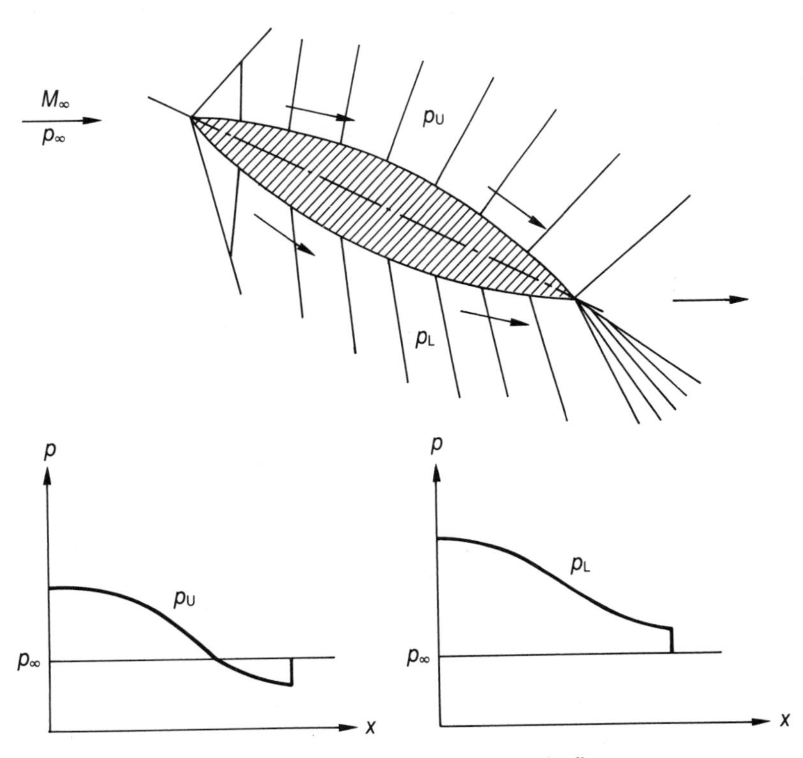

Figure 8.38 A curved symmetric aerofoil in a supersonic flow

supersonic aerofoil must possess a pointed nose, as shown in Figure 8.38. In this case, two oblique shock waves are attached to the nose of the aerofoil. The oblique shock wave on the lower surface of the aerofoil is stronger than that on the upper surface as long as the flow turning angle is larger on the lower side. As the angle of attack is increased, the nose half angle increases. Then, beyond a certain limit, the oblique shock wave becomes detached with a corresponding increase in the drag force.

Another possible aerofoil for supersonic flow is the diamond-shaped aerofoil, as shown in Figure 8.39. The flow on the upper surface is first expanded through a Prandtl–Meyer fan centred at point A, and then expanded by another one centred at point B. As long as the flow on the upper surface of the aerofoil must leave the aerofoil in the direction of the approaching stream, then the upper flow must be compressed to the free-stream pressure at point D. The flow on the lower side must first be compressed through an oblique shock wave attached to point A, which is followed by an expansion through a Prandtl–Meyer fan at point C. Finally, to turn back to the original flow direction, the lower flow must be expanded to the free-stream pressure through a system of expansion waves at point D. One should note that the flows

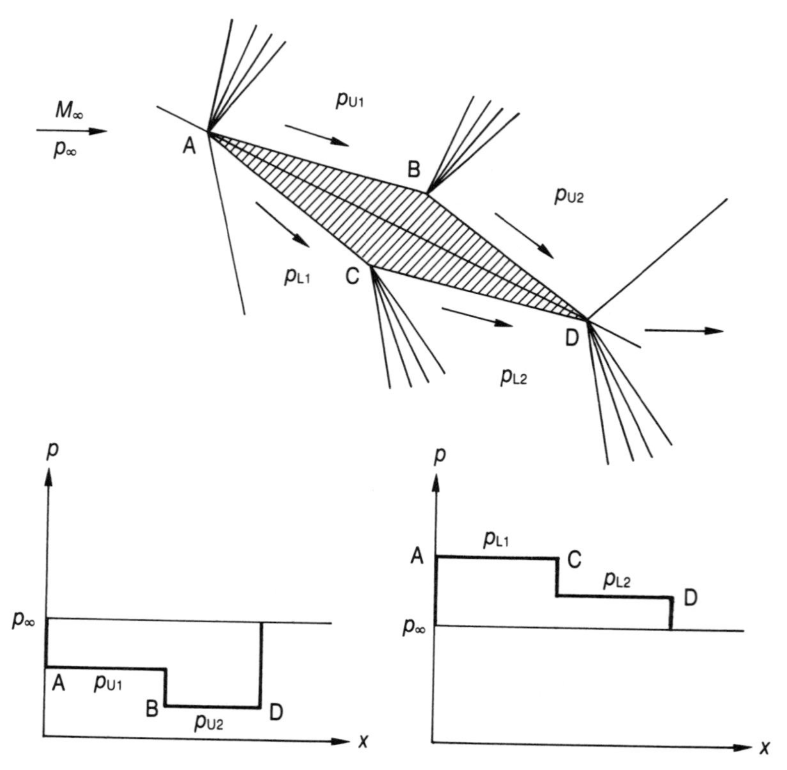

Figure 8.39 A diamond-shaped aerofoil in a supersonic flow

on the upper and lower sides become identical when the angle of attack is zero. In this case, both the lift and the drag forces are zero for frictionless flows.

Example 8.11

A two-dimensional diamond-shaped aerofoil is placed in a supersonic flow with a Mach number of 2.5, as shown in Figure 8.40. The angle of attack is 10°. If the chord length is 1 m, then determine the lift and drag coefficients. The wedge half angle is 5°.

Solution

The stagnation pressure of the free stream corresponding to $M_\infty = 2.5$ may be obtained from Appendix C as

$$p_\infty/p_{0\infty} = 0.058\,53 \quad \text{or} \quad p_{0\infty} = p_\infty/0.058\,53 = 17.09 p_\infty$$

For the expansion fan centred at point A of the aerofoil, the upstream Prandtl–Meyer angle for $M_\infty = 2.5$ may be obtained from Appendix H as $\nu_\infty = 39.12°$. Now, the Prandtl–Meyer angle after the first expansion fan is

$$\nu_{U1} = \nu_\infty + 5° = 39.12° + 5° = 44.12°$$

and that after the second expansion fan located at point B is

$$\nu_{U2} = \nu_{U1} + 10° = 44.12° + 10° = 54.12°$$

corresponding to Mach numbers of

$$M_{U1} = 2.72 \quad \text{and} \quad M_{U2} = 3.24$$

respectively, from Appendix H. One should note that the flow through an

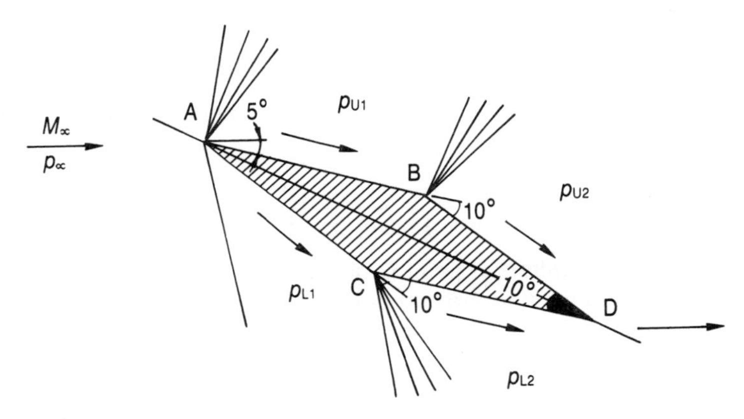

Figure 8.40 Sketch for Example 8.11

expansion fan is isentropic, so that

$$p_{0U2} = p_{0U1} = p_{0\infty} = 17.09 p_\infty$$

The stagnation pressure ratios corresponding to Mach numbers of $M_{U1} = 2.72$ and $M_{U2} = 3.24$ may now be obtained from Appendix C as

$$p_{U1}/p_{0U1} = 0.041\,65 \quad \text{or} \quad p_{U1} = (0.041\,65)(17.09 p_\infty) = 0.7118 p_\infty$$

and

$$p_{U2}/p_{0U2} = 0.019\,08 \quad \text{or} \quad p_{U2} = (0.019\,08)(17.09 p_\infty) = 0.3261 p_\infty$$

respectively. For the weak oblique shock wave, the shock-wave angle corresponding to $M_\infty = 2.5$ and $\delta_{L1} = 15^\circ$ may be obtained from Equation (8.8) as

$$\cot \delta_{L1} = \left(\frac{k+1}{2} \frac{M_\infty^2}{M_\infty^2 \sin^2 \theta_{\infty L1} - 1} - 1 \right) \tan \theta_{\infty L1}$$

or

$$\cot 15^\circ = \left(\frac{1.4+1}{2} \frac{(2.5)^2}{(2.5)^2 \sin^2 \theta_{\infty L1} - 1} - 1 \right) \tan \theta_{\infty L1}$$

which may then be solved for the shock-wave angle by trial and error as $\theta_{\infty L1} = 36.94^\circ$. Then the Mach number perpendicular to the oblique shock wave is

$$M_{\infty n} = M_\infty \sin \theta_{\infty L1} = (2.5)(\sin 36.94^\circ) = 1.503$$

and the corresponding downstream Mach number normal to the oblique shock-wave stagnation pressure and the pressure from Appendix D is

$$M_{L1n} = 0.7011$$

$$p_{0L1}/p_{0\infty} = 0.9298 \quad \text{or} \quad p_{0L1} = (0.9298)(17.09 p_\infty) = 15.89 p_\infty$$

and

$$p_{L1}/p_\infty = 2.458 \quad \text{or} \quad p_{L1} = 2.458 p_\infty$$

respectively. The normal component of the Mach number after the oblique shock wave is

$$M_{L1} = \frac{M_{L1n}}{\sin(\theta_{\infty L1} - \delta_{L1})} = \frac{0.7011}{\sin(36.94^\circ - 15^\circ)} = 1.876$$

For the expansion fan centred at point D of the aerofoil, the Prandtl–Meyer angle for $M_{L1} = 1.88$ may be obtained from Appendix H as $\nu_{L1} = 23.02^\circ$. Now the Prandtl–Meyer angle after the expansion fan is

$$\nu_{L2} = \nu_{L1} + 10^\circ = 23.02^\circ + 10^\circ = 33.02^\circ$$

which corresponds to a Mach number of $M_{L2} = 2.25$ from Appendix H. One should note that the flow through an expansion fan is isentropic, so that

$$p_{0L2} = p_{0L1} = 15.89 p_\infty$$

The stagnation pressure ratio corresponding to a Mach number of $M_{L2} = 2.25$ may now be obtained from Appendix C as

$$p_{L2}/p_{0L2} = 0.086\,48 \quad \text{or} \quad p_{L2} = (0.086\,48)(15.89 p_\infty) = 1.374 p_\infty$$

The length of each side of the diamond aerofoil is

$$l = \frac{c/2}{\cos 5^\circ} = \frac{0.5 \text{ m}}{\cos 5^\circ} = 0.5019 \text{ m}$$

The lift and drag forces may now be evaluated as

$$\begin{aligned} L &= (p_{L1} l)(\cos 15^\circ) + (p_{L2} l)(\cos 5^\circ) - (p_{U1} l)(\cos 5^\circ) - (p_{U2} l)(\cos 15^\circ) \\ &= (0.5019 \text{ m})\,[(2.458 p_\infty)(\cos 15^\circ) + (1.374 p_\infty)(\cos 5^\circ) \\ &\quad - (0.7118 p_\infty)(\cos 5^\circ) - (0.3261 p_\infty)(\cos 15^\circ)] = 1.365 p_\infty \end{aligned}$$

and

$$\begin{aligned} D &= (p_{L1} l)(\sin 15^\circ) + (p_{L2} l)(\sin 5^\circ) - (p_{U1} l)(\sin 5^\circ) - (p_{U2} l)(\sin 15^\circ) \\ &= (0.5019 \text{ m})\,[(2.458 p_\infty)(\sin 15^\circ) + (1.374 p_\infty)(\sin 5^\circ) \\ &\quad - (0.7118 p_\infty)(\sin 5^\circ) - (0.3261 p_\infty)(\sin 15^\circ)] = 0.3059 p_\infty \end{aligned}$$

respectively. Finally, the lift and drag coefficients are

$$C_L = \frac{L}{\frac{1}{2}\rho_\infty V_\infty^2 c} = \frac{L}{\frac{1}{2} k p_\infty M_\infty^2 c} = \frac{1.365 p_\infty}{(0.5)(1.4)(2.5)^2(1 \text{ m})\, p_\infty} = \mathbf{0.3120}$$

and

$$C_D = \frac{D}{\frac{1}{2}\rho_\infty V_\infty^2 c} = \frac{D}{\frac{1}{2} k p_\infty M_\infty^2 c} = \frac{0.3059 p_\infty}{(0.5)(1.4)(2.5)^2(1 \text{ m})\, p_\infty} = \mathbf{0.069\,92}$$

respectively.

FURTHER READING

Cambel, A. B. and Jennings, B. H. (1958) *Gas Dynamics*, Dover: New York.

Daneshyar, H. (1976) *One-Dimensional Compressible Flow*, Pergamon: Oxford.

Fox, R. W. and McDonald, A. T. (1985) *Introduction to Fluid Mechanics*, 3rd edn, Wiley: New York.

John, J. E. A. (1984) *Gas Dynamics*, 2nd edn, Allyn & Bacon: Boston, MA.

Liepman, H. W. and Roshko, A. (1957) *Elements of Gas Dynamics*, Wiley: New York.

Owczarek, J. A. (1964) *Fundamentals of Gas Dynamics*, International Textbook: Scranton, PA.
Shapiro, A. H. (1953) *The Dynamics and Thermodynamics of Fluid Flow*, Vol. 1, Ronald Press: New York.
Zucrow, M. J. and Hoffman, J. D. (1976) *Gas Dynamics*, Vol. 1, Wiley: New York.

PROBLEMS

8.1 A uniform supersonic air flow travelling at a Mach number of 2.0 passes over a concave corner, as shown in the figure below. An oblique shock wave, which makes an angle of 40° with the flow direction, is attached to the corner under the given conditions. If the pressure and temperature in the uniform flow are 40 kPa and 253 K, respectively, determine the pressure and temperature behind the wave, the downstream Mach number and the deflection angle. Also, find the maximum allowable deflection angle before the oblique shock wave detaches from the concave corner.

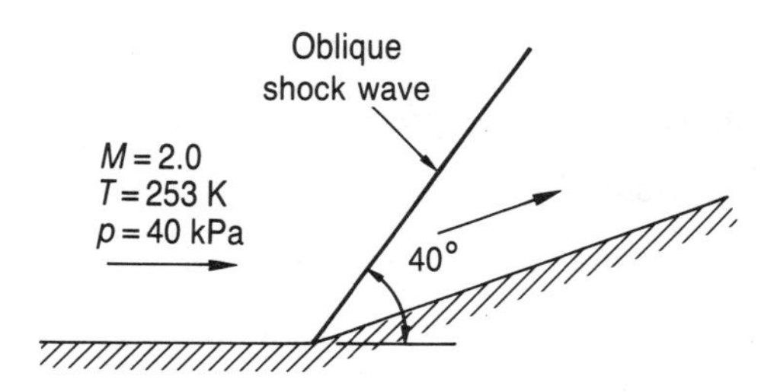

Problem 8.1

8.2 A uniform supersonic air flow travelling at a Mach number of 2.5 passes over a wedge, as shown in the figure below. An oblique shock wave, which

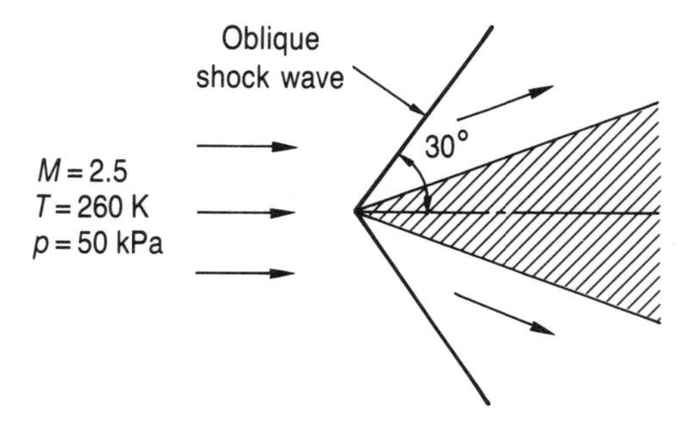

Problem 8.2

makes an angle of 30° with the flow direction, is attached to the wedge under the given conditions. If the pressure and temperature in the uniform flow are 50 kPa and 260 K, respectively, determine the pressure and temperature behind the wave, the downstream Mach number and the deflection angle. Also, find the maximum allowable deflection angle before the oblique shock wave detaches from the concave corner.

8.3 A uniform supersonic air flow travelling at a Mach number of 2.5 passes over a concave corner which deflects the flow by 10°, as shown in the figure below. If the pressure and temperature of the uniform flow are 50 kPa and 240 K, respectively, determine the shock-wave angle, Mach number, pressure and temperature downstream of the oblique shock wave.

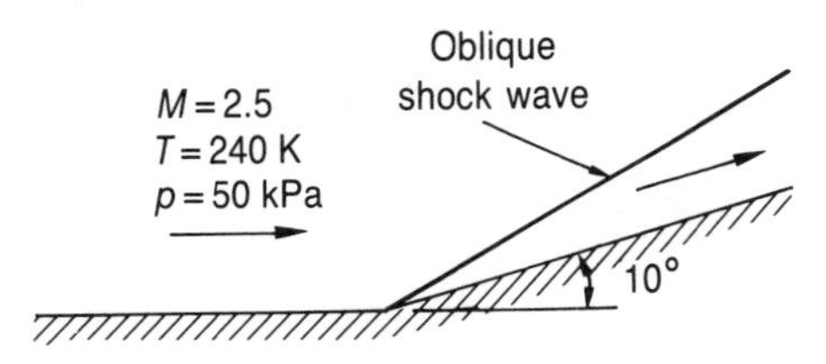

Problem 8.3

8.4 A uniform supersonic air flow travelling at a Mach number of 2.5 passes over a wedge which deflects the flow by 15°, as shown in the figure below. If the pressure and temperature of the uniform flow are 40 kPa and −10 °C, respectively, determine the shock-wave angle, Mach number, pressure and temperature downstream of the oblique shock wave.

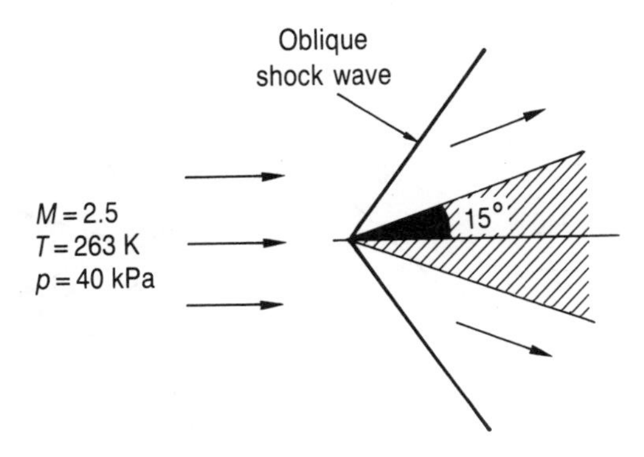

Problem 8.4

8.5 Air at a Mach number of 2.5 flows through an oblique shock wave which is inclined 30° to the flow direction, as shown in the figure below. The temperature and pressure of the uniform stream are 250 K and 50 kPa, respectively. Determine

a the pressure, temperature and Mach number after the incident shock wave,
b the pressure, temperature and Mach number after the reflected shock wave, and
c the angle at which the reflected shock wave is inclined to the flow direction.

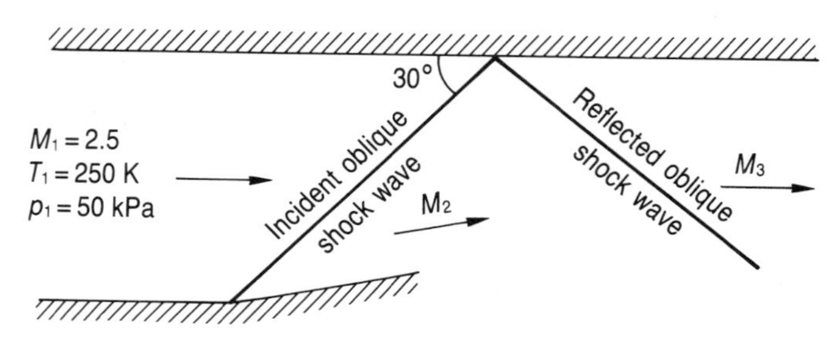

Problem 8.5

8.6 An oblique shock wave impinges on a free jet, as shown in the figure below. The jet of air is at a temperature of 240 K and moves with a Mach number of 3.0. Atmospheric pressure is 100 kPa. The angle of inclination of the oblique shock wave to the free jet is 30°. Determine the Mach number, pressure, temperature and angle of the flow to the horizontal direction

a in region 2, and
b in region 3.

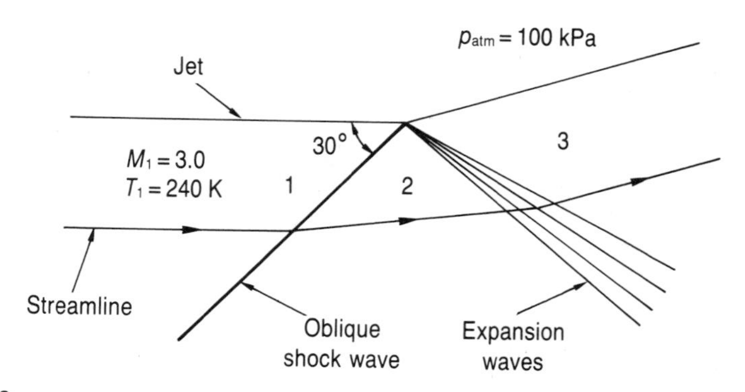

Problem 8.6

8.7 An oblique shock wave is incident on a solid boundary, as shown in the figure below. The boundary is to be turned through a certain angle so that there will be no reflected shock wave. Determine the angle, δ.

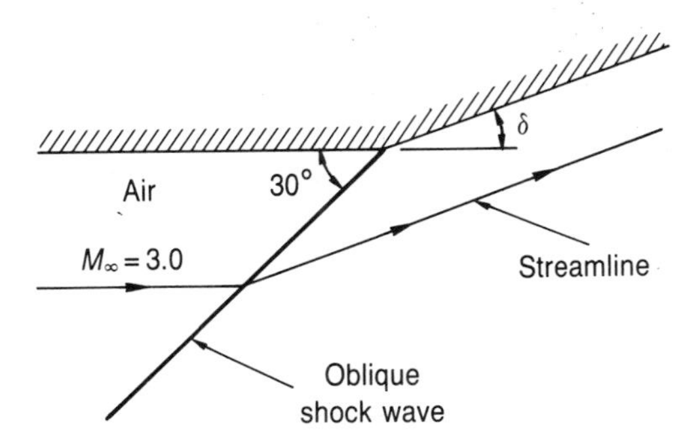

Problem 8.7

8.8 Two oblique shock waves, each making an angle of 30° with a uniform flow, intersect as shown in the figure below. The Mach number, pressure and temperature of the uniform flow are 2.5, 40 kPa and 240 K, respectively. Determine

a the pressure, temperature and Mach number after the incident shock wave,
b the pressure, temperature and Mach number after the reflected shock wave, and
c the angle at which the reflected shock wave is inclined to the flow direction.

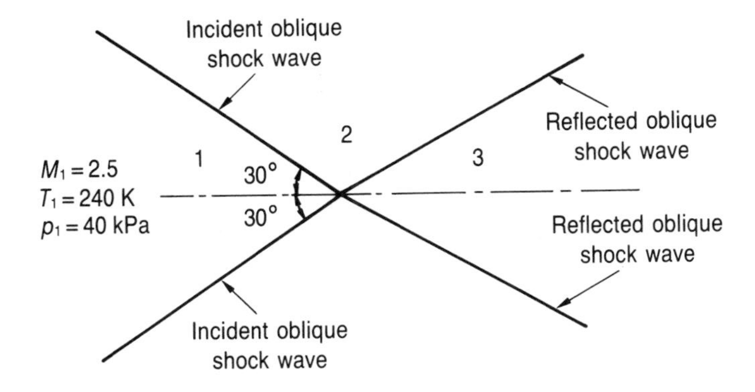

Problem 8.8

8.9 Air at a Mach number of 4.0 and a pressure of 100 kPa flows between two horizontal parallel plates, as shown in the figure below. At a certain cross-section, the upper wall turns downwards through an angle of 10°, while the lower one turns upwards through an angle of 20°. At these concave corners, two oblique shock waves are generated. These two shock waves then intersect at point O. Determine the pressure and Mach number in regions 4 and 5, and also find the net deflection of the flow.

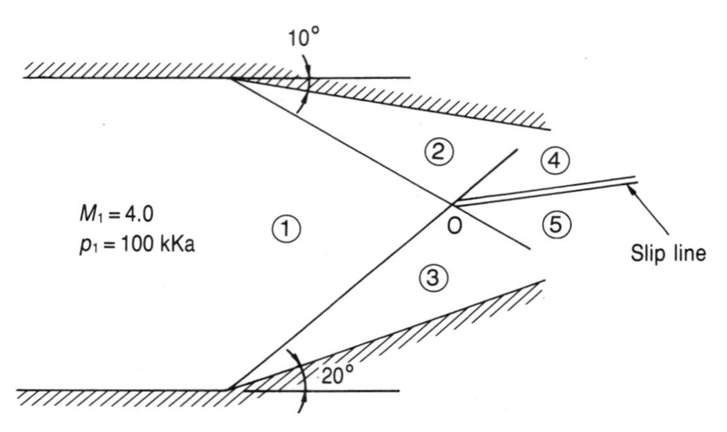

Problem 8.9

8.10 A uniform airstream at a Mach number of 1.5 expands through a 20° convex corner, as shown in the figure below. If the pressure and temperature of this airstream are 75 kPa and 250 K, respectively, determine the Mach number, pressure and temperature after the expansion and also determine the fan angle.

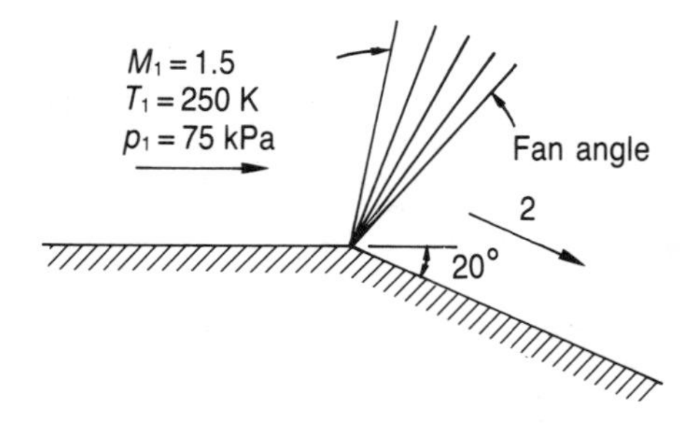

Problem 8.10

8.11 An overexpanded converging–diverging nozzle designed for an exit Mach number of 3.0 is shown in the figure below. The pressure and temperature of the air at the exit plane of the nozzle are 40 kPa and 250 K, respectively.

The nozzle discharges to a back pressure of 100 kPa. Determine the Mach number, pressure, temperature and flow direction in regions 2, 3 and 4.

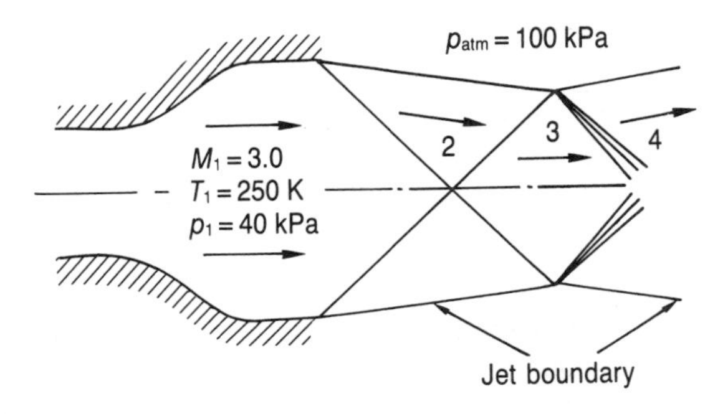

Problem 8.11

8.12 An overexpanded converging–diverging nozzle has throat and exit cross-sectional areas of 0.001 m^2 and 0.002 637 m^2, respectively, as shown in the figure below. Air is supplied from a large reservoir where the pressure and temperature are 750 kPa and 500 K, respectively. The nozzle discharges to a back pressure of 100 kPa. Determine the Mach number, pressure, temperature and flow direction in regions 2, 3 and 4.

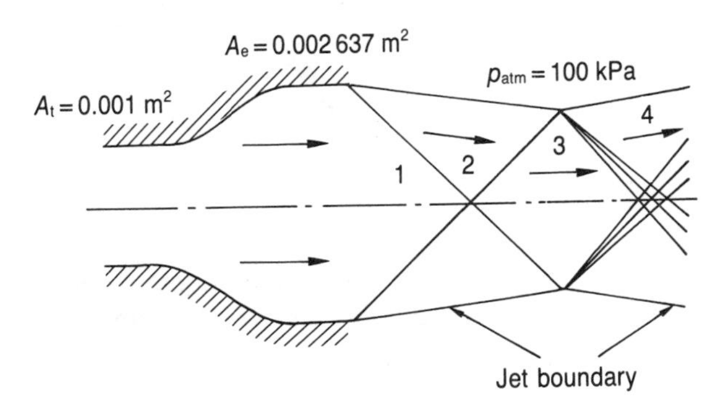

Problem 8.12

8.13 An underexpanded converging–diverging nozzle designed for an exit Mach number of 3.0 is shown in the figure below. The pressure and temperature of the air at the exit plane of the nozzle are 40 kPa and 250 K, respectively. If the back pressure is 20 kPa, determine the Mach number, temperature, and direction of the flow downstream of the expansion fan.

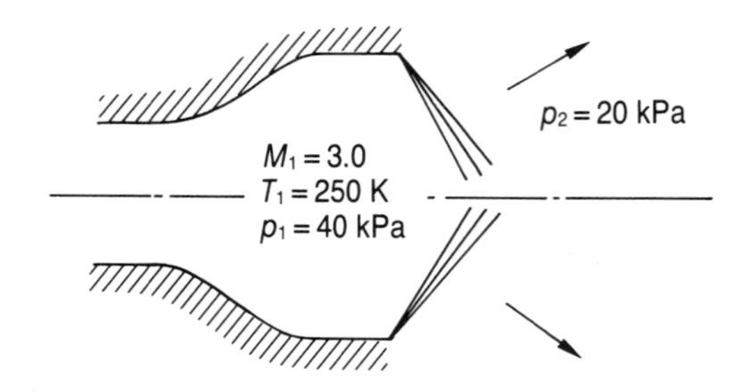

Problem 8.13

8.14 An underexpanded converging–diverging nozzle has throat and exit cross-sectional areas of 0.001 m^2 and 0.004 235 m^2, respectively, as shown in the figure below. Air is supplied from a large reservoir where the pressure and the temperature are 1500 kPa and 500 K, respectively. If the back pressure is 30 kPa, determine the Mach number, temperature and direction of the flow downstream of the expansion fan.

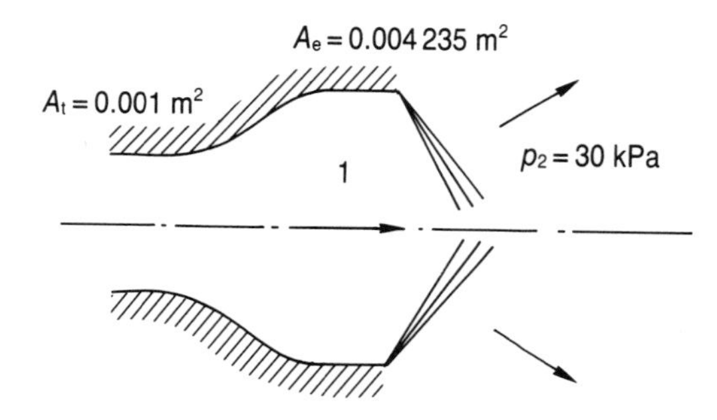

Problem 8.14

8.15 A spike-type oblique shock diffuser with one shock is operating at a Mach number of 2.6, as shown in the figure below. The temperature and

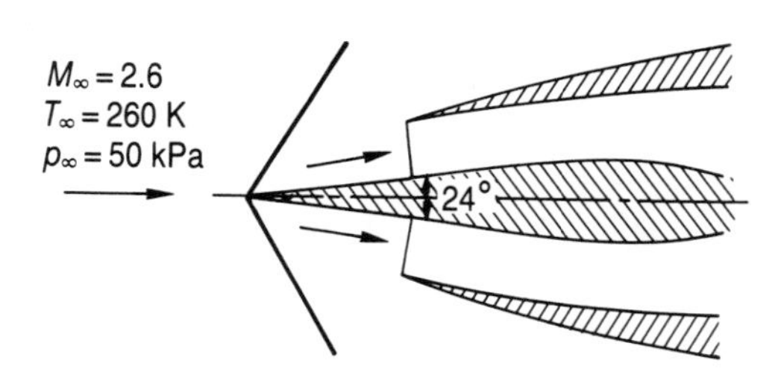

Problem 8.15

pressure of the approaching air are 260 K and 50 kPa, respectively. The cross-sectional area at the cowl inlet is 0.01 m^2. Determine the mass flow rate, pressure and temperature after the normal shock wave. The total spike angle is 24°.

8.16 A spike-type oblique shock diffuser with two shocks is operating at a Mach number of 2.2, as shown in the figure below. The temperature and pressure of the approaching air are 275 K and 75 kPa, respectively. The cross-sectional area at the cowl inlet is 0.02 m^2. Determine the mass flow rate, pressure and temperature after the normal shock wave. Each oblique shock wave turns the flow by 5°.

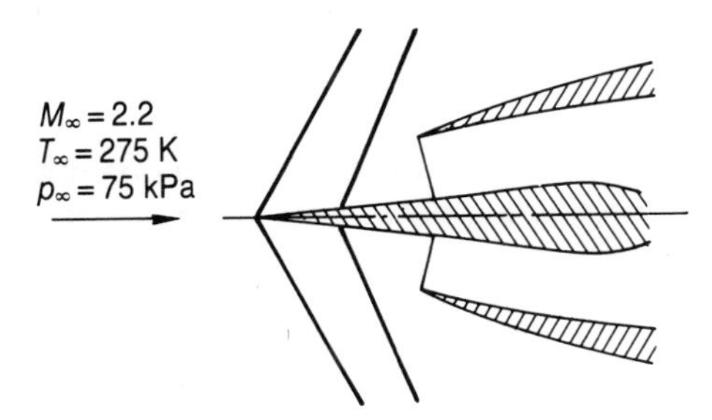

Problem 8.16

8.17 A spike-type oblique shock diffuser with one shock is operating at Mach number of 2.35 in the supercritical mode, as shown in the figure below. The temperature and pressure of the approaching air are 270 K and 65 kPa, respectively. The cross-sectional area at the cowl inlet is 0.01 m^2, while the cross-sectional area where the normal shock wave occurs is 0.0125 m^2. Determine the mass flow rate, pressure and temperature after the normal shock wave. The total spike angle is 30°.

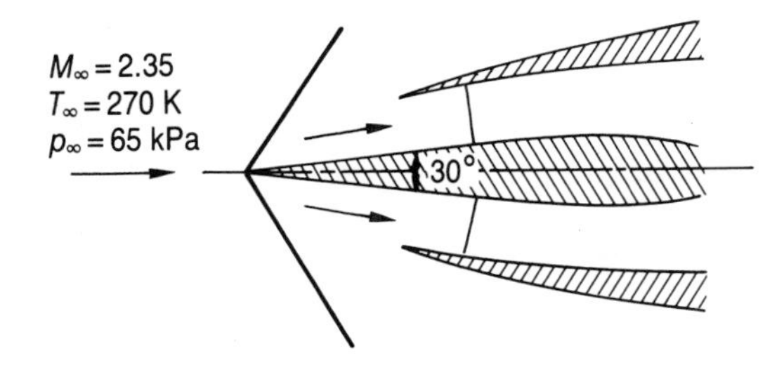

Problem 8.17

8.18 A flat plate aerofoil is placed in an airstream with a Mach number of 2.5, as shown in the figure below. If the angle of attack is 10°, determine the lift and drag coefficients.

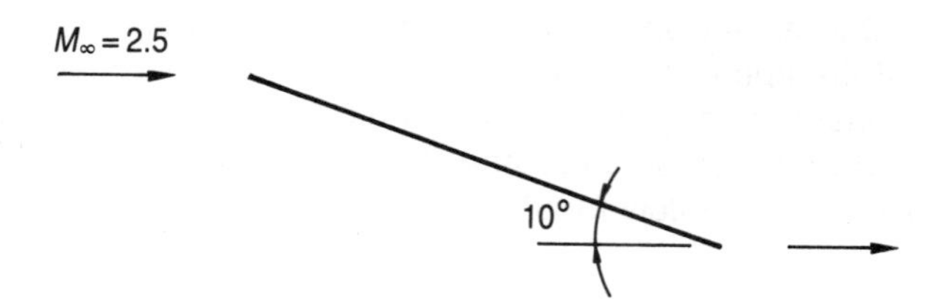

Problem 8.18

8.19 A two-dimensional aerofoil is placed in an airstream with a Mach number of 2.4, as shown in the figure below. If the angle of attack is 10°, determine the lift and drag coefficients.

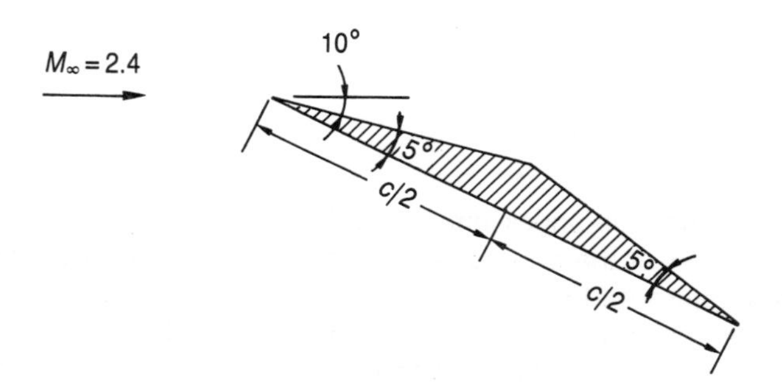

Problem 8.19

8.20 A flat plate aerofoil is placed in an airstream with a Mach number of 3.0, as shown in the figure below. Determine the lift and drag coefficients.

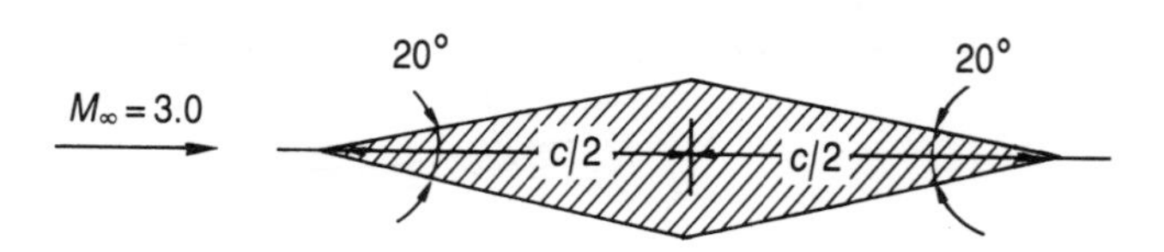

Problem 8.20

8.21 A two-dimensional aerofoil is placed in an airstream with a Mach number of 2.1, as shown in the figure below. If the angle of attack is 10°, determine the lift and drag coefficients.

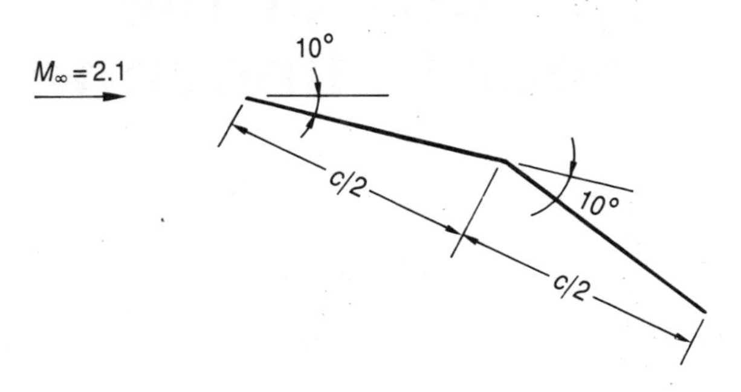

Problem 8.21

8.22 A flat plate aerofoil is placed in an airstream with a Mach number of 2.7, as shown in the figure below. If the angle of attack is 10°, determine the lift and drag coefficients.

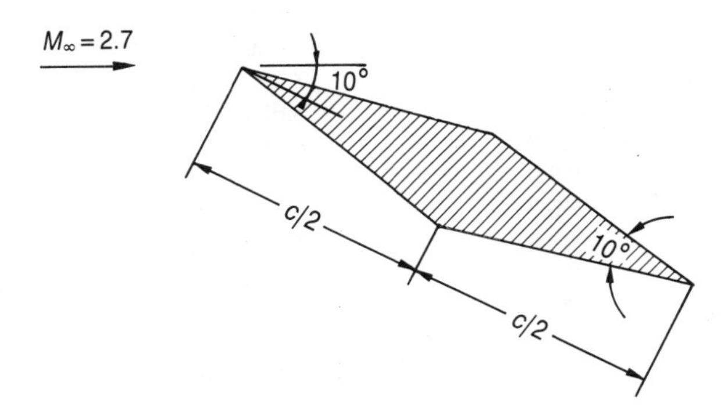

Problem 8.22

Appendix A
Properties of the standard atmosphere

Altitude (m)	*Temperature (K)*	*Pressure ($N\ m^{-2}$)*	*Density ($kg\ m^{-3}$)*
−1 000	294.7	1.139×10^5	1.347
−500	291.5	1.075×10^5	1.285
0	288.2	1.013×10^5	1.225
500	284.9	9.546×10^4	1.167
1 000	281.7	8.988×10^4	1.112
1 500	278.4	8.546×10^4	1.058
2 000	275.2	7.950×10^4	1.007
2 500	271.9	7.469×10^4	9.570×10^{-1}
3 000	268.7	7.012×10^4	9.023×10^{-1}
3 500	265.4	6.578×10^4	8.634×10^{-1}
4 000	262.2	6.166×10^4	8.192×10^{-1}
4 500	258.9	5.775×10^4	7.770×10^{-1}
5 000	255.7	5.405×10^4	7.364×10^{-1}
6 000	249.2	4.722×10^4	6.601×10^{-1}
7 000	242.7	4.111×10^4	5.900×10^{-1}
8 000	236.2	3.565×10^4	5.258×10^{-1}
9 000	229.7	3.080×10^4	4.671×10^{-1}
10 000	223.3	2.650×10^4	4.135×10^{-1}
11 000	216.8	2.270×10^4	3.648×10^{-1}
12 000	216.7	1.940×10^4	3.119×10^{-1}
13 000	216.7	1.658×10^4	2.666×10^{-1}
14 000	216.7	1.417×10^4	2.279×10^{-1}
15 000	216.7	1.211×10^4	1.948×10^{-1}
16 000	216.7	1.035×10^4	1.665×10^{-1}
17 000	216.7	8.850×10^3	1.423×10^{-1}
18 000	216.7	7.565×10^3	1.217×10^{-1}
19 000	216.7	6.468×10^3	1.040×10^{-1}
20 000	216.7	5.529×10^3	8.891×10^{-2}
21 000	217.6	4.729×10^3	7.572×10^{-2}
22 000	218.6	4.048×10^3	6.451×10^{-2}
23 000	219.6	3.467×10^3	5.501×10^{-2}
24 000	220.6	2.972×10^3	4.694×10^{-2}
25 000	221.6	2.549×10^3	4.008×10^{-2}

Appendix A *Continued*

Altitude (m)	*Temperature (K)*	*Pressure ($N\ m^{-2}$)*	*Density ($kg\ m^{-3}$)*
26 000	222.5	2.188×10^{3}	3.426×10^{-2}
27 000	223.5	1.880×10^{3}	2.930×10^{-2}
28 000	224.5	1.616×10^{3}	2.508×10^{-2}
29 000	225.5	1.390×10^{3}	2.148×10^{-2}
30 000	226.5	1.197×10^{3}	1.841×10^{-2}
31 000	227.6	1.031×10^{3}	1.579×10^{-2}
32 000	228.5	8.891×10^{2}	1.356×10^{-2}
33 000	231.0	7.673×10^{2}	1.157×10^{-2}
34 000	233.7	6.634×10^{2}	9.887×10^{-3}
35 000	236.5	5.746×10^{2}	8.463×10^{-3}
36 000	239.3	4.985×10^{2}	7.528×10^{-3}
37 000	242.1	4.333×10^{2}	6.236×10^{-3}
38 000	244.8	3.771×10^{2}	5.367×10^{-3}
39 000	247.6	3.288×10^{2}	4.627×10^{-3}
40 000	250.4	2.871×10^{2}	3.996×10^{-3}
42 000	255.9	2.200×10^{2}	2.995×10^{-3}
44 000	261.4	1.695×10^{2}	2.259×10^{-3}
46 000	266.9	1.313×10^{2}	1.714×10^{-3}
48 000	270.7	1.023×10^{2}	1.317×10^{-3}
50 000	270.7	7.978×10^{1}	1.027×10^{-3}
55 000	265.6	4.275×10^{1}	5.608×10^{-4}
60 000	255.8	2.246×10^{1}	3.059×10^{-4}
65 000	239.3	1.145×10^{1}	1.667×10^{-4}
70 000	219.7	5.251	8.754×10^{-5}
75 000	200.2	2.490	4.335×10^{-5}
80 000	180.7	1.037	1.999×10^{-5}
85 000	180.7	4.125×10^{-1}	7.955×10^{-6}
90 000	180.7	1.644×10^{-1}	3.170×10^{-6}
95 000	195.5	6.801×10^{-2}	1.211×10^{-6}
100 000	210.0	3.008×10^{-2}	4.974×10^{-7}

Appendix B
Thermodynamic properties of common gases at standard temperature and pressure (15 °C, 101.325 kPa)

Gas	*Chemical formula*	*Molecular mass, M*	*R* $(J\ kg^{-1}\ K^{-1})$	c_p $(J\ kg^{-1}\ K^{-1})$	c_v $(J\ kg^{-1}\ K^{-1})$	*k*
Air	–	28.96	287.1	1005	717.7	1.400
Carbon monoxide	CO	28.01	296.8	1043	745.8	1.398
Carbon dioxide	CO_2	44.01	188.9	844.9	655.9	1.288
Helium	He	4.000	2079	5233	3154	1.659
Hydrogen	H_2	2.016	4124	14 307	10 183	1.405
Methane	CH_4	16.04	518.3	2223	1705	1.304
Nitrogen	N_2	28.01	296.8	1039	742.0	1.400
Oxygen	O_2	32.00	259.8	917.6	657.8	1.395
Steam	H_2O	18.02	461.4	1864	1402	1.329

Appendix C
Isentropic flow of a perfect gas ($k = 1.4$)

M	M^*	p/p_0	T/T_0	ρ/ρ_0	A/A^*	I/I^*	$\dot{m}\sqrt{(RT_0)}/Ap_0$
0.00	0.000 00	1.0000	1.000	1.000	∞	∞	0.000 00
0.01	0.010 95	0.9999	1.000	0.9999	57.87	45.65	0.011 83
0.02	0.021 91	0.9997	0.9999	0.9998	28.94	22.83	0.023 66
0.03	0.032 86	0.9994	0.9998	0.9996	19.30	15.23	0.035 48
0.04	0.043 81	0.9989	0.9997	0.9992	14.48	11.43	0.047 28
0.05	0.054 76	0.9983	0.9995	0.9988	11.59	9.158	0.059 07
0.06	0.065 70	0.9975	0.9993	0.9982	9.666	7.643	0.070 84
0.07	0.076 64	0.9966	0.9990	0.9976	8.292	6.562	0.082 58
0.08	0.087 58	0.9955	0.9987	0.9968	7.262	5.753	0.094 29
0.09	0.098 51	0.9944	0.9984	0.9960	6.461	5.125	0.1060
0.10	0.1094	0.9930	0.9980	0.9950	5.822	4.624	0.1176
0.11	0.1204	0.9916	0.9976	0.9940	5.299	4.215	0.1292
0.12	0.1313	0.9900	0.9971	0.9928	4.864	3.875	0.1408
0.13	0.1422	0.9883	0.9966	0.9916	4.497	3.588	0.1523
0.14	0.1531	0.9864	0.9961	0.9903	4.182	3.343	0.1637
0.15	0.1639	0.9844	0.9955	0.9888	3.910	3.132	0.1751
0.16	0.1748	0.9823	0.9949	0.9873	3.673	2.947	0.1864
0.17	0.1857	0.9800	0.9943	0.9857	3.464	2.786	0.1977
0.18	0.1965	0.9776	0.9936	0.9840	3.278	2.642	0.2089
0.19	0.2074	0.9751	0.9928	0.9822	3.112	2.515	0.2200
0.20	0.2182	0.9725	0.9921	0.9803	2.964	2.400	0.2311
0.21	0.2290	0.9697	0.9913	0.9783	2.829	2.298	0.2420
0.22	0.2398	0.9668	0.9904	0.9762	2.708	2.205	0.2529
0.23	0.2506	0.9638	0.9895	0.9740	2.597	2.120	0.2637
0.24	0.2614	0.9607	0.9886	0.9718	2.496	2.043	0.2744
0.25	0.2722	0.9575	0.9877	0.9694	2.403	1.973	0.2850
0.26	0.2829	0.9541	0.9867	0.9670	2.317	1.909	0.2955
0.27	0.2936	0.9506	0.9856	0.9645	2.238	1.850	0.3059
0.28	0.3043	0.9470	0.9846	0.9619	2.166	1.795	0.3162
0.29	0.3150	0.9433	0.9835	0.9592	2.098	1.745	0.3264
0.30	0.3257	0.9395	0.9823	0.9564	2.035	1.698	0.3365
0.31	0.3364	0.9355	0.9811	0.9535	1.977	1.655	0.3464

continued

Appendix C *Continued*

M	M^*	p/p_0	T/T_0	ρ/ρ_0	A/A^*	I/I^*	$\dot{m}\sqrt{(RT_0)}/Ap_0$
0.32	0.3470	0.9315	0.9799	0.9506	1.922	1.614	0.3563
0.33	0.3576	0.9274	0.9787	0.9476	1.871	1.577	0.3660
0.34	0.3682	0.9231	0.9774	0.9445	1.823	1.542	0.3756
0.35	0.3788	0.9188	0.9761	0.9413	1.778	1.509	0.3851
0.36	0.3893	0.9143	0.9747	0.9380	1.736	1.479	0.3945
0.37	0.3999	0.9098	0.9733	0.9347	1.696	1.450	0.4037
0.38	0.4104	0.9052	0.9719	0.9313	1.659	1.424	0.4128
0.39	0.4209	0.9004	0.9705	0.9278	1.623	1.398	0.4218
0.40	0.4313	0.8956	0.9690	0.9243	1.590	1.375	0.4306
0.41	0.4418	0.8907	0.9675	0.9207	1.559	1.353	0.4393
0.42	0.4522	0.8857	0.9659	0.9170	1.529	1.332	0.4479
0.43	0.4626	0.8807	0.9643	0.9132	1.501	1.312	0.4563
0.44	0.4729	0.8755	0.9627	0.9094	1.474	1.294	0.4645
0.45	0.4833	0.8703	0.9611	0.9055	1.449	1.276	0.4727
0.46	0.4936	0.8650	0.9594	0.9016	1.425	1.260	0.4806
0.47	0.5038	0.8596	0.9577	0.8976	1.402	1.244	0.4885
0.48	0.5141	0.8541	0.9559	0.8935	1.380	1.230	0.4961
0.49	0.5243	0.8486	0.9542	0.8894	1.359	1.216	0.5037
0.50	0.5345	0.8430	0.9524	0.8852	1.340	1.203	0.5111
0.51	0.5447	0.8374	0.9506	0.8809	1.321	1.190	0.5183
0.52	0.5548	0.8317	0.9487	0.8766	1.303	1.179	0.5253
0.53	0.5649	0.8259	0.9468	0.8723	1.286	1.168	0.5323
0.54	0.5750	0.8201	0.9449	0.8679	1.270	1.157	0.5390
0.55	0.5851	0.8142	0.9430	0.8634	1.255	1.147	0.5456
0.56	0.5951	0.8082	0.9410	0.8589	1.240	1.138	0.5521
0.57	0.6051	0.8022	0.9390	0.8544	1.226	1.129	0.5584
0.58	0.6150	0.7962	0.9370	0.8498	1.213	1.121	0.5645
0.59	0.6249	0.7901	0.9349	0.8451	1.200	1.113	0.5705
0.60	0.6348	0.7840	0.9328	0.8405	1.188	1.105	0.5763
0.61	0.6447	0.7778	0.9307	0.8357	1.177	1.098	0.5819
0.62	0.6545	0.7716	0.9286	0.8310	1.166	1.091	0.5874
0.63	0.6643	0.7654	0.9265	0.8262	1.155	1.085	0.5928
0.64	0.6740	0.7591	0.9243	0.8213	1.145	1.079	0.5979
0.65	0.6837	0.7528	0.9221	0.8164	1.136	1.073	0.6030
0.66	0.6934	0.7465	0.9199	0.8115	1.127	1.068	0.6078
0.67	0.7031	0.7401	0.9176	0.8066	1.118	1.063	0.6125
0.68	0.7127	0.7338	0.9153	0.8016	1.110	1.058	0.6171
0.69	0.7223	0.7274	0.9131	0.7966	1.102	1.053	0.6215
0.70	0.7318	0.7209	0.9107	0.7916	1.094	1.049	0.6257
0.71	0.7413	0.7145	0.9084	0.7865	1.087	1.045	0.6298
0.72	0.7508	0.7080	0.9061	0.7814	1.081	1.041	0.6337
0.73	0.7602	0.7016	0.9037	0.7763	1.074	1.038	0.6374
0.74	0.7696	0.6951	0.9013	0.7712	1.068	1.034	0.6410
0.75	0.7789	0.6886	0.8989	0.7660	1.062	1.031	0.6445

Appendix C *Continued*

M	M^*	p/p_0	T/T_0	ρ/ρ_0	A/A^*	I/I^*	$\dot{m}\sqrt{(RT_0)}/Ap_0$
0.76	0.7883	0.6821	0.8964	0.7609	1.057	1.028	0.6478
0.77	0.7975	0.6756	0.8940	0.7557	1.052	1.026	0.6510
0.78	0.8068	0.6691	0.8915	0.7505	1.047	1.023	0.6540
0.79	0.8160	0.6625	0.8890	0.7452	1.043	1.021	0.6568
0.80	0.8251	0.6560	0.8865	0.7400	1.038	1.019	0.6595
0.81	0.8343	0.6495	0.8840	0.7347	1.034	1.016	0.6621
0.82	0.8433	0.6430	0.8815	0.7295	1.030	1.015	0.6645
0.83	0.8524	0.6365	0.8789	0.7242	1.027	1.013	0.6668
0.84	0.8614	0.6300	0.8763	0.7189	1.024	1.011	0.6689
0.85	0.8704	0.6235	0.8737	0.7136	1.021	1.010	0.6709
0.86	0.8793	0.6170	0.8711	0.7083	1.018	1.008	0.6727
0.87	0.8882	0.6106	0.8685	0.7030	1.015	1.007	0.6744
0.88	0.8970	0.6041	0.8659	0.6977	1.013	1.006	0.6760
0.89	0.9058	0.5977	0.8632	0.6924	1.011	1.005	0.6774
0.90	0.9146	0.5913	0.8606	0.6870	1.009	1.004	0.6787
0.91	0.9233	0.5849	0.8579	0.6817	1.007	1.003	0.6799
0.92	0.9320	0.5785	0.8552	0.6764	1.006	1.002	0.6809
0.93	0.9407	0.5721	0.8525	0.6711	1.004	1.002	0.6818
0.94	0.9493	0.5658	0.8498	0.6658	1.003	1.001	0.6826
0.95	0.9578	0.5595	0.8471	0.6604	1.002	1.001	0.6833
0.96	0.9663	0.5532	0.8444	0.6551	1.001	1.001	0.6838
0.97	0.9748	0.5469	0.8416	0.6498	1.001	1.000	0.6842
0.98	0.9832	0.5407	0.8389	0.6445	1.000	1.000	0.6845
0.99	0.9916	0.5345	0.8361	0.6392	1.000	1.000	0.6847
1.00	1.000	0.5283	0.8333	0.6339	1.000	1.000	0.6847
1.01	1.008	0.5221	0.8306	0.6287	1.000	1.000	0.6847
1.02	1.017	0.5160	0.8278	0.6234	1.000	1.000	0.6845
1.03	1.025	0.5099	0.8250	0.6181	1.001	1.000	0.6842
1.04	1.033	0.5039	0.8222	0.6129	1.001	1.001	0.6838
1.05	1.041	0.4979	0.8193	0.6077	1.002	1.001	0.6833
1.06	1.049	0.4919	0.8165	0.6024	1.003	1.001	0.6827
1.07	1.057	0.4860	0.8137	0.5972	1.004	1.002	0.6820
1.08	1.065	0.4800	0.8108	0.5920	1.005	1.002	0.6812
1.09	1.073	0.4742	0.8080	0.5869	1.006	1.003	0.6803
1.10	1.081	0.4684	0.8052	0.5817	1.008	1.003	0.6793
1.11	1.089	0.4626	0.8023	0.5766	1.010	1.004	0.6783
1.12	1.097	0.4568	0.7994	0.5714	1.011	1.004	0.6771
1.13	1.105	0.4511	0.7966	0.5663	1.013	1.005	0.6758
1.14	1.113	0.4455	0.7937	0.5612	1.015	1.006	0.6744
1.15	1.120	0.4398	0.7908	0.5562	1.017	1.006	0.6730
1.16	1.128	0.4343	0.7879	0.5511	1.020	1.007	0.6715
1.17	1.136	0.4287	0.7851	0.5461	1.022	1.008	0.6698

continued

Appendix C *Continued*

M	M^*	p/p_0	T/T_0	ρ/ρ_0	A/A^*	I/I^*	$\dot{m}\sqrt{(RT_0)}/Ap_0$
1.18	1.143	0.4232	0.7822	0.5411	1.025	1.009	0.6681
1.19	1.151	0.4178	0.7793	0.5361	1.028	1.010	0.6664
1.20	1.158	0.4124	0.7764	0.5311	1.030	1.011	0.6645
1.21	1.166	0.4070	0.7735	0.5262	1.033	1.012	0.6626
1.22	1.173	0.4017	0.7706	0.5213	1.037	1.013	0.6606
1.23	1.181	0.3964	0.7677	0.5164	1.040	1.014	0.6585
1.24	1.188	0.3912	0.7648	0.5115	1.043	1.015	0.6564
1.25	1.195	0.3861	0.7619	0.5067	1.047	1.016	0.6541
1.26	1.202	0.3809	0.7590	0.5019	1.050	1.017	0.6519
1.27	1.210	0.3759	0.7561	0.4971	1.054	1.018	0.6495
1.28	1.217	0.3708	0.7532	0.4923	1.058	1.019	0.6471
1.29	1.224	0.3658	0.7503	0.4876	1.062	1.021	0.6447
1.30	1.231	0.3609	0.7474	0.4829	1.066	1.022	0.6422
1.31	1.238	0.3560	0.7445	0.4782	1.071	1.023	0.6396
1.32	1.245	0.3512	0.7416	0.4736	1.075	1.024	0.6369
1.33	1.252	0.3464	0.7387	0.4690	1.080	1.025	0.6343
1.34	1.259	0.3417	0.7358	0.4644	1.084	1.027	0.6315
1.35	1.266	0.3370	0.7329	0.4598	1.089	1.028	0.6287
1.36	1.273	0.3323	0.7300	0.4553	1.094	1.029	0.6259
1.37	1.280	0.3277	0.7271	0.4508	1.099	1.031	0.6230
1.38	1.286	0.3232	0.7242	0.4463	1.104	1.032	0.6201
1.39	1.293	0.3187	0.7213	0.4418	1.109	1.033	0.6172
1.40	1.300	0.3142	0.7184	0.4374	1.115	1.035	0.6141
1.41	1.307	0.3098	0.7155	0.4330	1.120	1.036	0.6111
1.42	1.313	0.3055	0.7126	0.4287	1.126	1.037	0.6080
1.43	1.320	0.3012	0.7097	0.4244	1.132	1.039	0.6049
1.44	1.326	0.2969	0.7069	0.4201	1.138	1.040	0.6017
1.45	1.333	0.2927	0.7040	0.4158	1.144	1.042	0.5986
1.46	1.339	0.2886	0.7011	0.4116	1.150	1.043	0.5953
1.47	1.346	0.2845	0.6982	0.4074	1.156	1.044	0.5921
1.48	1.352	0.2804	0.6954	0.4032	1.163	1.046	0.5888
1.49	1.358	0.2764	0.6925	0.3991	1.169	1.047	0.5855
1.50	1.365	0.2724	0.6897	0.3950	1.176	1.049	0.5822
1.51	1.371	0.2685	0.6868	0.3909	1.183	1.050	0.5788
1.52	1.377	0.2646	0.6840	0.3869	1.190	1.052	0.5754
1.53	1.383	0.2608	0.6811	0.3829	1.197	1.053	0.5720
1.54	1.389	0.2570	0.6783	0.3789	1.204	1.055	0.5686
1.55	1.395	0.2533	0.6754	0.3750	1.212	1.056	0.5652
1.56	1.402	0.2496	0.6726	0.3710	1.219	1.058	0.5617
1.57	1.408	0.2459	0.6698	0.3672	1.227	1.059	0.5582
1.58	1.414	0.2423	0.6670	0.3633	1.234	1.060	0.5547
1.59	1.419	0.2388	0.6642	0.3595	1.242	1.062	0.5512
1.60	1.425	0.2353	0.6614	0.3557	1.250	1.063	0.5477
1.61	1.431	0.2318	0.6586	0.3520	1.258	1.065	0.5441

Appendix C *Continued*

M	M^*	p/p_0	T/T_0	ρ/ρ_0	A/A^*	I/I^*	$\dot{m}\sqrt{(RT_0)}/Ap_0$
1.62	1.437	0.2284	0.6558	0.3483	1.267	1.066	0.5406
1.63	1.443	0.2250	0.6530	0.3446	1.275	1.068	0.5370
1.64	1.449	0.2217	0.6502	0.3409	1.284	1.069	0.5335
1.65	1.454	0.2184	0.6475	0.3373	1.292	1.071	0.5299
1.66	1.460	0.2151	0.6447	0.3337	1.301	1.072	0.5263
1.67	1.466	0.2119	0.6419	0.3302	1.310	1.074	0.5227
1.68	1.471	0.2088	0.6392	0.3266	1.319	1.075	0.5191
1.69	1.477	0.2057	0.6364	0.3232	1.328	1.077	0.5155
1.70	1.482	0.2026	0.6337	0.3197	1.338	1.079	0.5119
1.71	1.488	0.1996	0.6310	0.3163	1.347	1.080	0.5083
1.72	1.493	0.1966	0.6283	0.3129	1.357	1.082	0.5047
1.73	1.499	0.1936	0.6256	0.3095	1.367	1.083	0.5011
1.74	1.504	0.1907	0.6229	0.3062	1.376	1.085	0.4975
1.75	1.510	0.1878	0.6202	0.3029	1.386	1.086	0.4939
1.76	1.515	0.1850	0.6175	0.2996	1.397	1.088	0.4903
1.77	1.520	0.1822	0.6148	0.2964	1.407	1.089	0.4866
1.78	1.526	0.1794	0.6121	0.2931	1.418	1.091	0.4830
1.79	1.531	0.1767	0.6095	0.2900	1.428	1.092	0.4794
1.80	1.536	0.1740	0.6068	0.2868	1.439	1.094	0.4758
1.81	1.541	0.1714	0.6041	0.2837	1.450	1.095	0.4723
1.82	1.546	0.1688	0.6015	0.2806	1.461	1.096	0.4687
1.83	1.551	0.1662	0.5989	0.2776	1.472	1.098	0.4651
1.84	1.556	0.1637	0.5963	0.2745	1.484	1.099	0.4615
1.85	1.561	0.1612	0.5936	0.2715	1.495	1.101	0.4580
1.86	1.566	0.1587	0.5910	0.2686	1.507	1.102	0.4544
1.87	1.571	0.1563	0.5885	0.2656	1.519	1.104	0.4509
1.88	1.576	0.1539	0.5859	0.2627	1.531	1.105	0.4473
1.89	1.581	0.1516	0.5833	0.2598	1.543	1.107	0.4438
1.90	1.586	0.1492	0.5807	0.2570	1.555	1.108	0.4403
1.91	1.591	0.1470	0.5782	0.2542	1.568	1.110	0.4368
1.92	1.596	0.1447	0.5756	0.2514	1.580	1.111	0.4333
1.93	1.600	0.1425	0.5731	0.2486	1.593	1.113	0.4298
1.94	1.605	0.1403	0.5705	0.2459	1.606	1.114	0.4263
1.95	1.610	0.1381	0.5680	0.2432	1.619	1.116	0.4229
1.96	1.615	0.1360	0.5655	0.2405	1.633	1.117	0.4194
1.97	1.619	0.1339	0.5630	0.2378	1.646	1.118	0.4160
1.98	1.624	0.1318	0.5605	0.2352	1.660	1.120	0.4126
1.99	1.628	0.1298	0.5580	0.2326	1.674	1.121	0.4092
2.00	1.633	0.1278	0.5556	0.2300	1.687	1.123	0.4058
2.01	1.638	0.1258	0.5531	0.2275	1.702	1.124	0.4024
2.02	1.642	0.1239	0.5506	0.2250	1.716	1.126	0.3990
2.03	1.646	0.1220	0.5482	0.2225	1.730	1.127	0.3957

continued

Appendix C *Continued*

M	M^*	p/p_0	T/T_0	ρ/ρ_0	A/A^*	I/I^*	$\dot{m}\sqrt{(RT_0)}/Ap_0$
2.04	1.651	0.1201	0.5458	0.2200	1.745	1.128	0.3924
2.05	1.655	0.1182	0.5433	0.2176	1.760	1.130	0.3891
2.06	1.660	0.1164	0.5409	0.2152	1.775	1.131	0.3858
2.07	1.664	0.1146	0.5385	0.2128	1.790	1.132	0.3825
2.08	1.668	0.1128	0.5361	0.2104	1.806	1.134	0.3792
2.09	1.673	0.1111	0.5337	0.2081	1.821	1.135	0.3760
2.10	1.677	0.1094	0.5313	0.2058	1.837	1.137	0.3728
2.11	1.681	0.1077	0.5290	0.2035	1.853	1.138	0.3695
2.12	1.685	0.1060	0.5266	0.2013	1.869	1.139	0.3664
2.13	1.689	0.1043	0.5243	0.1990	1.885	1.141	0.3632
2.14	1.694	0.1027	0.5219	0.1968	1.902	1.142	0.3600
2.15	1.698	0.1011	0.5196	0.1946	1.919	1.143	0.3569
2.16	1.702	0.099 56	0.5173	0.1925	1.935	1.145	0.3538
2.17	1.706	0.098 02	0.5150	0.1903	1.953	1.146	0.3507
2.18	1.710	0.096 50	0.5127	0.1882	1.970	1.147	0.3476
2.19	1.714	0.095 00	0.5104	0.1861	1.987	1.149	0.3446
2.20	1.718	0.093 52	0.5081	0.1841	2.005	1.150	0.3415
2.21	1.722	0.092 07	0.5059	0.1820	2.023	1.151	0.3385
2.22	1.726	0.090 64	0.5036	0.1800	2.041	1.153	0.3355
2.23	1.730	0.089 23	0.5014	0.1780	2.059	1.154	0.3325
2.24	1.734	0.087 85	0.4991	0.1760	2.078	1.155	0.3296
2.25	1.737	0.086 48	0.4969	0.1740	2.096	1.156	0.3266
2.26	1.741	0.085 14	0.4947	0.1721	2.115	1.158	0.3237
2.27	1.745	0.083 82	0.4925	0.1702	2.134	1.159	0.3208
2.28	1.749	0.082 52	0.4903	0.1683	2.154	1.160	0.3179
2.29	1.753	0.081 23	0.4881	0.1664	2.173	1.162	0.3151
2.30	1.756	0.079 97	0.4859	0.1646	2.193	1.163	0.3122
2.31	1.760	0.078 73	0.4837	0.1628	2.213	1.164	0.3094
2.32	1.764	0.077 51	0.4816	0.1609	2.233	1.165	0.3066
2.33	1.767	0.076 31	0.4794	0.1592	2.254	1.167	0.3038
2.34	1.771	0.075 12	0.4773	0.1574	2.274	1.168	0.3011
2.35	1.775	0.073 96	0.4752	0.1556	2.295	1.169	0.2983
2.36	1.778	0.072 81	0.4731	0.1539	2.316	1.170	0.2956
2.37	1.782	0.071 68	0.4709	0.1522	2.338	1.171	0.2929
2.38	1.785	0.070 57	0.4688	0.1505	2.359	1.173	0.2902
2.39	1.789	0.069 48	0.4668	0.1488	2.381	1.174	0.2876
2.40	1.792	0.068 40	0.4647	0.1472	2.403	1.175	0.2849
2.41	1.796	0.067 34	0.4626	0.1456	2.425	1.176	0.2823
2.42	1.799	0.066 30	0.4606	0.1439	2.448	1.177	0.2797
2.43	1.802	0.065 27	0.4585	0.1424	2.471	1.179	0.2772
2.44	1.806	0.064 26	0.4565	0.1408	2.494	1.180	0.2746
2.45	1.809	0.063 27	0.4544	0.1392	2.517	1.181	0.2721
2.46	1.813	0.062 29	0.4524	0.1377	2.540	1.182	0.2695
2.47	1.816	0.061 33	0.4504	0.1362	2.564	1.183	0.2671

Appendix C *Continued*

M	M^*	p/p_0	T/T_0	ρ/ρ_0	A/A^*	I/I^*	$\dot{m}\sqrt{(RT_0)}/Ap_0$
2.48	1.819	0.060 38	0.4484	0.1346	2.588	1.184	0.2646
2.49	1.822	0.059 45	0.4464	0.1332	2.612	1.186	0.2621
2.50	1.826	0.058 53	0.4444	0.1317	2.637	1.187	0.2597
2.51	1.829	0.057 62	0.4425	0.1302	2.661	1.188	0.2573
2.52	1.832	0.056 74	0.4405	0.1288	2.686	1.189	0.2549
2.53	1.835	0.055 86	0.4386	0.1274	2.712	1.190	0.2525
2.54	1.839	0.055 00	0.4366	0.1260	2.737	1.191	0.2502
2.55	1.842	0.054 15	0.4347	0.1246	2.763	1.192	0.2478
2.56	1.845	0.053 32	0.4328	0.1232	2.789	1.193	0.2455
2.57	1.848	0.052 50	0.4309	0.1218	2.815	1.195	0.2432
2.58	1.851	0.051 69	0.4289	0.1205	2.842	1.196	0.2409
2.59	1.854	0.050 90	0.4271	0.1192	2.869	1.197	0.2387
2.60	1.857	0.050 12	0.4252	0.1179	2.896	1.198	0.2364
2.61	1.860	0.049 35	0.4233	0.1166	2.923	1.199	0.2342
2.62	1.863	0.048 59	0.4214	0.1153	2.951	1.200	0.2320
2.63	1.866	0.047 84	0.4196	0.1140	2.979	1.201	0.2298
2.64	1.869	0.047 11	0.4177	0.1128	3.007	1.202	0.2277
2.65	1.872	0.046 39	0.4159	0.1115	3.036	1.203	0.2255
2.66	1.875	0.045 68	0.4141	0.1103	3.065	1.204	0.2234
2.67	1.878	0.044 98	0.4122	0.1091	3.094	1.205	0.2213
2.68	1.881	0.044 29	0.4104	0.1079	3.123	1.206	0.2192
2.69	1.884	0.043 62	0.4086	0.1067	3.153	1.207	0.2172
2.70	1.887	0.042 95	0.4068	0.1056	3.183	1.208	0.2151
2.71	1.889	0.042 29	0.4051	0.1044	3.213	1.209	0.2131
2.72	1.892	0.041 65	0.4033	0.1033	3.244	1.210	0.2111
2.73	1.895	0.041 02	0.4015	0.1022	3.275	1.211	0.2091
2.74	1.898	0.040 39	0.3998	0.1010	3.306	1.212	0.2071
2.75	1.901	0.039 78	0.3980	0.099 94	3.338	1.213	0.2052
2.76	1.903	0.039 17	0.3963	0.098 85	3.370	1.214	0.2032
2.77	1.906	0.038 58	0.3945	0.097 78	3.402	1.215	0.2013
2.78	1.909	0.037 99	0.3928	0.096 71	3.434	1.216	0.1994
2.79	1.911	0.037 42	0.3911	0.095 66	3.467	1.217	0.1975
2.80	1.914	0.036 85	0.3894	0.094 63	3.500	1.218	0.1956
2.81	1.917	0.036 29	0.3877	0.093 60	3.534	1.219	0.1938
2.82	1.919	0.035 74	0.3860	0.092 59	3.567	1.220	0.1919
2.83	1.922	0.035 20	0.3844	0.091 58	3.601	1.221	0.1901
2.84	1.925	0.034 67	0.3827	0.090 59	3.636	1.222	0.1883
2.85	1.927	0.034 15	0.3810	0.089 62	3.671	1.223	0.1865
2.86	1.930	0.033 63	0.3794	0.088 65	3.706	1.224	0.1848
2.87	1.932	0.033 12	0.3777	0.087 69	3.741	1.225	0.1830
2.88	1.935	0.032 63	0.3761	0.086 75	3.777	1.226	0.1813
2.89	1.937	0.032 13	0.3745	0.085 81	3.813	1.227	0.1796

continued

Appendix C *Continued*

M	M^*	p/p_0	T/T_0	ρ/ρ_0	A/A^*	I/I^*	$\dot{m}\sqrt{(RT_0)}/Ap_0$
2.90	1.940	0.031 65	0.3729	0.084 89	3.850	1.228	0.1779
2.91	1.942	0.031 18	0.3712	0.083 98	3.887	1.229	0.1762
2.92	1.945	0.030 71	0.3696	0.083 07	3.924	1.229	0.1745
2.93	1.947	0.030 25	0.3681	0.082 18	3.961	1.230	0.1729
2.94	1.950	0.029 80	0.3665	0.081 30	3.999	1.231	0.1712
2.95	1.952	0.029 35	0.3649	0.080 43	4.038	1.232	0.1696
2.96	1.954	0.028 91	0.3633	0.079 57	4.076	1.233	0.1680
2.97	1.957	0.028 48	0.3618	0.078 72	4.115	1.234	0.1664
2.98	1.959	0.028 05	0.3602	0.077 88	4.155	1.235	0.1648
2.99	1.962	0.027 64	0.3587	0.077 05	4.194	1.236	0.1632
3.00	1.964	0.027 22	0.3571	0.076 23	4.235	1.237	0.1617
3.01	1.966	0.026 82	0.3556	0.075 41	4.275	1.237	0.1602
3.02	1.969	0.026 42	0.3541	0.074 61	4.316	1.238	0.1586
3.03	1.971	0.026 03	0.3526	0.073 82	4.357	1.239	0.1571
3.04	1.973	0.025 64	0.3511	0.073 03	4.399	1.240	0.1557
3.05	1.975	0.025 26	0.3496	0.072 26	4.441	1.241	0.1542
3.06	1.978	0.024 89	0.3481	0.071 49	4.483	1.242	0.1527
3.07	1.980	0.024 52	0.3466	0.070 74	4.526	1.243	0.1513
3.08	1.982	0.024 16	0.3452	0.069 99	4.570	1.243	0.1498
3.09	1.984	0.023 80	0.3437	0.069 25	4.613	1.244	0.1484
3.10	1.987	0.023 45	0.3422	0.068 52	4.657	1.245	0.1470
3.11	1.989	0.023 10	0.3408	0.067 79	4.702	1.246	0.1456
3.12	1.991	0.022 76	0.3393	0.067 08	4.747	1.247	0.1443
3.13	1.993	0.022 43	0.3379	0.066 37	4.792	1.247	0.1429
3.14	1.995	0.022 10	0.3365	0.065 68	4.838	1.248	0.1415
3.15	1.997	0.021 77	0.3351	0.064 99	4.884	1.249	0.1402
3.16	2.000	0.021 46	0.3337	0.064 30	4.930	1.250	0.1389
3.17	2.002	0.021 14	0.3323	0.063 63	4.977	1.251	0.1376
3.18	2.004	0.020 83	0.3309	0.062 96	5.025	1.251	0.1363
3.19	2.006	0.020 53	0.3295	0.062 31	5.073	1.252	0.1350
3.20	2.008	0.020 23	0.3281	0.061 65	5.121	1.253	0.1337
3.21	2.010	0.019 93	0.3267	0.061 01	5.170	1.254	0.1325
3.22	2.012	0.019 64	0.3253	0.060 37	5.219	1.254	0.1312
3.23	2.014	0.019 36	0.3240	0.059 75	5.268	1.255	0.1300
3.24	2.016	0.019 08	0.3226	0.059 12	5.319	1.256	0.1287
3.25	2.018	0.018 80	0.3213	0.058 51	5.369	1.257	0.1275
3.26	2.020	0.018 53	0.3199	0.057 90	5.420	1.258	0.1263
3.27	2.022	0.018 26	0.3186	0.057 30	5.472	1.258	0.1251
3.28	2.024	0.017 99	0.3173	0.056 71	5.523	1.259	0.1240
3.29	2.026	0.017 73	0.3160	0.056 12	5.576	1.260	0.1228
3.30	2.028	0.017 48	0.3147	0.055 54	5.629	1.260	0.1217
3.31	2.030	0.017 22	0.3134	0.054 97	5.682	1.261	0.1205
3.32	2.032	0.016 98	0.3121	0.054 40	5.736	1.262	0.1194
3.33	2.034	0.016 73	0.3108	0.053 84	5.790	1.263	0.1183

Appendix C *Continued*

M	M^*	p/p_0	T/T_0	ρ/ρ_0	A/A^*	I/I^*	$\dot{m}\sqrt{(RT_0)}/Ap_0$
3.34	2.035	0.016 49	0.3095	0.053 29	5.845	1.263	0.1172
3.35	2.037	0.016 25	0.3082	0.052 74	5.900	1.264	0.1161
3.36	2.039	0.016 02	0.3069	0.052 20	5.956	1.265	0.1150
3.37	2.041	0.015 79	0.3057	0.051 66	6.012	1.265	0.1139
3.38	2.043	0.015 57	0.3044	0.051 13	6.069	1.266	0.1128
3.39	2.045	0.015 34	0.3032	0.050 61	6.126	1.267	0.1118
3.40	2.047	0.015 12	0.3019	0.050 09	6.184	1.268	0.1107
3.41	2.048	0.014 91	0.3007	0.049 58	6.242	1.268	0.1097
3.42	2.050	0.014 70	0.2995	0.049 08	6.301	1.269	0.1087
3.43	2.052	0.014 49	0.2982	0.048 58	6.360	1.270	0.1077
3.44	2.054	0.014 28	0.2970	0.048 08	6.420	1.270	0.1067
3.45	2.056	0.014 08	0.2958	0.047 59	6.480	1.271	0.1057
3.46	2.057	0.013 88	0.2946	0.047 11	6.541	1.272	0.1047
3.47	2.059	0.013 68	0.2934	0.046 63	6.602	1.272	0.1037
3.48	2.061	0.013 49	0.2922	0.046 16	6.664	1.273	0.1027
3.49	2.062	0.013 30	0.2910	0.045 69	6.727	1.274	0.1018
3.50	2.064	0.013 11	0.2899	0.045 23	6.790	1.274	0.1008
3.51	2.066	0.012 93	0.2887	0.044 78	6.853	1.275	0.099 91
3.52	2.068	0.012 74	0.2875	0.044 33	6.917	1.276	0.098 99
3.53	2.069	0.012 56	0.2864	0.043 88	6.982	1.276	0.098 07
3.54	2.071	0.012 39	0.2852	0.043 44	7.047	1.277	0.097 17
3.55	2.073	0.012 21	0.2841	0.043 00	7.113	1.278	0.096 27
3.56	2.074	0.012 04	0.2829	0.042 57	7.179	1.278	0.095 38
3.57	2.076	0.011 88	0.2818	0.042 14	7.246	1.279	0.094 50
3.58	2.078	0.011 71	0.2806	0.041 72	7.313	1.279	0.093 63
3.59	2.079	0.011 55	0.2795	0.041 31	7.381	1.280	0.092 76
3.60	2.081	0.011 38	0.2784	0.040 89	7.450	1.281	0.091 91
3.61	2.082	0.011 23	0.2773	0.040 49	7.519	1.281	0.091 06
3.62	2.084	0.011 07	0.2762	0.040 08	7.589	1.282	0.090 23
3.63	2.086	0.010 92	0.2751	0.039 68	7.659	1.283	0.089 40
3.64	2.087	0.010 76	0.2740	0.039 29	7.730	1.283	0.088 58
3.65	2.089	0.010 62	0.2729	0.038 90	7.802	1.284	0.087 76
3.66	2.090	0.010 47	0.2718	0.038 52	7.874	1.284	0.086 96
3.67	2.092	0.010 32	0.2707	0.038 13	7.947	1.285	0.086 16
3.68	2.093	0.010 18	0.2697	0.037 76	8.020	1.286	0.085 37
3.69	2.095	0.010 04	0.2686	0.037 39	8.094	1.286	0.084 59
3.70	2.096	0.009 903	0.2675	0.037 02	8.169	1.287	0.083 82
3.71	2.098	0.009 767	0.2665	0.036 65	8.244	1.287	0.083 05
3.72	2.099	0.009 633	0.2654	0.036 29	8.320	1.288	0.082 30
3.73	2.101	0.009 500	0.2644	0.035 94	8.397	1.288	0.081 55
3.74	2.102	0.009 370	0.2633	0.035 58	8.474	1.289	0.080 80
3.75	2.104	0.009 242	0.2623	0.035 24	8.552	1.290	0.080 07

continued

Appendix C *Continued*

M	M^*	p/p_0	T/T_0	ρ/ρ_0	A/A^*	I/I^*	$\dot{m}\sqrt{(RT_0)}/Ap_0$
3.76	2.105	0.009 116	0.2613	0.034 89	8.630	1.290	0.079 34
3.77	2.107	0.008 991	0.2602	0.034 55	8.709	1.291	0.078 62
3.78	2.108	0.008 869	0.2592	0.034 21	8.789	1.291	0.077 91
3.79	2.110	0.008 748	0.2582	0.033 88	8.869	1.292	0.077 20
3.80	2.111	0.008 629	0.2572	0.033 55	8.951	1.292	0.076 50
3.81	2.113	0.008 512	0.2562	0.033 22	9.032	1.293	0.075 81
3.82	2.114	0.008 396	0.2552	0.032 90	9.115	1.293	0.075 12
3.83	2.115	0.008 283	0.2542	0.032 58	9.198	1.294	0.074 44
3.84	2.117	0.008 171	0.2532	0.032 27	9.282	1.295	0.073 77
3.85	2.118	0.008 060	0.2522	0.031 95	9.366	1.295	0.073 11
3.86	2.120	0.007 951	0.2513	0.031 65	9.451	1.296	0.072 45
3.87	2.121	0.007 844	0.2503	0.031 34	9.537	1.296	0.071 80
3.88	2.122	0.007 739	0.2493	0.031 04	9.624	1.297	0.071 15
3.89	2.124	0.007 635	0.2484	0.030 74	9.711	1.297	0.070 51
3.90	2.125	0.007 532	0.2474	0.030 44	9.799	1.298	0.069 88
3.91	2.126	0.007 431	0.2464	0.030 15	9.888	1.298	0.069 25
3.92	2.128	0.007 332	0.2455	0.029 86	9.977	1.299	0.068 63
3.93	2.129	0.007 233	0.2446	0.029 58	10.07	1.299	0.068 02
3.94	2.130	0.007 137	0.2436	0.029 29	10.16	1.300	0.067 41
3.95	2.132	0.007 042	0.2427	0.029 02	10.25	1.300	0.066 81
3.96	2.133	0.006 948	0.2418	0.028 74	10.34	1.301	0.066 21
3.97	2.134	0.006 855	0.2408	0.028 46	10.44	1.301	0.065 62
3.98	2.136	0.006 764	0.2399	0.028 19	10.53	1.302	0.065 03
3.99	2.137	0.006 675	0.2390	0.027 93	10.62	1.302	0.064 45
4.00	2.138	0.006 586	0.2381	0.027 66	10.72	1.303	0.063 88
4.01	2.139	0.006 499	0.2372	0.027 40	10.81	1.303	0.063 31
4.02	2.141	0.006 413	0.2363	0.027 14	10.91	1.304	0.062 75
4.03	2.142	0.006 328	0.2354	0.026 88	11.01	1.304	0.062 20
4.04	2.143	0.006 245	0.2345	0.026 63	11.11	1.305	0.061 64
4.05	2.144	0.006 163	0.2336	0.026 38	11.21	1.305	0.061 10
4.06	2.146	0.006 082	0.2327	0.026 13	11.31	1.306	0.060 56
4.07	2.147	0.006 002	0.2319	0.025 89	11.41	1.306	0.060 02
4.08	2.148	0.005 923	0.2310	0.025 64	11.51	1.307	0.059 49
4.09	2.149	0.005 845	0.2301	0.025 40	11.61	1.307	0.058 97
4.10	2.150	0.005 769	0.2293	0.025 16	11.71	1.308	0.058 45
4.11	2.152	0.005 694	0.2284	0.024 93	11.82	1.308	0.057 94
4.12	2.153	0.005 619	0.2275	0.024 70	11.92	1.309	0.057 43
4.13	2.154	0.005 546	0.2267	0.024 47	12.03	1.309	0.056 92
4.14	2.155	0.005 474	0.2258	0.024 24	12.14	1.310	0.056 42
4.15	2.156	0.005 403	0.2250	0.024 01	12.24	1.310	0.055 93
4.16	2.158	0.005 333	0.2242	0.023 79	12.35	1.311	0.055 44
4.17	2.159	0.005 264	0.2233	0.023 57	12.46	1.311	0.054 96
4.18	2.160	0.005 195	0.2225	0.023 35	12.57	1.311	0.054 48
4.19	2.161	0.005 128	0.2217	0.023 13	12.68	1.312	0.054 00

Appendix C *Continued*

M	M^*	p/p_0	T/T_0	ρ/ρ_0	A/A^*	I/I^*	$\dot{m}\sqrt{(RT_0)}/Ap_0$
4.20	2.162	0.005 062	0.2208	0.022 92	12.79	1.312	0.053 53
4.21	2.163	0.004 997	0.2200	0.022 71	12.90	1.313	0.053 06
4.22	2.164	0.004 932	0.2192	0.022 50	13.02	1.313	0.052 60
4.23	2.166	0.004 869	0.2184	0.022 29	13.13	1.314	0.052 14
4.24	2.167	0.004 806	0.2176	0.022 09	13.25	1.314	0.051 69
4.25	2.168	0.004 745	0.2168	0.021 89	13.36	1.315	0.051 24
4.26	2.169	0.004 684	0.2160	0.021 68	13.48	1.315	0.050 80
4.27	2.170	0.004 624	0.2152	0.021 49	13.60	1.315	0.050 36
4.28	2.171	0.004 565	0.2144	0.021 29	13.72	1.316	0.049 93
4.29	2.172	0.004 507	0.2136	0.021 10	13.83	1.316	0.049 49
4.30	2.173	0.004 449	0.2129	0.020 90	13.95	1.317	0.049 07
4.31	2.174	0.004 393	0.2121	0.020 71	14.08	1.317	0.048 64
4.32	2.175	0.004 337	0.2113	0.020 52	14.20	1.318	0.048 23
4.33	2.176	0.004 282	0.2105	0.020 34	14.32	1.318	0.047 81
4.34	2.177	0.004 228	0.2098	0.020 15	14.45	1.318	0.047 40
4.35	2.179	0.004 174	0.2090	0.019 97	14.57	1.319	0.046 99
4.36	2.180	0.004 121	0.2082	0.019 79	14.70	1.319	0.046 59
4.37	2.181	0.004 069	0.2075	0.019 61	14.82	1.320	0.046 19
4.38	2.182	0.004 018	0.2067	0.019 43	14.95	1.320	0.045 80
4.39	2.183	0.003 967	0.2060	0.019 26	15.08	1.320	0.045 41
4.40	2.184	0.003 918	0.2053	0.019 09	15.21	1.321	0.045 02
4.41	2.185	0.003 868	0.2045	0.018 92	15.34	1.321	0.044 63
4.42	2.186	0.003 820	0.2038	0.018 75	15.47	1.322	0.044 25
4.43	2.187	0.003 772	0.2030	0.018 58	15.61	1.322	0.043 88
4.44	2.188	0.003 725	0.2023	0.018 41	15.74	1.322	0.043 51
4.45	2.189	0.003 678	0.2016	0.018 25	15.87	1.323	0.043 14
4.46	2.190	0.003 632	0.2009	0.018 08	16.01	1.323	0.042 77
4.47	2.191	0.003 587	0.2002	0.017 92	16.15	1.324	0.042 41
4.48	2.192	0.003 543	0.1994	0.017 76	16.28	1.324	0.042 05
4.49	2.193	0.003 499	0.1987	0.017 61	16.42	1.324	0.041 69
4.50	2.194	0.003 455	0.1980	0.017 45	16.56	1.325	0.041 34
4.51	2.195	0.003 412	0.1973	0.017 29	16.70	1.325	0.040 99
4.52	2.196	0.003 370	0.1966	0.017 14	16.84	1.325	0.040 65
4.53	2.196	0.003 329	0.1959	0.016 99	16.99	1.326	0.040 31
4.54	2.197	0.003 288	0.1952	0.016 84	17.13	1.326	0.039 97
4.55	2.198	0.003 247	0.1945	0.016 69	17.28	1.327	0.039 63
4.56	2.199	0.003 207	0.1938	0.016 54	17.42	1.327	0.039 30
4.57	2.200	0.003 168	0.1932	0.016 40	17.57	1.327	0.038 97
4.58	2.201	0.003 129	0.1925	0.016 25	17.72	1.328	0.038 65
4.59	2.202	0.003 090	0.1918	0.016 11	17.87	1.328	0.038 32
4.60	2.203	0.003 053	0.1911	0.015 97	18.02	1.328	0.038 00
4.61	2.204	0.003 015	0.1905	0.015 83	18.17	1.329	0.037 69

continued

Appendix C *Continued*

M	M^*	p/p_0	T/T_0	ρ/ρ_0	A/A^*	I/I^*	$\dot{m}\sqrt{(RT_0)}/Ap_0$
4.62	2.205	0.002 978	0.1898	0.015 69	18.32	1.329	0.037 37
4.63	2.206	0.002 942	0.1891	0.015 56	18.48	1.330	0.037 06
4.64	2.207	0.002 906	0.1885	0.015 42	18.63	1.330	0.036 75
4.65	2.208	0.002 871	0.1878	0.015 29	18.79	1.330	0.036 45
4.66	2.208	0.002 836	0.1872	0.015 15	18.94	1.331	0.036 15
4.67	2.209	0.002 802	0.1865	0.015 02	19.10	1.331	0.035 85
4.68	2.210	0.002 768	0.1859	0.014 89	19.26	1.331	0.035 55
4.69	2.211	0.002 734	0.1852	0.014 76	19.42	1.332	0.035 26
4.70	2.212	0.002 701	0.1846	0.014 64	19.58	1.332	0.034 97
4.71	2.213	0.002 669	0.1839	0.014 51	19.75	1.332	0.034 68
4.72	2.214	0.002 636	0.1833	0.014 38	19.91	1.333	0.034 39
4.73	2.215	0.002 605	0.1827	0.014 26	20.07	1.333	0.034 11
4.74	2.215	0.002 573	0.1820	0.014 14	20.24	1.333	0.033 83
4.75	2.216	0.002 543	0.1814	0.014 02	20.41	1.334	0.033 55
4.76	2.217	0.002 512	0.1808	0.013 90	20.58	1.334	0.033 28
4.77	2.218	0.002 482	0.1802	0.013 78	20.75	1.334	0.033 00
4.78	2.219	0.002 452	0.1795	0.013 66	20.92	1.335	0.032 73
4.79	2.220	0.002 423	0.1789	0.013 54	21.09	1.335	0.032 47
4.80	2.220	0.002 394	0.1783	0.013 43	21.26	1.335	0.032 20
4.81	2.221	0.002 366	0.1777	0.013 31	21.44	1.336	0.031 94
4.82	2.222	0.002 338	0.1771	0.013 20	21.61	1.336	0.031 68
4.83	2.223	0.002 310	0.1765	0.013 09	21.79	1.336	0.031 42
4.84	2.224	0.002 282	0.1759	0.012 98	21.97	1.337	0.031 17
4.85	2.224	0.002 255	0.1753	0.012 87	22.15	1.337	0.030 91
4.86	2.225	0.002 229	0.1747	0.012 76	22.33	1.337	0.030 66
4.87	2.226	0.002 202	0.1741	0.012 65	22.51	1.338	0.030 41
4.88	2.227	0.002 176	0.1735	0.012 54	22.70	1.338	0.030 17
4.89	2.228	0.002 151	0.1729	0.012 44	22.88	1.338	0.029 93
4.90	2.228	0.002 126	0.1724	0.012 33	23.07	1.339	0.029 68
4.91	2.229	0.002 101	0.1718	0.012 23	23.25	1.339	0.029 44
4.92	2.230	0.002 076	0.1712	0.012 13	23.44	1.339	0.029 21
4.93	2.231	0.002 052	0.1706	0.012 02	23.63	1.340	0.028 97
4.94	2.232	0.002 028	0.1700	0.011 92	23.82	1.340	0.028 74
4.95	2.232	0.002 004	0.1695	0.011 82	24.02	1.340	0.028 51
4.96	2.233	0.001 981	0.1689	0.011 73	24.21	1.340	0.028 28
4.97	2.234	0.001 957	0.1683	0.011 63	24.41	1.341	0.028 06
4.98	2.235	0.001 935	0.1678	0.011 53	24.60	1.341	0.027 83
4.99	2.235	0.001 912	0.1672	0.011 44	24.80	1.341	0.027 61
5.00	2.236	0.001 890	0.1667	0.011 34	25.00	1.342	0.027 39
∞	2.450	0.000 000	0.0000	0.000 00	∞	1.429	0.000 00

Appendix D
Flow of a perfect gas across a normal shock wave ($k = 1.4$)

M_x	M_y	$\frac{p_y}{p_x}$	$\frac{T_y}{T_x}$	$\frac{\rho_y}{\rho_x} = \frac{V_x}{V_y}$	$\frac{p_{0y}}{p_{0x}} = \frac{A_x^*}{A_y^*}$	$\frac{s_y - s_x}{R}$
1.00	1.0000	1.000	1.000	1.000	1.0000	0.000 000 000 0
1.01	0.9901	1.023	1.007	1.017	1.0000	0.000 000 852 2
1.02	0.9805	1.047	1.013	1.033	1.0000	0.000 010 05
1.03	0.9712	1.071	1.020	1.050	1.0000	0.000 032 91
1.04	0.9620	1.095	1.026	1.067	0.9999	0.000 076 72
1.05	0.9531	1.120	1.033	1.084	0.9999	0.000 147 1
1.06	0.9444	1.144	1.039	1.101	0.9998	0.000 249 3
1.07	0.9360	1.169	1.046	1.118	0.9996	0.000 388 5
1.08	0.9277	1.194	1.052	1.135	0.9994	0.000 569 3
1.09	0.9196	1.219	1.059	1.152	0.9992	0.000 796 2
1.10	0.9118	1.245	1.065	1.169	0.9989	0.001 072
1.11	0.9041	1.271	1.071	1.186	0.9986	0.001 402
1.12	0.8966	1.297	1.078	1.203	0.9982	0.001 789
1.13	0.8892	1.323	1.084	1.221	0.9978	0.000 235
1.14	0.8820	1.350	1.090	1.238	0.9973	0.002 743
1.15	0.8750	1.376	1.097	1.255	0.9967	0.003 316
1.16	0.8682	1.403	1.103	1.272	0.9961	0.003 956
1.17	0.8615	1.430	1.109	1.290	0.9953	0.004 666
1.18	0.8549	1.458	1.115	1.307	0.9946	0.005 446
1.19	0.8485	1.485	1.122	1.324	0.9937	0.006 300
1.20	0.8422	1.513	1.128	1.342	0.9928	0.007 228
1.21	0.8360	1.541	1.134	1.359	0.9918	0.008 232
1.22	0.8300	1.570	1.141	1.376	0.9907	0.009 313
1.23	0.8241	1.598	1.147	1.394	0.9896	0.010 47
1.24	0.8183	1.627	1.153	1.411	0.9884	0.011 71
1.25	0.8126	1.656	1.159	1.429	0.9871	0.013 03
1.26	0.8071	1.686	1.166	1.446	0.9857	0.014 43
1.27	0.8016	1.715	1.172	1.463	0.9842	0.015 91

continued

Appendix D *Continued*

M_x	M_y	$\frac{p_y}{p_x}$	$\frac{T_y}{T_x}$	$\frac{\rho_y}{\rho_x} = \frac{V_x}{V_y}$	$\frac{p_{0y}}{p_{0x}} = \frac{A_x^*}{A_y^*}$	$\frac{s_y - s_x}{R}$
1.28	0.7963	1.745	1.178	1.481	0.9827	0.017 47
1.29	0.7911	1.775	1.185	1.498	0.9811	0.019 11
1.30	0.7860	1.805	1.191	1.516	0.9794	0.020 84
1.31	0.7809	1.835	1.197	1.533	0.9776	0.022 65
1.32	0.7760	1.866	1.204	1.551	0.9758	0.024 55
1.33	0.7712	1.897	1.210	1.568	0.9738	0.026 52
1.34	0.7664	1.928	1.216	1.585	0.9718	0.028 59
1.35	0.7618	1.960	1.223	1.603	0.9697	0.030 73
1.36	0.7572	1.991	1.229	1.620	0.9676	0.032 96
1.37	0.7527	2.023	1.235	1.638	0.9653	0.035 27
1.38	0.7483	2.055	1.242	1.655	0.9630	0.037 67
1.39	0.7440	2.087	1.248	1.672	0.9607	0.040 14
1.40	0.7397	2.120	1.255	1.690	0.9582	0.042 70
1.41	0.7355	2.153	1.261	1.707	0.9557	0.045 35
1.42	0.7314	2.186	1.268	1.724	0.9531	0.048 07
1.43	0.7274	2.219	1.274	1.742	0.9504	0.050 88
1.44	0.7235	2.253	1.281	1.759	0.9476	0.053 77
1.45	0.7196	2.286	1.287	1.776	0.9448	0.056 74
1.46	0.7157	2.320	1.294	1.793	0.9420	0.059 80
1.47	0.7120	2.354	1.300	1.811	0.9390	0.062 93
1.48	0.7083	2.389	1.307	1.828	0.9360	0.066 14
1.49	0.7047	2.423	1.314	1.845	0.9329	0.069 43
1.50	0.7011	2.458	1.320	1.862	0.9298	0.072 80
1.51	0.6976	2.493	1.327	1.879	0.9266	0.076 25
1.52	0.6941	2.529	1.334	1.896	0.9233	0.079 77
1.53	0.6907	2.564	1.340	1.913	0.9200	0.083 38
1.54	0.6874	2.600	1.347	1.930	0.9166	0.087 06
1.55	0.6841	2.636	1.354	1.947	0.9132	0.090 81
1.56	0.6809	2.673	1.361	1.964	0.9097	0.094 64
1.57	0.6777	2.709	1.367	1.981	0.9062	0.098 55
1.58	0.6746	2.746	1.374	1.998	0.9026	0.1025
1.59	0.6715	2.783	1.381	2.015	0.8989	0.1066
1.60	0.6684	2.820	1.388	2.032	0.8952	0.1107
1.61	0.6655	2.857	1.395	2.049	0.8915	0.1149
1.62	0.6625	2.895	1.402	2.065	0.8877	0.1192
1.63	0.6596	2.933	1.409	2.082	0.8838	0.1235
1.64	0.6568	2.971	1.416	2.099	0.8799	0.1279
1.65	0.6540	3.010	1.423	2.115	0.8760	0.1324
1.66	0.6512	3.048	1.430	2.132	0.8720	0.1369
1.67	0.6485	3.087	1.437	2.148	0.8680	0.1416
1.68	0.6458	3.126	1.444	2.165	0.8639	0.1462
1.69	0.6431	3.165	1.451	2.181	0.8599	0.1510

Appendix D *Continued*

M_x	M_y	$\frac{p_y}{p_x}$	$\frac{T_y}{T_x}$	$\frac{\rho_y}{\rho_x} = \frac{V_x}{V_y}$	$\frac{p_{0y}}{p_{0x}} = \frac{A_x^*}{A_y^*}$	$\frac{s_y - s_x}{R}$
1.70	0.6405	3.205	1.458	2.198	0.8557	0.1558
1.71	0.6380	3.245	1.466	2.214	0.8516	0.1607
1.72	0.6355	3.285	1.473	2.230	0.8474	0.1656
1.73	0.6330	3.325	1.480	2.247	0.8431	0.1706
1.74	0.6305	3.366	1.487	2.263	0.8389	0.1757
1.75	0.6281	3.406	1.495	2.279	0.8346	0.1808
1.76	0.6257	3.447	1.502	2.295	0.8302	0.1860
1.77	0.6234	3.488	1.509	2.311	0.8259	0.1913
1.78	0.6210	3.530	1.517	2.327	0.8215	0.1966
1.79	0.6188	3.571	1.524	2.343	0.8171	0.2020
1.80	0.6165	3.613	1.532	2.359	0.8127	0.2074
1.81	0.6143	3.655	1.539	2.375	0.8082	0.2129
1.82	0.6121	3.698	1.547	2.391	0.8038	0.2185
1.83	0.6099	3.740	1.554	2.407	0.7993	0.2241
1.84	0.6078	3.783	1.562	2.422	0.7948	0.2297
1.85	0.6057	3.826	1.569	2.438	0.7902	0.2354
1.86	0.6036	3.870	1.577	2.454	0.7857	0.2412
1.87	0.6016	3.913	1.585	2.469	0.7811	0.2470
1.88	0.5996	3.957	1.592	2.485	0.7765	0.2529
1.89	0.5976	4.001	1.600	2.500	0.7720	0.2588
1.90	0.5956	4.045	1.608	2.516	0.7674	0.2648
1.91	0.5937	4.089	1.616	2.531	0.7627	0.2708
1.92	0.5918	4.134	1.624	2.546	0.7581	0.2769
1.93	0.5899	4.179	1.631	2.562	0.7535	0.2830
1.94	0.5880	4.224	1.639	2.577	0.7488	0.2892
1.95	0.5862	4.270	1.647	2.592	0.7442	0.2955
1.96	0.5844	4.315	1.655	2.607	0.7395	0.3017
1.97	0.5826	4.361	1.663	2.622	0.7349	0.3080
1.98	0.5808	4.407	1.671	2.637	0.7302	0.3144
1.99	0.5791	4.453	1.679	2.652	0.7255	0.3208
2.00	0.5774	4.500	1.687	2.667	0.7209	0.3273
2.01	0.5757	4.547	1.696	2.681	0.7162	0.3338
2.02	0.5740	4.594	1.704	2.696	0.7115	0.3403
2.03	0.5723	4.641	1.712	2.711	0.7069	0.3469
2.04	0.5707	4.689	1.720	2.725	0.7022	0.3536
2.05	0.5691	4.736	1.729	2.740	0.6975	0.3602
2.06	0.5675	4.784	1.737	2.755	0.6928	0.3670
2.07	0.5659	4.832	1.745	2.769	0.6882	0.3737
2.08	0.5643	4.881	1.754	2.783	0.6835	0.3805
2.09	0.5628	4.929	1.762	2.798	0.6789	0.3873

continued

Appendix D *Continued*

M_x	M_y	$\frac{p_y}{p_x}$	$\frac{T_y}{T_x}$	$\frac{\rho_y}{\rho_x}=\frac{V_x}{V_y}$	$\frac{p_{0y}}{p_{0x}}=\frac{A_x^*}{A_y^*}$	$\frac{s_y-s_x}{R}$
2.10	0.5613	4.978	1.770	2.812	0.6742	0.3942
2.11	0.5598	5.027	1.779	2.826	0.6696	0.4011
2.12	0.5583	5.077	1.787	2.840	0.6649	0.4081
2.13	0.5568	5.126	1.796	2.854	0.6603	0.4151
2.14	0.5554	5.176	1.805	2.868	0.6557	0.4221
2.15	0.5540	5.226	1.813	2.882	0.6511	0.4292
2.16	0.5525	5.277	1.822	2.896	0.6464	0.4363
2.17	0.5511	5.327	1.831	2.910	0.6419	0.4434
2.18	0.5498	5.378	1.839	2.924	0.6373	0.4506
2.19	0.5484	5.429	1.848	2.938	0.6327	0.4578
2.20	0.5471	5.480	1.857	2.951	0.6281	0.4650
2.21	0.5457	5.531	1.866	2.965	0.6236	0.4723
2.22	0.5444	5.583	1.875	2.978	0.6191	0.4796
2.23	0.5431	5.635	1.883	2.992	0.6145	0.4869
2.24	0.5418	5.687	1.892	3.005	0.6100	0.4943
2.25	0.5406	5.740	1.901	3.019	0.6055	0.5017
2.26	0.5393	5.792	1.910	3.032	0.6011	0.5091
2.27	0.5381	5.845	1.919	3.045	0.5966	0.5165
2.28	0.5368	5.898	1.929	3.058	0.5921	0.5240
2.29	0.5356	5.951	1.938	3.071	0.5877	0.5315
2.30	0.5344	6.005	1.947	3.085	0.5833	0.5391
2.31	0.5332	6.059	1.956	3.098	0.5789	0.5466
2.32	0.5321	6.113	1.965	3.110	0.5745	0.5542
2.33	0.5309	6.167	1.974	3.123	0.5702	0.5618
2.34	0.5297	6.222	1.984	3.136	0.5658	0.5695
2.35	0.5286	6.276	1.993	3.149	0.5615	0.5772
2.36	0.5275	6.331	2.002	3.162	0.5572	0.5849
2.37	0.5264	6.386	2.012	3.174	0.5529	0.5926
2.38	0.5253	6.442	2.021	3.187	0.5486	0.6003
2.39	0.5242	6.497	2.031	3.199	0.5444	0.6081
2.40	0.5231	6.553	2.040	3.212	0.5401	0.6159
2.41	0.5221	6.609	2.050	3.224	0.5359	0.6237
2.42	0.5210	6.666	2.059	3.237	0.5317	0.6316
2.43	0.5200	6.722	2.069	3.249	0.5276	0.6395
2.44	0.5189	6.779	2.079	3.261	0.5234	0.6473
2.45	0.5179	6.836	2.088	3.273	0.5193	0.6553
2.46	0.5169	6.894	2.098	3.285	0.5152	0.6632
2.47	0.5159	6.951	2.108	3.298	0.5111	0.6711
2.48	0.5149	7.009	2.118	3.310	0.5071	0.6791
2.49	0.5140	7.067	2.128	3.321	0.5030	0.6871
2.50	0.5130	7.125	2.137	3.333	0.4990	0.6951
2.51	0.5120	7.183	2.147	3.345	0.4950	0.7032

Appendix D *Continued*

M_x	M_y	$\frac{p_y}{p_x}$	$\frac{T_y}{T_x}$	$\frac{\rho_y}{\rho_x} = \frac{V_x}{V_y}$	$\frac{p_{0y}}{p_{0x}} = \frac{A_x^*}{A_y^*}$	$\frac{s_y - s_x}{R}$
2.52	0.5111	7.242	2.157	3.357	0.4911	0.7112
2.53	0.5102	7.301	2.167	3.369	0.4871	0.7193
2.54	0.5092	7.360	2.177	3.380	0.4832	0.7274
2.55	0.5083	7.420	2.187	3.392	0.4793	0.7355
2.56	0.5074	7.479	2.198	3.403	0.4754	0.7436
2.57	0.5065	7.539	2.208	3.415	0.4715	0.7517
2.58	0.5056	7.599	2.218	3.426	0.4677	0.7599
2.59	0.5047	7.659	2.228	3.438	0.4639	0.7681
2.60	0.5039	7.720	2.238	3.449	0.4601	0.7763
2.61	0.5030	7.781	2.249	3.460	0.4564	0.7845
2.62	0.5022	7.842	2.259	3.471	0.4526	0.7927
2.63	0.5013	7.903	2.269	3.483	0.4489	0.8009
2.64	0.5005	7.965	2.280	3.494	0.4452	0.8092
2.65	0.4996	8.026	2.290	3.505	0.4416	0.8174
2.66	0.4988	8.088	2.301	3.516	0.4379	0.8257
2.67	0.4980	8.150	2.311	3.527	0.4343	0.8340
2.68	0.4972	8.213	2.322	3.537	0.4307	0.8423
2.69	0.4964	8.275	2.332	3.548	0.4271	0.8507
2.70	0.4956	8.338	2.343	3.559	0.4236	0.8590
2.71	0.4949	8.401	2.354	3.570	0.4201	0.8673
2.72	0.4941	8.465	2.364	3.580	0.4166	0.8757
2.73	0.4933	8.528	2.375	3.591	0.4131	0.8841
2.74	0.4926	8.592	2.386	3.601	0.4097	0.8925
2.75	0.4918	8.656	2.397	3.612	0.4062	0.9008
2.76	0.4911	8.721	2.407	3.622	0.4028	0.9093
2.77	0.4903	8.785	2.418	3.633	0.3994	0.9177
2.78	0.4896	8.850	2.429	3.643	0.3961	0.9261
2.79	0.4889	8.915	2.440	3.653	0.3928	0.9345
2.80	0.4882	8.980	2.451	3.664	0.3895	0.9430
2.81	0.4875	9.045	2.462	3.674	0.3862	0.9514
2.82	0.4868	9.111	2.473	3.684	0.3829	0.9599
2.83	0.4861	9.177	2.484	3.694	0.3797	0.9684
2.84	0.4854	9.243	2.496	3.704	0.3765	0.9769
2.85	0.4847	9.310	2.507	3.714	0.3733	0.9854
2.86	0.4840	9.376	2.518	3.724	0.3701	0.9939
2.87	0.4833	9.443	2.529	3.734	0.3670	1.002
2.88	0.4827	9.510	2.540	3.743	0.3639	1.011
2.89	0.4820	9.577	2.552	3.753	0.3608	1.019
2.90	0.4814	9.645	2.563	3.763	0.3577	1.028
2.91	0.4807	9.713	2.575	3.773	0.3547	1.037

continued

M_x	M_y	$\frac{p_y}{p_x}$	$\frac{T_y}{T_x}$	$\frac{\rho_y}{\rho_x} = \frac{V_x}{V_y}$	$\frac{p_{0y}}{p_{0x}} = \frac{A_x^*}{A_y^*}$	$\frac{s_y - s_x}{R}$
2.92	0.4801	9.781	2.586	3.782	0.3517	1.045
2.93	0.4795	9.849	2.598	3.792	0.3487	1.054
2.94	0.4788	9.918	2.609	3.801	0.3457	1.062
2.95	0.4782	9.986	2.621	3.811	0.3428	1.071
2.96	0.4776	10.06	2.632	3.820	0.3398	1.079
2.97	0.4770	10.12	2.644	3.829	0.3369	1.088
2.98	0.4764	10.19	2.656	3.839	0.3340	1.097
2.99	0.4758	10.26	2.667	3.848	0.3312	1.105
3.00	0.4752	10.33	2.679	3.857	0.3283	1.114
3.01	0.4746	10.40	2.691	3.866	0.3255	1.122
3.02	0.4740	10.47	2.703	3.875	0.3227	1.131
3.03	0.4734	10.54	2.714	3.884	0.3200	1.140
3.04	0.4729	10.62	2.726	3.893	0.3172	1.148
3.05	0.4723	10.69	2.738	3.902	0.3145	1.157
3.06	0.4717	10.76	2.750	3.911	0.3118	1.165
3.07	0.4712	10.83	2.762	3.920	0.3091	1.174
3.08	0.4706	10.90	2.774	3.929	0.3065	1.183
3.09	0.4701	10.97	2.786	3.938	0.3038	1.191
3.10	0.4695	11.04	2.799	3.947	0.3012	1.200
3.11	0.4690	11.12	2.811	3.955	0.2986	1.209
3.12	0.4685	11.19	2.823	3.964	0.2960	1.217
3.13	0.4679	11.26	2.835	3.973	0.2935	1.226
3.14	0.4674	11.34	2.848	3.981	0.2910	1.235
3.15	0.4669	11.41	2.860	3.990	0.2885	1.243
3.16	0.4664	11.48	2.872	3.998	0.2860	1.252
3.17	0.4659	11.56	2.885	4.006	0.2835	1.261
3.18	0.4654	11.63	2.897	4.015	0.2811	1.269
3.19	0.4648	11.71	2.909	4.023	0.2786	1.278
3.20	0.4643	11.78	2.922	4.031	0.2762	1.287
3.21	0.4639	11.85	2.935	4.040	0.2738	1.295
3.22	0.4634	11.93	2.947	4.048	0.2715	1.304
3.23	0.4629	12.01	2.960	4.056	0.2691	1.313
3.24	0.4624	12.08	2.972	4.064	0.2668	1.321
3.25	0.4619	12.16	2.985	4.072	0.2645	1.330
3.26	0.4614	12.23	2.998	4.080	0.2622	1.339
3.27	0.4610	12.31	3.011	4.088	0.2600	1.347
3.28	0.4605	12.38	3.023	4.096	0.2577	1.356
3.29	0.4600	12.46	3.036	4.104	0.2555	1.365
3.30	0.4596	12.54	3.049	4.112	0.2533	1.373
3.31	0.4591	12.62	3.062	4.120	0.2511	1.382
3.32	0.4587	12.69	3.075	4.128	0.2489	1.391
3.33	0.4582	12.77	3.088	4.135	0.2468	1.399

Appendix D *Continued*

M_x	M_y	$\frac{p_y}{p_x}$	$\frac{T_y}{T_x}$	$\frac{\rho_y}{\rho_x} = \frac{V_x}{V_y}$	$\frac{p_{0y}}{p_{0x}} = \frac{A_x^*}{A_y^*}$	$\frac{s_y - s_x}{R}$
3.34	0.4578	12.85	3.101	4.143	0.2446	1.408
3.35	0.4573	12.93	3.114	4.151	0.2425	1.417
3.36	0.4569	13.00	3.127	4.158	0.2404	1.425
3.37	0.4565	13.08	3.141	4.166	0.2383	1.434
3.38	0.4560	13.16	3.154	4.173	0.2363	1.443
3.39	0.4556	13.24	3.167	4.181	0.2342	1.451
3.40	0.4552	13.32	3.180	4.188	0.2322	1.460
3.41	0.4548	13.40	3.194	4.196	0.2302	1.469
3.42	0.4544	13.48	3.207	4.203	0.2282	1.477
3.43	0.4540	13.56	3.220	4.211	0.2263	1.486
3.44	0.4535	13.64	3.234	4.218	0.2243	1.495
3.45	0.4531	13.72	3.247	4.225	0.2224	1.503
3.46	0.4527	13.80	3.261	4.232	0.2205	1.512
3.47	0.4523	13.88	3.274	4.240	0.2186	1.521
3.48	0.4519	13.96	3.288	4.247	0.2167	1.529
3.49	0.4515	14.04	3.301	4.254	0.2148	1.538
3.50	0.4512	14.13	3.315	4.261	0.2129	1.547
3.51	0.4508	14.21	3.329	4.268	0.2111	1.555
3.52	0.4504	14.29	3.342	4.275	0.2093	1.564
3.53	0.4500	14.37	3.356	4.282	0.2075	1.573
3.54	0.4496	14.45	3.370	4.289	0.2057	1.581
3.55	0.4492	14.54	3.384	4.296	0.2039	1.590
3.56	0.4489	14.62	3.398	4.303	0.2022	1.599
3.57	0.4485	14.70	3.412	4.309	0.2004	1.607
3.58	0.4481	14.79	3.426	4.316	0.1987	1.616
3.59	0.4478	14.87	3.440	4.323	0.1970	1.625
3.60	0.4474	14.95	3.454	4.330	0.1953	1.633
3.61	0.4471	15.04	3.468	4.336	0.1936	1.642
3.62	0.4467	15.12	3.482	4.343	0.1920	1.650
3.63	0.4463	15.21	3.496	4.350	0.1903	1.659
3.64	0.4460	15.29	3.510	4.356	0.1887	1.668
3.65	0.4456	15.38	3.525	4.363	0.1871	1.676
3.66	0.4453	15.46	3.539	4.369	0.1855	1.685
3.67	0.4450	15.55	3.553	4.376	0.1839	1.693
3.68	0.4446	15.63	3.567	4.382	0.1823	1.702
3.69	0.4443	15.72	3.582	4.388	0.1807	1.711
3.70	0.4439	15.80	3.596	4.395	0.1792	1.719
3.71	0.4436	15.89	3.611	4.401	0.1777	1.728
3.72	0.4433	15.98	3.625	4.408	0.1761	1.736
3.73	0.4430	16.07	3.640	4.414	0.1746	1.745

continued

Appendix D *Continued*

M_x	M_y	$\frac{p_y}{p_x}$	$\frac{T_y}{T_x}$	$\frac{\rho_y}{\rho_x} = \frac{V_x}{V_y}$	$\frac{p_{0y}}{p_{0x}} = \frac{A_x^*}{A_y^*}$	$\frac{s_y - s_x}{R}$
3.74	0.4426	16.15	3.654	4.420	0.1731	1.754
3.75	0.4423	16.24	3.669	4.426	0.1717	1.762
3.76	0.4420	16.33	3.684	4.432	0.1702	1.771
3.77	0.4417	16.42	3.698	4.439	0.1687	1.779
3.78	0.4414	16.50	3.713	4.445	0.1673	1.788
3.79	0.4410	16.59	3.728	4.451	0.1659	1.796
3.80	0.4407	16.68	3.743	4.457	0.1645	1.805
3.81	0.4404	16.77	3.757	4.463	0.1631	1.814
3.82	0.4401	16.86	3.772	4.469	0.1617	1.822
3.83	0.4398	16.95	3.787	4.475	0.1603	1.831
3.84	0.4395	17.04	3.802	4.481	0.1589	1.839
3.85	0.4392	17.13	3.817	4.487	0.1576	1.848
3.86	0.4389	17.22	3.832	4.492	0.1563	1.856
3.87	0.4386	17.31	3.847	4.498	0.1549	1.865
3.88	0.4383	17.40	3.862	4.504	0.1536	1.873
3.89	0.4380	17.49	3.878	4.510	0.1523	1.882
3.90	0.4377	17.58	3.893	4.516	0.1510	1.890
3.91	0.4375	17.67	3.908	4.521	0.1497	1.899
3.92	0.4372	17.76	3.923	4.527	0.1485	1.907
3.93	0.4369	17.85	3.939	4.533	0.1472	1.916
3.94	0.4366	17.94	3.954	4.538	0.1460	1.924
3.95	0.4363	18.04	3.969	4.544	0.1448	1.933
3.96	0.4360	18.13	3.985	4.549	0.1435	1.941
3.97	0.4358	18.22	4.000	4.555	0.1423	1.950
3.98	0.4355	18.31	4.016	4.560	0.1411	1.958
3.99	0.4352	18.41	4.031	4.566	0.1399	1.967
4.00	0.4350	18.50	4.047	4.571	0.1388	1.975
4.01	0.4347	18.59	4.062	4.577	0.1376	1.983
4.02	0.4344	18.69	4.078	4.582	0.1364	1.992
4.03	0.4342	18.78	4.094	4.588	0.1353	2.000
4.04	0.4339	18.88	4.110	4.593	0.1342	2.009
4.05	0.4336	18.97	4.125	4.598	0.1330	2.017
4.06	0.4334	19.06	4.141	4.604	0.1319	2.026
4.07	0.4331	19.16	4.157	4.609	0.1308	2.034
4.08	0.4329	19.25	4.173	4.614	0.1297	2.042
4.09	0.4326	19.35	4.189	4.619	0.1286	2.051
4.10	0.4324	19.44	4.205	4.624	0.1276	2.059
4.11	0.4321	19.54	4.221	4.630	0.1265	2.068
4.12	0.4319	19.64	4.237	4.635	0.1254	2.076
4.13	0.4316	19.73	4.253	4.640	0.1244	2.084
4.14	0.4314	19.83	4.269	4.645	0.1234	2.093
4.15	0.4311	19.93	4.285	4.650	0.1223	2.101

Appendix D *Continued*

M_x	M_y	$\dfrac{p_y}{p_x}$	$\dfrac{T_y}{T_x}$	$\dfrac{\rho_y}{\rho_x} = \dfrac{V_x}{V_y}$	$\dfrac{p_{0y}}{p_{0x}} = \dfrac{A_x^*}{A_y^*}$	$\dfrac{s_y - s_x}{R}$
4.16	0.4309	20.02	4.301	4.655	0.1213	2.109
4.17	0.4306	20.12	4.318	4.660	0.1203	2.118
4.18	0.4304	20.22	4.334	4.665	0.1193	2.126
4.19	0.4302	20.32	4.350	4.670	0.1183	2.134
4.20	0.4299	20.41	4.367	4.675	0.1173	2.143
4.21	0.4297	20.51	4.383	4.680	0.1164	2.151
4.22	0.4295	20.61	4.399	4.685	0.1154	2.159
4.23	0.4292	20.71	4.416	4.690	0.1144	2.168
4.24	0.4290	20.81	4.432	4.694	0.1135	2.176
4.25	0.4288	20.91	4.449	4.699	0.1126	2.184
4.26	0.4286	21.01	4.465	4.704	0.1116	2.193
4.27	0.4283	21.11	4.482	4.709	0.1107	2.201
4.28	0.4281	21.20	4.499	4.713	0.1098	2.209
4.29	0.4279	21.30	4.515	4.718	0.1089	2.217
4.30	0.4277	21.40	4.532	4.723	0.1080	2.226
4.31	0.4275	21.51	4.549	4.728	0.1071	2.234
4.32	0.4272	21.61	4.566	4.732	0.1062	2.242
4.33	0.4270	21.71	4.583	4.737	0.1054	2.250
4.34	0.4268	21.81	4.600	4.741	0.1045	2.259
4.35	0.4266	21.91	4.616	4.746	0.1036	2.267
4.36	0.4264	22.01	4.633	4.751	0.1028	2.275
4.37	0.4262	22.11	4.650	4.755	0.1020	2.283
4.38	0.4260	22.22	4.668	4.760	0.1011	2.291
4.39	0.4258	22.32	4.685	4.764	0.1003	2.300
4.40	0.4255	22.42	4.702	4.768	0.099 48	2.308
4.41	0.4253	22.52	4.719	4.773	0.098 67	2.316
4.42	0.4251	22.63	4.736	4.777	0.097 87	2.324
4.43	0.4249	22.73	4.753	4.782	0.097 07	2.332
4.44	0.4247	22.83	4.771	4.786	0.096 28	2.340
4.45	0.4245	22.94	4.788	4.790	0.095 50	2.349
4.46	0.4243	23.04	4.805	4.795	0.094 73	2.357
4.47	0.4241	23.14	4.823	4.799	0.093 96	2.365
4.48	0.4239	23.25	4.840	4.803	0.093 20	2.373
4.49	0.4237	23.35	4.858	4.808	0.092 44	2.381
4.50	0.4236	23.46	4.875	4.812	0.091 70	2.389
4.51	0.4234	23.56	4.893	4.816	0.090 96	2.397
4.52	0.4232	23.67	4.910	4.820	0.090 22	2.405
4.53	0.4230	23.77	4.928	4.824	0.089 50	2.414
4.54	0.4228	23.88	4.946	4.829	0.088 78	2.422
4.55	0.4226	23.99	4.963	4.833	0.088 06	2.430

continued

Appendix D *Continued*

M_x	M_y	$\frac{p_y}{p_x}$	$\frac{T_y}{T_x}$	$\frac{\rho_y}{\rho_x} = \frac{V_x}{V_y}$	$\frac{p_{0y}}{p_{0x}} = \frac{A_x^*}{A_y^*}$	$\frac{s_y - s_x}{R}$
4.56	0.4224	24.09	4.981	4.837	0.087 35	2.438
4.57	0.4222	24.20	4.999	4.841	0.086 65	2.446
4.58	0.4220	24.31	5.017	4.845	0.085 96	2.454
4.59	0.4219	24.41	5.034	4.849	0.085 27	2.462
4.60	0.4217	24.52	5.052	4.853	0.084 59	2.470
4.61	0.4215	24.63	5.070	4.857	0.083 91	2.478
4.62	0.4213	24.74	5.088	4.861	0.083 24	2.486
4.63	0.4211	24.84	5.106	4.865	0.082 57	2.494
4.64	0.4210	24.95	5.124	4.869	0.081 92	2.502
4.65	0.4208	25.06	5.142	4.873	0.081 26	2.510
4.66	0.4206	25.17	5.161	4.877	0.080 62	2.518
4.67	0.4204	25.28	5.179	4.881	0.079 98	2.526
4.68	0.4203	25.39	5.197	4.885	0.079 34	2.534
4.69	0.4201	25.50	5.215	4.889	0.078 71	2.542
4.70	0.4199	25.60	5.233	4.893	0.078 09	2.550
4.71	0.4197	25.71	5.252	4.896	0.077 47	2.558
4.72	0.4196	25.82	5.270	4.900	0.076 85	2.566
4.73	0.4194	25.94	5.289	4.904	0.076 25	2.574
4.74	0.4192	26.05	5.307	4.908	0.075 64	2.582
4.75	0.4191	26.16	5.325	4.912	0.075 05	2.590
4.76	0.4189	26.27	5.344	4.915	0.074 45	2.598
4.77	0.4187	26.38	5.363	4.919	0.073 87	2.605
4.78	0.4186	26.49	5.381	4.923	0.073 29	2.613
4.79	0.4184	26.60	5.400	4.926	0.072 71	2.621
4.80	0.4183	26.71	5.418	4.930	0.072 14	2.629
4.81	0.4181	26.83	5.437	4.934	0.071 57	2.637
4.82	0.4179	26.94	5.456	4.937	0.071 01	2.645
4.83	0.4178	27.05	5.475	4.941	0.070 46	2.653
4.84	0.4176	27.16	5.493	4.945	0.069 91	2.661
4.85	0.4175	27.28	5.512	4.948	0.069 36	2.668
4.86	0.4173	27.39	5.531	4.952	0.068 82	2.676
4.87	0.4172	27.50	5.550	4.955	0.068 28	2.684
4.88	0.4170	27.62	5.569	4.959	0.067 75	2.692
4.89	0.4169	27.73	5.588	4.962	0.067 22	2.700
4.90	0.4167	27.84	5.607	4.966	0.066 70	2.708
4.91	0.4165	27.96	5.626	4.969	0.066 18	2.715
4.92	0.4164	28.07	5.646	4.973	0.065 67	2.723
4.93	0.4162	28.19	5.665	4.976	0.065 16	2.731
4.94	0.4161	28.30	5.684	4.980	0.064 65	2.739
4.95	0.4160	28.42	5.703	4.983	0.064 15	2.746
4.96	0.4158	28.54	5.722	4.987	0.063 66	2.754
4.97	0.4157	28.65	5.742	4.990	0.063 17	2.762

Appendix D *Continued*

M_x	M_y	$\frac{p_y}{p_x}$	$\frac{T_y}{T_x}$	$\frac{\rho_y}{\rho_x} = \frac{V_x}{V_y}$	$\frac{p_{0y}}{p_{0x}} = \frac{A_x^*}{A_y^*}$	$\frac{s_y - s_x}{R}$
4.98	0.4155	28.77	5.761	4.993	0.062 68	2.770
4.99	0.4154	28.88	5.781	4.997	0.062 20	2.777
5.00	0.4152	29.00	5.800	5.000	0.061 72	2.785
∞	0.3780	∞	∞	6.000	0.000 00	∞

Appendix E
Flow of a perfect gas on the Fanno line ($k = 1.4$)

M	$\frac{p}{p^*}$	$\frac{T}{T^*}$	$\frac{\rho}{\rho^*} = \frac{V^*}{V}$	$\frac{p_0}{p_0^*}$	$\frac{I}{I^*}$	$\frac{s^* - s}{R}$	$\frac{4\bar{f}L_{max}}{D}$
0.00	∞	1.200	∞	∞	∞	∞	∞
0.01	109.5	1.200	91.29	57.87	45.65	4.058	7134.0
0.02	54.77	1.200	45.65	28.94	22.83	3.365	1779.0
0.03	36.51	1.200	30.43	19.30	15.23	2.960	787.1
0.04	27.38	1.200	22.83	14.48	11.44	2.673	440.4
0.05	21.90	1.199	18.26	11.59	9.158	2.450	280.0
0.06	18.25	1.199	15.22	9.666	7.643	2.269	193.0
0.07	15.64	1.199	13.05	8.292	6.562	2.115	140.7
0.08	13.68	1.198	11.42	7.262	5.753	1.983	106.7
0.09	12.16	1.198	10.15	6.461	5.125	1.866	83.50
0.10	10.94	1.198	9.138	5.822	4.624	1.762	66.92
0.11	9.947	1.197	8.309	5.299	4.215	1.668	54.69
0.12	9.116	1.197	7.618	4.864	3.875	1.582	45.41
0.13	8.412	1.196	7.034	4.497	3.588	1.503	38.21
0.14	7.809	1.195	6.533	4.182	3.343	1.431	32.51
0.15	7.287	1.195	6.099	3.910	3.132	1.364	27.93
0.16	6.829	1.194	5.720	3.673	2.947	1.301	24.20
0.17	6.425	1.193	5.385	3.464	2.786	1.242	21.12
0.18	6.066	1.192	5.088	3.278	2.642	1.187	18.54
0.19	5.745	1.191	4.822	3.112	2.515	1.135	16.38
0.20	5.455	1.190	4.583	2.964	2.400	1.086	14.53
0.21	5.194	1.190	4.366	2.829	2.298	1.040	12.96
0.22	4.955	1.188	4.169	2.708	2.205	0.9961	11.60
0.23	4.738	1.187	3.990	2.597	2.120	0.9543	10.42
0.24	4.538	1.186	3.825	2.496	2.043	0.9145	9.386
0.25	4.355	1.185	3.674	2.403	1.973	0.8766	8.483
0.26	4.185	1.184	3.535	2.317	1.909	0.8404	7.688
0.27	4.028	1.183	3.406	2.238	1.850	0.8058	6.983
0.28	3.882	1.181	3.286	2.166	1.795	0.7727	6.357
0.29	3.746	1.180	3.174	2.098	1.745	0.7409	5.799
0.30	3.619	1.179	3.070	2.035	1.698	0.7105	5.299

Appendix E *Continued*

M	$\frac{p}{p^*}$	$\frac{T}{T^*}$	$\frac{\rho}{\rho^*} = \frac{V^*}{V}$	$\frac{p_0}{p_0^*}$	$\frac{I}{I^*}$	$\frac{s^* - s}{R}$	$\frac{4\bar{f}L_{max}}{D}$
0.31	3.500	1.177	2.973	1.977	1.655	0.6813	4.851
0.32	3.389	1.176	2.882	1.922	1.614	0.6533	4.447
0.33	3.284	1.174	2.796	1.871	1.577	0.6263	4.082
0.34	3.185	1.173	2.716	1.823	1.542	0.6004	3.752
0.35	3.092	1.171	2.640	1.778	1.509	0.5755	3.452
0.36	3.004	1.170	2.568	1.736	1.479	0.5515	3.180
0.37	2.921	1.168	2.501	1.696	1.450	0.5283	2.932
0.38	2.842	1.166	2.437	1.659	1.424	0.5060	2.705
0.39	2.767	1.165	2.376	1.623	1.398	0.4845	2.498
0.40	2.696	1.163	2.318	1.590	1.375	0.4638	2.308
0.41	2.628	1.161	2.264	1.559	1.353	0.4438	2.134
0.42	2.563	1.159	2.212	1.529	1.332	0.4246	1.974
0.43	2.502	1.157	2.162	1.501	1.312	0.4059	1.827
0.44	2.443	1.155	2.114	1.474	1.294	0.3880	1.692
0.45	2.386	1.153	2.069	1.449	1.276	0.3706	1.566
0.46	2.333	1.151	2.026	1.425	1.260	0.3539	1.451
0.47	2.281	1.149	1.985	1.402	1.244	0.3378	1.344
0.48	2.231	1.147	1.945	1.380	1.230	0.3222	1.245
0.49	2.184	1.145	1.907	1.359	1.216	0.3071	1.154
0.50	2.138	1.143	1.871	1.340	1.203	0.2926	1.069
0.51	2.094	1.141	1.836	1.321	1.190	0.2785	0.9904
0.52	2.052	1.138	1.802	1.303	1.179	0.2650	0.9174
0.53	2.011	1.136	1.770	1.286	1.168	0.2519	0.8496
0.54	1.972	1.134	1.739	1.270	1.157	0.2393	0.7866
0.55	1.934	1.132	1.709	1.255	1.147	0.2271	0.7281
0.56	1.898	1.129	1.680	1.240	1.138	0.2154	0.6736
0.57	1.862	1.127	1.653	1.226	1.129	0.2040	0.6229
0.58	1.828	1.124	1.626	1.213	1.121	0.1931	0.5757
0.59	1.795	1.122	1.600	1.200	1.113	0.1826	0.5317
0.60	1.763	1.119	1.575	1.188	1.105	0.1724	0.4908
0.61	1.733	1.117	1.551	1.177	1.098	0.1627	0.4527
0.62	1.703	1.114	1.528	1.166	1.091	0.1533	0.4172
0.63	1.674	1.112	1.505	1.155	1.085	0.1442	0.3841
0.64	1.646	1.109	1.484	1.145	1.079	0.1355	0.3533
0.65	1.618	1.107	1.463	1.136	1.073	0.1272	0.3246
0.66	1.592	1.104	1.442	1.127	1.068	0.1192	0.2979
0.67	1.566	1.101	1.422	1.118	1.063	0.1114	0.2730
0.68	1.541	1.098	1.403	1.110	1.058	0.1041	0.2498
0.69	1.517	1.096	1.385	1.102	1.053	0.096 97	0.2282
0.70	1.493	1.093	1.367	1.094	1.049	0.090 18	0.2081

continued

Appendix E *Continued*

M	$\frac{p}{p^*}$	$\frac{T}{T^*}$	$\frac{\rho}{\rho^*} = \frac{V^*}{V}$	$\frac{p_0}{p_0^*}$	$\frac{I}{I^*}$	$\frac{s^* - s}{R}$	$\frac{4\bar{f}L_{max}}{D}$
0.71	1.471	1.090	1.349	1.087	1.045	0.083 69	0.1895
0.72	1.448	1.087	1.332	1.081	1.041	0.077 49	0.1722
0.73	1.427	1.084	1.315	1.074	1.038	0.071 57	0.1561
0.74	1.405	1.082	1.299	1.068	1.034	0.065 92	0.1411
0.75	1.385	1.079	1.284	1.062	1.031	0.060 55	0.1273
0.76	1.365	1.076	1.269	1.057	1.028	0.055 43	0.1145
0.77	1.345	1.073	1.254	1.052	1.026	0.050 58	0.1026
0.78	1.326	1.070	1.240	1.047	1.023	0.045 98	0.091 67
0.79	1.307	1.067	1.226	1.043	1.021	0.041 63	0.081 58
0.80	1.289	1.064	1.212	1.038	1.019	0.037 52	0.072 29
0.81	1.272	1.061	1.199	1.034	1.016	0.033 65	0.063 76
0.82	1.254	1.058	1.186	1.030	1.015	0.030 01	0.055 93
0.83	1.237	1.055	1.173	1.027	1.013	0.026 60	0.048 78
0.84	1.221	1.052	1.161	1.024	1.011	0.023 42	0.042 26
0.85	1.205	1.048	1.149	1.021	1.010	0.020 46	0.036 33
0.86	1.189	1.045	1.137	1.018	1.008	0.017 71	0.030 97
0.87	1.173	1.042	1.126	1.015	1.007	0.015 18	0.026 13
0.88	1.158	1.039	1.115	1.013	1.006	0.012 86	0.021 80
0.89	1.144	1.036	1.104	1.011	1.005	0.010 74	0.017 93
0.90	1.129	1.033	1.093	1.009	1.004	0.008 823	0.014 51
0.91	1.115	1.029	1.083	1.007	1.003	0.007 105	0.011 51
0.92	1.101	1.026	1.073	1.006	1.002	0.005 581	0.008 914
0.93	1.088	1.023	1.063	1.004	1.002	0.004 249	0.006 687
0.94	1.074	1.020	1.053	1.003	1.001	0.003 103	0.004 816
0.95	1.061	1.017	1.044	1.002	1.001	0.002 143	0.003 278
0.96	1.049	1.013	1.035	1.001	1.001	0.001 364	0.002 057
0.97	1.036	1.010	1.026	1.001	1.000	0.000 762 5	0.001 135
0.98	1.024	1.007	1.017	1.000	1.000	0.000 337 0	0.000 494 8
0.99	1.012	1.003	1.008	1.000	1.000	0.000 083 75	0.000 121 3
1.00	1.000	1.000	1.000	1.000	1.000	0.000 000 00	0.000 000 0
1.01	0.9884	0.9967	0.9918	1.000	1.000	0.000 082 97	0.000 116 8
1.02	0.9771	0.9933	0.9837	1.000	1.000	0.000 329 3	0.000 458 8
1.03	0.9660	0.9900	0.9758	1.001	1.000	0.000 737 6	0.001 013
1.04	0.9551	0.9866	0.9681	1.001	1.001	0.001 304	0.001 769
1.05	0.9443	0.9832	0.9605	1.002	1.001	0.002 027	0.002 714
1.06	0.9338	0.9798	0.9531	1.003	1.001	0.002 903	0.003 838
1.07	0.9235	0.9764	0.9458	1.004	1.002	0.003 930	0.005 131
1.08	0.9133	0.9730	0.9387	1.005	1.002	0.005 106	0.006 585
1.09	0.9034	0.9696	0.9317	1.006	1.003	0.006 428	0.008 189
1.10	0.8936	0.9662	0.9249	1.008	1.003	0.007 894	0.009 935
1.11	0.8840	0.9628	0.9182	1.010	1.004	0.009 502	0.011 82
1.12	0.8745	0.9593	0.9116	1.011	1.004	0.011 25	0.013 82

M	$\frac{p}{p^*}$	$\frac{T}{T^*}$	$\frac{\rho}{\rho^*}=\frac{V^*}{V}$	$\frac{p_0}{p_0^*}$	$\frac{I}{I^*}$	$\frac{s^*-s}{R}$	$\frac{4\bar{f}L_{max}}{D}$
1.13	0.8652	0.9559	0.9051	1.013	1.005	0.013 13	0.015 95
1.14	0.8561	0.9524	0.8988	1.015	1.006	0.015 15	0.018 19
1.15	0.8471	0.9490	0.8926	1.017	1.006	0.017 30	0.020 53
1.16	0.8383	0.9455	0.8865	1.020	1.007	0.019 59	0.022 98
1.17	0.8296	0.9421	0.8806	1.022	1.008	0.022 00	0.025 52
1.18	0.8210	0.9386	0.8747	1.025	1.009	0.024 54	0.028 14
1.19	0.8126	0.9351	0.8690	1.028	1.010	0.027 20	0.030 85
1.20	0.8044	0.9317	0.8633	1.030	1.011	0.029 99	0.033 64
1.21	0.7962	0.9282	0.8578	1.033	1.012	0.032 89	0.036 50
1.22	0.7882	0.9247	0.8524	1.037	1.013	0.035 92	0.039 43
1.23	0.7803	0.9212	0.8470	1.040	1.014	0.039 06	0.042 42
1.24	0.7726	0.9178	0.8418	1.043	1.015	0.042 32	0.045 47
1.25	0.7649	0.9143	0.8367	1.047	1.016	0.045 69	0.048 58
1.26	0.7574	0.9108	0.8316	1.050	1.017	0.049 18	0.051 74
1.27	0.7500	0.9073	0.8266	1.054	1.018	0.052 77	0.054 95
1.28	0.7427	0.9038	0.8218	1.058	1.019	0.056 47	0.058 20
1.29	0.7356	0.9003	0.8170	1.062	1.021	0.060 28	0.061 50
1.30	0.7285	0.8969	0.8123	1.066	1.022	0.064 20	0.064 83
1.31	0.7215	0.8934	0.8076	1.071	1.023	0.068 22	0.068 20
1.32	0.7147	0.8899	0.8031	1.075	1.024	0.072 34	0.071 61
1.33	0.7079	0.8864	0.7986	1.080	1.025	0.076 56	0.075 04
1.34	0.7012	0.8829	0.7942	1.084	1.027	0.080 88	0.078 50
1.35	0.6947	0.8794	0.7899	1.089	1.028	0.085 30	0.081 99
1.36	0.6882	0.8760	0.7856	1.094	1.029	0.089 81	0.085 50
1.37	0.6818	0.8725	0.7814	1.099	1.031	0.094 42	0.089 04
1.38	0.6755	0.8690	0.7773	1.104	1.032	0.099 12	0.092 59
1.39	0.6693	0.8655	0.7733	1.109	1.033	0.1039	0.096 16
1.40	0.6632	0.8621	0.7693	1.115	1.035	0.1088	0.099 74
1.41	0.6572	0.8586	0.7654	1.120	1.036	0.1138	0.1033
1.42	0.6512	0.8551	0.7615	1.126	1.037	0.1188	0.1070
1.43	0.6454	0.8517	0.7577	1.132	1.039	0.1240	0.1106
1.44	0.6396	0.8482	0.7540	1.138	1.040	0.1292	0.1142
1.45	0.6339	0.8448	0.7503	1.144	1.042	0.1345	0.1178
1.46	0.6282	0.8413	0.7467	1.150	1.043	0.1399	0.1215
1.47	0.6227	0.8379	0.7432	1.156	1.044	0.1454	0.1251
1.48	0.6172	0.8344	0.7397	1.163	1.046	0.1509	0.1287
1.49	0.6118	0.8310	0.7362	1.169	1.047	0.1566	0.1324
1.50	0.6065	0.8276	0.7328	1.176	1.049	0.1623	0.1361
1.51	0.6012	0.8242	0.7295	1.183	1.050	0.1680	0.1397
1.52	0.5960	0.8207	0.7262	1.190	1.052	0.1739	0.1433

continued

Appendix E *Continued*

M	$\frac{p}{p^*}$	$\frac{T}{T^*}$	$\frac{\rho}{\rho^*} = \frac{V^*}{V}$	$\frac{p_0}{p_0^*}$	$\frac{I}{I^*}$	$\frac{s^* - s}{R}$	$\frac{4\bar{f}L_{max}}{D}$
1.53	0.5909	0.8173	0.7229	1.197	1.053	0.1798	0.1470
1.54	0.5858	0.8139	0.7198	1.204	1.055	0.1858	0.1506
1.55	0.5808	0.8105	0.7166	1.212	1.056	0.1919	0.1543
1.56	0.5759	0.8071	0.7135	1.219	1.058	0.1981	0.1579
1.57	0.5710	0.8038	0.7105	1.227	1.059	0.2043	0.1615
1.58	0.5662	0.8004	0.7074	1.234	1.060	0.2106	0.1651
1.59	0.5615	0.7970	0.7045	1.242	1.062	0.2169	0.1688
1.60	0.5568	0.7937	0.7016	1.250	1.063	0.2233	0.1724
1.61	0.5522	0.7903	0.6987	1.258	1.065	0.2298	0.1760
1.62	0.5476	0.7869	0.6958	1.267	1.066	0.2364	0.1795
1.63	0.5431	0.7836	0.6930	1.275	1.068	0.2430	0.1831
1.64	0.5386	0.7803	0.6903	1.284	1.069	0.2496	0.1867
1.65	0.5342	0.7770	0.6876	1.292	1.071	0.2564	0.1902
1.66	0.5299	0.7736	0.6849	1.301	1.072	0.2631	0.1938
1.67	0.5256	0.7703	0.6823	1.310	1.074	0.2700	0.1973
1.68	0.5213	0.7670	0.6796	1.319	1.075	0.2769	0.2008
1.69	0.5171	0.7637	0.6771	1.328	1.077	0.2839	0.2043
1.70	0.5130	0.7605	0.6745	1.338	1.079	0.2909	0.2078
1.71	0.5089	0.7572	0.6721	1.347	1.080	0.2980	0.2113
1.72	0.5048	0.7539	0.6696	1.357	1.082	0.3051	0.2147
1.73	0.5008	0.7507	0.6672	1.367	1.083	0.3123	0.2182
1.74	0.4969	0.7474	0.6648	1.376	1.085	0.3195	0.2216
1.75	0.4929	0.7442	0.6624	1.386	1.086	0.3268	0.2250
1.76	0.4891	0.7410	0.6601	1.397	1.088	0.3341	0.2284
1.77	0.4853	0.7377	0.6578	1.407	1.089	0.3415	0.2318
1.78	0.4815	0.7345	0.6555	1.418	1.091	0.3489	0.2352
1.79	0.4778	0.7313	0.6533	1.428	1.092	0.3564	0.2385
1.80	0.4741	0.7282	0.6511	1.439	1.094	0.3639	0.2419
1.81	0.4704	0.7250	0.6489	1.450	1.095	0.3715	0.2452
1.82	0.4668	0.7218	0.6467	1.461	1.096	0.3791	0.2485
1.83	0.4632	0.7187	0.6446	1.472	1.098	0.3868	0.2518
1.84	0.4597	0.7155	0.6425	1.484	1.099	0.3945	0.2551
1.85	0.4562	0.7124	0.6404	1.495	1.101	0.4023	0.2583
1.86	0.4528	0.7093	0.6384	1.507	1.102	0.4101	0.2616
1.87	0.4494	0.7061	0.6364	1.519	1.104	0.4179	0.2648
1.88	0.4460	0.7030	0.6344	1.531	1.105	0.4258	0.2680
1.89	0.4427	0.6999	0.6324	1.543	1.107	0.4337	0.2712
1.90	0.4394	0.6969	0.6305	1.555	1.108	0.4416	0.2743
1.91	0.4361	0.6938	0.6286	1.568	1.110	0.4496	0.2775
1.92	0.4329	0.6907	0.6267	1.580	1.111	0.4577	0.2806
1.93	0.4297	0.6877	0.6248	1.593	1.113	0.4657	0.2837
1.94	0.4265	0.6847	0.6230	1.606	1.114	0.4739	0.2868

Appendix E *Continued*

M	$\frac{p}{p^*}$	$\frac{T}{T^*}$	$\frac{\rho}{\rho^*} = \frac{V^*}{V}$	$\frac{p_0}{p_0^*}$	$\frac{I}{I^*}$	$\frac{s^* - s}{R}$	$\frac{4\bar{f}L_{max}}{D}$
1.95	0.4234	0.6816	0.6211	1.619	1.116	0.4820	0.2899
1.96	0.4203	0.6786	0.6193	1.633	1.117	0.4902	0.2929
1.97	0.4172	0.6756	0.6176	1.646	1.118	0.4984	0.2960
1.98	0.4142	0.6726	0.6158	1.660	1.120	0.5066	0.2990
1.99	0.4112	0.6696	0.6141	1.674	1.121	0.5149	0.3020
2.00	0.4082	0.6667	0.6124	1.687	1.123	0.5232	0.3050
2.01	0.4053	0.6637	0.6107	1.702	1.124	0.5316	0.3080
2.02	0.4024	0.6608	0.6090	1.716	1.126	0.5400	0.3109
2.03	0.3995	0.6578	0.6074	1.730	1.127	0.5484	0.3138
2.04	0.3967	0.6549	0.6057	1.745	1.128	0.5568	0.3168
2.05	0.3939	0.6520	0.6041	1.760	1.130	0.5653	0.3197
2.06	0.3911	0.6491	0.6025	1.775	1.131	0.5738	0.3225
2.07	0.3883	0.6462	0.6010	1.790	1.132	0.5823	0.3254
2.08	0.3856	0.6433	0.5994	1.806	1.134	0.5909	0.3282
2.09	0.3829	0.6405	0.5979	1.821	1.135	0.5995	0.3310
2.10	0.3802	0.6376	0.5963	1.837	1.137	0.6081	0.3339
2.11	0.3776	0.6348	0.5948	1.853	1.138	0.6167	0.3366
2.12	0.3750	0.6320	0.5934	1.869	1.139	0.6254	0.3394
2.13	0.3724	0.6291	0.5919	1.885	1.141	0.6341	0.3422
2.14	0.3698	0.6263	0.5905	1.902	1.142	0.6428	0.3449
2.15	0.3673	0.6235	0.5890	1.919	1.143	0.6516	0.3476
2.16	0.3648	0.6208	0.5876	1.935	1.145	0.6603	0.3503
2.17	0.3623	0.6180	0.5862	1.953	1.146	0.6691	0.3530
2.18	0.3598	0.6152	0.5848	1.970	1.147	0.6779	0.3556
2.19	0.3574	0.6125	0.5835	1.987	1.149	0.6868	0.3583
2.20	0.3549	0.6098	0.5821	2.005	1.150	0.6956	0.3609
2.21	0.3525	0.6070	0.5808	2.023	1.151	0.7045	0.3635
2.22	0.3502	0.6043	0.5794	2.041	1.153	0.7134	0.3661
2.23	0.3478	0.6016	0.5781	2.059	1.154	0.7223	0.3687
2.24	0.3455	0.5989	0.5768	2.078	1.155	0.7313	0.3712
2.25	0.3432	0.5963	0.5756	2.096	1.156	0.7402	0.3738
2.26	0.3409	0.5936	0.5743	2.115	1.158	0.7492	0.3763
2.27	0.3387	0.5910	0.5731	2.134	1.159	0.7582	0.3788
2.28	0.3364	0.5883	0.5718	2.154	1.160	0.7672	0.3813
2.29	0.3342	0.5857	0.5706	2.173	1.162	0.7763	0.3838
2.30	0.3320	0.5831	0.5694	2.193	1.163	0.7853	0.3862
2.31	0.3298	0.5805	0.5682	2.213	1.164	0.7944	0.3887
2.32	0.3277	0.5779	0.5670	2.233	1.165	0.8035	0.3911
2.33	0.3255	0.5753	0.5658	2.254	1.167	0.8126	0.3935
2.34	0.3234	0.5728	0.5647	2.274	1.168	0.8217	0.3959

continued

Appendix E *Continued*

M	$\frac{p}{p^*}$	$\frac{T}{T^*}$	$\frac{\rho}{\rho^*} = \frac{V^*}{V}$	$\frac{p_0}{p_0^*}$	$\frac{I}{I^*}$	$\frac{s^* - s}{R}$	$\frac{4\bar{f}L_{max}}{D}$
2.35	0.3213	0.5702	0.5635	2.295	1.169	0.8309	0.3983
2.36	0.3193	0.5677	0.5624	2.316	1.170	0.8400	0.4006
2.37	0.3172	0.5651	0.5613	2.338	1.171	0.8492	0.4030
2.38	0.3152	0.5626	0.5602	2.359	1.173	0.8584	0.4053
2.39	0.3131	0.5601	0.5591	2.381	1.174	0.8675	0.4076
2.40	0.3111	0.5576	0.5580	2.403	1.175	0.8768	0.4099
2.41	0.3092	0.5551	0.5569	2.425	1.176	0.8860	0.4122
2.42	0.3072	0.5527	0.5558	2.448	1.177	0.8952	0.4144
2.43	0.3053	0.5502	0.5548	2.471	1.179	0.9045	0.4167
2.44	0.3033	0.5478	0.5537	2.494	1.180	0.9137	0.4189
2.45	0.3014	0.5453	0.5527	2.517	1.181	0.9230	0.4211
2.46	0.2995	0.5429	0.5517	2.540	1.182	0.9323	0.4233
2.47	0.2976	0.5405	0.5507	2.564	1.183	0.9416	0.4255
2.48	0.2958	0.5381	0.5497	2.588	1.184	0.9509	0.4277
2.49	0.2939	0.5357	0.5487	2.612	1.186	0.9602	0.4298
2.50	0.2921	0.5333	0.5477	2.637	1.187	0.9695	0.4320
2.51	0.2903	0.5310	0.5468	2.661	1.188	0.9799	0.4341
2.52	0.2885	0.5286	0.5458	2.686	1.189	0.9882	0.4362
2.53	0.2867	0.5263	0.5448	2.712	1.190	0.9976	0.4383
2.54	0.2850	0.5239	0.5439	2.737	1.191	1.007	0.4404
2.55	0.2832	0.5216	0.5430	2.763	1.192	1.016	0.4425
2.56	0.2815	0.5193	0.5421	2.789	1.193	1.026	0.4445
2.57	0.2798	0.5170	0.5411	2.815	1.195	1.035	0.4466
2.58	0.2781	0.5147	0.5402	2.842	1.196	1.044	0.4486
2.59	0.2764	0.5125	0.5393	2.869	1.197	1.054	0.4506
2.60	0.2747	0.5102	0.5385	2.896	1.198	1.063	0.4526
2.61	0.2731	0.5080	0.5376	2.923	1.199	1.073	0.4546
2.62	0.2714	0.5057	0.5367	2.951	1.200	1.082	0.4565
2.63	0.2698	0.5035	0.5359	2.979	1.201	1.092	0.4585
2.64	0.2682	0.5013	0.5350	3.007	1.202	1.101	0.4604
2.65	0.2666	0.4991	0.5342	3.036	1.203	1.111	0.4624
2.66	0.2650	0.4969	0.5333	3.065	1.204	1.120	0.4643
2.67	0.2634	0.4947	0.5325	3.094	1.205	1.129	0.4662
2.68	0.2619	0.4925	0.5317	3.123	1.206	1.139	0.4681
2.69	0.2603	0.4904	0.5309	3.153	1.207	1.148	0.4700
2.70	0.2588	0.4882	0.5301	3.183	1.208	1.158	0.4718
2.71	0.2573	0.4861	0.5293	3.213	1.209	1.167	0.4737
2.72	0.2558	0.4839	0.5285	3.244	1.210	1.177	0.4755
2.73	0.2543	0.4818	0.5277	3.275	1.211	1.186	0.4773
2.74	0.2528	0.4797	0.5269	3.306	1.212	1.196	0.4791
2.75	0.2513	0.4776	0.5262	3.338	1.213	1.205	0.4809
2.76	0.2498	0.4755	0.5254	3.370	1.214	1.215	0.4827

Appendix E *Continued*

M	$\frac{p}{p^*}$	$\frac{T}{T^*}$	$\frac{\rho}{\rho^*}=\frac{V^*}{V}$	$\frac{p_0}{p_0^*}$	$\frac{I}{I^*}$	$\frac{s^*-s}{R}$	$\frac{4\bar{f}L_{max}}{D}$
2.77	0.2484	0.4735	0.5247	3.402	1.215	1.224	0.4845
2.78	0.2470	0.4714	0.5239	3.434	1.216	1.234	0.4863
2.79	0.2455	0.4693	0.5232	3.467	1.217	1.243	0.4880
2.80	0.2441	0.4673	0.5225	3.500	1.218	1.253	0.4898
2.81	0.2427	0.4653	0.5217	3.534	1.219	1.262	0.4915
2.82	0.2414	0.4632	0.5210	3.567	1.220	1.272	0.4932
2.83	0.2400	0.4612	0.5203	3.601	1.221	1.281	0.4949
2.84	0.2386	0.4592	0.5196	3.636	1.222	1.291	0.4966
2.85	0.2373	0.4572	0.5189	3.671	1.223	1.300	0.4983
2.86	0.2359	0.4552	0.5182	3.706	1.224	1.310	0.5000
2.87	0.2346	0.4533	0.5175	3.741	1.225	1.319	0.5016
2.88	0.2333	0.4513	0.5169	3.777	1.226	1.329	0.5033
2.89	0.2320	0.4494	0.5162	3.813	1.227	1.338	0.5049
2.90	0.2307	0.4474	0.5155	3.850	1.228	1.348	0.5065
2.91	0.2294	0.4455	0.5149	3.887	1.229	1.358	0.5081
2.92	0.2281	0.4436	0.5142	3.924	1.229	1.367	0.5097
2.93	0.2268	0.4417	0.5136	3.961	1.230	1.377	0.5113
2.94	0.2256	0.4398	0.5129	3.999	1.231	1.386	0.5129
2.95	0.2243	0.4379	0.5123	4.038	1.232	1.396	0.5145
2.96	0.2231	0.4360	0.5116	4.076	1.233	1.405	0.5160
2.97	0.2218	0.4341	0.5110	4.115	1.234	1.415	0.5176
2.98	0.2206	0.4323	0.5104	4.155	1.235	1.424	0.5191
2.99	0.2194	0.4304	0.5098	4.194	1.236	1.434	0.5206
3.00	0.2182	0.4286	0.5092	4.235	1.237	1.443	0.5222
3.01	0.2170	0.4267	0.5086	4.275	1.237	1.453	0.5237
3.02	0.2158	0.4249	0.5080	4.316	1.238	1.462	0.5252
3.03	0.2147	0.4231	0.5074	4.357	1.239	1.472	0.5266
3.04	0.2135	0.4213	0.5068	4.399	1.240	1.481	0.5281
3.05	0.2124	0.4195	0.5062	4.441	1.241	1.491	0.5296
3.06	0.2112	0.4177	0.5056	4.483	1.242	1.500	0.5310
3.07	0.2101	0.4159	0.5051	4.526	1.243	1.510	0.5325
3.08	0.2090	0.4142	0.5045	4.570	1.243	1.519	0.5339
3.09	0.2078	0.4124	0.5039	4.613	1.244	1.529	0.5354
3.10	0.2067	0.4107	0.5034	4.657	1.245	1.538	0.5368
3.11	0.2056	0.4089	0.5028	4.702	1.246	1.548	0.5382
3.12	0.2045	0.4072	0.5023	4.747	1.247	1.557	0.5396
3.13	0.2034	0.4055	0.5017	4.792	1.247	1.567	0.5410
3.14	0.2024	0.4038	0.5012	4.838	1.248	1.576	0.5424
3.15	0.2013	0.4021	0.5007	4.884	1.249	1.586	0.5437
3.16	0.2002	0.4004	0.5001	4.930	1.250	1.595	0.5451

continued

Appendix E *Continued*

M	$\frac{p}{p^*}$	$\frac{T}{T^*}$	$\frac{\rho}{\rho^*}=\frac{V^*}{V}$	$\frac{p_0}{p_0^*}$	$\frac{I}{I^*}$	$\frac{s^*-s}{R}$	$\frac{4\bar{f}L_{max}}{D}$
3.17	0.1992	0.3987	0.4996	4.977	1.251	1.605	0.5464
3.18	0.1981	0.3970	0.4991	5.025	1.251	1.614	0.5478
3.19	0.1971	0.3954	0.4986	5.073	1.252	1.624	0.5491
3.20	0.1961	0.3937	0.4980	5.121	1.253	1.633	0.5504
3.21	0.1951	0.3921	0.4975	5.170	1.254	1.643	0.5518
3.22	0.1940	0.3904	0.4970	5.219	1.254	1.652	0.5531
3.23	0.1930	0.3888	0.4965	5.268	1.255	1.662	0.5544
3.24	0.1920	0.3872	0.4960	5.319	1.256	1.671	0.5557
3.25	0.1911	0.3855	0.4955	5.369	1.257	1.681	0.5569
3.26	0.1901	0.3839	0.4951	5.420	1.258	1.690	0.5582
3.27	0.1891	0.3823	0.4946	5.472	1.258	1.700	0.5595
3.28	0.1881	0.3807	0.4941	5.523	1.259	1.709	0.5607
3.29	0.1872	0.3792	0.4936	5.576	1.260	1.718	0.5620
3.30	0.1862	0.3776	0.4931	5.629	1.260	1.728	0.5632
3.31	0.1853	0.3760	0.4927	5.682	1.261	1.737	0.5645
3.32	0.1843	0.3745	0.4922	5.736	1.262	1.747	0.5657
3.33	0.1834	0.3729	0.4917	5.790	1.263	1.756	0.5669
3.34	0.1825	0.3714	0.4913	5.845	1.263	1.766	0.5681
3.35	0.1815	0.3699	0.4908	5.900	1.264	1.775	0.5693
3.36	0.1806	0.3683	0.4904	5.956	1.265	1.784	0.5705
3.37	0.1797	0.3668	0.4899	6.012	1.265	1.794	0.5717
3.38	0.1788	0.3653	0.4895	6.069	1.266	1.803	0.5729
3.39	0.1779	0.3638	0.4891	6.126	1.267	1.813	0.5740
3.40	0.1770	0.3623	0.4886	6.184	1.268	1.822	0.5752
3.41	0.1762	0.3608	0.4882	6.242	1.268	1.831	0.5764
3.42	0.1753	0.3594	0.4878	6.301	1.269	1.841	0.5775
3.43	0.1744	0.3579	0.4873	6.360	1.270	1.850	0.5786
3.44	0.1736	0.3564	0.4869	6.420	1.270	1.859	0.5798
3.45	0.1727	0.3550	0.4865	6.480	1.271	1.869	0.5809
3.46	0.1718	0.3535	0.4861	6.541	1.272	1.878	0.5820
3.47	0.1710	0.3521	0.4857	6.602	1.272	1.887	0.5831
3.48	0.1702	0.3507	0.4853	6.664	1.273	1.897	0.5842
3.49	0.1693	0.3492	0.4849	6.727	1.274	1.906	0.5853
3.50	0.1685	0.3478	0.4845	6.790	1.274	1.915	0.5864
3.51	0.1677	0.3464	0.4841	6.853	1.275	1.925	0.5875
3.52	0.1669	0.3450	0.4837	6.917	1.276	1.934	0.5886
3.53	0.1661	0.3436	0.4833	6.982	1.276	1.943	0.5897
3.54	0.1653	0.3422	0.4829	7.047	1.277	1.953	0.5907
3.55	0.1645	0.3409	0.4825	7.113	1.278	1.962	0.5918
3.56	0.1637	0.3395	0.4821	7.179	1.278	1.971	0.5928
3.57	0.1629	0.3381	0.4817	7.246	1.279	1.980	0.5939
3.58	0.1621	0.3368	0.4813	7.313	1.279	1.990	0.5949

Appendix E *Continued*

M	$\frac{p}{p^*}$	$\frac{T}{T^*}$	$\frac{\rho}{\rho^*}=\frac{V^*}{V}$	$\frac{p_0}{p_0^*}$	$\frac{I}{I^*}$	$\frac{s^*-s}{R}$	$\frac{4\bar{f}L_{max}}{D}$
3.59	0.1613	0.3354	0.4810	7.381	1.280	1.999	0.5959
3.60	0.1606	0.3341	0.4806	7.450	1.281	2.008	0.5970
3.61	0.1598	0.3327	0.4802	7.519	1.281	2.017	0.5980
3.62	0.1590	0.3314	0.4799	7.589	1.282	2.027	0.5990
3.63	0.1583	0.3301	0.4795	7.659	1.283	2.036	0.6000
3.64	0.1575	0.3288	0.4791	7.730	1.283	2.045	0.6010
3.65	0.1568	0.3275	0.4788	7.802	1.284	2.054	0.6020
3.66	0.1560	0.3262	0.4784	7.874	1.284	2.064	0.6030
3.67	0.1553	0.3249	0.4781	7.947	1.285	2.073	0.6039
3.68	0.1546	0.3236	0.4777	8.020	1.286	2.082	0.6049
3.69	0.1539	0.3223	0.4774	8.094	1.286	2.091	0.6059
3.70	0.1531	0.3210	0.4770	8.169	1.287	2.100	0.6068
3.71	0.1524	0.3198	0.4767	8.244	1.287	2.110	0.6078
3.72	0.1517	0.3185	0.4763	8.320	1.288	2.119	0.6087
3.73	0.1510	0.3172	0.4760	8.397	1.288	2.128	0.6097
3.74	0.1503	0.3160	0.4757	8.474	1.289	2.137	0.6106
3.75	0.1496	0.3148	0.4753	8.552	1.290	2.146	0.6115
3.76	0.1489	0.3135	0.4750	8.630	1.290	2.155	0.6125
3.77	0.1482	0.3123	0.4747	8.709	1.291	2.164	0.6134
3.78	0.1475	0.3111	0.4743	8.789	1.291	2.174	0.6143
3.79	0.1469	0.3099	0.4740	8.869	1.292	2.183	0.6152
3.80	0.1462	0.3086	0.4737	8.951	1.292	2.192	0.6161
3.81	0.1455	0.3074	0.4734	9.032	1.293	2.201	0.6170
3.82	0.1449	0.3062	0.4730	9.115	1.293	2.210	0.6179
3.83	0.1442	0.3051	0.4727	9.198	1.294	2.219	0.6188
3.84	0.1436	0.3039	0.4724	9.282	1.295	2.228	0.6197
3.85	0.1429	0.3027	0.4721	9.366	1.295	2.237	0.6206
3.86	0.1423	0.3015	0.4718	9.451	1.296	2.246	0.6214
3.87	0.1416	0.3003	0.4715	9.537	1.296	2.255	0.6223
3.88	0.1410	0.2992	0.4712	9.624	1.297	2.264	0.6231
3.89	0.1403	0.2980	0.4709	9.711	1.297	2.273	0.6240
3.90	0.1397	0.2969	0.4706	9.799	1.298	2.282	0.6248
3.91	0.1391	0.2957	0.4703	9.888	1.298	2.291	0.6257
3.92	0.1385	0.2946	0.4700	9.977	1.299	2.300	0.6265
3.93	0.1378	0.2935	0.4697	10.07	1.299	2.309	0.6274
3.94	0.1372	0.2923	0.4694	10.16	1.300	2.318	0.6282
3.95	0.1366	0.2912	0.4691	10.25	1.300	2.327	0.6290
3.96	0.1360	0.2901	0.4688	10.34	1.301	2.336	0.6298
3.97	0.1354	0.2890	0.4686	10.44	1.301	2.345	0.6307
3.98	0.1348	0.2879	0.4683	10.53	1.302	2.354	0.6315

continued

Appendix E *Continued*

M	$\frac{p}{p^*}$	$\frac{T}{T^*}$	$\frac{\rho}{\rho^*} = \frac{V^*}{V}$	$\frac{p_0}{p_0^*}$	$\frac{I}{I^*}$	$\frac{s^* - s}{R}$	$\frac{4\bar{f}L_{max}}{D}$
3.99	0.1342	0.2868	0.4680	10.62	1.302	2.363	0.6323
4.00	0.1336	0.2857	0.4677	10.72	1.303	2.372	0.6331
4.01	0.1330	0.2846	0.4674	10.81	1.303	2.381	0.6339
4.02	0.1325	0.2835	0.4672	10.91	1.304	2.390	0.6347
4.03	0.1319	0.2825	0.4669	11.01	1.304	2.399	0.6354
4.04	0.1313	0.2814	0.4666	11.11	1.305	2.408	0.6362
4.05	0.1307	0.2803	0.4663	11.21	1.305	2.417	0.6370
4.06	0.1302	0.2793	0.4661	11.31	1.306	2.425	0.6378
4.07	0.1296	0.2782	0.4658	11.41	1.306	2.434	0.6385
4.08	0.1290	0.2772	0.4655	11.51	1.307	2.443	0.6393
4.09	0.1285	0.2761	0.4653	11.61	1.307	2.452	0.6401
4.10	0.1279	0.2751	0.4650	11.71	1.308	2.461	0.6408
4.11	0.1274	0.2741	0.4648	11.82	1.308	2.470	0.6416
4.12	0.1268	0.2730	0.4645	11.92	1.309	2.479	0.6423
4.13	0.1263	0.2720	0.4642	12.03	1.309	2.487	0.6430
4.14	0.1257	0.2710	0.4640	12.14	1.310	2.496	0.6438
4.15	0.1252	0.2700	0.4637	12.24	1.310	2.505	0.6445
4.16	0.1247	0.2690	0.4635	12.35	1.311	2.514	0.6452
4.17	0.1241	0.2680	0.4632	12.46	1.311	2.523	0.6460
4.18	0.1236	0.2670	0.4630	12.57	1.311	2.531	0.6467
4.19	0.1231	0.2660	0.4627	12.68	1.312	2.540	0.6474
4.20	0.1226	0.2650	0.4625	12.79	1.312	2.549	0.6481
4.21	0.1221	0.2640	0.4623	12.90	1.313	2.558	0.6488
4.22	0.1215	0.2631	0.4620	13.02	1.313	2.566	0.6495
4.23	0.1210	0.2621	0.4618	13.13	1.314	2.575	0.6502
4.24	0.1205	0.2611	0.4615	13.25	1.314	2.584	0.6509
4.25	0.1200	0.2602	0.4613	13.36	1.315	2.592	0.6516
4.26	0.1195	0.2592	0.4611	13.48	1.315	2.601	0.6523
4.27	0.1190	0.2583	0.4608	13.60	1.315	2.610	0.6530
4.28	0.1185	0.2573	0.4606	13.72	1.316	2.618	0.6536
4.29	0.1180	0.2564	0.4604	13.83	1.316	2.627	0.6543
4.30	0.1175	0.2554	0.4601	13.95	1.317	2.636	0.6550
4.31	0.1170	0.2545	0.4599	14.08	1.317	2.644	0.6557
4.32	0.1166	0.2536	0.4597	14.20	1.318	2.653	0.6563
4.33	0.1161	0.2526	0.4595	14.32	1.318	2.662	0.6570
4.34	0.1156	0.2517	0.4592	14.45	1.318	2.670	0.6576
4.35	0.1151	0.2508	0.4590	14.57	1.319	2.679	0.6583
4.36	0.1147	0.2499	0.4588	14.70	1.319	2.688	0.6589
4.37	0.1142	0.2490	0.4586	14.82	1.320	2.696	0.6596
4.38	0.1137	0.2481	0.4584	14.95	1.320	2.705	0.6602
4.39	0.1133	0.2472	0.4582	15.08	1.320	2.713	0.6609
4.40	0.1128	0.2463	0.4579	15.21	1.321	2.722	0.6615

Appendix E *Continued*

M	$\frac{p}{p^*}$	$\frac{T}{T^*}$	$\frac{\rho}{\rho^*} = \frac{V^*}{V}$	$\frac{p_0}{p_0^*}$	$\frac{I}{I^*}$	$\frac{s^* - s}{R}$	$\frac{4\bar{f}L_{max}}{D}$
4.41	0.1123	0.2454	0.4577	15.34	1.321	2.731	0.6621
4.42	0.1119	0.2445	0.4575	15.47	1.322	2.739	0.6627
4.43	0.1114	0.2437	0.4573	15.61	1.322	2.748	0.6634
4.44	0.1110	0.2428	0.4571	15.74	1.322	2.756	0.6640
4.45	0.1105	0.2419	0.4569	15.87	1.323	2.765	0.6646
4.46	0.1101	0.2410	0.4567	16.01	1.323	2.773	0.6652
4.47	0.1096	0.2402	0.4565	16.15	1.324	2.782	0.6658
4.48	0.1092	0.2393	0.4563	16.28	1.324	2.790	0.6664
4.49	0.1088	0.2385	0.4561	16.42	1.324	2.799	0.6670
4.50	0.1083	0.2376	0.4559	16.56	1.325	2.807	0.6676
4.51	0.1079	0.2368	0.4557	16.70	1.325	2.816	0.6682
4.52	0.1075	0.2359	0.4555	16.84	1.325	2.824	0.6688
4.53	0.1070	0.2351	0.4553	16.99	1.326	2.832	0.6694
4.54	0.1066	0.2343	0.4551	17.13	1.326	2.841	0.6700
4.55	0.1062	0.2334	0.4549	17.28	1.327	2.849	0.6706
4.56	0.1058	0.2326	0.4547	17.42	1.327	2.858	0.6712
4.57	0.1054	0.2318	0.4545	17.57	1.327	2.866	0.6717
4.58	0.1049	0.2310	0.4543	17.72	1.328	2.875	0.6723
4.59	0.1045	0.2302	0.4541	17.87	1.328	2.883	0.6729
4.60	0.1041	0.2294	0.4539	18.02	1.328	2.891	0.6734
4.61	0.1037	0.2286	0.4537	18.17	1.329	2.900	0.6740
4.62	0.1033	0.2278	0.4536	18.32	1.329	2.908	0.6746
4.63	0.1029	0.2270	0.4534	18.48	1.330	2.916	0.6751
4.64	0.1025	0.2262	0.4532	18.63	1.330	2.925	0.6757
4.65	0.1021	0.2254	0.4530	18.79	1.330	2.933	0.6762
4.66	0.1017	0.2246	0.4528	18.94	1.331	2.941	0.6768
4.67	0.1013	0.2238	0.4526	19.10	1.331	2.950	0.6773
4.68	0.1009	0.2230	0.4525	19.26	1.331	2.958	0.6779
4.69	0.1005	0.2223	0.4523	19.42	1.332	2.966	0.6784
4.70	0.1001	0.2215	0.4521	19.58	1.332	2.975	0.6790
4.71	0.099 75	0.2207	0.4519	19.75	1.332	2.983	0.6795
4.72	0.099 36	0.2200	0.4517	19.91	1.333	2.991	0.6800
4.73	0.098 98	0.2192	0.4516	20.07	1.333	2.999	0.6805
4.74	0.098 60	0.2184	0.4514	20.24	1.333	3.008	0.6811
4.75	0.098 23	0.2177	0.4512	20.41	1.334	3.016	0.6816
4.76	0.097 85	0.2169	0.4510	20.58	1.334	3.024	0.6821
4.77	0.097 48	0.2162	0.4509	20.75	1.334	3.032	0.6826
4.78	0.097 11	0.2155	0.4507	20.92	1.335	3.041	0.6831
4.79	0.096 74	0.2147	0.4505	21.09	1.335	3.049	0.6837
4.80	0.096 37	0.2140	0.4504	21.26	1.335	3.057	0.6842

continued

Appendix E *Continued*

M	$\frac{p}{p^*}$	$\frac{T}{T^*}$	$\frac{\rho}{\rho^*} = \frac{V^*}{V}$	$\frac{p_0}{p_0^*}$	$\frac{I}{I^*}$	$\frac{s^* - s}{R}$	$\frac{4\bar{f}L_{max}}{D}$
4.81	0.096 01	0.2132	0.4502	21.44	1.336	3.065	0.6847
4.82	0.095 64	0.2125	0.4500	21.61	1.336	3.073	0.6852
4.83	0.095 28	0.2118	0.4499	21.79	1.336	3.082	0.6857
4.84	0.094 92	0.2111	0.4497	21.97	1.337	3.090	0.6862
4.85	0.094 57	0.2104	0.4495	22.15	1.337	3.098	0.6867
4.86	0.094 21	0.2096	0.4494	22.33	1.337	3.106	0.6872
4.87	0.093 86	0.2089	0.4492	22.51	1.338	3.114	0.6877
4.88	0.093 51	0.2082	0.4491	22.70	1.338	3.122	0.6881
4.89	0.093 16	0.2075	0.4489	22.88	1.338	3.130	0.6886
4.90	0.092 81	0.2068	0.4487	23.07	1.339	3.138	0.6891
4.91	0.092 47	0.2061	0.4486	23.25	1.339	3.146	0.6896
4.92	0.092 12	0.2054	0.4484	23.44	1.339	3.155	0.6901
4.93	0.091 78	0.2047	0.4483	23.63	1.340	3.163	0.6905
4.94	0.091 44	0.2041	0.4481	23.82	1.340	3.171	0.6910
4.95	0.091 10	0.2034	0.4480	24.02	1.340	3.179	0.6915
4.96	0.090 77	0.2027	0.4478	24.21	1.340	3.187	0.6920
4.97	0.090 43	0.2020	0.4477	24.41	1.341	3.195	0.6924
4.98	0.090 10	0.2013	0.4475	24.60	1.341	3.203	0.6929
4.99	0.089 77	0.2007	0.4474	24.80	1.341	3.211	0.6933
5.00	0.089 44	0.2000	0.4472	25.00	1.342	3.219	0.6938
∞	0.000 00	0.0000	0.4083	∞	1.429	∞	0.8215

Appendix F
Isothermal flow of a perfect gas ($k = 1.4$)

M	$\frac{p}{p^{*t}} = \frac{\rho}{\rho^{*t}} = \frac{V^{*t}}{V}$	$\frac{p_0}{p_0^{*t}}$	$\frac{T_0}{T_0^{*t}}$	$\frac{I}{I^{*t}}$	$\frac{s^{*t} - s}{R}$	$\frac{4\bar{f}L_{max}}{D}$
0.00	∞	∞	0.8750	∞	∞	∞
0.01	84.52	52.97	0.8750	42.26	4.4369	7133.0
0.02	42.26	26.49	0.8751	21.14	3.7438	1777.0
0.03	28.17	17.67	0.8752	14.10	3.3383	786.0
0.04	21.13	13.26	0.8753	10.59	3.0506	439.3
0.05	16.90	10.61	0.8754	8.481	2.8275	279.1
0.06	14.09	8.849	0.8756	7.078	2.6452	192.1
0.07	12.07	7.592	0.8759	6.078	2.4910	139.8
0.08	10.56	6.650	0.8761	5.330	2.3575	105.9
0.09	9.391	5.918	0.8764	4.749	2.2397	82.70
0.10	8.452	5.333	0.8768	4.285	2.1343	66.16
0.11	7.683	4.856	0.8771	3.907	2.0390	53.95
0.12	7.043	4.458	0.8775	3.592	1.9520	44.70
0.13	6.501	4.122	0.8780	3.328	1.8720	37.52
0.14	6.037	3.835	0.8784	3.101	1.7979	31.85
0.15	5.634	3.587	0.8789	2.906	1.7289	27.29
0.16	5.282	3.370	0.8795	2.736	1.6643	23.57
0.17	4.971	3.179	0.8801	2.586	1.6037	20.51
0.18	4.695	3.010	0.8807	2.454	1.5466	17.95
0.19	4.448	2.859	0.8813	2.336	1.4925	15.80
0.20	4.226	2.723	0.8820	2.231	1.4412	13.98
0.21	4.025	2.601	0.8827	2.137	1.3924	12.41
0.22	3.842	2.490	0.8835	2.051	1.3459	11.07
0.23	3.675	2.389	0.8843	1.973	1.3014	9.900
0.24	3.521	2.297	0.8851	1.903	1.2589	8.883
0.25	3.381	2.213	0.8859	1.838	1.2181	7.993
0.26	3.251	2.135	0.8868	1.779	1.1788	7.209
0.27	3.130	2.063	0.8878	1.725	1.1411	6.516
0.28	3.018	1.997	0.8887	1.675	1.1047	5.901
0.29	2.914	1.936	0.8897	1.629	1.0696	5.354

continued

Appendix F *Continued*

M	$\frac{p}{p^{*t}} = \frac{\rho}{\rho^{*t}} = \frac{V^{*t}}{V}$	$\frac{p_0}{p_0^{*t}}$	$\frac{T_0}{T_0^{*t}}$	$\frac{I}{I^{*t}}$	$\frac{s^{*t} - s}{R}$	$\frac{4\bar{f}L_{max}}{D}$
0.30	2.817	1.879	0.8908	1.586	1.0357	4.865
0.31	2.726	1.826	0.8918	1.547	1.0029	4.427
0.32	2.641	1.777	0.8929	1.510	0.9712	4.033
0.33	2.561	1.731	0.8941	1.476	0.9404	3.678
0.34	2.486	1.687	0.8952	1.444	0.9106	3.358
0.35	2.415	1.647	0.8964	1.414	0.8816	3.068
0.36	2.348	1.609	0.8977	1.387	0.8534	2.805
0.37	2.284	1.573	0.8990	1.361	0.8260	2.566
0.38	2.224	1.540	0.9003	1.337	0.7993	2.348
0.39	2.167	1.508	0.9016	1.314	0.7734	2.149
0.40	2.113	1.478	0.9030	1.293	0.7481	1.968
0.41	2.061	1.450	0.9044	1.273	0.7234	1.802
0.42	2.012	1.424	0.9059	1.255	0.6993	1.651
0.43	1.965	1.399	0.9074	1.237	0.6757	1.512
0.44	1.921	1.375	0.9089	1.221	0.6527	1.384
0.45	1.878	1.352	0.9104	1.205	0.6303	1.267
0.46	1.837	1.331	0.9120	1.191	0.6083	1.159
0.47	1.798	1.311	0.9137	1.177	0.5868	1.060
0.48	1.761	1.292	0.9153	1.164	0.5657	0.9687
0.49	1.725	1.274	0.9170	1.152	0.5451	0.8847
0.50	1.690	1.256	0.9187	1.141	0.5249	0.8073
0.51	1.657	1.240	0.9205	1.130	0.5051	0.7360
0.52	1.625	1.225	0.9223	1.120	0.4857	0.6702
0.53	1.595	1.210	0.9242	1.111	0.4666	0.6096
0.54	1.565	1.196	0.9260	1.102	0.4480	0.5536
0.55	1.537	1.183	0.9279	1.094	0.4296	0.5021
0.56	1.509	1.170	0.9299	1.086	0.4116	0.4545
0.57	1.483	1.158	0.9319	1.079	0.3939	0.4107
0.58	1.457	1.147	0.9339	1.072	0.3765	0.3703
0.59	1.433	1.136	0.9359	1.065	0.3594	0.3332
0.60	1.409	1.126	0.9380	1.059	0.3426	0.2990
0.61	1.386	1.116	0.9401	1.054	0.3261	0.2675
0.62	1.363	1.107	0.9423	1.048	0.3098	0.2386
0.63	1.342	1.098	0.9445	1.044	0.2938	0.2121
0.64	1.321	1.090	0.9467	1.039	0.2781	0.1878
0.65	1.300	1.082	0.9489	1.035	0.2626	0.1655
0.66	1.281	1.075	0.9512	1.031	0.2473	0.1452
0.67	1.261	1.068	0.9536	1.027	0.2322	0.1267
0.68	1.243	1.062	0.9559	1.024	0.2174	0.1099
0.69	1.225	1.055	0.9583	1.021	0.2028	0.094 63
0.70	1.207	1.050	0.9608	1.018	0.1885	0.080 85
0.71	1.190	1.044	0.9632	1.015	0.1743	0.068 44

Appendix F *Continued*

M	$\frac{p}{p^{*t}} = \frac{\rho}{\rho^{*t}} = \frac{V^{*t}}{V}$	$\frac{p_0}{p_0^{*t}}$	$\frac{T_0}{T_0^{*t}}$	$\frac{I}{I^{*t}}$	$\frac{s^{*t} - s}{R}$	$\frac{4\bar{f}L_{max}}{D}$
0.72	1.174	1.039	0.9657	1.013	0.1603	0.057 33
0.73	1.158	1.034	0.9683	1.011	0.1465	0.047 43
0.74	1.142	1.030	0.9708	1.009	0.1329	0.038 66
0.75	1.127	1.026	0.9734	1.007	0.1195	0.030 95
0.76	1.112	1.022	0.9761	1.006	0.1062	0.024 24
0.77	1.098	1.018	0.9788	1.004	0.093 13	0.018 48
0.78	1.084	1.015	0.9815	1.003	0.080 23	0.013 59
0.79	1.067	1.012	0.9842	1.002	0.067 49	0.009 532
0.80	1.056	1.009	0.9870	1.002	0.054 91	0.006 257
0.81	1.043	1.007	0.9898	1.001	0.042 49	0.003 715
0.82	1.031	1.005	0.9927	1.001	0.030 22	0.001 863
0.83	1.018	1.003	0.9956	1.000	0.018 09	0.000 662 8
0.84	1.006	1.001	0.9985	1.000	0.006 117	0.000 075 15
0.85	0.9943	0.9993	1.001	1.000	−0.005 717	0.000 065 12
0.86	0.9827	0.9981	1.004	1.000	−0.017 41	0.000 599 5
0.87	0.9714	0.9970	1.008	1.000	−0.028 97	0.001 647
0.88	0.9604	0.9962	1.011	1.001	−0.040 40	0.003 179
0.89	0.9496	0.9956	1.014	1.001	−0.051 70	0.005 167
0.90	0.9391	0.9953	1.017	1.002	−0.062 88	0.007 585
0.91	0.9287	0.9951	1.020	1.003	−0.073 93	0.010 41
0.92	0.9186	0.9952	1.023	1.004	−0.084 86	0.013 62
0.93	0.9088	0.9954	1.026	1.005	−0.095 67	0.017 19
0.94	0.8991	0.9958	1.030	1.006	−0.1064	0.021 10
0.95	0.8896	0.9965	1.033	1.007	−0.1169	0.025 34
0.96	0.8804	0.9973	1.036	1.008	−0.1274	0.029 88
0.97	0.8713	0.9983	1.040	1.010	−0.1378	0.034 71
0.98	0.8624	0.9996	1.043	1.011	−0.1480	0.039 81
0.99	0.8537	1.0010	1.047	1.013	−0.1582	0.045 16
1.00	0.8452	1.0025	1.050	1.014	−0.1682	0.050 76
1.01	0.8368	1.004	1.054	1.016	−0.1782	0.056 58
1.02	0.8286	1.006	1.057	1.018	−0.1880	0.062 63
1.03	0.8205	1.008	1.061	1.020	−0.1978	0.068 87
1.04	0.8126	1.011	1.064	1.022	−0.2075	0.075 31
1.05	0.8049	1.013	1.068	1.024	−0.2170	0.081 93
1.06	0.7973	1.016	1.072	1.026	−0.2265	0.088 72
1.07	0.7899	1.019	1.075	1.028	−0.2359	0.095 67
1.08	0.7826	1.022	1.079	1.030	−0.2452	0.1028
1.09	0.7754	1.025	1.083	1.033	−0.2544	0.1100
1.10	0.7683	1.028	1.087	1.035	−0.2635	0.1174
1.11	0.7614	1.031	1.091	1.037	−0.2726	0.1249

continued

Appendix F *Continued*

M	$\frac{p}{p^{*t}} = \frac{\rho}{\rho^{*t}} = \frac{V^{*t}}{V}$	$\frac{p_0}{p_0^{*t}}$	$\frac{T_0}{T_0^{*t}}$	$\frac{I}{I^{*t}}$	$\frac{s^{*t} - s}{R}$	$\frac{4\bar{f}L_{max}}{D}$
1.12	0.7546	1.035	1.095	1.040	−0.2816	0.1326
1.13	0.7479	1.039	1.098	1.042	−0.2905	0.1403
1.14	0.7414	1.043	1.102	1.045	−0.2993	0.1482
1.15	0.7349	1.047	1.106	1.048	−0.3080	0.1561
1.16	0.7286	1.051	1.110	1.051	−0.3167	0.1641
1.17	0.7224	1.056	1.115	1.053	−0.3252	0.1723
1.18	0.7162	1.061	1.119	1.056	−0.3338	0.1805
1.19	0.7102	1.065	1.123	1.059	−0.3422	0.1888
1.20	0.7043	1.070	1.127	1.062	−0.3506	0.1972
1.21	0.6985	1.075	1.131	1.065	−0.3589	0.2056
1.22	0.6927	1.081	1.135	1.068	−0.3671	0.2141
1.23	0.6871	1.086	1.140	1.071	−0.3753	0.2226
1.24	0.6816	1.092	1.144	1.074	−0.3833	0.2312
1.25	0.6761	1.097	1.148	1.078	−0.3914	0.2399
1.26	0.6708	1.103	1.153	1.081	−0.3993	0.2486
1.27	0.6655	1.110	1.157	1.084	−0.4073	0.2574
1.28	0.6603	1.116	1.162	1.087	−0.4151	0.2662
1.29	0.6552	1.122	1.166	1.091	−0.4229	0.2750
1.30	0.6501	1.129	1.171	1.094	−0.4306	0.2839
1.31	0.6452	1.136	1.175	1.098	−0.4383	0.2928
1.32	0.6403	1.142	1.180	1.101	−0.4459	0.3017
1.33	0.6355	1.150	1.185	1.105	−0.4534	0.3106
1.34	0.6307	1.157	1.189	1.108	−0.4609	0.3196
1.35	0.6260	1.164	1.194	1.112	−0.4683	0.3286
1.36	0.6214	1.172	1.199	1.115	−0.4757	0.3376
1.37	0.6169	1.180	1.203	1.119	−0.4830	0.3467
1.38	0.6124	1.187	1.208	1.123	−0.4903	0.3557
1.39	0.6080	1.196	1.213	1.126	−0.4975	0.3648
1.40	0.6037	1.204	1.218	1.130	−0.5047	0.3739
1.41	0.5994	1.212	1.223	1.134	−0.5118	0.3829
1.42	0.5952	1.221	1.228	1.138	−0.5189	0.3920
1.43	0.5910	1.230	1.233	1.142	−0.5259	0.4011
1.44	0.5869	1.239	1.238	1.145	−0.5329	0.4102
1.45	0.5829	1.248	1.243	1.149	−0.5398	0.4193
1.46	0.5789	1.257	1.248	1.153	−0.5467	0.4284
1.47	0.5749	1.267	1.253	1.157	−0.5535	0.4376
1.48	0.5711	1.276	1.258	1.161	−0.5603	0.4467
1.49	0.5672	1.286	1.264	1.165	−0.5670	0.4558
1.50	0.5634	1.296	1.269	1.169	−0.5737	0.4649
1.51	0.5597	1.306	1.274	1.173	−0.5803	0.4740
1.52	0.5560	1.317	1.279	1.177	−0.5869	0.4831
1.53	0.5524	1.327	1.285	1.181	−0.5935	0.4921

Appendix F *Continued*

M	$\frac{p}{p^{*t}} = \frac{\rho}{\rho^{*t}} = \frac{V^{*t}}{V}$	$\frac{p_0}{p_0^{*t}}$	$\frac{T_0}{T_0^{*t}}$	$\frac{I}{I^{*t}}$	$\frac{s^{*t} - s}{R}$	$\frac{4f\bar{L}_{max}}{D}$
1.54	0.5488	1.338	1.290	1.185	−0.6000	0.5012
1.55	0.5453	1.349	1.295	1.190	−0.6065	0.5103
1.56	0.5418	1.360	1.301	1.194	−0.6129	0.5194
1.57	0.5383	1.372	1.306	1.198	−0.6193	0.5284
1.58	0.5349	1.383	1.312	1.202	−0.6257	0.5374
1.59	0.5315	1.395	1.317	1.206	−0.6320	0.5465
1.60	0.5282	1.407	1.323	1.211	−0.6382	0.5555
1.61	0.5249	1.419	1.329	1.215	−0.6445	0.5645
1.62	0.5217	1.431	1.334	1.219	−0.6507	0.5735
1.63	0.5185	1.444	1.340	1.224	−0.6568	0.5825
1.64	0.5153	1.457	1.346	1.228	−0.6629	0.5914
1.65	0.5122	1.470	1.351	1.232	−0.6690	0.6004
1.66	0.5091	1.483	1.357	1.237	−0.6751	0.6093
1.67	0.5061	1.496	1.363	1.241	−0.6811	0.6182
1.68	0.5031	1.510	1.369	1.245	−0.6870	0.6271
1.69	0.5001	1.524	1.375	1.250	−0.6930	0.6360
1.70	0.4971	1.538	1.381	1.254	−0.6989	0.6449
1.71	0.4942	1.552	1.387	1.259	−0.7047	0.6537
1.72	0.4914	1.567	1.393	1.263	−0.7106	0.6626
1.73	0.4885	1.581	1.399	1.268	−0.7164	0.6714
1.74	0.4857	1.596	1.405	1.272	−0.7221	0.6802
1.75	0.4829	1.611	1.411	1.277	−0.7279	0.6889
1.76	0.4802	1.627	1.417	1.281	−0.7335	0.6977
1.77	0.4775	1.642	1.423	1.286	−0.7392	0.7064
1.78	0.4748	1.658	1.429	1.290	−0.7448	0.7151
1.79	0.4722	1.674	1.436	1.295	−0.7505	0.7238
1.80	0.4695	1.691	1.442	1.300	−0.7560	0.7325
1.81	0.4669	1.707	1.448	1.304	−0.7616	0.7412
1.82	0.4644	1.724	1.455	1.309	−0.7671	0.7498
1.83	0.4618	1.741	1.461	1.314	−0.7726	0.7584
1.84	0.4593	1.758	1.467	1.318	−0.7780	0.7670
1.85	0.4568	1.776	1.474	1.323	−0.7834	0.7755
1.86	0.4544	1.794	1.480	1.328	−0.7888	0.7841
1.87	0.4520	1.812	1.487	1.332	−0.7942	0.7926
1.88	0.4496	1.830	1.494	1.337	−0.7995	0.8011
1.89	0.4472	1.849	1.500	1.342	−0.8048	0.8096
1.90	0.4448	1.868	1.507	1.346	−0.8101	0.8180
1.91	0.4425	1.887	1.513	1.351	−0.8153	0.8265
1.92	0.4402	1.906	1.520	1.356	−0.8206	0.8349
1.93	0.4379	1.926	1.527	1.361	−0.8258	0.8433

continued

Appendix F *Continued*

M	$\frac{p}{p^{*t}} = \frac{\rho}{\rho^{*t}} = \frac{V^{*t}}{V}$	$\frac{p_0}{p_0^{*t}}$	$\frac{T_0}{T_0^{*t}}$	$\frac{I}{I^{*t}}$	$\frac{s^{*t} - s}{R}$	$\frac{4\bar{f}L_{max}}{D}$
1.94	0.4356	1.946	1.534	1.366	−0.8309	0.8516
1.95	0.4334	1.966	1.540	1.370	−0.8361	0.8600
1.96	0.4312	1.987	1.547	1.375	−0.8412	0.8683
1.97	0.4290	2.008	1.554	1.380	−0.8463	0.8766
1.98	0.4268	2.029	1.561	1.385	−0.8513	0.8849
1.99	0.4247	2.050	1.568	1.390	−0.8564	0.8931
2.00	0.4226	2.072	1.575	1.395	−0.8614	0.9013
2.01	0.4205	2.094	1.582	1.399	−0.8664	0.9095
2.02	0.4184	2.116	1.589	1.404	−0.8713	0.9177
2.03	0.4163	2.139	1.596	1.409	−0.8763	0.9259
2.04	0.4143	2.162	1.603	1.414	−0.8812	0.9340
2.05	0.4123	2.185	1.610	1.419	−0.8861	0.9421
2.06	0.4103	2.209	1.617	1.424	−0.8909	0.9502
2.07	0.4083	2.233	1.624	1.429	−0.8958	0.9583
2.08	0.4063	2.257	1.632	1.434	−0.9006	0.9663
2.09	0.4044	2.281	1.639	1.439	−0.9054	0.9743
2.10	0.4025	2.306	1.646	1.444	−0.9102	0.9823
2.11	0.4005	2.332	1.654	1.449	−0.9149	0.9903
2.12	0.3987	2.357	1.661	1.454	−0.9197	0.9982
2.13	0.3968	2.383	1.669	1.459	−0.9244	1.006
2.14	0.3949	2.409	1.676	1.464	−0.9290	1.014
2.15	0.3931	2.436	1.684	1.469	−0.9337	1.022
2.16	0.3913	2.463	1.691	1.474	−0.9383	1.030
2.17	0.3895	2.490	1.699	1.479	−0.9430	1.038
2.18	0.3877	2.518	1.707	1.484	−0.9476	1.045
2.19	0.3859	2.546	1.714	1.489	−0.9521	1.053
2.20	0.3842	2.574	1.722	1.494	−0.9567	1.061
2.21	0.3824	2.603	1.730	1.499	−0.9612	1.069
2.22	0.3807	2.632	1.737	1.504	−0.9657	1.076
2.23	0.3790	2.662	1.745	1.509	−0.9702	1.084
2.24	0.3773	2.691	1.753	1.514	−0.9747	1.092
2.25	0.3756	2.722	1.761	1.519	−0.9792	1.099
2.26	0.3740	2.753	1.769	1.524	−0.9836	1.107
2.27	0.3723	2.784	1.777	1.529	−0.9880	1.115
2.28	0.3707	2.815	1.785	1.534	−0.9924	1.122
2.29	0.3691	2.847	1.793	1.539	−0.9968	1.130
2.30	0.3675	2.879	1.801	1.544	−1.001	1.137
2.31	0.3659	2.912	1.809	1.550	−1.005	1.145
2.32	0.3643	2.945	1.817	1.555	−1.010	1.152
2.33	0.3627	2.979	1.825	1.560	−1.014	1.160
2.34	0.3612	3.013	1.833	1.565	−1.018	1.167
2.35	0.3596	3.047	1.841	1.570	−1.023	1.175

Appendix F *Continued*

M	$\frac{p}{p^{*t}} = \frac{\rho}{\rho^{*t}} = \frac{V^{*t}}{V}$	$\frac{p_0}{p_0^{*t}}$	$\frac{T_0}{T_0^{*t}}$	$\frac{I}{I^{*t}}$	$\frac{s^{*t} - s}{R}$	$\frac{4\bar{f}L_{max}}{D}$
2.36	0.3581	3.082	1.850	1.575	−1.027	1.182
2.37	0.3566	3.118	1.858	1.580	−1.031	1.189
2.38	0.3551	3.153	1.866	1.586	−1.035	1.197
2.39	0.3536	3.190	1.875	1.591	−1.040	1.204
2.40	0.3521	3.226	1.883	1.596	−1.044	1.211
2.41	0.3507	3.263	1.891	1.601	−1.048	1.219
2.42	0.3492	3.301	1.900	1.606	−1.052	1.226
2.43	0.3478	3.339	1.908	1.612	−1.056	1.233
2.44	0.3464	3.378	1.917	1.617	−1.060	1.240
2.45	0.3450	3.417	1.925	1.622	−1.064	1.248
2.46	0.3436	3.456	1.934	1.627	−1.068	1.255
2.47	0.3422	3.496	1.943	1.632	−1.072	1.262
2.48	0.3408	3.537	1.951	1.638	−1.076	1.269
2.49	0.3394	3.578	1.960	1.643	−1.081	1.276
2.50	0.3381	3.620	1.969	1.648	−1.085	1.283
2.51	0.3367	3.662	1.978	1.653	−1.089	1.290
2.52	0.3354	3.704	1.986	1.659	−1.092	1.297
2.53	0.3341	3.747	1.995	1.664	−1.096	1.305
2.54	0.3327	3.791	2.004	1.669	−1.100	1.312
2.55	0.3314	3.835	2.013	1.674	−1.104	1.319
2.56	0.3301	3.880	2.022	1.680	−1.108	1.325
2.57	0.3289	3.925	2.031	1.685	−1.112	1.332
2.58	0.3276	3.971	2.040	1.690	−1.116	1.339
2.59	0.3263	4.018	2.049	1.695	−1.120	1.346
2.60	0.3251	4.065	2.058	1.701	−1.124	1.353
2.61	0.3238	4.112	2.067	1.706	−1.128	1.360
2.62	0.3226	4.160	2.076	1.711	−1.131	1.367
2.63	0.3214	4.209	2.085	1.717	−1.135	1.374
2.64	0.3201	4.258	2.095	1.722	−1.139	1.381
2.65	0.3189	4.308	2.104	1.727	−1.143	1.387
2.66	0.3177	4.359	2.113	1.733	−1.147	1.394
2.67	0.3165	4.410	2.123	1.738	−1.150	1.401
2.68	0.3154	4.462	2.132	1.743	−1.154	1.408
2.69	0.3142	4.514	2.141	1.749	−1.158	1.414
2.70	0.3130	4.567	2.151	1.754	−1.161	1.421
2.71	0.3119	4.621	2.160	1.759	−1.165	1.428
2.72	0.3107	4.675	2.170	1.765	−1.169	1.434
2.73	0.3096	4.730	2.179	1.770	−1.173	1.441
2.74	0.3085	4.786	2.189	1.775	−1.176	1.448
2.75	0.3073	4.842	2.198	1.781	−1.180	1.454

continued

M	$\frac{p}{p^{*t}} = \frac{\rho}{\rho^{*t}} = \frac{V^{*t}}{V}$	$\frac{p_0}{p_0^{*t}}$	$\frac{T_0}{T_0^{*t}}$	$\frac{I}{I^{*t}}$	$\frac{s^{*t} - s}{R}$	$\frac{4\bar{f}L_{max}}{D}$
2.76	0.3062	4.899	2.208	1.786	−1.183	1.461
2.77	0.3051	4.956	2.218	1.791	−1.187	1.467
2.78	0.3040	5.015	2.227	1.797	−1.191	1.474
2.79	0.3029	5.074	2.237	1.802	−1.194	1.480
2.80	0.3018	5.133	2.247	1.807	−1.198	1.487
2.81	0.3008	5.194	2.257	1.813	−1.201	1.493
2.82	0.2997	5.255	2.267	1.818	−1.205	1.500
2.83	0.2986	5.316	2.277	1.824	−1.209	1.506
2.84	0.2976	5.379	2.286	1.829	−1.212	1.513
2.85	0.2965	5.442	2.296	1.834	−1.216	1.519
2.86	0.2955	5.506	2.306	1.840	−1.219	1.525
2.87	0.2945	5.571	2.316	1.845	−1.223	1.532
2.88	0.2935	5.637	2.327	1.851	−1.226	1.538
2.89	0.2924	5.703	2.337	1.856	−1.229	1.545
2.90	0.2914	5.770	2.347	1.861	−1.233	1.551
2.91	0.2904	5.838	2.357	1.867	−1.236	1.557
2.92	0.2894	5.906	2.367	1.872	−1.240	1.563
2.93	0.2884	5.976	2.377	1.878	−1.243	1.570
2.94	0.2875	6.046	2.388	1.883	−1.247	1.576
2.95	0.2865	6.117	2.398	1.888	−1.250	1.582
2.96	0.2855	6.189	2.408	1.894	−1.253	1.588
2.97	0.2846	6.262	2.419	1.899	−1.257	1.595
2.98	0.2836	6.335	2.429	1.905	−1.260	1.601
2.99	0.2827	6.410	2.440	1.910	−1.264	1.607
3.00	0.2817	6.485	2.450	1.916	−1.267	1.613
3.01	0.2808	6.561	2.461	1.921	−1.270	1.619
3.02	0.2799	6.638	2.471	1.927	−1.273	1.625
3.03	0.2789	6.716	2.482	1.932	−1.277	1.631
3.04	0.2780	6.794	2.492	1.937	−1.280	1.637
3.05	0.2771	6.874	2.503	1.943	−1.283	1.644
3.06	0.2762	6.955	2.514	1.948	−1.287	1.650
3.07	0.2753	7.036	2.524	1.954	−1.290	1.656
3.08	0.2744	7.118	2.535	1.959	−1.293	1.662
3.09	0.2735	7.202	2.546	1.965	−1.296	1.668
3.10	0.2726	7.286	2.557	1.970	−1.300	1.674
3.11	0.2718	7.371	2.568	1.976	−1.303	1.680
3.12	0.2709	7.457	2.579	1.981	−1.306	1.686
3.13	0.2700	7.544	2.589	1.987	−1.309	1.691
3.14	0.2692	7.632	2.600	1.992	−1.312	1.697
3.15	0.2683	7.722	2.611	1.998	−1.316	1.703
3.16	0.2675	7.812	2.622	2.003	−1.319	1.709
3.17	0.2666	7.903	2.634	2.009	−1.322	1.715

Appendix F *Continued*

M	$\frac{p}{p^{*t}} = \frac{\rho}{\rho^{*t}} = \frac{V^{*t}}{V}$	$\frac{p_0}{p_0^{*t}}$	$\frac{T_0}{T_0^{*t}}$	$\frac{I}{I^{*t}}$	$\frac{s^{*t} - s}{R}$	$\frac{4\bar{f}L_{max}}{D}$
3.18	0.2658	7.995	2.645	2.014	−1.325	1.721
3.19	0.2649	8.088	2.656	2.020	−1.328	1.727
3.20	0.2641	8.182	2.667	2.025	−1.331	1.733
3.21	0.2633	8.277	2.678	2.031	−1.335	1.738
3.22	0.2625	8.374	2.689	2.036	−1.338	1.744
3.23	0.2617	8.471	2.701	2.042	−1.341	1.750
3.24	0.2609	8.569	2.712	2.047	−1.344	1.756
3.25	0.2600	8.669	2.723	2.053	−1.347	1.761
3.26	0.2592	8.769	2.735	2.058	−1.350	1.767
3.27	0.2585	8.871	2.746	2.064	−1.353	1.773
3.28	0.2577	8.974	2.758	2.069	−1.356	1.779
3.29	0.2569	9.078	2.769	2.075	−1.359	1.784
3.30	0.2561	9.183	2.781	2.080	−1.362	1.790
3.31	0.2553	9.289	2.792	2.086	−1.365	1.796
3.32	0.2546	9.397	2.804	2.091	−1.368	1.801
3.33	0.2538	9.505	2.816	2.097	−1.371	1.807
3.34	0.2530	9.615	2.827	2.102	−1.374	1.812
3.35	0.2523	9.726	2.839	2.108	−1.377	1.818
3.36	0.2515	9.838	2.851	2.114	−1.380	1.824
3.37	0.2508	9.952	2.862	2.119	−1.383	1.829
3.38	0.2500	10.07	2.874	2.125	−1.386	1.835
3.39	0.2493	10.18	2.886	2.130	−1.389	1.840
3.40	0.2486	10.30	2.898	2.136	−1.392	1.846
3.41	0.2478	10.42	2.910	2.141	−1.395	1.851
3.42	0.2471	10.54	2.922	2.147	−1.398	1.857
3.43	0.2464	10.66	2.934	2.152	−1.401	1.862
3.44	0.2457	10.78	2.946	2.158	−1.404	1.868
3.45	0.2450	10.90	2.958	2.164	−1.407	1.873
3.46	0.2443	11.03	2.970	2.169	−1.410	1.879
3.47	0.2436	11.15	2.982	2.175	−1.412	1.884
3.48	0.2429	11.28	2.994	2.180	−1.415	1.890
3.49	0.2422	11.41	3.007	2.186	−1.418	1.895
3.50	0.2415	11.54	3.019	2.191	−1.421	1.900
3.51	0.2408	11.67	3.031	2.197	−1.424	1.906
3.52	0.2401	11.81	3.043	2.203	−1.427	1.911
3.53	0.2394	11.94	3.056	2.208	−1.430	1.916
3.54	0.2387	12.08	3.068	2.214	−1.432	1.922
3.55	0.2381	12.21	3.080	2.219	−1.435	1.927
3.56	0.2374	12.35	3.093	2.225	−1.438	1.932
3.57	0.2367	12.49	3.105	2.230	−1.441	1.938

continued

Appendix F *Continued*

M	$\frac{p}{p^{*t}} = \frac{\rho}{\rho^{*t}} = \frac{V^{*t}}{V}$	$\frac{p_0}{p_0^{*t}}$	$\frac{T_0}{T_0^{*t}}$	$\frac{I}{I^{*t}}$	$\frac{s^{*t} - s}{R}$	$\frac{4\bar{f}L_{max}}{D}$
3.58	0.2361	12.63	3.118	2.236	−1.444	1.943
3.59	0.2354	12.78	3.130	2.242	−1.446	1.948
3.60	0.2348	12.92	3.143	2.247	−1.449	1.953
3.61	0.2341	13.07	3.156	2.253	−1.452	1.959
3.62	0.2335	13.22	3.168	2.258	−1.455	1.964
3.63	0.2328	13.37	3.181	2.264	−1.457	1.969
3.64	0.2322	13.52	3.194	2.270	−1.460	1.974
3.65	0.2315	13.67	3.206	2.275	−1.463	1.980
3.66	0.2309	13.82	3.219	2.281	−1.466	1.985
3.67	0.2303	13.98	3.232	2.286	−1.468	1.990
3.68	0.2297	14.14	3.245	2.292	−1.471	1.995
3.69	0.2290	14.29	3.258	2.298	−1.474	2.000
3.70	0.2284	14.45	3.271	2.303	−1.477	2.005
3.71	0.2278	14.62	3.284	2.309	−1.479	2.010
3.72	0.2272	14.78	3.297	2.314	−1.482	2.016
3.73	0.2266	14.95	3.310	2.320	−1.485	2.021
3.74	0.2260	15.11	3.323	2.326	−1.487	2.026
3.75	0.2254	15.28	3.336	2.331	−1.490	2.031
3.76	0.2248	15.45	3.349	2.337	−1.493	2.036
3.77	0.2242	15.62	3.362	2.342	−1.495	2.041
3.78	0.2236	15.80	3.375	2.348	−1.498	2.046
3.79	0.2230	15.97	3.389	2.354	−1.501	2.051
3.80	0.2224	16.15	3.402	2.359	−1.503	2.056
3.81	0.2218	16.33	3.415	2.365	−1.506	2.061
3.82	0.2212	16.51	3.429	2.371	−1.508	2.066
3.83	0.2207	16.70	3.442	2.376	−1.511	2.071
3.84	0.2201	16.88	3.455	2.382	−1.514	2.076
3.85	0.2195	17.07	3.469	2.387	−1.516	2.081
3.86	0.2190	17.26	3.482	2.393	−1.519	2.086
3.87	0.2184	17.45	3.496	2.399	−1.521	2.091
3.88	0.2178	17.64	3.510	2.404	−1.524	2.096
3.89	0.2173	17.83	3.523	2.410	−1.527	2.100
3.90	0.2167	18.03	3.537	2.416	−1.529	2.105
3.91	0.2162	18.23	3.550	2.421	−1.532	2.110
3.92	0.2156	18.43	3.564	2.427	−1.534	2.115
3.93	0.2151	18.63	3.578	2.433	−1.537	2.120
3.94	0.2145	18.83	3.592	2.438	−1.539	2.125
3.95	0.2140	19.04	3.605	2.444	−1.542	2.130
3.96	0.2134	19.25	3.619	2.449	−1.544	2.135
3.97	0.2129	19.46	3.633	2.455	−1.547	2.139
3.98	0.2124	19.67	3.647	2.461	−1.550	2.144
3.99	0.2118	19.89	3.661	2.466	−1.552	2.149

Appendix F *Continued*

M	$\frac{p}{p^{*t}} = \frac{\rho}{\rho^{*t}} = \frac{V^{*t}}{V}$	$\frac{p_0}{p_0^{*t}}$	$\frac{T_0}{T_0^{*t}}$	$\frac{I}{I^{*t}}$	$\frac{s^{*t} - s}{R}$	$\frac{4\bar{f}L_{max}}{D}$
4.00	0.2113	20.10	3.675	2.472	−1.555	2.154
4.01	0.2108	20.32	3.689	2.478	−1.557	2.158
4.02	0.2102	20.54	3.703	2.483	−1.560	2.163
4.03	0.2097	20.77	3.717	2.489	−1.562	2.168
4.04	0.2092	20.99	3.731	2.495	−1.564	2.173
4.05	0.2087	21.22	3.745	2.500	−1.567	2.177
4.06	0.2082	21.45	3.760	2.506	−1.569	2.182
4.07	0.2077	21.68	3.774	2.512	−1.572	2.187
4.08	0.2071	21.92	3.788	2.517	−1.574	2.192
4.09	0.2066	22.15	3.802	2.523	−1.577	2.196
4.10	0.2061	22.39	3.817	2.529	−1.579	2.201
4.11	0.2056	22.63	3.831	2.534	−1.582	2.206
4.12	0.2051	22.88	3.846	2.540	−1.584	2.210
4.13	0.2046	23.12	3.860	2.546	−1.587	2.215
4.14	0.2041	23.37	3.874	2.551	−1.589	2.220
4.15	0.2037	23.62	3.889	2.557	−1.591	2.224
4.16	0.2032	23.87	3.903	2.563	−1.594	2.229
4.17	0.2027	24.13	3.918	2.568	−1.596	2.233
4.18	0.2022	24.39	3.933	2.574	−1.599	2.238
4.19	0.2017	24.65	3.947	2.580	−1.601	2.243
4.20	0.2012	24.91	3.962	2.585	−1.603	2.247
4.21	0.2007	25.18	3.977	2.591	−1.606	2.252
4.22	0.2003	25.44	3.991	2.597	−1.608	2.256
4.23	0.1998	25.71	4.006	2.602	−1.610	2.261
4.24	0.1993	25.99	4.021	2.608	−1.613	2.265
4.25	0.1989	26.26	4.036	2.614	−1.615	2.270
4.26	0.1984	26.54	4.051	2.619	−1.618	2.274
4.27	0.1979	26.82	4.066	2.625	−1.620	2.279
4.28	0.1975	27.11	4.081	2.631	−1.622	2.283
4.29	0.1970	27.39	4.096	2.637	−1.625	2.288
4.30	0.1965	27.68	4.111	2.642	−1.627	2.292
4.31	0.1961	27.97	4.126	2.648	−1.629	2.297
4.32	0.1956	28.27	4.141	2.654	−1.631	2.301
4.33	0.1952	28.56	4.156	2.659	−1.634	2.306
4.34	0.1947	28.86	4.171	2.665	−1.636	2.310
4.35	0.1943	29.17	4.186	2.671	−1.638	2.315
4.36	0.1938	29.47	4.202	2.676	−1.641	2.319
4.37	0.1934	29.78	4.217	2.682	−1.643	2.323
4.38	0.1930	30.09	4.232	2.688	−1.645	2.328
4.39	0.1925	30.41	4.248	2.693	−1.648	2.332

continued

Appendix F *Continued*

M	$\frac{p}{p^{*t}} = \frac{\rho}{\rho^{*t}} = \frac{V^{*t}}{V}$	$\frac{p_0}{p_0^{*t}}$	$\frac{T_0}{T_0^{*t}}$	$\frac{I}{I^{*t}}$	$\frac{s^{*t} - s}{R}$	$\frac{4\bar{f}L_{max}}{D}$
4.40	0.1921	30.72	4.263	2.699	−1.650	2.337
4.41	0.1916	31.04	4.278	2.705	−1.652	2.341
4.42	0.1912	31.37	4.294	2.711	−1.654	2.345
4.43	0.1908	31.69	4.309	2.716	−1.657	2.350
4.44	0.1904	32.02	4.325	2.722	−1.659	2.354
4.45	0.1899	32.36	4.340	2.728	−1.661	2.358
4.46	0.1895	32.69	4.356	2.733	−1.663	2.363
4.47	0.1891	33.03	4.372	2.739	−1.666	2.367
4.48	0.1887	33.37	4.387	2.745	−1.668	2.371
4.49	0.1882	33.71	4.403	2.750	−1.670	2.376
4.50	0.1878	34.06	4.419	2.756	−1.672	2.380
4.51	0.1874	34.41	4.435	2.762	−1.675	2.384
4.52	0.1870	34.77	4.450	2.768	−1.677	2.388
4.53	0.1866	35.12	4.466	2.773	−1.679	2.393
4.54	0.1862	35.48	4.482	2.779	−1.681	2.397
4.55	0.1857	35.85	4.498	2.785	−1.683	2.401
4.56	0.1853	36.22	4.514	2.790	−1.686	2.405
4.57	0.1849	36.59	4.530	2.796	−1.688	2.410
4.58	0.1845	36.96	4.546	2.802	−1.690	2.414
4.59	0.1841	37.34	4.562	2.808	−1.692	2.418
4.60	0.1837	37.72	4.578	2.813	−1.694	2.422
4.61	0.1833	38.10	4.594	2.819	−1.696	2.427
4.62	0.1829	38.49	4.610	2.825	−1.699	2.431
4.63	0.1825	38.88	4.626	2.830	−1.701	2.435
4.64	0.1821	39.27	4.643	2.836	−1.703	2.439
4.65	0.1818	39.67	4.659	2.842	−1.705	2.443
4.66	0.1814	40.07	4.675	2.848	−1.707	2.447
4.67	0.1810	40.48	4.692	2.853	−1.709	2.452
4.68	0.1806	40.89	4.708	2.859	−1.712	2.456
4.69	0.1802	41.30	4.724	2.865	−1.714	2.460
4.70	0.1798	41.72	4.741	2.870	−1.716	2.464
4.71	0.1794	42.14	4.757	2.876	−1.718	2.468
4.72	0.1791	42.56	4.774	2.882	−1.720	2.472
4.73	0.1787	42.99	4.790	2.888	−1.722	2.476
4.74	0.1783	43.42	4.807	2.893	−1.724	2.480
4.75	0.1779	43.85	4.823	2.899	−1.726	2.484
4.76	0.1776	44.29	4.840	2.905	−1.728	2.488
4.77	0.1772	44.73	4.857	2.911	−1.731	2.493
4.78	0.1768	45.18	4.873	2.916	−1.733	2.497
4.79	0.1764	45.63	4.890	2.922	−1.735	2.501
4.80	0.1761	46.08	4.907	2.928	−1.737	2.505
4.81	0.1757	46.54	4.924	2.933	−1.739	2.509

Appendix F *Continued*

M	$\frac{p}{p^{*t}} = \frac{\rho}{\rho^{*t}} = \frac{V^{*t}}{V}$	$\frac{p_0}{p_0^{*t}}$	$\frac{T_0}{T_0^{*t}}$	$\frac{I}{I^{*t}}$	$\frac{s^{*t} - s}{R}$	$\frac{4\bar{f}L_{max}}{D}$
4.82	0.1753	47.00	4.941	2.939	−1.741	2.513
4.83	0.1750	47.47	4.958	2.945	−1.743	2.517
4.84	0.1746	47.94	4.974	2.951	−1.745	2.521
4.85	0.1743	48.42	4.991	2.956	−1.747	2.525
4.86	0.1739	48.89	5.008	2.962	−1.749	2.529
4.87	0.1735	49.38	5.025	2.968	−1.751	2.533
4.88	0.1732	49.86	5.043	2.974	−1.753	2.537
4.89	0.1728	50.35	5.060	2.979	−1.755	2.541
4.90	0.1725	50.85	5.077	2.985	−1.757	2.545
4.91	0.1721	51.35	5.094	2.991	−1.760	2.549
4.92	0.1718	51.85	5.111	2.997	−1.762	2.553
4.93	0.1714	52.36	5.128	3.002	−1.764	2.557
4.94	0.1711	52.87	5.146	3.008	−1.766	2.560
4.95	0.1707	53.39	5.163	3.014	−1.768	2.564
4.96	0.1704	53.91	5.180	3.020	−1.770	2.568
4.97	0.1701	54.44	5.198	3.025	−1.772	2.572
4.98	0.1697	54.97	5.215	3.031	−1.774	2.576
4.99	0.1694	55.50	5.233	3.037	−1.776	2.580
5.00	0.1690	56.04	5.250	3.043	−1.778	2.584
∞	0.0000	∞	∞	∞	∞	∞

Appendix G
Flow of a perfect gas on the Rayleigh line ($k = 1.4$)

M	$\frac{p}{p^*}$	$\frac{T}{T^*}$	$\frac{\rho}{\rho^*} = \frac{V^*}{V}$	$\frac{p_0}{p_0^*}$	$\frac{T_0}{T_0^*}$	$\frac{s^* - s}{R}$
0.00	2.400	0.000 000 0	∞	1.268	0.000 000 0	∞
0.01	2.400	0.000 578 8	4167.0	1.268	0.000 479 9	26.98
0.02	2.399	0.002 301	1042.0	1.268	0.001 918	22.14
0.03	2.397	0.005 171	463.6	1.267	0.004 310	19.30
0.04	2.395	0.009 175	261.0	1.266	0.007 648	17.29
0.05	2.392	0.014 30	167.3	1.266	0.011 92	15.74
0.06	2.388	0.020 53	116.32	1.265	0.017 12	14.47
0.07	2.384	0.027 84	85.62	1.264	0.023 22	13.40
0.08	2.379	0.036 21	65.69	1.262	0.030 22	12.48
0.09	2.373	0.045 62	52.02	1.261	0.038 08	11.67
0.10	2.367	0.056 02	42.25	1.259	0.046 78	10.95
0.11	2.360	0.067 39	35.02	1.257	0.056 30	10.30
0.12	2.353	0.079 70	29.52	1.255	0.066 61	9.709
0.13	2.345	0.092 90	25.24	1.253	0.077 68	9.169
0.14	2.336	0.1070	21.84	1.251	0.089 47	8.672
0.15	2.327	0.1218	19.10	1.249	0.1020	8.213
0.16	2.317	0.1374	16.86	1.246	0.1151	7.787
0.17	2.307	0.1538	15.00	1.243	0.1289	7.389
0.18	2.296	0.1708	13.44	1.241	0.1432	7.017
0.19	2.285	0.1884	12.13	1.238	0.1581	6.668
0.20	2.273	0.2066	11.00	1.235	0.1736	6.340
0.21	2.260	0.2253	10.03	1.231	0.1894	6.031
0.22	2.248	0.2445	9.192	1.228	0.2057	5.739
0.23	2.235	0.2641	8.460	1.225	0.2224	5.464
0.24	2.221	0.2841	7.817	1.221	0.2395	5.202
0.25	2.207	0.3044	7.250	1.218	0.2568	4.955
0.26	2.193	0.3250	6.747	1.214	0.2745	4.719
0.27	2.178	0.3457	6.299	1.210	0.2923	4.496
0.28	2.163	0.3667	5.898	1.206	0.3104	4.283
0.29	2.147	0.3877	5.538	1.203	0.3286	4.080
0.30	2.131	0.4089	5.213	1.199	0.3469	3.887

Appendix G *Continued*

M	$\frac{p}{p^*}$	$\frac{T}{T^*}$	$\frac{\rho}{\rho^*} = \frac{V^*}{V}$	$\frac{p_0}{p_0^*}$	$\frac{T_0}{T_0^*}$	$\frac{s^* - s}{R}$
0.31	2.115	0.4300	4.919	1.195	0.3653	3.703
0.32	2.099	0.4512	4.652	1.190	0.3837	3.527
0.33	2.083	0.4723	4.409	1.186	0.4021	3.359
0.34	2.066	0.4933	4.188	1.182	0.4206	3.199
0.35	2.049	0.5141	3.985	1.178	0.4389	3.046
0.36	2.031	0.5348	3.798	1.174	0.4572	2.899
0.37	2.014	0.5553	3.627	1.169	0.4754	2.759
0.38	1.996	0.5755	3.469	1.165	0.4935	2.625
0.39	1.979	0.5955	3.323	1.161	0.5113	2.497
0.40	1.961	0.6152	3.187	1.157	0.5290	2.374
0.41	1.943	0.6345	3.062	1.152	0.5465	2.256
0.42	1.925	0.6535	2.945	1.148	0.5638	2.144
0.43	1.906	0.6721	2.837	1.144	0.5808	2.036
0.44	1.888	0.6903	2.736	1.139	0.5975	1.933
0.45	1.870	0.7080	2.641	1.135	0.6139	1.834
0.46	1.852	0.7254	2.552	1.131	0.6301	1.740
0.47	1.833	0.7423	2.470	1.127	0.6459	1.649
0.48	1.815	0.7587	2.392	1.122	0.6613	1.562
0.49	1.796	0.7747	2.319	1.118	0.6766	1.479
0.50	1.778	0.7901	2.250	1.114	0.6914	1.400
0.51	1.759	0.8051	2.185	1.110	0.7058	1.324
0.52	1.741	0.8196	2.124	1.106	0.7199	1.251
0.53	1.723	0.8335	2.067	1.102	0.7336	1.181
0.54	1.704	0.8469	2.012	1.098	0.7470	1.115
0.55	1.686	0.8599	1.961	1.094	0.7599	1.051
0.56	1.668	0.8723	1.912	1.090	0.7725	0.9898
0.57	1.650	0.8842	1.866	1.086	0.7847	0.9315
0.58	1.632	0.8955	1.822	1.083	0.7965	0.8758
0.59	1.614	0.9064	1.780	1.079	0.8079	0.8226
0.60	1.596	0.9167	1.741	1.075	0.8189	0.7717
0.61	1.578	0.9265	1.703	1.072	0.8296	0.7232
0.62	1.560	0.9358	1.667	1.068	0.8398	0.6770
0.63	1.543	0.9447	1.633	1.065	0.8497	0.6328
0.64	1.525	0.9530	1.601	1.061	0.8592	0.5908
0.65	1.508	0.9608	1.570	1.058	0.8683	0.5507
0.66	1.491	0.9682	1.540	1.055	0.8771	0.5126
0.67	1.474	0.9750	1.512	1.052	0.8855	0.4763
0.68	1.457	0.9814	1.484	1.049	0.8935	0.4419
0.69	1.440	0.9874	1.458	1.046	0.9012	0.4092
0.70	1.423	0.9929	1.434	1.043	0.9085	0.3781

continued

Appendix G *Continued*

M	$\frac{p}{p^*}$	$\frac{T}{T^*}$	$\frac{\rho}{\rho^*} = \frac{V^*}{V}$	$\frac{p_0}{p_0^*}$	$\frac{T_0}{T_0^*}$	$\frac{s^* - s}{R}$
0.71	1.407	0.9980	1.410	1.040	0.9155	0.3486
0.72	1.391	1.003	1.387	1.038	0.9221	0.3207
0.73	1.375	1.007	1.365	1.035	0.9284	0.2943
0.74	1.359	1.011	1.344	1.033	0.9344	0.2694
0.75	1.343	1.014	1.324	1.030	0.9401	0.2459
0.76	1.327	1.017	1.305	1.028	0.9455	0.2237
0.77	1.311	1.020	1.286	1.026	0.9505	0.2028
0.78	1.296	1.022	1.268	1.023	0.9553	0.1832
0.79	1.281	1.024	1.251	1.021	0.9598	0.1649
0.80	1.266	1.026	1.234	1.019	0.9639	0.1477
0.81	1.251	1.027	1.218	1.017	0.9679	0.1316
0.82	1.236	1.028	1.203	1.016	0.9715	0.1167
0.83	1.222	1.028	1.188	1.014	0.9749	0.1028
0.84	1.207	1.029	1.174	1.012	0.9781	0.089 95
0.85	1.193	1.029	1.160	1.011	0.9810	0.078 10
0.86	1.179	1.028	1.147	1.010	0.9836	0.067 22
0.87	1.165	1.028	1.134	1.008	0.9861	0.057 27
0.88	1.152	1.027	1.121	1.007	0.9883	0.048 22
0.89	1.138	1.026	1.109	1.006	0.9903	0.040 04
0.90	1.125	1.025	1.098	1.005	0.9921	0.032 70
0.91	1.111	1.023	1.086	1.004	0.9937	0.026 18
0.92	1.098	1.021	1.076	1.003	0.9951	0.020 44
0.93	1.086	1.019	1.065	1.002	0.9963	0.015 47
0.94	1.073	1.017	1.055	1.002	0.9973	0.011 24
0.95	1.060	1.015	1.045	1.001	0.9981	0.007 714
0.96	1.048	1.012	1.035	1.001	0.9988	0.004 881
0.97	1.036	1.009	1.026	1.000	0.9993	0.002 715
0.98	1.024	1.006	1.017	1.000	0.9997	0.001 193
0.99	1.012	1.003	1.008	1.000	0.9999	0.000 294 8
1.00	1.000	1.000	1.000	1.000	1.0000	0.000 000 0
1.01	0.9884	0.9966	0.9918	1.000	0.9999	0.000 288 2
1.02	0.9770	0.9930	0.9838	1.000	0.9997	0.001 141
1.03	0.9657	0.9894	0.9761	1.000	0.9994	0.002 539
1.04	0.9546	0.9855	0.9686	1.001	0.9989	0.004 466
1.05	0.9436	0.9816	0.9613	1.001	0.9984	0.006 902
1.06	0.9327	0.9776	0.9542	1.002	0.9977	0.009 833
1.07	0.9221	0.9734	0.9473	1.002	0.9969	0.013 24
1.08	0.9115	0.9691	0.9406	1.003	0.9960	0.017 11
1.09	0.9011	0.9648	0.9340	1.004	0.9950	0.021 43
1.10	0.8909	0.9603	0.9277	1.005	0.9939	0.026 18
1.11	0.8808	0.9558	0.9215	1.006	0.9927	0.031 35
1.12	0.8708	0.9512	0.9155	1.007	0.9915	0.036 92

Appendix G *Continued*

M	$\frac{p}{p^*}$	$\frac{T}{T^*}$	$\frac{\rho}{\rho^*} = \frac{V^*}{V}$	$\frac{p_0}{p_0^*}$	$\frac{T_0}{T_0^*}$	$\frac{s^* - s}{R}$
1.13	0.8609	0.9465	0.9096	1.008	0.9901	0.042 88
1.14	0.8512	0.9417	0.9039	1.010	0.9887	0.049 22
1.15	0.8417	0.9369	0.8984	1.011	0.9872	0.055 93
1.16	0.8322	0.9320	0.8930	1.012	0.9856	0.062 98
1.17	0.8229	0.9270	0.8877	1.014	0.9840	0.070 39
1.18	0.8137	0.9220	0.8826	1.016	0.9823	0.078 12
1.19	0.8047	0.9169	0.8776	1.018	0.9805	0.086 17
1.20	0.7958	0.9118	0.8727	1.019	0.9787	0.094 53
1.21	0.7870	0.9067	0.8679	1.021	0.9768	0.1032
1.22	0.7783	0.9015	0.8633	1.023	0.9749	0.1121
1.23	0.7697	0.8963	0.8587	1.026	0.9729	0.1214
1.24	0.7613	0.8911	0.8543	1.028	0.9709	0.1309
1.25	0.7529	0.8858	0.8500	1.030	0.9689	0.1406
1.26	0.7447	0.8805	0.8458	1.033	0.9668	0.1506
1.27	0.7366	0.8752	0.8417	1.035	0.9646	0.1609
1.28	0.7287	0.8699	0.8376	1.038	0.9624	0.1714
1.29	0.7208	0.8645	0.8337	1.041	0.9602	0.1821
1.30	0.7130	0.8592	0.8299	1.044	0.9580	0.1930
1.31	0.7054	0.8538	0.8261	1.047	0.9557	0.2041
1.32	0.6978	0.8484	0.8225	1.050	0.9534	0.2155
1.33	0.6904	0.8430	0.8189	1.053	0.9511	0.2270
1.34	0.6830	0.8377	0.8154	1.056	0.9487	0.2388
1.35	0.6758	0.8323	0.8120	1.059	0.9464	0.2507
1.36	0.6686	0.8269	0.8086	1.063	0.9440	0.2628
1.37	0.6616	0.8215	0.8053	1.066	0.9416	0.2750
1.38	0.6546	0.8161	0.8021	1.070	0.9391	0.2875
1.39	0.6478	0.8108	0.7990	1.074	0.9367	0.3001
1.40	0.6410	0.8054	0.7959	1.078	0.9343	0.3128
1.41	0.6344	0.8000	0.7929	1.082	0.9318	0.3257
1.42	0.6278	0.7947	0.7900	1.086	0.9293	0.3387
1.43	0.6213	0.7894	0.7871	1.090	0.9268	0.3519
1.44	0.6149	0.7840	0.7843	1.094	0.9243	0.3652
1.45	0.6086	0.7787	0.7815	1.098	0.9218	0.3787
1.46	0.6024	0.7735	0.7788	1.103	0.9193	0.3922
1.47	0.5962	0.7682	0.7762	1.107	0.9168	0.4059
1.48	0.5902	0.7629	0.7736	1.112	0.9143	0.4197
1.49	0.5842	0.7577	0.7710	1.117	0.9118	0.4336
1.50	0.5783	0.7525	0.7685	1.122	0.9093	0.4476
1.51	0.5725	0.7473	0.7661	1.126	0.9068	0.4617
1.52	0.5668	0.7422	0.7637	1.132	0.9042	0.4759

continued

Appendix G *Continued*

M	$\frac{p}{p^*}$	$\frac{T}{T^*}$	$\frac{\rho}{\rho^*}=\frac{V^*}{V}$	$\frac{p_0}{p_0^*}$	$\frac{T_0}{T_0^*}$	$\frac{s^*-s}{R}$
1.53	0.5611	0.7370	0.7613	1.137	0.9017	0.4902
1.54	0.5555	0.7319	0.7590	1.142	0.8992	0.5046
1.55	0.5500	0.7268	0.7568	1.147	0.8967	0.5191
1.56	0.5446	0.7217	0.7545	1.153	0.8942	0.5336
1.57	0.5392	0.7167	0.7524	1.158	0.8917	0.5482
1.58	0.5339	0.7117	0.7502	1.164	0.8892	0.5630
1.59	0.5287	0.7067	0.7481	1.170	0.8867	0.5777
1.60	0.5236	0.7017	0.7461	1.176	0.8842	0.5926
1.61	0.5185	0.6968	0.7441	1.182	0.8817	0.6075
1.62	0.5135	0.6919	0.7421	1.188	0.8792	0.6225
1.63	0.5085	0.6870	0.7402	1.194	0.8768	0.6375
1.64	0.5036	0.6822	0.7383	1.200	0.8743	0.6527
1.65	0.4988	0.6774	0.7364	1.207	0.8718	0.6678
1.66	0.4940	0.6726	0.7345	1.213	0.8694	0.6830
1.67	0.4894	0.6678	0.7327	1.220	0.8670	0.6983
1.68	0.4847	0.6631	0.7310	1.226	0.8645	0.7136
1.69	0.4801	0.6584	0.7292	1.233	0.8621	0.7290
1.70	0.4756	0.6538	0.7275	1.240	0.8597	0.7444
1.71	0.4712	0.6491	0.7258	1.247	0.8573	0.7598
1.72	0.4668	0.6445	0.7242	1.254	0.8549	0.7753
1.73	0.4624	0.6400	0.7226	1.262	0.8526	0.7908
1.74	0.4581	0.6355	0.7210	1.269	0.8502	0.8064
1.75	0.4539	0.6310	0.7194	1.277	0.8478	0.8220
1.76	0.4497	0.6265	0.7178	1.284	0.8455	0.8376
1.77	0.4456	0.6221	0.7163	1.292	0.8432	0.8532
1.78	0.4415	0.6176	0.7148	1.300	0.8409	0.8689
1.79	0.4375	0.6133	0.7134	1.308	0.8386	0.8846
1.80	0.4335	0.6089	0.7119	1.316	0.8363	0.9003
1.81	0.4296	0.6046	0.7105	1.324	0.8340	0.9161
1.82	0.4257	0.6004	0.7091	1.332	0.8317	0.9318
1.83	0.4219	0.5961	0.7078	1.341	0.8295	0.9476
1.84	0.4181	0.5919	0.7064	1.349	0.8273	0.9634
1.85	0.4144	0.5877	0.7051	1.358	0.8250	0.9792
1.86	0.4107	0.5836	0.7038	1.367	0.8228	0.9951
1.87	0.4071	0.5795	0.7025	1.376	0.8206	1.011
1.88	0.4035	0.5754	0.7012	1.385	0.8185	1.027
1.89	0.3999	0.5714	0.7000	1.394	0.8163	1.043
1.90	0.3964	0.5673	0.6988	1.403	0.8141	1.059
1.91	0.3930	0.5634	0.6975	1.413	0.8120	1.074
1.92	0.3895	0.5594	0.6964	1.422	0.8099	1.090
1.93	0.3862	0.5555	0.6952	1.432	0.8078	1.106
1.94	0.3828	0.5516	0.6940	1.442	0.8057	1.122

Appendix G *Continued*

M	$\frac{p}{p^*}$	$\frac{T}{T^*}$	$\frac{\rho}{\rho^*} = \frac{V^*}{V}$	$\frac{p_0}{p_0^*}$	$\frac{T_0}{T_0^*}$	$\frac{s^* - s}{R}$
1.95	0.3795	0.5477	0.6929	1.452	0.8036	1.138
1.96	0.3763	0.5439	0.6918	1.462	0.8015	1.154
1.97	0.3731	0.5401	0.6907	1.472	0.7995	1.170
1.98	0.3699	0.5364	0.6896	1.482	0.7974	1.186
1.99	0.3667	0.5326	0.6885	1.493	0.7954	1.202
2.00	0.3636	0.5289	0.6875	1.503	0.7934	1.218
2.01	0.3606	0.5253	0.6865	1.514	0.7914	1.233
2.02	0.3575	0.5216	0.6854	1.525	0.7894	1.249
2.03	0.3545	0.5180	0.6844	1.536	0.7874	1.265
2.04	0.3516	0.5144	0.6835	1.547	0.7855	1.281
2.05	0.3487	0.5109	0.6825	1.558	0.7835	1.297
2.06	0.3458	0.5074	0.6815	1.569	0.7816	1.313
2.07	0.3429	0.5039	0.6806	1.581	0.7797	1.329
2.08	0.3401	0.5004	0.6796	1.592	0.7778	1.345
2.09	0.3373	0.4970	0.6787	1.604	0.7759	1.361
2.10	0.3345	0.4936	0.6778	1.616	0.7741	1.376
2.11	0.3318	0.4902	0.6769	1.628	0.7722	1.392
2.12	0.3291	0.4868	0.6760	1.640	0.7704	1.408
2.13	0.3265	0.4835	0.6752	1.653	0.7685	1.424
2.14	0.3238	0.4802	0.6743	1.665	0.7667	1.440
2.15	0.3212	0.4770	0.6735	1.678	0.7649	1.455
2.16	0.3186	0.4737	0.6726	1.691	0.7631	1.471
2.17	0.3161	0.4705	0.6718	1.704	0.7614	1.487
2.18	0.3136	0.4673	0.6710	1.717	0.7596	1.503
2.19	0.3111	0.4642	0.6702	1.730	0.7579	1.519
2.20	0.3086	0.4611	0.6694	1.743	0.7561	1.534
2.21	0.3062	0.4580	0.6686	1.757	0.7544	1.550
2.22	0.3038	0.4549	0.6679	1.771	0.7527	1.566
2.23	0.3014	0.4518	0.6671	1.785	0.7510	1.581
2.24	0.2991	0.4488	0.6664	1.799	0.7493	1.597
2.25	0.2968	0.4458	0.6656	1.813	0.7477	1.613
2.26	0.2945	0.4428	0.6649	1.827	0.7460	1.628
2.27	0.2922	0.4399	0.6642	1.842	0.7444	1.644
2.28	0.2899	0.4370	0.6635	1.856	0.7428	1.659
2.29	0.2877	0.4341	0.6628	1.871	0.7411	1.675
2.30	0.2855	0.4312	0.6621	1.886	0.7395	1.690
2.31	0.2833	0.4284	0.6614	1.901	0.7380	1.706
2.32	0.2812	0.4256	0.6607	1.916	0.7364	1.722
2.33	0.2791	0.4228	0.6601	1.932	0.7348	1.737
2.34	0.2769	0.4200	0.6594	1.948	0.7333	1.752

continued

Appendix G *Continued*

M	$\frac{p}{p^*}$	$\frac{T}{T^*}$	$\frac{\rho}{\rho^*} = \frac{V^*}{V}$	$\frac{p_0}{p_0^*}$	$\frac{T_0}{T_0^*}$	$\frac{s^* - s}{R}$
2.35	0.2749	0.4172	0.6588	1.963	0.7317	1.768
2.36	0.2728	0.4145	0.6581	1.979	0.7302	1.783
2.37	0.2708	0.4118	0.6575	1.996	0.7287	1.799
2.38	0.2688	0.4091	0.6569	2.012	0.7272	1.814
2.39	0.2668	0.4065	0.6563	2.028	0.7257	1.829
2.40	0.2648	0.4038	0.6557	2.045	0.7242	1.845
2.41	0.2628	0.4012	0.6551	2.062	0.7227	1.860
2.42	0.2609	0.3986	0.6545	2.079	0.7213	1.875
2.43	0.2590	0.3961	0.6539	2.096	0.7198	1.891
2.44	0.2571	0.3935	0.6533	2.114	0.7184	1.906
2.45	0.2552	0.3910	0.6527	2.131	0.7170	1.921
2.46	0.2534	0.3885	0.6522	2.149	0.7156	1.936
2.47	0.2515	0.3860	0.6516	2.167	0.7142	1.951
2.48	0.2497	0.3836	0.6511	2.185	0.7128	1.967
2.49	0.2479	0.3811	0.6505	2.203	0.7114	1.982
2.50	0.2462	0.3787	0.6500	2.222	0.7101	1.997
2.51	0.2444	0.3763	0.6495	2.241	0.7087	2.012
2.52	0.2427	0.3739	0.6489	2.259	0.7074	2.027
2.53	0.2409	0.3716	0.6484	2.279	0.7060	2.042
2.54	0.2392	0.3692	0.6479	2.298	0.7047	2.057
2.55	0.2375	0.3669	0.6474	2.317	0.7034	2.072
2.56	0.2359	0.3646	0.6469	2.337	0.7021	2.087
2.57	0.2342	0.3623	0.6464	2.357	0.7008	2.102
2.58	0.2326	0.3601	0.6459	2.377	0.6995	2.117
2.59	0.2310	0.3578	0.6454	2.397	0.6983	2.131
2.60	0.2294	0.3556	0.6450	2.418	0.6970	2.146
2.61	0.2278	0.3534	0.6445	2.438	0.6957	2.161
2.62	0.2262	0.3512	0.6440	2.459	0.6945	2.176
2.63	0.2246	0.3491	0.6436	2.480	0.6933	2.191
2.64	0.2231	0.3469	0.6431	2.502	0.6921	2.205
2.65	0.2216	0.3448	0.6427	2.523	0.6908	2.220
2.66	0.2201	0.3427	0.6422	2.545	0.6896	2.235
2.67	0.2186	0.3406	0.6418	2.567	0.6885	2.249
2.68	0.2171	0.3385	0.6413	2.589	0.6873	2.264
2.69	0.2156	0.3364	0.6409	2.612	0.6861	2.279
2.70	0.2142	0.3344	0.6405	2.634	0.6849	2.293
2.71	0.2127	0.3324	0.6401	2.657	0.6838	2.308
2.72	0.2113	0.3304	0.6397	2.680	0.6826	2.322
2.73	0.2099	0.3284	0.6392	2.704	0.6815	2.337
2.74	0.2085	0.3264	0.6388	2.727	0.6804	2.351
2.75	0.2071	0.3244	0.6384	2.751	0.6793	2.366
2.76	0.2058	0.3225	0.6380	2.775	0.6781	2.380

Appendix G *Continued*

M	$\frac{p}{p^*}$	$\frac{T}{T^*}$	$\frac{\rho}{\rho^*}=\frac{V^*}{V}$	$\frac{p_0}{p_0^*}$	$\frac{T_0}{T_0^*}$	$\frac{s^*-s}{R}$
2.77	0.2044	0.3205	0.6376	2.799	0.6770	2.394
2.78	0.2030	0.3186	0.6372	2.823	0.6760	2.409
2.79	0.2017	0.3167	0.6369	2.848	0.6749	2.423
2.80	0.2004	0.3149	0.6365	2.873	0.6738	2.437
2.81	0.1991	0.3130	0.6361	2.898	0.6727	2.452
2.82	0.1978	0.3111	0.6357	2.924	0.6717	2.466
2.83	0.1965	0.3093	0.6354	2.949	0.6706	2.480
2.84	0.1953	0.3075	0.6350	2.975	0.6696	2.494
2.85	0.1940	0.3057	0.6346	3.001	0.6685	2.508
2.86	0.1927	0.3039	0.6343	3.028	0.6675	2.522
2.87	0.1915	0.3021	0.6339	3.054	0.6665	2.537
2.88	0.1903	0.3004	0.6336	3.081	0.6655	2.551
2.89	0.1891	0.2986	0.6332	3.108	0.6645	2.565
2.90	0.1879	0.2969	0.6329	3.136	0.6635	2.579
2.91	0.1867	0.2951	0.6325	3.164	0.6625	2.593
2.92	0.1855	0.2934	0.6322	3.191	0.6615	2.607
2.93	0.1843	0.2918	0.6319	3.220	0.6606	2.621
2.94	0.1832	0.2901	0.6315	3.248	0.6596	2.634
2.95	0.1820	0.2884	0.6312	3.277	0.6586	2.648
2.96	0.1809	0.2868	0.6309	3.306	0.6577	2.662
2.97	0.1798	0.2851	0.6306	3.335	0.6568	2.676
2.98	0.1787	0.2835	0.6303	3.365	0.6558	2.690
2.99	0.1776	0.2819	0.6299	3.394	0.6549	2.704
3.00	0.1765	0.2803	0.6296	3.424	0.6540	2.717
3.01	0.1754	0.2787	0.6293	3.455	0.6531	2.731
3.02	0.1743	0.2771	0.6290	3.485	0.6522	2.745
3.03	0.1732	0.2756	0.6287	3.516	0.6513	2.758
3.04	0.1722	0.2740	0.6284	3.548	0.6504	2.772
3.05	0.1711	0.2725	0.6281	3.579	0.6495	2.786
3.06	0.1701	0.2709	0.6278	3.611	0.6486	2.799
3.07	0.1691	0.2694	0.6275	3.643	0.6477	2.813
3.08	0.1681	0.2679	0.6273	3.675	0.6469	2.826
3.09	0.1670	0.2664	0.6270	3.708	0.6460	2.840
3.10	0.1660	0.2650	0.6267	3.741	0.6452	2.853
3.11	0.1651	0.2635	0.6264	3.774	0.6443	2.867
3.12	0.1641	0.2620	0.6261	3.808	0.6435	2.880
3.13	0.1631	0.2606	0.6259	3.841	0.6426	2.893
3.14	0.1621	0.2592	0.6256	3.876	0.6418	2.907
3.15	0.1612	0.2577	0.6253	3.910	0.6410	2.920
3.16	0.1602	0.2563	0.6251	3.945	0.6402	2.933

continued

Appendix G *Continued*

M	$\frac{p}{p^*}$	$\frac{T}{T^*}$	$\frac{\rho}{\rho^*} = \frac{V^*}{V}$	$\frac{p_0}{p_0^*}$	$\frac{T_0}{T_0^*}$	$\frac{s^* - s}{R}$
3.17	0.1593	0.2549	0.6248	3.980	0.6394	2.947
3.18	0.1583	0.2535	0.6245	4.015	0.6386	2.960
3.19	0.1574	0.2522	0.6243	4.051	0.6378	2.973
3.20	0.1565	0.2508	0.6240	4.087	0.6370	2.986
3.21	0.1556	0.2494	0.6238	4.123	0.6362	3.000
3.22	0.1547	0.2481	0.6235	4.160	0.6354	3.013
3.23	0.1538	0.2467	0.6233	4.197	0.6347	3.026
3.24	0.1529	0.2454	0.6230	4.234	0.6339	3.039
3.25	0.1520	0.2441	0.6228	4.272	0.6331	3.052
3.26	0.1511	0.2428	0.6225	4.310	0.6324	3.065
3.27	0.1503	0.2415	0.6223	4.348	0.6316	3.078
3.28	0.1494	0.2402	0.6221	4.387	0.6309	3.091
3.29	0.1486	0.2389	0.6218	4.426	0.6301	3.104
3.30	0.1477	0.2377	0.6216	4.465	0.6294	3.117
3.31	0.1469	0.2364	0.6214	4.505	0.6287	3.130
3.32	0.1461	0.2352	0.6211	4.545	0.6280	3.143
3.33	0.1452	0.2339	0.6209	4.586	0.6272	3.155
3.34	0.1444	0.2327	0.6207	4.626	0.6265	3.168
3.35	0.1436	0.2315	0.6205	4.667	0.6258	3.181
3.36	0.1428	0.2303	0.6202	4.709	0.6251	3.194
3.37	0.1420	0.2290	0.6200	4.751	0.6244	3.207
3.38	0.1412	0.2279	0.6198	4.793	0.6237	3.219
3.39	0.1404	0.2267	0.6196	4.835	0.6230	3.232
3.40	0.1397	0.2255	0.6194	4.878	0.6224	3.245
3.41	0.1389	0.2243	0.6192	4.922	0.6217	3.257
3.42	0.1381	0.2232	0.6190	4.965	0.6210	3.270
3.43	0.1374	0.2220	0.6187	5.009	0.6203	3.282
3.44	0.1366	0.2209	0.6185	5.054	0.6197	3.295
3.45	0.1359	0.2197	0.6183	5.098	0.6190	3.308
3.46	0.1351	0.2186	0.6181	5.144	0.6184	3.320
3.47	0.1344	0.2175	0.6179	5.189	0.6177	3.333
3.48	0.1337	0.2164	0.6177	5.235	0.6171	3.345
3.49	0.1329	0.2153	0.6175	5.281	0.6164	3.357
3.50	0.1322	0.2142	0.6173	5.328	0.6158	3.370
3.51	0.1315	0.2131	0.6172	5.375	0.6152	3.382
3.52	0.1308	0.2120	0.6170	5.423	0.6145	3.395
3.53	0.1301	0.2110	0.6168	5.471	0.6139	3.407
3.54	0.1294	0.2099	0.6166	5.519	0.6133	3.419
3.55	0.1287	0.2088	0.6164	5.568	0.6127	3.432
3.56	0.1280	0.2078	0.6162	5.617	0.6121	3.444
3.57	0.1274	0.2068	0.6160	5.666	0.6115	3.456
3.58	0.1267	0.2057	0.6158	5.716	0.6109	3.468

Appendix G *Continued*

M	$\frac{p}{p^*}$	$\frac{T}{T^*}$	$\frac{\rho}{\rho^*}=\frac{V^*}{V}$	$\frac{p_0}{p_0^*}$	$\frac{T_0}{T_0^*}$	$\frac{s^*-s}{R}$
3.59	0.1260	0.2047	0.6157	5.767	0.6103	3.480
3.60	0.1254	0.2037	0.6155	5.817	0.6097	3.493
3.61	0.1247	0.2027	0.6153	5.869	0.6091	3.505
3.62	0.1241	0.2017	0.6151	5.920	0.6085	3.517
3.63	0.1234	0.2007	0.6150	5.972	0.6080	3.529
3.64	0.1228	0.1997	0.6148	6.025	0.6074	3.541
3.65	0.1221	0.1987	0.6146	6.078	0.6068	3.553
3.66	0.1215	0.1977	0.6144	6.131	0.6062	3.565
3.67	0.1209	0.1968	0.6143	6.185	0.6057	3.577
3.68	0.1202	0.1958	0.6141	6.239	0.6051	3.589
3.69	0.1196	0.1949	0.6139	6.294	0.6046	3.601
3.70	0.1190	0.1939	0.6138	6.349	0.6040	3.613
3.71	0.1184	0.1930	0.6136	6.404	0.6035	3.625
3.72	0.1178	0.1920	0.6134	6.460	0.6029	3.637
3.73	0.1172	0.1911	0.6133	6.517	0.6024	3.648
3.74	0.1166	0.1902	0.6131	6.574	0.6018	3.660
3.75	0.1160	0.1893	0.6130	6.631	0.6013	3.672
3.76	0.1154	0.1884	0.6128	6.689	0.6008	3.684
3.77	0.1148	0.1875	0.6126	6.748	0.6003	3.696
3.78	0.1143	0.1866	0.6125	6.806	0.5997	3.707
3.79	0.1137	0.1857	0.6123	6.866	0.5992	3.719
3.80	0.1131	0.1848	0.6122	6.926	0.5987	3.731
3.81	0.1126	0.1839	0.6120	6.986	0.5982	3.742
3.82	0.1120	0.1830	0.6119	7.047	0.5977	3.754
3.83	0.1114	0.1822	0.6117	7.108	0.5972	3.766
3.84	0.1109	0.1813	0.6116	7.170	0.5967	3.777
3.85	0.1103	0.1805	0.6114	7.232	0.5962	3.789
3.86	0.1098	0.1796	0.6113	7.295	0.5957	3.800
3.87	0.1093	0.1788	0.6112	7.358	0.5952	3.812
3.88	0.1087	0.1779	0.6110	7.422	0.5947	3.823
3.89	0.1082	0.1771	0.6109	7.486	0.5942	3.835
3.90	0.1077	0.1763	0.6107	7.551	0.5937	3.846
3.91	0.1071	0.1754	0.6106	7.616	0.5933	3.858
3.92	0.1066	0.1746	0.6104	7.682	0.5928	3.869
3.93	0.1061	0.1738	0.6103	7.748	0.5923	3.880
3.94	0.1056	0.1730	0.6102	7.815	0.5918	3.892
3.95	0.1051	0.1722	0.6100	7.882	0.5914	3.903
3.96	0.1046	0.1714	0.6099	7.950	0.5909	3.914
3.97	0.1041	0.1706	0.6098	8.018	0.5904	3.926
3.98	0.1036	0.1699	0.6096	8.087	0.5900	3.937

continued

M	$\frac{p}{p^*}$	$\frac{T}{T^*}$	$\frac{\rho}{\rho^*}=\frac{V^*}{V}$	$\frac{p_0}{p_0^*}$	$\frac{T_0}{T_0^*}$	$\frac{s^*-s}{R}$
3.99	0.1031	0.1691	0.6095	8.157	0.5895	3.948
4.00	0.1026	0.1683	0.6094	8.227	0.5891	3.960
4.01	0.1021	0.1675	0.6092	8.297	0.5886	3.971
4.02	0.1016	0.1668	0.6091	8.369	0.5882	3.982
4.03	0.1011	0.1660	0.6090	8.440	0.5878	3.993
4.04	0.1006	0.1653	0.6089	8.513	0.5873	4.004
4.05	0.1002	0.1645	0.6087	8.585	0.5869	4.015
4.06	0.099 68	0.1638	0.6086	8.659	0.5864	4.026
4.07	0.099 21	0.1630	0.6085	8.733	0.5860	4.038
4.08	0.098 75	0.1623	0.6084	8.807	0.5856	4.049
4.09	0.098 28	0.1616	0.6082	8.882	0.5852	4.060
4.10	0.097 82	0.1609	0.6081	8.958	0.5847	4.071
4.11	0.097 37	0.1601	0.6080	9.034	0.5843	4.082
4.12	0.096 91	0.1594	0.6079	9.111	0.5839	4.093
4.13	0.096 46	0.1587	0.6078	9.188	0.5835	4.104
4.14	0.096 02	0.1580	0.6076	9.266	0.5831	4.114
4.15	0.095 57	0.1573	0.6075	9.345	0.5827	4.125
4.16	0.095 13	0.1566	0.6074	9.424	0.5823	4.136
4.17	0.094 70	0.1559	0.6073	9.504	0.5818	4.147
4.18	0.094 26	0.1552	0.6072	9.585	0.5814	4.158
4.19	0.093 83	0.1546	0.6071	9.666	0.5810	4.169
4.20	0.093 40	0.1539	0.6070	9.747	0.5807	4.180
4.21	0.092 97	0.1532	0.6068	9.830	0.5803	4.190
4.22	0.092 55	0.1525	0.6067	9.912	0.5799	4.201
4.23	0.092 13	0.1519	0.6066	9.996	0.5795	4.212
4.24	0.091 71	0.1512	0.6065	10.08	0.5791	4.223
4.25	0.091 30	0.1506	0.6064	10.17	0.5787	4.233
4.26	0.090 89	0.1499	0.6063	10.25	0.5783	4.244
4.27	0.090 48	0.1493	0.6062	10.34	0.5779	4.255
4.28	0.090 07	0.1486	0.6061	10.42	0.5776	4.265
4.29	0.089 67	0.1480	0.6060	10.51	0.5772	4.276
4.30	0.089 27	0.1473	0.6059	10.60	0.5768	4.287
4.31	0.088 87	0.1467	0.6058	10.69	0.5764	4.297
4.32	0.088 47	0.1461	0.6057	10.78	0.5761	4.308
4.33	0.088 08	0.1455	0.6056	10.87	0.5757	4.318
4.34	0.087 69	0.1448	0.6055	10.96	0.5754	4.329
4.35	0.087 30	0.1442	0.6054	11.05	0.5750	4.339
4.36	0.086 91	0.1436	0.6053	11.14	0.5746	4.350
4.37	0.086 53	0.1430	0.6052	11.23	0.5743	4.360
4.38	0.086 15	0.1424	0.6051	11.33	0.5739	4.371
4.39	0.085 77	0.1418	0.6050	11.42	0.5736	4.381
4.40	0.085 40	0.1412	0.6049	11.52	0.5732	4.391

Appendix G *Continued*

M	$\frac{p}{p^*}$	$\frac{T}{T^*}$	$\frac{\rho}{\rho^*} = \frac{V^*}{V}$	$\frac{p_0}{p_0^*}$	$\frac{T_0}{T_0^*}$	$\frac{s^* - s}{R}$
4.41	0.085 02	0.1406	0.6048	11.61	0.5729	4.402
4.42	0.084 65	0.1400	0.6047	11.71	0.5725	4.412
4.43	0.084 28	0.1394	0.6046	11.80	0.5722	4.423
4.44	0.083 92	0.1388	0.6045	11.90	0.5718	4.433
4.45	0.083 56	0.1383	0.6044	12.00	0.5715	4.443
4.46	0.083 19	0.1377	0.6043	12.10	0.5712	4.453
4.47	0.082 84	0.1371	0.6042	12.20	0.5708	4.464
4.48	0.082 48	0.1365	0.6041	12.30	0.5705	4.474
4.49	0.082 12	0.1360	0.6040	12.40	0.5702	4.484
4.50	0.081 77	0.1354	0.6039	12.50	0.5698	4.494
4.51	0.081 42	0.1348	0.6038	12.60	0.5695	4.505
4.52	0.081 07	0.1343	0.6037	12.71	0.5692	4.515
4.53	0.080 73	0.1337	0.6036	12.81	0.5688	4.525
4.54	0.080 39	0.1332	0.6035	12.92	0.5685	4.535
4.55	0.080 04	0.1326	0.6035	13.02	0.5682	4.545
4.56	0.079 70	0.1321	0.6034	13.13	0.5679	4.555
4.57	0.079 37	0.1316	0.6033	13.24	0.5676	4.565
4.58	0.079 03	0.1310	0.6032	13.34	0.5673	4.575
4.59	0.078 70	0.1305	0.6031	13.45	0.5669	4.585
4.60	0.078 37	0.1300	0.6030	13.56	0.5666	4.595
4.61	0.078 04	0.1294	0.6029	13.67	0.5663	4.606
4.62	0.077 71	0.1289	0.6029	13.78	0.5660	4.615
4.63	0.077 39	0.1284	0.6028	13.90	0.5657	4.625
4.64	0.077 07	0.1279	0.6027	14.01	0.5654	4.635
4.65	0.076 75	0.1274	0.6026	14.12	0.5651	4.645
4.66	0.076 43	0.1268	0.6025	14.24	0.5648	4.655
4.67	0.076 11	0.1263	0.6024	14.35	0.5645	4.665
4.68	0.075 80	0.1258	0.6024	14.47	0.5642	4.675
4.69	0.075 48	0.1253	0.6023	14.58	0.5639	4.685
4.70	0.075 17	0.1248	0.6022	14.70	0.5636	4.695
4.71	0.074 86	0.1243	0.6021	14.82	0.5633	4.705
4.72	0.074 56	0.1238	0.6020	14.94	0.5630	4.714
4.73	0.074 25	0.1234	0.6020	15.06	0.5628	4.724
4.74	0.073 95	0.1229	0.6019	15.18	0.5625	4.734
4.75	0.073 65	0.1224	0.6018	15.30	0.5622	4.744
4.76	0.073 35	0.1219	0.6017	15.42	0.5619	4.753
4.77	0.073 05	0.1214	0.6016	15.55	0.5616	4.763
4.78	0.072 75	0.1209	0.6016	15.67	0.5613	4.773
4.79	0.072 46	0.1205	0.6015	15.80	0.5611	4.783
4.80	0.072 17	0.1200	0.6014	15.92	0.5608	4.792

continued

Appendix G *Continued*

M	$\frac{p}{p^*}$	$\frac{T}{T^*}$	$\frac{\rho}{\rho^*} = \frac{V^*}{V}$	$\frac{p_0}{p_0^*}$	$\frac{T_0}{T_0^*}$	$\frac{s^* - s}{R}$
4.81	0.071 88	0.1195	0.6013	16.05	0.5605	4.802
4.82	0.071 59	0.1191	0.6013	16.18	0.5602	4.812
4.83	0.071 30	0.1186	0.6012	16.31	0.5600	4.821
4.84	0.071 01	0.1181	0.6011	16.44	0.5597	4.831
4.85	0.070 73	0.1177	0.6010	16.57	0.5594	4.840
4.86	0.070 45	0.1172	0.6010	16.70	0.5591	4.850
4.87	0.070 17	0.1168	0.6009	16.83	0.5589	4.860
4.88	0.069 89	0.1163	0.6008	16.96	0.5586	4.869
4.89	0.069 61	0.1159	0.6008	17.10	0.5584	4.879
4.90	0.069 34	0.1154	0.6007	17.23	0.5581	4.888
4.91	0.069 06	0.1150	0.6006	17.37	0.5578	4.898
4.92	0.068 79	0.1145	0.6005	17.51	0.5576	4.907
4.93	0.068 52	0.1141	0.6005	17.64	0.5573	4.917
4.94	0.068 25	0.1137	0.6004	17.78	0.5571	4.926
4.95	0.067 98	0.1132	0.6003	17.92	0.5568	4.935
4.96	0.067 72	0.1128	0.6003	18.06	0.5566	4.945
4.97	0.067 45	0.1124	0.6002	18.20	0.5563	4.954
4.98	0.067 19	0.1120	0.6001	18.35	0.5561	4.964
4.99	0.066 93	0.1115	0.6001	18.49	0.5558	4.973
5.00	0.066 67	0.1111	0.6000	18.63	0.5556	4.982
∞	0.000 00	0.0000	0.5833	∞	0.4898	∞

Appendix H
Prandtl–Meyer flow of a perfect gas ($k = 1.4$)

M	μ (°)	ν (°)	M	μ (°)	ν (°)
1.00	90.00	0.000 00	1.32	49.25	6.721
1.01	81.93	0.044 73	1.33	48.75	7.000
1.02	78.64	0.1257	1.34	48.27	7.280
1.03	76.14	0.2294	1.35	47.79	7.561
1.04	74.06	0.3510	1.36	47.33	7.844
1.05	72.25	0.4874	1.37	46.88	8.128
1.06	70.63	0.6367	1.38	46.44	8.413
1.07	69.16	0.7973	1.39	46.01	8.699
1.08	67.81	0.9680	1.40	45.58	8.986
1.09	66.55	1.148	1.41	45.17	9.276
1.10	65.38	1.336	1.42	44.77	9.565
1.11	64.28	1.532	1.43	44.37	9.855
1.12	63.23	1.735	1.44	43.98	10.15
1.13	62.25	1.944	1.45	43.60	10.44
1.14	61.31	2.160	1.46	43.23	10.73
1.15	60.41	2.381	1.47	42.86	11.02
1.16	59.55	2.607	1.48	42.51	11.32
1.17	58.73	2.839	1.49	42.16	11.61
1.18	57.94	3.074	1.50	41.81	11.91
1.19	57.18	3.314	1.51	41.47	12.20
1.20	56.44	3.558	1.52	41.14	12.49
1.21	55.74	3.806	1.53	40.81	12.79
1.22	55.05	4.057	1.54	40.49	13.09
1.23	54.39	4.312	1.55	40.18	13.38
1.24	53.75	4.569	1.56	39.87	13.68
1.25	53.13	4.830	1.57	39.56	13.97
1.26	52.53	5.093	1.58	39.27	14.27
1.27	51.94	5.359	1.59	38.97	14.56
1.28	51.38	5.627	1.60	38.68	14.86
1.29	50.82	5.898	1.61	38.40	15.16
1.30	50.28	6.170	1.62	38.12	15.45
1.31	49.76	6.445	1.63	37.84	15.75

continued

Appendix H *Continued*

M	μ (°)	ν (°)	M	μ (°)	ν (°)
1.64	37.57	16.04	2.07	28.89	28.29
1.65	37.31	16.34	2.08	28.74	28.56
1.66	37.04	16.63	2.09	28.59	28.83
1.67	36.78	16.93	2.10	28.44	29.10
1.68	36.53	17.22	2.11	28.29	29.36
1.69	36.28	17.52	2.12	28.14	29.63
1.70	36.03	17.81	2.13	28.00	29.90
1.71	35.79	18.10	2.14	27.86	30.16
1.72	35.55	18.40	2.15	27.72	30.43
1.73	35.31	18.69	2.16	27.58	30.69
1.74	35.08	18.98	2.17	27.44	30.95
1.75	34.85	19.27	2.18	27.30	31.21
1.76	34.62	19.56	2.19	27.17	31.47
1.77	34.40	19.86	2.20	27.04	31.73
1.78	34.18	20.15	2.21	26.90	31.99
1.79	33.96	20.44	2.22	26.77	32.25
1.80	33.75	20.73	2.23	26.64	32.51
1.81	33.54	21.01	2.24	26.51	32.76
1.82	33.33	21.30	2.25	26.39	33.02
1.83	33.12	21.59	2.26	26.26	33.27
1.84	32.92	21.88	2.27	26.14	33.53
1.85	32.72	22.16	2.28	26.01	33.78
1.86	32.52	22.45	2.29	25.89	34.03
1.87	32.33	22.73	2.30	25.77	34.28
1.88	32.13	23.02	2.31	25.65	34.53
1.89	31.94	23.30	2.32	25.53	34.78
1.90	31.76	23.59	2.33	25.42	35.03
1.91	31.57	23.87	2.34	25.30	35.28
1.92	31.39	24.15	2.35	25.18	35.53
1.93	31.21	24.43	2.36	25.07	35.77
1.94	31.03	24.71	2.37	24.96	36.02
1.95	30.85	24.99	2.38	24.85	36.26
1.96	30.68	25.27	2.39	24.73	36.50
1.97	30.51	25.55	2.40	24.62	36.75
1.98	30.33	25.83	2.41	24.52	36.99
1.99	30.17	26.10	2.42	24.41	37.23
2.00	30.00	26.38	2.43	24.30	37.47
2.01	29.84	26.66	2.44	24.19	37.71
2.02	29.67	26.93	2.45	24.09	37.95
2.03	29.51	27.20	2.46	23.99	38.18
2.04	29.35	27.48	2.47	23.88	38.42
2.05	29.20	27.75	2.48	23.78	38.66
2.06	29.04	28.02	2.49	23.68	38.89

Appendix H *Continued*

M	μ (°)	ν (°)	M	μ (°)	ν (°)
2.50	23.58	39.12	2.91	20.10	47.99
2.51	23.48	39.36	2.92	20.03	48.19
2.52	23.38	39.59	2.93	19.96	48.39
2.53	23.28	39.82	2.94	19.89	48.59
2.54	23.18	40.05	2.95	19.81	48.78
2.55	23.09	40.28	2.96	19.75	48.98
2.56	22.99	40.51	2.97	19.68	49.18
2.57	22.90	40.74	2.98	19.61	49.37
2.58	22.81	40.96	2.99	19.54	49.56
2.59	22.71	41.19	3.00	19.47	49.76
2.60	22.62	41.41	3.01	19.40	49.95
2.61	22.53	41.64	3.02	19.34	50.14
2.62	22.44	41.86	3.03	19.27	50.33
2.63	22.35	42.09	3.04	19.20	50.52
2.64	22.26	42.31	3.05	19.14	50.71
2.65	22.17	42.53	3.06	19.07	50.90
2.66	22.08	42.75	3.07	19.01	51.09
2.67	22.00	42.97	3.08	18.95	51.28
2.68	21.91	43.19	3.09	18.88	51.46
2.69	21.82	43.40	3.10	18.82	51.65
2.70	21.74	43.62	3.11	18.76	51.83
2.71	21.65	43.84	3.12	18.69	52.02
2.72	21.57	44.05	3.13	18.63	52.20
2.73	21.49	44.27	3.14	18.57	52.39
2.74	21.41	44.48	3.15	18.51	52.57
2.75	21.32	44.69	3.16	18.45	52.75
2.76	21.24	44.91	3.17	18.39	52.93
2.77	21.16	45.12	3.18	18.33	53.11
2.78	21.08	45.33	3.19	18.27	53.29
2.79	21.00	45.54	3.20	18.21	53.47
2.80	20.92	45.75	3.21	18.15	53.65
2.81	20.85	45.95	3.22	18.09	53.83
2.82	20.77	46.16	3.23	18.03	54.00
2.83	20.69	46.37	3.24	17.98	54.18
2.84	20.62	46.57	3.25	17.92	54.35
2.85	20.54	46.78	3.26	17.86	54.53
2.86	20.47	46.98	3.27	17.81	54.70
2.87	20.39	47.19	3.28	17.75	54.88
2.88	20.32	47.39	3.29	17.70	55.05
2.89	20.24	47.59	3.30	17.64	55.22
2.90	20.17	47.79	3.31	17.58	55.39

continued

Appendix H *Continued*

M	μ (°)	ν (°)	M	μ (°)	ν (°)
3.32	17.53	55.56	3.75	15.47	62.33
3.33	17.48	55.73	3.76	15.42	62.47
3.34	17.42	55.90	3.77	15.38	62.61
3.35	17.37	56.07	3.78	15.34	62.76
3.36	17.31	56.24	3.79	15.30	62.90
3.37	17.26	56.41	3.80	15.26	63.04
3.38	17.21	56.58	3.81	15.22	63.19
3 39	17.16	56.74	3.82	15.18	63.33
3.40	17.10	56.91	3.83	15.14	63.47
3.41	17.05	57.07	3.84	15.09	63.61
3.42	17.00	57.24	3.85	15.05	63.75
3.43	16.95	57.40	3.86	15.01	63.89
3.44	16.90	57.56	3.87	14.98	64.03
3.45	16.85	57.73	3.88	14.94	64.16
3.46	16.80	57.89	3.89	14.90	64.30
3.47	16.75	58.05	3.90	14.86	64.44
3.48	16.70	58.21	3.91	14.82	64.58
3.49	16.65	58.37	3.92	14.78	64.71
3.50	16.60	58.53	3.93	14.74	64.85
3.51	16.55	58.69	3.94	14.70	64.98
3.52	16.50	58.85	3.95	14.66	65.12
3.53	16.46	59.00	3.96	14.63	65.25
3.54	16.41	59.16	3.97	14.59	65.39
3.55	16.36	59.32	3.98	14.55	65.52
3.56	16.31	59.47	3.99	14.51	65.65
3.57	16.27	59.63	4.00	14.48	65.78
3.58	16.22	59.78	4.01	14.44	65.92
3.59	16.17	59.94	4.02	14.40	66.05
3.60	16.13	60.09	4.03	14.37	66.18
3.61	16.08	60.24	4.04	14.33	66.31
3.62	16.04	60.40	4.05	14.29	66.44
3.63	15.99	60.55	4.06	14.26	66.57
3.64	15.95	60.70	4.07	14.22	66.70
3.65	15.90	60.85	4.08	14.19	66.83
3.66	15.86	61.00	4.09	14.15	66.95
3.67	15.81	61.15	4.10	14.12	67.08
3.68	15.77	61.30	4.11	14.08	67.21
3.69	15.72	61.45	4.12	14.05	67.34
3.70	15.68	61.60	4.13	14.01	67.46
3.71	15.64	61.74	4.14	13.98	67.59
3.72	15.59	61.89	4.15	13.94	67.71
3.73	15.55	62.04	4.16	13.91	67.84
3.74	15.51	62.18	4.17	13.88	67.96

Appendix H *Continued*

M	μ (°)	ν (°)	M	μ (°)	ν (°)
4.18	13.84	68.09	4.60	12.56	72.92
4.19	13.81	68.21	4.61	12.53	73.03
4.20	13.77	68.33	4.62	12.50	73.13
4.21	13.74	68.46	4.63	12.47	73.24
4.22	13.71	68.58	4.64	12.45	73.34
4.23	13.67	68.70	4.65	12.42	73.45
4.24	13.64	68.82	4.66	12.39	73.55
4.25	13.61	68.94	4.67	12.36	73.66
4.26	13.58	69.06	4.68	12.34	73.76
4.27	13.54	69.18	4.69	12.31	73.87
4.28	13.51	69.30	4.70	12.28	73.97
4.29	13.48	69.42	4.71	12.26	74.07
4.30	13.45	69.54	4.72	12.23	74.18
4.31	13.42	69.66	4.73	12.21	74.28
4.32	13.38	69.78	4.74	12.18	74.38
4.33	13.35	69.89	4.75	12.15	74.48
4.34	13.32	70.01	4.76	12.13	74.58
4.35	13.29	70.13	4.77	12.10	74.69
4.36	13.26	70.24	4.78	12.08	74.79
4.37	13.23	70.36	4.79	12.05	74.89
4.38	13.20	70.48	4.80	12.02	74.99
4.39	13.17	70.59	4.81	12.00	75.09
4.40	13.14	70.71	4.82	11.97	75.19
4.41	13.11	70.82	4.83	11.95	75.28
4.42	13.08	70.93	4.84	11.92	75.38
4.43	13.05	71.05	4.85	11.90	75.48
4.44	13.02	71.16	4.86	11.87	75.58
4.45	12.99	71.27	4.87	11.85	75.68
4.46	12.96	71.39	4.88	11.82	75.78
4.47	12.93	71.50	4.89	11.80	75.87
4.48	12.90	71.61	4.90	11.78	75.97
4.49	12.87	71.72	4.91	11.75	76.07
4.50	12.84	71.83	4.92	11.73	76.16
4.51	12.81	71.94	4.93	11.70	76.26
4.52	12.78	72.05	4.94	11.68	76.35
4.53	12.75	72.16	4.95	11.66	76.45
4.54	12.72	72.27	4.96	11.63	76.54
4.55	12.70	72.38	4.97	11.61	76.64
4.56	12.67	72.49	4.98	11.58	76.73
4.57	12.64	72.60	4.99	11.56	76.83
4.58	12.61	72.70	5.00	11.54	76.92
4.59	12.58	72.81	∞	0.00	130.4

Answers to problems

CHAPTER 1

1.1 93.75 $m\,s^{-1}$
1.2 1.545 $kg\,s^{-1}$
1.3 (a) compressible; (b) unsteady
1.4 7455 s
1.5 −0.4044 $kg\,s^{-1}\,m^{-3}$
1.6 −12.5 N
1.7 $(p_{atm} - p_e)A_e - \rho_e A_e V_e^2$
1.8 (a) 7 $m\,s^{-1}$; (b) 36 kPa
1.9 129 kN

CHAPTER 2

2.1 (a) 9000 J; (b) 4944 J; (c) 4438 J
2.2 (b) 21 000 J; (c) 42.86 per cent
2.3 42 kJ, 147 kJ
2.4 344.9 kW
2.5 0.047 39 $kg\,s^{-1}$
2.6 1.777 kg, 490 K
2.7 40.04 kJ
2.8 340.3 K, 0.009 34 kg
2.9 (a) 328.1 K; (b) 6.635 kg; (c) 437.2 K, 412.8 kPa; (d) 3361 $J\,K^{-1}$

CHAPTER 3

3.1 334.3 $m\,s^{-1}$
3.2 398.1 K
3.3 (a) 304.4 $m\,s^{-1}$; (b) 317.2 $m\,s^{-1}$
3.4 2.381
3.5 0.8050, imaginary

3.6 1873 m s^{-1}
3.7 0.5, 170.1 m s^{-1}
3.8 4, 1373 m s^{-1}, 14.48°
3.9 506 m s^{-1}, 4.419 s
3.10 12.96 s
3.11 2447 m
3.12 456.9 m s^{-1}
3.13 (a) 1.414; (b) 500 m s^{-1},(c) 6000 m; (d) 311.1 K
3.15 1.3 kPa, 1 K

CHAPTER 4

4.1 (b) 0.093 99 per cent
4.2 225.8 kPa, 334.4 K, 156.1 kPa, 334.4 K
4.3 1.56, 469.1 m s^{-1}
4.4 (a) 385.6 m s^{-1}; (b) 352 m s^{-1}; (c) 0.5332, 0.5682; (d) 862.2 m s^{-1}, ∞, 2.449;

(e) $$\frac{V^2}{(862.2)^2} + \frac{a^2}{(385.6)^2} = 1$$

4.5 364.3 K, 144.2 kPa, 267.9 m s^{-1}, 1.379 kg m^{-3}, 3.694 kg s^{-1}
4.6 0.89, 302.1 K, 310.2 m s^{-1}, 2.683 kg s^{-1}
4.7 0.26, 381.6 kPa, 296 K, 89.67 m s^{-1}
4.8 330.1 K, 0.549, 162.8 kPa, 1.718 kg s^{-1}
4.9 0.3020, 1.064 kg s^{-1}
4.10 0.5124, 1.742 kg s^{-1}
4.11 0.46, 103.3 kg m^{-2} s^{-1}
4.12 0.40, 392.5 K, 187.1 kPa, 0.2638 kg s^{-1}
4.13 270.2 kPa, 3.601, 395.1 kg m^{-2} s^{-1}
4.14 172.6 m s^{-1}, 141.8 kPa, 0.4118 m^2
4.15 1.0, 666.6 K, 317 kPa, 517.6 m s^{-1}, 0.4286 kg s^{-1}
4.16 100 kPa, 0.90, 344.2 K, 334.8 m s^{-1}, 0.3388 kg s^{-1}
4.17 1.0, 343.4 kPa, 291.7 K, 0.001 94 m^2, 2.724 kg s^{-1}
4.18 100 kPa, 0.78, 356.6 K, 0.004 017 m^2, 1.163 kg s^{-1}
4.19 1.0, 158.5 kPa, 333.3 K, 0.003 785 m^2, 2.294 kg s^{-1}
4.20 100 kPa, 0.78, 312 K, 0.002 339 m^2, 0.7239 kg s^{-1}
4.21 188.2 kPa
4.22 135.1 kPa, 0.022 08 m^2
4.23 0.4485 kg s^{-1}, 0.001 374 m^2
4.24 6.552 kg s^{-1}
4.25 1.714 s
4.26 19.1 s
4.27 11.63 s
4.28 2.20, 152.4 K, 18.7 kPa, 18.7 kPa, 0.4655 kg s^{-1}

4.29 2.20, 213.4 K, 22.19 kPa, 22.19 kPa, 0.4667 kg s^{-1}
4.30 100 kPa, 0.37, 340.7 K, 0.2101 kg s^{-1}
4.31 100 kPa, 0.37, 394.2 K, 0.1948 kg s^{-1}
4.32 1.60, 330.7 K, 211.8 kPa, 1301 kg m^{-2} s^{-1}
4.33 1.85, 114.7 kPa, 249.3 K, 938.7 kg m^{-2}s^{-1}
4.34 0.9983 kg s^{-1}
4.35 (a) 1.85, 80.6 kPa, 207.8 K, 722.4 kg m^{-2} s^{-1}, 534.7 m s^{-1}; (b) 507.3 m s^{-1}, 80.60 kPa, 222 K, 1.70, 642.4 kg m^{-2} s^{-1}; (c) 0.9487; (d) 0.8893
4.36 (a) 2.20, 65.46 kPa, 254.1 K, 0.8202 kg s^{-1}, 703.1 m s^{-1}; (b) 0.8038 kg s^{-1}, 65.46 kPa, 698.2 m s^{-1}, 257.5 K, 2.17; (c) 0.9861; (d) 0.993

4.37 (a) $$\left(1-\frac{1}{\eta}\,\frac{\frac{1}{2}(k-1)M^2}{1+\frac{1}{2}(k-1)M^2}\right)^{k/(k-1)}$$

(b) $$\sqrt{(k)}M\sqrt{\left(1+\frac{k-1}{2}M^2\right)}\left(1-\frac{1}{\eta}\,\frac{\frac{1}{2}(k-1)M^2}{1+\frac{1}{2}(k-1)M^2}\right)^{k/(k-1)}$$

4.38 (a) 31.12 kN; (b) 2033 N s kg^{-1}, (c) 1.556
4.39 (a) 23.90 kN, (b) 2507 N s kg^{-1}, (c) 1.707
4.40 0.7509

CHAPTER 5

5.1 176.6 kPa, 374.4 K, 247.5 kPa, 412.5 K, 276.1 m s^{-1}
5.2 669.8 K, 246.6 m s^{-1}, 319.8 J kg^{-1} K^{-1}
5.3 196.8 kPa, 597.9 kPa
5.4 (a) 15.38; (b) 6.156
5.5 −284.1 N
5.7 (a) 732.9 m s^{-1}; (b) 951 kPa, 762 K; (c) 2.708 MPa, 1028 K
5.8 448.8 m s^{-1}
5.9 (a) 378.6 m s^{-1}, 438.7 m s^{-1}; (b) 6.468 s
5.10 306.6 m s^{-1}
5.11 (a) 407.3 m s^{-1}; (b) 10.40; (c) 5.027; (d) 6469 kPa, 1400 K
5.12 7587 kPa
5.13 (a) 425 m s^{-1}; (b) 2244 kPa, 873.9 K
5.14 (a) 187.1 kPa; (b) 18.70 kPa; (c) 102.5 kPa; (d) 200 kPa $> p_b >$ 187.1 kPa, 187.1 kPa $> p_b >$ 102.5 kPa, 102.5 kPa $> p_b >$ 18.70 kPa, $p_b <$ 18.70 kPa
5.15 (a) 229.6 $> p_b >$ 219.8 kPa, 219.8 kPa $> p_b >$ 103.0 kPa, 103 kPa $> p_b >$ 15.71 kPa, $p_b <$ 15.71 kPa; (b) 127 kPa
5.16 (a) 280.7 kPa $> p_b >$ 153.8 kPa, 153.8 kPa $> p_b >$ 28.06 kPa, $p_b <$ 28.06 kPa; (b) 0.6964 kg s^{-1}
5.17 (a) 0.1527 kg s^{-1}; (b) 0.0004 m^2, 1.00; (c) 2.33, 14.60 kPa
5.18 1.60, 105.9 kPa, 330.7 K, 650.5 kg m^{-2} s^{-1}

5.19 (a) 134.1 kPa; (b) 1.372 kg s^{-1}; (c) 9.798 × 10^{-4} m^2; (d) 417.3 kPa; (e) 417.3 kPa
5.20 304.3 kPa, 103.4 kPa, 400 K
5.21 414.3 m s^{-1}
5.22 0.4507*L*, 0.7536
5.23 111.0 kPa, 1.85, 0.6057, 1.5
5.24 320 kPa, 0.50
5.25 (a) 0.5431; (b) 153.7 kPa; (c) 291 kPa
5.26 2.073 s
5.27 (a) 268.6 kPa, 141.9 kPa, 12.46 kPa, 0.4996; (b) 118.6 kPa
5.28 (a) 3.267 kg s^{-1}; (b) 0.010 21 m^2
5.29 (a) 2.20; (b) 0.6368 m^2, (c) Yes, possible; (d) 55.95 kPa; (e) 96.07 kPa; (f) 0.84 m^2, (g) 69.02 kPa; (h) 100 kPa $> p_b >$ 96.07 kPa, 96.07 kPa $> p_b >$ 55.95 kPa, $p_b <$ 55.95 kPa
5.30 (a) 0.8557; (b) 0.8952; (c) 1.83, 1.472
5.31 21.47 kW, 63.48 kW
5.32 (a) 2.65; (b) 1.70; (c) 85.47 kg s^{-1}; (d) 54.83 kg s^{-1}
5.33 (a) 0.069 87 m^2; (b) 0.059 79 m^2
5.34 (a) 1.474; (b) 1.340; (c) 445.4 kPa, 528.3 kPa
5.35 392.5 m s^{-1}
5.36 (a) 487.7 m s^{-1} (b) 255.3 m s^{-1}, 123.8 K
5.37 (a) 534.8 m s^{-1}; (b) 0.6395; (c) 130 kPa, 404.1 K
5.38 (a) 635.6 m s^{-1}; (b) 371.5 m s^{-1}, 0.8582; (c) 18.7 kPa, 466.2 K; (d) 338.3 m s^{-1}; (e) 0.004 529 s
5.39 (a) 635.6 m s^{-1}; (b) 371.5 m s^{-1}; (c) 0.8582; (d) 0.003 355 s; (e) 338.3 m s^{-1}; (g) 4.01 m
5.40 (a) 347.3 m s^{-1}; (b) 0.057 59 s; (c) 98 m s^{-1}; (d) 229.7 m s^{-1}
5.41 89.5 m s^{-1}, −310.7 m s^{-1}
5.42 48.16

CHAPTER 6

6.1 0.66, 296.0 K, 135.5 kPa, 0.2565 kg s^{-1}
6.2 1.33, 313.5 K, 221.4 kPa, 0.3648 kg s^{-1}
6.3 271.2 K, 316.1 K, 74.04 J kg^{-1} K^{-1}
6.4 0.93, 57.33 kPa, 343.3 K
6.5 1.88, 677.4 m s^{-1}, 5.113 kg s^{-1}
6.6 572.4 kPa, 44.1 J kg^{-1} K^{-1}
6.7 (a) 0.3720; (b) 1.00, 256.9 K, 34.24 kPa; (c) −3.705 kN
6.8 362.5 K, 29.1 N, 46.6 J kg^{-1} K^{-1}
6.9 0.65, 0.40, 0.19, 0.18
6.10 2.504 kg s^{-1}
6.11 0.4494

6.12 94.62 kPa, 0.4646 kg s^{-1}
6.13 200 kPa > p_b > 83.16 kPa, 83.16 kPa > p_b > 0 kPa
6.14 (a) 0.70, 1.00; (b) 193.2 kPa, 291.7 K; (c) 0.3876 kg s^{-1}
6.15 (a) 0.45, 1.00; (b) 167.7 kPa, 260.2 K; (c) 0.9120 kg s^{-1}
6.16 (a) 0.50, 0.53; (b) 119.0 kPa, 378.6 K; (c) 0.2843 kg s^{-1}
6.17 (a) 0.51, 0.71; (b) 105.0 kPa; 286.6 K, (c) 0.6055 kg s^{-1}
6.18 (a) 3.197 m; (b) 102.4 kPa; (c) 37.79 kPa; (d) 87.67 kPa;
(e) 200 kPa > p_b > 102.4 kPa, 102.4 kPa > p_b > 87.67 kPa > p_b > 37.79 kPa,
37.79 kPa > p_b > 0
6.19 0.41, 209 kPa 290.3 K, 0.4415 kg s^{-1}
6.20 1.46, 37.79 kPa, 280.4 K, 0.2886 kg s^{-1}
6.21 1.57, 234.4 K, 45.41 kPa, 1.625 kg s^{-1}
6.22 0.64, 323.5 K, 262 kPa, 3.249 kg s^{-1}
6.23 (a) 0.6458 m; (b) 106 kPa; (c) 0.47 m; (d) 153.5 kPa;
(e) 300 kPa > p_b > 153.5 kPa, 153.5 kPa > p_b > 106 kPa, 106 kPa > p_b > 0
6.24 1.00, 291.7 K, 60.02 kPa, 1.234 kg s^{-1}
6.25 0.88, 61 kPa, 259.8 K, 0.018 39 kg s^{-1}
6.26 0.43, 289.3 K, 165 kPa, 0.5721 kg s^{-1}
6.27 (a) 0.66, 280.1 K, 71.28 kPa; (b) 1.52 N; (c) 0.096 39 kg s^{-1}; (d) 406.9 W
6.28 (a) 1.09, 350 K, 183 kPa; (b) 10.7 N; (c) 0.2340 kg s^{-1}; (d) − 9.548 kW
6.29 (a) 0.8452, 300 K, 101.8 kPa; (b) 15.18 N; (c) 0.1089 kg s^{-1}; (d) 3.535 kW
6.30 (a) 0.8452, 400 K, 177.5 kPa; (b) 42.3 N; (c) 0.3702 kg s^{-1}; (d) − 45.47 kW

CHAPTER 7

7.1 247.8 kJ kg^{-1}, 283.5 kPa
7.2 0.51, 1239 K, 1303 K, 1.601 MW
7.3 1.48, 594 m s^{-1}, 156.6 kPa, 76.58 kJ kg^{-1}
7.4 255.9 kJ kg^{-1}, 639.2 J kg^{-1} K^{-1}, −17.2 kPa
7.5 834.2 kW, 221.6 J kg^{-1} K^{-1}, −13.3 kPa
7.6 648.5 K, 0.47, 642.6 J kg^{-1} K^{-1}, − 24 kPa
7.7 2.204 kg s^{-1}
7.8 0.7944, 67.94 kW
7.9 0.3699, 576.4 N
7.10 408.6 kJ kg^{-1}

CHAPTER 8

8.1 71 kPa, 299.8 K, 1.61, 10.57°, 22.97°
8.2 82.8 kPa, 301.3 K, 2.17, 8.01°, 29.8°
8.3 31.85°, 2.085, 93.3 kPa, 289 K
8.4 36.94°, 1.876, 98.32 kPa, 347.2 K

8.5 (a) 82.8 kPa, 289.8 K, 2.169; (b) 130.0 kPa, 330.7 K, 1.874; (c) 26.3°
8.6 (a) 2.367, 245.8 kPa, 316.8 K, 12.77°; (b) 2.95, 100 kPa, 245.3 K, 25.53°
8.7 12.77°
8.8 (a) 66.24 kPa, 278.2 K, 2.169; (b) 104.0 kPa, 317.4 K, 1.874; (c) 26.3°
8.9 1008 kPa, 2.204, 1024 kPa, 2.116, 9.5°
8.10 2.21, 25.35 kPa, 183.4 K, 34.91°
8.11 2.356, 100 kPa, 331.2 K, −13°, 1.835, 215.3 kPa, 417.6 K, 0°, 2.33, 100 kPa, 335.6 K, 13.15°
8.12 1.934, 100 kPa, 286.6 K, −13.61°, 1.440, 202.3 kPa, 354 K, 0°, 1.91, 100 kPa, 289.1 K, 13.72°
8.13 3.47, 205.4 K, 8.29°
8.14 3.21, 163.4 K, 3.89°
8.15 8.628 kg s^{-1}, 525.7 kPa, 575.1 K
8.16 18.98 kg s^{-1}, 495.5 kPa, 502.8 K
8.17 10.06 kg s^{-1}, 466.6 kPa, 533.5 K
8.18 0.3104, 0.054 74
8.19 0.4931, 0.1352
8.20 0, 0.045 43
8.21 0.5876, 0.1804
8.22 0.4425, 0.1338

Index